Biopesticides in Environment and Food Security

Issues and Strategies

About the Editors

Opender Koul, Fellow of the National Academy of Agricultural Sciences and the Indian Academy of Entomology, is an insect toxicologist/physiologist/chemical ecologist and currently the Director of the Insect Biopesticide Research Centre, Jalandhar, India and Adjunct Professor, Kasetsart University, Bangkok, Thailand. He has been a Visiting Scientist at the University of Kanazawa, Japan (1985-1986), University of British Columbia, Canada (1988-1992), and Institute of Plant Protection, Poznan, Poland (2001). His extensive research experience concerns insect-plant interactions, spanning toxicological, physiological and agricultural aspects. He has been honoured with an Indian National Science Academy Medal (INSA), the Kothari Scientific Research Institute Award, KEC Science Society Award, Recognition Award of National Academy of Agricultural Sciences of India (2005); Prof. R.C. Saxena Memorial Medal (2009); and Late P.P. Singhal Memorial Award (2010) for outstanding contributions in the field of Insect Toxicology/ Physiology and Plant Protection. Dr. Koul was also nominated for the Recognition Award of ESA (USA) in 2006. Dr. Koul is on panel of experts in many committees and leading international and national journals. He has also been an informal consultant to BOSTID, NRC of USA at ICIPE, Nairobi, UNIDO Expert on Biopesticides in China (2006) and Nigeria (2007), Biopesticides Expert, Tsukuba, Japan (2009) and Member External Advisory Board (ADAPPT Net Work for African, Caribbean and Pacific Group of States, European Development Fund), UK, 2009. He has authored over 200 research papers and articles, and is the author/editor of 20 books on different aspects of biopesticides.

G. S. Dhaliwal is currently Visiting Professor, Insect Biopesticide Research Centre, Jalandhar, India. He obtained Ph.D. from the Indian Agricultural Research Institute, New Delhi in 1972 and remained Post-doctoral Fellow at the International Rice Research Institute, Los Banos, Philippines, for two years. A member of the World Food Prize Nominating Academy, he is Fellow of the National Environmental Science academy (NESA), Society of Plant Protection Sciences, Society of Pesticide Sciences India, Plant Protection Association of India, Indian Society for the Advancement of Insect Science and Entomological Society of India. Honoured with Best Scientist Award of NESA and Lifetime Achievement Award of International Allelopathy Foun-dation, he has been Founder President of the Indian Society for the Advancement of Insect Science and Society of Biopesticide Sciences India, and President of the Indian Ecological Society, He has authored more than 50 book chapters, 250 research papers and review articles, and is author/ editor of more than 40 books on different aspects of pest management, environment and sustainable agriculture.

Sucheta Khokhar is currently Dean, College of Agriculture, and served as Head, Department of Entomology (until June 2010) at CCS Haryana Agricultural University, Hisar, She obtained her Ph.D. from CCS Haryana Agricultural University, Hisar, in 1979 and was a FAO Post-doctoral Training Fellow at Ohio State University, USA in 1999 in the Diagnostic Clinics for Insects and Other Pathogens. A Fellow of the Applied Zoological Research Association (AZRA), CRRI, Cuttack, she is the recipient of Dr. Mrs. Jagadeswari Rao Women Scientist Award 1999 and 2000 bestowed by AZRA. She has established the new identity of 35 insects from Haryana and adjoining areas, and discovered 10 new species of insects. She has published more than 70 research papers in national and international journals of scientific repute and contributed 10 chapters in the books and 3 review papers. She is a co-editor of two books/compendia. She is life member of several scientific societies. She has guided more than 12 post-graduate students and taught taxonomy, morphology, classification and biosystematics courses of Entomology to under-graduate and post-graduate students at CCS Haryana Agricultural University, Hisar and Indian Agricultural Research institute, New Delhi.

Ram Singh did his Ph.D. from the Indian Agricultural Research Institute, New Delhi, in 1985. He was awarded the European Commission Post-doctoral Fellowship (1991-92) and FAO Post-doctoral Training Fellowship (1998) when he worked on various aspects of insect-plant interactions and integrated pest management at the Horticulture Research International, Wellesbourne, United Kingdom. A Fellow of the Royal Entomological Society, London, he is the recipient of University Gold Medal and Aspee Gold Medal, from HAU, Hisar and ICAR award from New Delhi. He has published more than 150 research papers in national and international journals of scientific repute. He has also contributed 17 chapters in the books and 11 review papers. Dr. Singh has co-edited/authored four books and four compendia. He is life member of several scientific societies and also the Fellow of Plant Protection Association of India. He has presented his work at several international conferences in India, Netherlands, Scotland and United Kingdom.

Biopesticides in Environment and Food Security

Issues and Strategies

Editors

Opender Koul **G.S. Dhaliwal**
Sucheta Khokhar **Ram Singh**

Published by:

Scientific Publishers (India)
5 A, New Pali Road, P.O. Box 91
Jodhpur 342 001 (India)

Branch Office
Scientific Publishers (India)
4806/24, Ansari Road, Daryaganj
New Delhi - 110 002 (India)

E-mail: info@scientificpub.com
Website: www.scientificpub.com

ISBN: 978-81-7233-797-1

Typeset by Rajesh Ojha
Printed in India

PREFACE

The current global population of about 7 billion is likely to increase to 9.5 billion by 2050. The population growth will tax food supplies and outstrip our efforts in many areas. It is thus clear that the farmers all over the world will need to produce at least 50% more food than that was produced in 2000 to feed the expanding population. The efforts to increase food production in the 20th century have caused many environmental, economic and social problems. Meeting the food supply needs of an ever-increasing world population is both critical and at the same time stressing the natural resources needed for crop production. In spite of the significant progress over the last four decades towards food security in several regions of the world, around 850 million people remain malnourished and about 2 billion hectares of arable land has been degraded. Therefore, we face a gigantic task of ensuring environment and food security in the twenty-first century.

The large scale use of pesticides during the green revolution era has caused several adverse effects like environmental pollution, pesticide resistance, pest resurgence, and emergence of secondary pest problems. Therefore, the quest for shifting the focus on to more reliable, sustainable and environment friendly agents of pest control has recently started gaining momentum. Biopesticides are expected to fill in this slot and play a prominent role in future integrated pest management programmes. While the overall market for pesticides is showing a decline in average annual growth rate of 11%, biopesticide market is growing rapidly, increasing from US$ 672 million in 2005 to over US$ 1 billion in 2010, at an annual growth rate of 9.9%. Biopesticides accounted for 2.5% of the total global pesticide market in 2005, which increased to 4.5% in 2010. During the next five years, an annual growth rate of 10 to 15% has been projected. Several constraints such as high cost, variable performance under diverse use conditions, inadequate quality reinforcement, lack of knowledge and awareness, lack of interest in their promotion by big entrepreneurs, and time consuming and cumbersome regulatory system have hindered the desired pace of progress in their adoption and use. The present book focuses on these multifarious issues related to the development of biopesticides and outlines a road map for future.

In all, there are 21 chapters in the book dealing with various aspects of the biopesticides. The first chapter gives a broad overview of the salient developments of various categories of biopesticides and sets the stage for the book. Next three chapters discuss various aspects of semiochemicals to show their potential, possibilities of technology development including nanotechnology. Chapter 5 discusses deployment of transgenic crops for pest management and their ecological implications in terms of environmental safety. Genetic improvement of bioagents in recent years has been discussed in chapters 7 and 8 to bring forth the possibilities of having new products in future. Plant proteinase inhibitors are extensively studied as a class of biopesticides, however, in recent years there has been emphasis on molecular basis of interaction of such inhibitors. The details are comprehensively dealt in chapter 9. Pest management strategies are very important for the success of any biopesticide, whether for crop protection or post-harvest pest management. Chapters 10, 11 and 12 deal with these aspects along with the registration requirements, which are important component of any IPM strategy to be used at farmers' level. These requirements have been discussed in chapter 12. There are many organisms that have in recent past shown lot of potential as biocontrol agents, like acarines, rhizomicroorganisms, endophytic fungi, and fungi as disease management organisms. All these aspects make the contents of chapters 14 to 17. The potential and constraints of botanical biopesticides in pest management have been extensively discussed in Chapter 18. Chapters 19 to 21 discuss some country wide studies of biopesticides, like India, Canada, Africa and Thailand. On the whole, this volume gives comprehensive state-of-art information about recent trends in biopesticides and at the same time evaluates the issues and strategies.

We express our sincere thanks to the Department of Biotechnology (DBT), Government of India, Council of Scientific and Industrial Research (CSIR), Agricultural and Processed Food Product Export Development Authority (APEDA) and Indian National Science Academy (INSA) for the financial support. We are thankful to Dr. K. S. Khokhar, Vice-Chancellor, CCS Haryana Agricultural University, Hisar, for his expert guidance and support. Thanks are also due to Dr. Arvind Kumar, Deputy Director General (Education) and Dr. C. Devakumar, Assistant Director General (Education), Indian Council of Agricultural Research, New Delhi, for their encouragement and cooperation. We are also grateful to Dr. O.P. Tokey, Dean, Postgraduate Studies and Dr. R.K. Saini, Head, Department of Entomology, CCS Haryana Agricultural University, Hisar, for their ever-willing help from time to time. We are also thankful to all the contributors, particularly those from Australia, Canada, Germany, Malavi, Mauritius, New Zealand, Thailand, USA and Zimbabwe, for submitting their chapters within the stipulated period. We thank Dr. Vikas Jindal, Assistant Entomologist, Punjab Agricultural University, Ludhiana, for going through certain chapters of the book and making useful suggestions. We also thank Dr. Birpal Kaur and Shikha for their help in the preparation of the index. Our special thanks are due to M/s Scientific Publishers (India), Jodhpur for bringing out the book speedily and in an excellent form.

Opender Koul
G. S. Dhaliwal
Sucheta Khokhar
Ram Singh

CONTRIBUTORS

V. Ambethgar
Tamil Nadu Agricultural University
Regional Research Station
Vridhachalam-606001, Tamil Nadu, India

Ruisheng An
Department of Entomology
The Ohio State University
OARDC, Wooster, OH-44691
USA

Harit K. Bal
Department of Entomology
The Ohio State University
OARDC, Wooster, OH-44691
USA

O. M. Bambawale
National Centre for Integrated Pest Management
IARI Campus
New Delhi-110012, India

Steven R. Belmain
Natural Resources Institute
University of Greenwich
Kent ME4 4TB, UK

Someshwar Bhagat
National Centre for Integrated Pest Management
IARI Campus
New Delhi-110012, India

Vasakorn Bullangpoti
Zoology Department, Faculty of Science
Kasetsart University
Bangkok-10900, Thailand

Surendra K. Dara
University of California
Cooperative Extension
San Luis Obispo, CA-93401
USA

Suseelendra Desai
Central Research Institute for Dryland Agriculture
Santoshnagar, Hyderabad-500059, India

G. S. Dhaliwal
Insect Biopesticide Research Centre, 30 Parkash Nagar
Jalandhar-144003, India

M. K. Dhillon
Division of Entomology
Indian Agricultural Research Institute, New Delhi-110012
India

S. Facknath
Faculty of Agriculture
University of Mauritius
Reduit, Mauritius

Ashok P. Giri
Plant Molecular Biology Unit
Division of Biochemical Sciences
National Chemical Laboratory
Pune-411008, India

Sreeraj Gopi
Asthagiri Herbal Research Foundation
J162-A, Perangudi Industrial Estate, Perangudi,
Chennai-600096, India

Parwinder S. Grewal
Department of Entomology
The Ohio State University
OARDC, Wooster, OH-44691
USA

Rachna Gulati
Directorate of Research
CCS Haryana Agricultural
University, Hisar-125004
India

Vidya S. Gupta
Plant Molecular Biology Unit
Division of Biochemical Sciences
National Chemical Laboratory
Pune-411008, India

G. M. Gurr
Charles Sturt University
PO Box 883, Orange, NSW 2800
Australia

Detlef F. Hein
Illinois Natural History Survey
Institute of Natural Resource
Sustainability, University of
Illinois, 1816 Oak Street
Champaign, IL-61820, USA

Hans E. Hummel
Justus-Liebig University of
Giessen, Karl-Glockner-Strasse
21C, D-35394, Giessen, Germany

Murray B. Isman
Faculty of Land and Food
Systems, University of British
Columbia, 2357 Main Mall
Suite 248, Vancouver, BC
Canada V6T 1Z4

Saleem Javeed
National Bureau of
Agriculturally Important Insects
H A Farm Post, Hebbal
Bangaluru-560024, India

Vikas Jindal
Department of Entomology
Punjab Agricultural University
Ludhiana-141004, India

Rakesh S. Joshi
Plant Molecular Biology Unit
Division of Biochemical Sciences
National Chemical Laboratory
Pune-411008, India

John Kamanula
Department of Chemistry
Mzuzu University, Luwinga
Muzuzu 2, Malawi

S. Kannan
Asthagiri Herbal Research
Foundation, J162-A
Perangudi Industrial Estate
Perangudi, Chennai-600096
India

Jaspal Kaur
Department of Plant Pathology
Punjab Agricultural University
Ludhiana-141004, India

Neha Khandelwal
Plant Molecular Biology Unit
Division of Biochemical Sciences
National Chemical Laboratory
Pune-411008, India

Sucheta Khokhar
College of Agriculture
CCS Haryana Agricultural
University, Hisar-125004, India

Opender Koul
Insect Biopesticide Research
Centre, 30 Parkash Nagar
Jalandhar-144003, India

Deva Kumar
Indian Council of Agricultural
Rersearch, KAB II,
New Delhi-110012, India

B. Lalljee
Consultancy and Contract
Research Centre
Univeristy of Mauritius
Reduit, Mauritius

Brighton M. Mvumi
Faculty of Agriculture
University of Zimbabwe, Mount
Pleasant, Harare, Zimbabwe

M. Nagesh
National Bureau of
Agriculturally Important Insects
H A Farm Post, Hebbal
Bangaluru-560024, India

S. Narasimhan
Asthagiri Herbal Research
Foundation, J162-A
Perangudi Industrial Estate,
Perangudi, Chennai-600096,
India

Stephen P. Nyirenda
Lunyangwa Research Station
Department of Agricultural
Research Services
Mzuzu, Malawi

Thomas W. Phillips
Department of Entomology
123 W. Waters Hall
Kansas State University
Manhattan, KS-66502, USA

Y. R. Sarma
FAO Consultant
Aramam, M 10-5, KSHB Colony
Calicut (Kozhikode)-673009,
India

Rita Seffrin
Faculty of Land and Food
Systems, University of British
Columbia, 2357 Main Mall
Suite 248, Vancouver, BC,
Canada V6T 1Z4

H. C. Sharma
International Crops Research
Institute for the Semi-Arid
Tropics (ICRISAT)
Patancheru-502324
Andhra Pradesh, India

M. Simpson
Charles Sturt University
PO Box 883, Orange, NSW 2800
Australia

Gudeta Sileshi
World Agroforestry Centre
(ICRAF), Chitedze Agricultural
Research Station, Lilongwe
Malawi

Ram Singh
Department of Entomology
CCS Haryana Agricultural
University, Hisar-125004, India

Rama S. Singh
Department of Plant Pathology
Punjab Agricultural University
Ludhiana-141004, India

Phosiso Sola
Southern Alliance for Indigenous
Resources, Belvedere, Harare,
Zimbabwe

Philip C. Stevenson
Natural Resources Institute
University of Greenwich
Kent ME4 4TB, UK

R. Thippeswamy
National Bureau of
Agriculturally Important Insects
H A Farm Post, Hebbal
Bangaluru-560024, India

Seema Wahab
Department of Biotechnology
Government of India
Block 2, 6th Floor
C G O Complex, Lodhi Road
New Delhi-11003, India

Suresh Walia
Division of Agricultural
Chemicals, Indian Agricultural
Research Institute, New Delhi-
110012, india

S. D. Wratten
Lincoln University
National Center for Advanced
Bio-Protection Technologies
P. O. Box 84, Lincoln 7647
New Zealand

CONTENTS

1

Biopesticides: Springboard to Environment and Food Security

G.S. DHALIWAL[1*], OPENDER KOUL[1], SUCHETA KHOKHAR[2] AND RAM SINGH[2]

[1]Insect Biopesticide Research Centre, 30 Parkash Nagar
Jalandhar-144003, India
[2] Department of Entomology, CCS Haryana Agricultural University
Hisar-125004, India
*E-mail: gurmail_d@yahoo.in

INTRODUCTION

In the history of the world, no century can match the population growth of the one that has recently come to close. The 20th century began with 1.6 billion, and at the end of the century; these two numbers had simply reversed to 6.1 billion. The world's population now grows by about one billion every 12 years. The global population reached 6.89 billion in mid-2010 and is likely to reach 9.48 billion in 2050. Much of this increase would occur in developing countries particularly sub-Saharan Africa and Asia. During 2010, China occupied the first position with a population of 1.34 billion followed by India with a population of 1.19 billion. However, India is projected to be the world's most populous nation at 1.75 billion in 2050, overtaking the current leader China, which is forecast to hit 1.44 billion (PRB, 2010). Such demographic changes would have profound implications for the economy, environment, health and quality of life of people. The population growth will tax food supplies and outstrip nations' efforts in many areas. It is thus clear that world's farmers will need to produce at least 50% more food than in 2000 to feed the expanding population. The efforts to increase food production in 20th century have caused many environmental, economic and social problems. Meeting the food supply needs of an ever-increasing world population is both critical and at the same time, stressing the natural resources needed for crop production. In spite of the significant progress over the last four decades towards food security in several regions of the world, around 850 million people remain malnourished. Therefore, we face a gigantic task of ensuring environment and food security in the twenty-first century (FAO, 2006; Dhaliwal and Koul, 2010). In this context, there is a strong need to promote ecofriendly technologies in agriculture and biopesticides are expected to play a prominent role in sustainable crop production (Koul *et al.*, 2009).

BIOPESTICIDE UMBRELLA

In a broader sense, biopesticides include pesticides of biological origin. There is no unanimity in the definitions of biopesticides, which vary from a very narrow to a very wide scope. For example, Van Latum and Gerrits (1999) interpreted biopesticides as simply the plant-derived substances. Similarly, Lisansky (1989) defined biopesticides as preparations or formulations manufactured to be used in the control or eradication of pests, weeds or disease organisms in which the active ingredient or principle is based on a living microorganism or is derived from one without significant purification or modification. These narrow definitions stand in contrast to the wider one, which includes all organic substances having a protective effect on plants, whether found in nature or chemically synthesized (Frost and Neubert, 1991).

Recently, some rational attempts have been made to define biopesticides. Zechendorf (1995) included substances such as plant extracts, hormones, pheromones and toxins from organic origin in the scope of biopesticides. Copping and Menn (2000) referred the term biopesticides to encompass many aspects of pest control such as microbial, entomophagous nematodes, plant-derived pesticides, secondary metabolites from microorganisms, pheromones and genes used to transform crops to express resistance to pests. More recently, the encouragement of natural enemies (parasitoids, predators, microbes, etc.) and the use of biorational compounds (semiochemicals, insect growth regulators, plant-derived materials, etc.), and transgenic crop varieties in pest management, have been considered to constitute the biopesticide umbrella (Dhaliwal *et al.*, 2010b).

PREDATORS AND PARASITOIDS

Predators and parasitoids are natural enemies, which attack various life stages of insects. The effect of these natural enemies ranges from a temporary or minor effect to the death of the host insect pest. The insects that prey other insects and mites occur in many insect orders, the most prominent being Coleoptera, Neuroptera, Hymenoptera, Diptera, Hemiptera and Odonata. In addition, there are several species of mites and spiders feeding on a wide range of insects and mites. The majority of the parasitoids utilized in biological control of insect pests belong to two orders, namely Hymenoptera (especially wasps of super families Chalcidoidea, Ichneumonoidea and Proctotrupoidea), and Diptera (flies, particularly of the family Tachinidae). It has been estimated that only 15% of all entomophages have so far been identified. More than 125 species of natural enemies are available at global level for augmentative biological control (Thacker, 2002).

The first well planned and successful biological control attempt involved the control of cottony cushion scale, *Icerya purchasi*, in California (USA) during 1887-88, by importing the Vedalia beetle, *Rodolia cardinalis*, from Adelaide (Australia). Complete control of the pest was achieved within 18 months and even after 100 years, the scale is still under check. Following the success of this programme, the beetle has been imported to control this scale in over 50 other countries. The recent outstanding example is the control of cassava mealy bug, *Phenacoccus manihoti*, by a tiny wasp, *Epidinocarsis lopezi* in Africa. The mealy bug was devastating the cassava plant in late 1970s, destroying as much as 80% of the crop in some areas and widespread famine was a real possibility. *E. lopezi* was imported from Paraguay and first released in Nigeria in 1981. Subsequently, the wasp was sent to other countries and by the end of 1986, it was established over an area of more than 750,000 km^2 in 16 African countries. Over the next 7 years, the cassava mealy bug problem was effectively eliminated from 30 countries. In view of the outstanding success of the programme, Dr Hans R. Herren was awarded the World Food Prize in 1995 (Bellotti *et al.*, 1999; Dhaliwal and Koul, 2007).

Recent records of the introduction of entomophagous insects reveal that there have been 5634 releases of 2119 species of entomophagous arthropods to control 597 pest species upto March, 2004. However, complete control was obtained in only 212 cases and substantial control in 419 cases. The primary target pests in classical biological control programmes belonged to Homoptera (2420 releases), followed by Lepidoptera (1722), Coleoptera (596), Diptera (533),

Hymenoptera (137) and Heteroptera (96); the Homoptera have been the most successful group controlled (429 releases or 17.7% were successful). The primary agents in classical biological control programmes belonged to Coleoptera (1168), Diptera (457) and Hymenoptera (3887); the success rates recorded were 9.2, 8.1 and 12.3% for these groups, respectively (Gurr *et al.,* 2007).

MICROBIAL PATHOGENS

The exploitation of disease causing organisms (microbial pathogens) to reduce the population of insect pests below the damaging levels is referred to as microbial control. So far over 3000 microorganisms have been reported to cause diseases in insects. Nevertheless, scientists familiar with specific pathogen groups agree that a very large number of insect pathogens remain undiscovered or unidentified. More than 100 bacteria have been identified as arthropod pathogens, among which *Bacillus thuringiensis* (Bt) has received maximum attention as microbial control agent. The crystal proteins produced by subspecies of *kurstaki* and *aizawai* are effective against most species of Lepidoptera. Since the discovery of the dipteran specific subspecies *israelensis,* the recorded host range of Bt has expanded considerably with the addition of more than 100 susceptible species of mosquitoes and black flies. The more recently discovered subspecies *tenebrionis* and *sandiego* produce toxins specific in their activity against coleopterans. Viruses have been isolated from more than 1000 species of insects from at least 13 different insect orders. World over, about 525 insect species in 52 families and 8 orders are known to be infected by nuclear polyhedrosis virus (NPV) and of this large number of species, 455 belong to Lepidoptera. Among the lepidopterans, noctuids comprise of the largest family with 107 species. There are over 72,000 described species of fungi and the total number of species may be as high as 1.5 million. Until now, over 800 species of entomopathogenic fungi have been identified. The two major groups of entomopathogenic nematodes are *Steinernema* (55 species) and *Heterorhabditis* (12 species). More than 1000 species of protozoa pathogenic to insects have been described and many more remain to be discovered (Dhaliwal and Koul, 2010).

The world's largest programme for the use of an entomopathogen to control a pest on a single crop involved the management of soybean caterpillar, *Anticarsia gemmatalis (Hubner)* by its nucleopolyhdrovirus (AgNPV). In pilot scale field trials with AgNPV, reductions over 80% of *A. gemmatalis* larval populations were obtained.

Approximately 2000 ha were treated in 1982-83, which increased to 500,000 ha in 1986-87, and to 1 million ha in 1989-90. With the commercialization of AgNPV by private companies, use of the biopesticide, increased to about 1.5 million ha in 1995. The virus is presently used on 2.0 million ha soybean in Brazil and it is expected that its use may increase to 4.0 million ha per year within 5 years. The use of AgNPV in Brazil has generated substantial economic, ecological and social benefits. The current annual savings at the soybean-grower level are over US$ 14 million. Since implementation of the programme in 1982-83 up to 2004-05, it is estimated that the cumulative use of AgNPV reached approximately 23,000,000 ha, which represents savings to farmers of about US$ 161 million. Most importantly, since the beginning of the programme, more than 23 million litres of chemical pesticides were not sprayed in the environment, resulting in considerable ecological benefits to the society. In addition, handpicking of the virus-killed larvae during the soybean season provided an additional source of income to resource-poor families, which contributed to their socio-economic improvement. With recent achievements in commercial production of AgMNPV, it is expected that the use of the biological insecticides may reach 4.0 million ha/year within 5 years (Moscardi, 2007).

BIORATIONAL PESTICIDES

Biorational pesticides were initially defined as naturally occurring substances (e.g. insect pheromones and hormones) or microbial agents that are somewhat species specific (Falcon, 1985). Subsequently, the chemicals that affect insect behaviour, growth or reproduction for the suppression of insect populations were often referred to as biorational pesticides (Dhaliwal, 2005). However, a broader definition that includes all chemicals that affect the growth, development, biology and ecology of pest species differently from that of beneficial species seems to be more appropriate (Graves *et al.*, 1999). According to Hara (2000), biorational or reduced risk pesticides are synthetic or natural compounds that effectively control insect pests, but have low toxicity to non-target organisms (such as humans, animals and natural enemies) and the environment.

Insect Based Biorationals

Two groups of biorationals have gained prominence recently and are based on insect behaviour modifiers (behaviour modifying chemicals or semiochemicals (pheromones and allelochemicals), and growth and development regulators (insect growth regulators). There was a rapid growth in the identification of insect pheromones during 1970s and by the end of 1980s pheromones and pheromone mimics were known for about 1000 species of insects. To day, more than 1500 moth sex pheromones and hundreds of other pheromones have been identified, including sex and aggregation pheromones from beetles and other groups. The first trials that involved assaying pheromones for pest control were carried out in mid-1960s. Since then, hundreds of other pheromones have been identified and there are more than 50 (in over 300 different formulations) that can be used in pest control programmes. Over 30 target species have been controlled successfully by mating disruption. The worldwide sales of semiochemical products is about US$ 70-80 million or 1% of the agrochemical market. The fact that producers rely on the deployment of pheromones for pest control on 150,000 to 200,000 ha worldwide justifies continued efforts to develop and implement management programmes based on the use of these biorationals (Gut *et al.,* 2004).

Plant-derived Biorationals

It has been estimated that there are at least 250,000 species of plants in the world, of which only 10% have been examined chemically. So far, more than 6500 plant species have been screened for anti-insect properties and of these, nearly 2500 species belonging to 235 plant families exhibit measurable to considerable pest control activity. Various isolated compounds from these plants indicate that approximately 350 compounds are insecticidal and more than 900 isolates are feeding deterrents alone (Koul, 2005). Among the various pesticidal plants, neem (*Azadirachta indica* A. Juss.) has been exploited for pest management at global level. Today, neem trees are found in nearly 80 countries worldwide and the global estimate is about 91 million trees. During the last three decades, neem has been established as a multipurpose tree, but its recognition as a source of valuable plant chemicals, especially for insecticidal, insect repellent, antifeedant and growth regulatory properties of neem kernel extracts, has attracted worldwide attention. The most important components identified in neem are the tetranortriterpenoids, the azadirachtins. These occur at concentrations of 0.1-0.9% in seed core and it has been established that 30–60 g azadirachtin per ha is sufficient to combat and repel key pests of various crops. Inspite of high selectivity, neem derivatives are known to affect more than 500 species of insects belonging to different orders (Koul and Wahab, 2004).

Microbe-derived Biorationals

Several important classes of biorationals have been produced mostly from soil microorganisms by fermentation procedures. Avermectins are a group of macrocyclic lactones isolated from fermentation of the soil actinomycete microorganisms, *Streptomyces avermitilis.* Avermectins are among the most potent antihelmintic, insecticidal and acaricidal compounds known. These compounds exhibit a broad spectrum of activity against agricultural pests such as aphids, mites, caterpillars, and other pests that feed on crops and ornamental plants. They are also used in veterinary medicine for treatment of internal and external parasites, and mites of pests and livestock. Abamectin is a commercial product recommended for use on ornamentals, cotton, citrus, pome, nut crops, vegetables, potatoes and many other crops to control the motile stages of a wide range of arthropods, including spider mites, leaf miners, ants, cockroaches and selected lepidopteran species. The spinosyns, a new class of insecticidal macrocyclic lactones, derived by fermentation of the soil actinomycetes, *Saccharopolyspora spinosa,* are active against many pests of crops, ornamentals, forestry, greenhouse, garden and households. The most effective product of this group is spinosad, which has shown contact and stomach activity against Coleoptera, Diptera, Hymenoptera, Lepidoptera, Siphonaptera and Thysanoptera. It is particularly effective against lepidopterans such as heliothine complex and is one of the most powerful chemicals for controlling thrips and leaf miners (Horowitz and Ishaaya, 2004).

Most of the biorational pesticides are preferable to conventional pesticides because of their specificity to target pests, their effectiveness at low rates and their non-persistent characteristics in the environment. In view of their specificity, the market potential of these chemicals is limited and the industry is not keen to invest on the development and use of new chemicals, which requires huge investment in terms of resources and time. Furthermore, many products available in the market are not of uniform quality and give inconsistent results. However, these problems are not insurmountable and are being taken care of. Many of the recent biorationals have exhibited broad spectrum of activity, and suitable standards for quality and performance have been laid out for many biorational compounds. Therefore, the biorational pesticides are becoming more reliable and look set for playing a vital role in ecofriendly management of crop pests in future (Dhaliwal and Koul, 2010).

TRANSGENIC CROPS

The introduction of gene technology has added a new dimension to efforts for increasing world food supplies. The global losses due to insect pests after the introduction of transgenic crops (and other advances in crop protection) have declined from 13.6 to 10.8%, and the corresponding losses in Indian agriculture have declined from 23.3 to 17.5% (Dhaliwal *et al.*, 2010a). The global area under transgenic crops has increased from 1.7 million ha in 1996 to 148 million ha in 2010. This represents 87-fold increase, indicating the fastest adoption of technology in the recent history of agriculture. Of the 15.4 million farmers growing transgenic crops in 29 countries, over 90% (14.4 million) are small and resource-poor farmers from developing countries. It is estimated that by 2015, transgenic crops would be grown on 200 million ha in about 40 countries or even more. Thus, transgenic crops have an enormous potential for contributing to the Millennium Development Goal (MDG) of reducing poverty by 50% by 2015 (James, 2010). Transgenic crops have played an important role by contributing to food security and environmental protection. During 1996-2008, the total crop production gain of 4 principal transgenic crops (soybean, maize, cotton and canola) was 167.1 million tones. The global accumulated economic benefits due to transgenic crops during 1996-2008 were US$ 51.9 billion; $26.1 billion for developing and $25.8 billion for developed countries. The accumulative reduction in the use of pesticides for the period 1996-2008 was estimated at 356 million kg of active ingredient (a.i.). During 2008 alone, there was a pesticide reduction of 34.6 million kg of a.i. in the environment. Similarly, biotech agriculture contributed to reduction in emission of 14.4 billion kg of carbon dioxide in 2008, which is equivalent to removing 7 million cars from the road (Brooks and Barfoot, 2010).

The public concern about the biosafety of transgenic plants has occupied the centre stage. There is an on-going conflict between the proponents and critics of the application of gene technology in agriculture. The proponents of transgenic crops argue that biotech agriculture is necessary to feed the world. They claim that it is more environment friendly than chemical pesticides to control weeds and animal pests. In addition, gene technology is a natural extension of conventional breeding techniques with an added advantage of making precise genetic changes and may, therefore, be used to solve the problems caused by intensive agriculture. In spite of the large-scale adoption of biotech agriculture at global level, there is a great concern about the potential risks associated with transgenic plants. The critics argue that biotech agriculture would introduce new ecological risks and decrease the biodiversity of plant cultivars. Food safety risks associated with transgenic plants include the spread of antibiotic resistance, changes in the nutrient composition of the plant, and production of toxic proteins and allergens. There is a potential danger that just as the farmers became trapped in 'pesticide treadmills' during the green revolution era, they may be trapped in 'gene treadmills' in the current gene revolution era. Thus, there is a strong need to narrow the gap between the proponents and critics of application of gene technology. More involvement of scientific experts is needed for improving the public awareness with regard to transgenic technology and the perceived risks thereof. The debate over food and environmental security, regulatory oversight and welfare of the farming community would predominate as the technology moves forward with new crops and new adopting countries (Dhaliwal and Koul, 2010).

CONCLUSIONS

The growing demand for pesticide residue free food for the expanding population has brought the biopesticides to the forefront of pest management. While the overall market for pesticides is showing a decline in annual growth rate of 1.1%, the biopesticide market is growing rapidly at an annual growth rate of 9.9%. The different groups of biopesticides have proved their potential against specific pests in specific situations. At present, more than 1000 commercial biopesticide formulations with more than 225 registered biopesticide active ingredients are known to be available in the global market. The annual growth rate in biopesticide market is estimated to increase to 20% or even more by the middle of the twenty-first century. In order to make biopesticides more effective and sustainable, a detailed understanding of mechanisms of their action, ecology of insect-plant interactions and the impact on non-target organisms is essential. There is also need to standardize biopesticide formulations and consistently monitor their quality. The risks and benefits of individual transgenic crops on case by case basis be examined thoroughly in order to make this technology available to resource-poor farmers, in developing countries to meet the ever-increasing demand for food in future.

REFERENCES

Bellotti, A.C., Smith, L. and Lapointe, S.L. (1999) Recent advances in cassava pest management. *Annu. Rev. Entomol.*, **44**, 343–370.

Brooks, G. and Barfoot, P. (2010) *GM Crops: Global Socio-economic and Environmental Impacts.* P.G. Economics Ltd., Dorchester, UK.

Copping, L.G. and Menn, J.J. (2000) Biopesticides: A review of their action, applications and efficacy. *Pest Manage. Sci.*, **56**, 651–676.

Dhaliwal, G.S. (2005) Prospects of biorational approaches to pest management. In: B.S. Chhillar, K.K. Dahiya and J.S. Rana (eds.), *Advances in Biorational Approaches to Pest Management,* CCS Haryana Agricultural University, Hisar, pp.18–28.

Dhaliwal, G.S. and Koul, O. (2007) *Biopesticides and Pest Management: Conventional and Biotechnological Approaches.* Kalyani Publishers, New Delhi.

Dhaliwal, G.S. and Koul, O. (2010) *Quest for Pest Management: From Green Revolution to Gene Revolution.* Kalyani Publishers, New Delhi.

Dhaliwal, G.S., Jindal, V. and Dhawan, A.K. (2010a) Insect pest problems and crop losses: Changing trends. *Indian J. Ecol.,* **37**, 1–7.

Dhaliwal, G.S., Koul, O., Khokhar, S. and Singh, R. (2010b) Biopesticides: An ecosafe pathway to food and environment security. In: O. Koul, G.S. Dhaliwal, S. Khokhar and R. Singh (eds.), *Biopesticides in Food and Environment Security.* Abstracts Third Conference on Biopesticides: Emerging Trends (BET 2010), October 20-22, Hisar, India. Society of Biopesticide Sciences India, Jalandhar, India, pp. 3–6.

Falcon, L.A. (1985) Development and use of microbial insecticides. In: M.A. Hoy and D.C. Herzog (eds.), *Biological Control in Agricultural IPM Systems,* Academic Press, New York, pp. 229–242.

FAO (2006) *The State of Food Insecurity in the World 2006.* Food and Agriculture Organization of the United Nations, Rome, Italy.

Frost, M. and Neubert, S. (1991) *Bio-Pestizide.* Pestizid Aktions-Netzweck, Hamburg.

Graves, J.B., Leonard, B.R. and Ottea, J.A. (1999) Chemical approaches to managing arthropod pests. In: J.R. Ruberson (ed.), *Handbook of Pest Management.* Marcel Dekker, Inc., New York, pp. 449–486.

Gurr, G.M., Price, P.W., Urrutia, M., Wade, M., Wratten, S.D. and Simmons, A.T. (2007) Ecology of predator-prey and parasitoid-host system: Its role in integrated pest management. In: O. Koul and G.W. Cuperus (eds.), *Ecologically Based Integrated Pest Management,* CAB International, Wallingford, UK, pp. 222–248.

Gut, L.J., Stelinsky, L.L., Thomson, D.R. and Miller, J.R. (2004) Behaviour-modifying chemicals: Prospects and constraints in IPM. In: O. Koul, G.S. Dhaliwal and G.W. Cuperus (eds.), *Integrated Pest Management: Potential, Constraints and Challenges,* CAB International, Wallingford, UK. pp. 73–121.

Hara, A.H. (2000) Finding alternating ways to control alien pests – bait 2: New insecticides introduced to fight old pests. *Hawaii Landscape,* **4**, 5.

Horowitz, A.R. and Ishaaya, I. (2004) Biorational insecticides-Mechanisms, selectivity and importance in pest management. In: A.R. Horowitz and I. Ishaaya (eds.), *Insect Pest Management: Field and Protected Crops,* Springer-Verlag, Berlin, Germany, pp. 1–28.

James, C. (2010) *Global Status of Commercialized Biotech / GM Crops: 2010.* ISAAA Brief No. 42. International Service for the Acquisition of Agri-biotech Applications, Ithaca, New York.

Koul, O. (2005) *Insect Antifeedants.* CRC Press, Boca Raton, Florida, USA.

Koul, O. and Wahab, S. (2004) *Neem: Today and in the New Millennium.* Kluwer Academic Publishers, Dordrecht, The Netherlands.

Koul, O., Dhaliwal, G.S. and Kaul, V.K. (2009) *Sustainable Crop Protection: Biopesticde Strategies.* Kalyani Publishers, New Delhi.

Lisansky, S.G. (1989) Biopesticides. *Agbiotech News Int.* **1**, 349.

Moscardi, F. (2007) A nucleopolyhedrovirus for control of velvetbean cater-pillar in Brazilian soybeans. In: C. Vincent, M.S Goettel and G. Lazarovits (eds.), *Biological Control: A Global Perspective,* CAB International, Wallingford, UK, pp. 344–352.

PRB (2010) *2010 World Population Data Sheet.* Population Reference Bureau, Washigton, DC.

Thacker, J.R.M. (2002) *An Introduction to Arthropod Pest Control.* Cambridge University Press, Cambridge, UK.

Van Lantum, E.B.J. and Gerrits, R. (1999) *Bio-Pesticides in Developing Countries.* African Centre for Technology Studies, Nairobi.

Zechendorf, B. (1995) Biopesticides as integrated pest management strategy in agriculture. In: R. Reuveni (ed.), *Novel Approaches to Integrated Pest Management,* CRC Press, Inc., Boca Raton, Florida, USA, pp. 231–259.

2

A Novel Approach to Enhancing Biological Control: Attracting Predators and Parasitoids Using Herbivore Induced Plant Volatiles (HIPVs) and Nectar 'Rewards'

G.M. GURR[1*], M. SIMPSON[1] AND S.D. WRATTEN[2]

[1] EH Graham Centre for Agricultural Innovation, (Industry & Investment NSW and Charles Sturt University) Charles Sturt University, Faculty of Science, School of Agricultural & Wine Sciences, Leeds Parade, Orange, NSW 2800,
[2] Lincoln University, National Centre for Advanced Bio-Protection Technologies, PO Box 84 Lincoln 7647, New Zealand.
*E-mail: ggurr@csu.edu.au

INTRODUCTION

Plants have evolved a wide range of mechanisms to protect themselves from herbivores. Spines, tough skins and toxins are amongst the most familiar, but plants also have more subtle approaches. Defences can be 'switched on' when the metabolic cost is justified by a current need for defence against herbivory. Laboratory work over two decades has shown that plants under attack by arthropod herbivores produce volatile chemicals that attract predators (Dicke and Sabelis, 1998) and parasitoid wasps (Turlings *et al.*, 1990). Several of the compounds responsible for this effect have been identified, synthesised and used in slow-release dispensers or as sprays. Under field conditions they result in elevated catches of biological control agents (James, 2005). Remarkably, application to plants of some herbivore induced plant volatiles (HIPVs) or related compounds such as jasmonic acid involved in induced plant defence metabolic pathways can also induce the production of a natural blend of HIPVs (James, 2003). Such findings suggest that applying synthetic HIPVs to crops may attract – both directly and indirectly – the natural enemies that could protect crops from pest damage. A criticism of this approach, however, is that natural enemies may rapidly leave the crop or starve unless their attraction is carefully synchronised with periods of relatively high pest abundance. This would be unacceptable in many agricultural systems, where pest densities need to be maintained at low levels. Fortunately, nature suggests a solution to this problem. Other than the production of HIPVs, induced plant defence mechanisms include elevated production of nectar from extrafloral nectaries (Arimura *et al.*, 2005), structures on the leaves, stems or stipules and able to attract parasitoids and predators of pests. Unlike nectar from flowers, which is available only during the pollination period, extrafloral nectar is available for much of the growing season. Extrafloral nectar can serve as an alternative food for omnivorous predators during periods of prey scarcity. In the case of parasitoid adults it can greatly increase longevity, allowing survival even when host herbivores are relatively rare. Although the production of extrafloral nectar might have a useful role in cropping systems, most crop species do not have such nectaries. Amongst those that do, such as cotton, there is also the problem that some pests are able to exploit extrafloral nectar (James and Price, 2004). This can increase pest fitness and plant damage. So, how to provide nectar 'rewards' in a targeted fashion so as to deny benefit to pests?

Biological control can be elegant, sustainable and pollution-free but its overall success rate has been low, especially for arthropod biological control agents against arthropod pests (Greathead and Greathead, 2000). A reason for this low success is that many such agents are omnivorous, even to the extent that adults may be dependent upon plant foods whilst immatures prey upon or parasitise arthropods (Polis and Strong, 1996). Despite a prescient warning published in *Science* over sixty years ago (Wolcott, 1942) the importance of this life-history omnivory has often been overlooked by biological control practitioners. Releasing a promising agent into an environment depauperate in appropriate plant foods, can limit its survival and fecundity and its ultimate impact on pest populations. In conservation biological control, researchers and practitioners seek to boost the plant diversity of farm habitats to provide shelter and food resources for biological control agents. Providing 'rewards' to biological control agents in the form of nectar and pollen can markedly increase success in otherwise inhospitable monocultures (Wäckers *et al.*, 2007). Use of flowering ground covers and field margin strips is popular but shares the same potential risk as naturally-produced extrafloral nectar as they may be used by pest species. Selectivity, where prior screening has identified plant species that produce nectar benefiting natural enemies but of no value to the target pest (Baggen and Gurr, 1998; Baggen *et al.*, 1999) has, therefore, become a major theme in conservation biological control. Many 'selective' foods plants are now known. Additionally, nectar 'rewards' will have little effect if predators and parasitoids are rare in targeted crops. Biological control agents are often more common in adjacent non-crop vegetation (Thies and Tscharntke, 1999) but the stimulus for them to colonise a monocultural crop may be inadequate, even with added nectar sources. There is also the need to re-locate pests' natural enemies after they have been disturbed by agricultural practices such as spraying and harvesting.

Applying exogenous HIPVs to crops offers scope to attract biological control agents to crops early in the season to prevent pest build-up. The simultaneous presence of 'selective' food plants then provides a nectar 'reward' to these biological control agents retaining them in the crop and maximising their fecundity, searching ability and adult longevity. 'Attract and reward' technologies are based on natural, induced plant defence mechanisms and when combined in agriculture represent a synergy between two previously separate approaches to biological control and could make a powerful contribution to future sustainable pest management.

The foregoing mini-review suggests a hypothesis: that the 'attract' and 'reward' approaches could result in synergistic effects that lead to higher abundance of natural enemies. The following section constitutes a case study evaluating the actual potential of this approach. This set of results complements very recent reports of other aspects of that Australian study (Simpson *et al.*, 2011a,b).

CASE STUDY

Two field experiments were conducted in the central west region of New South Wales (NSW), Australia in sweet-corn (*Zea mays* L.) and broccoli (*Brassica oleracea* L.) A randomised block design was used for all experiments, all treatments were replicated four times with a plot spacing of a least 15 m (James, 2005) to minimise inter-plot interference.

A canola oil-based formulation (Synertrol® (Organic Crop Protectants Pty. Ltd. Lilyfield, Australia) was enhanced with several synthetic HIPVs at 1.0% v/v. The formulations were diluted in water at 5ml/ 1L and applied with a hand-sprayer to the foliage until runoff. A preceding study evaluated several synthetic HIPVs at three concentrations in the same crops (Simpson *et al.*, 2011a) and formed the basis for the HIPVs and dosage used in this current study. Additionally, a HIPV mix was included because plants emit HIPV blends in response to herbivorous damage (Yu *et al.*, 2008). The HIPVs were spray-applied, because previous studies demonstrated success via this technique (James *in press*; Simpson *et al.* unpublished data) and of its mechanization advantage which is of agronomic importance.

A floral reward strategy was used rather than an artificial food supplement which may need to be applied repeatedly. Buckwheat (cv. Ikeda) (*Fagopyrum esculentum*) was chosen, because previous studies that used this plant, demonstrated increased abundance and fitness of natural enemies, parasitism rates and lower pest abundance (Berndt *et al.*, 2002; Lavandero *et al.*, 2005; Lee and Heimpel, 2005; Begum *et al.*, 2006; Irvin *et al.*, 2006; Witting-Bissinger *et al.*, 2008). It was also chosen because of its quick germination and short sowing to flowering time (Bowie *et al.*, 1995) which was of agronomic importance.

The first experiment used sweet corn located on a farm near the town of The Lagoon, NSW, Australia (33°32′ 51.84″S, 149°36′51. 27.75″E). The experiment included 10 treatments: methyl salicylate (MeSA), methyl jasmonate (MeJA), methyl anthranilate (MeA), a mix of MeSA, MeJA, MeA (each HIPV was added to Synertrol® equally making a 1.0% formulation); and a water control, each with and without buckwheat. Plots comprised an average of 12 corn plants in one row. Buckwheat was added to reward treatment plots as a 1 × 2 m row immediately alongside the corn plants. Buckwheat seeds were hand-sown at 100g/plot into cultivated ground on 27 November, 2008, one day after the sweet-corn was sown. HIPV treatments were spray-applied to sweet-corn plants on 9 February, 2009. At this stage the buckwheat in all of the 'reward' plots was in bloom and the sweet-corn was at silking growth stage which is a critical phase for *Helicoverpa* spp. larvae attack. Abundance of insects was measured with non-attractive sticky traps (21 × 14.9 cm, two traps/treatment plot), made of clear acetate sheets coated with a thin layer of a clear insect trapping adhesive (The Tanglefoot Company®, Grand Rapids, MI 49504, USA). Traps were located within 15cm of the corn plants to the side of the buckwheat and were replaced on day one, four, 10, 18 and

22 after spraying. Trapped insects were examined with a dissecting microscope and identified using Naumann (1991) and Stevens *et al.* (2007).

The second experiment was with broccoli. The study site was located on a farm near the town of Oberon, NSW, Australia, (33° 39′46.19″S, 149°50′21.46″E). The experiment included 10 treatments: MeSA, cis-3-hexenyl acetate (HA), benzaldehyde (Be) and a mix of MeSA, HA and Be (each HIPV was added equally to Synertrol® making a 2.0% formulation) and a water control, each with and without reward. Plots comprised of four broccoli plants in one row. Buckwheat was established in the reward treatment plots as a 0.60 × 2 m row immediately adjacent to the broccoli plants. Buckwheat seeds were hand-sown at 50g/plot into cultivated ground on 15 January 2009 one day after the broccoli seedlings were planted. HIPV treatments were spray-applied to broccoli plants on 2 March 2009 when the buckwheat in all of the reward plots was in bloom. The broccoli plants were about six weeks old and at the vegetative growth stage with side shoots being developed. Two traps were located within 15 cm of the broccoli plants to the side of the buckwheat and were replaced on day two, four, 11, 18 and 25 after spraying. Arthropod monitoring and identifications were as previously described.

GenStat (12th Edition, Payne *et al.* (2009)) was used for data analysis. Insect data from sticky traps were converted to daily captures due to the uneven time periods of trap replacement. Raw data were square root transformed $\sqrt{(x+0.5)}$ due to non-normal distribution and to ensure residuals were homogenous. Restricted Maximum Likelihood procedure which account for spatial and temporal variability was used for the analysis with attract (HIPV), reward (buckwheat) and time as factors to test for main effects of these, also for interactions. Treatment means within each significant factor and/ or interactions were differentiated using the least significant difference (l.s.d.) at 5% generated from the repeated measures model.

In the sweetcorn experiment, two parasitoid families responded to the treatments with a significant HIPV*reward response. In the HIPV treatments without buckwheat reward plots, Eulophidae numbers were higher for MeA, MeJA and the HIPV mix compared to the water control (F = 4.44, d.f.=4, 150, P=0.002, Fig. 2.1a). The presence of buckwheat led to a significant increase in eulophids in the water control treatment. The same also applied to the MeJA and MeSA treatments. For one particular parasitoid family, Encyrtidae, there were again significant (F = 3.16, df = 4, 150; P = 0.016) HIPV* reward responses. The buckwheat had a significant effect for MeJA and the HIPV mix treatments (Fig. 2.1b). In the former treatment, the buckwheat reward increased encyrtid numbers but the converse was true in the HIPV mix (Fig. 2.1b).

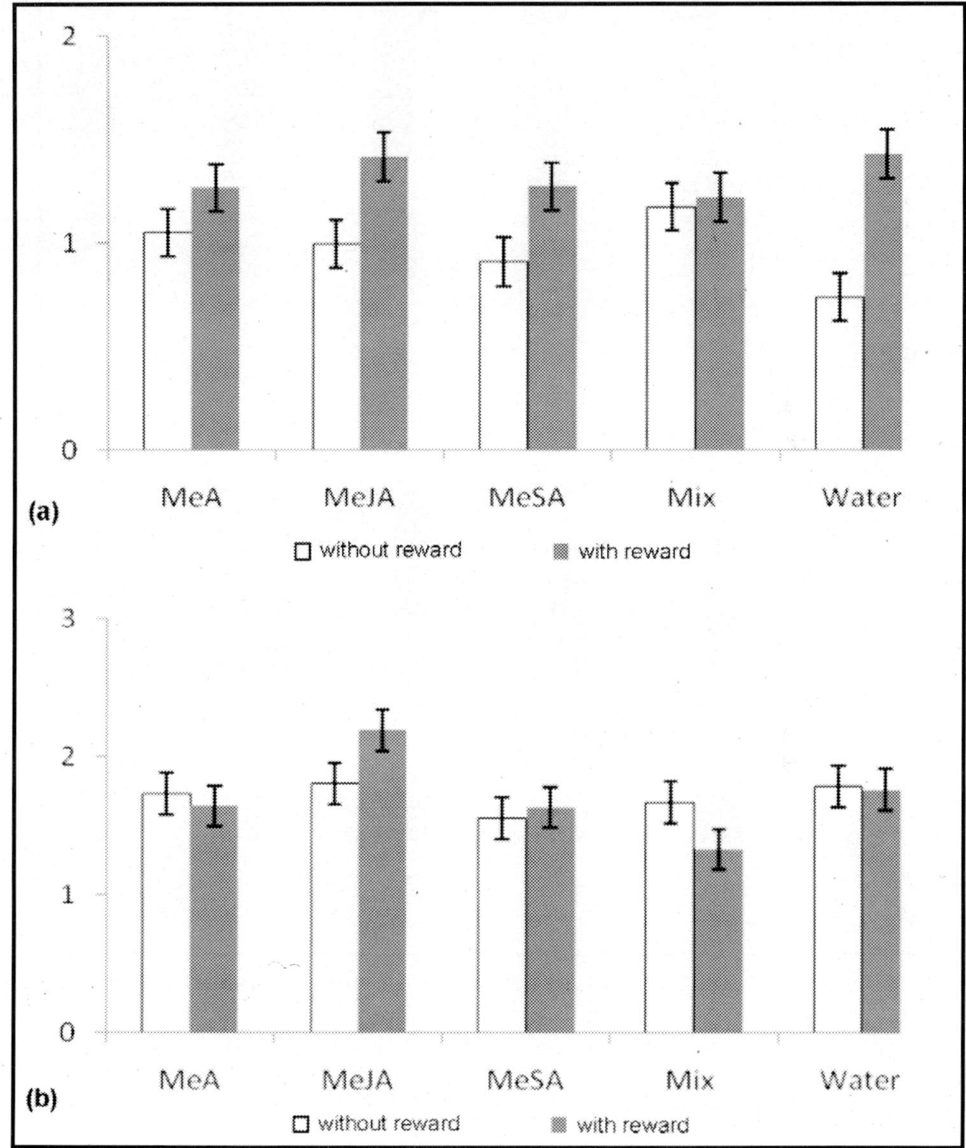

Fig. 2.1. Mean (√x+.0.5) daily trap catches of parasitoids on non-attracting sticky traps adjacent to sweet-corn plants treated with synthetic herbivore-induced plant volatile compounds and with or without a reward of buckwheat flowers (all time periods pooled); (a) Eulophidae, (b) Encyrtidae

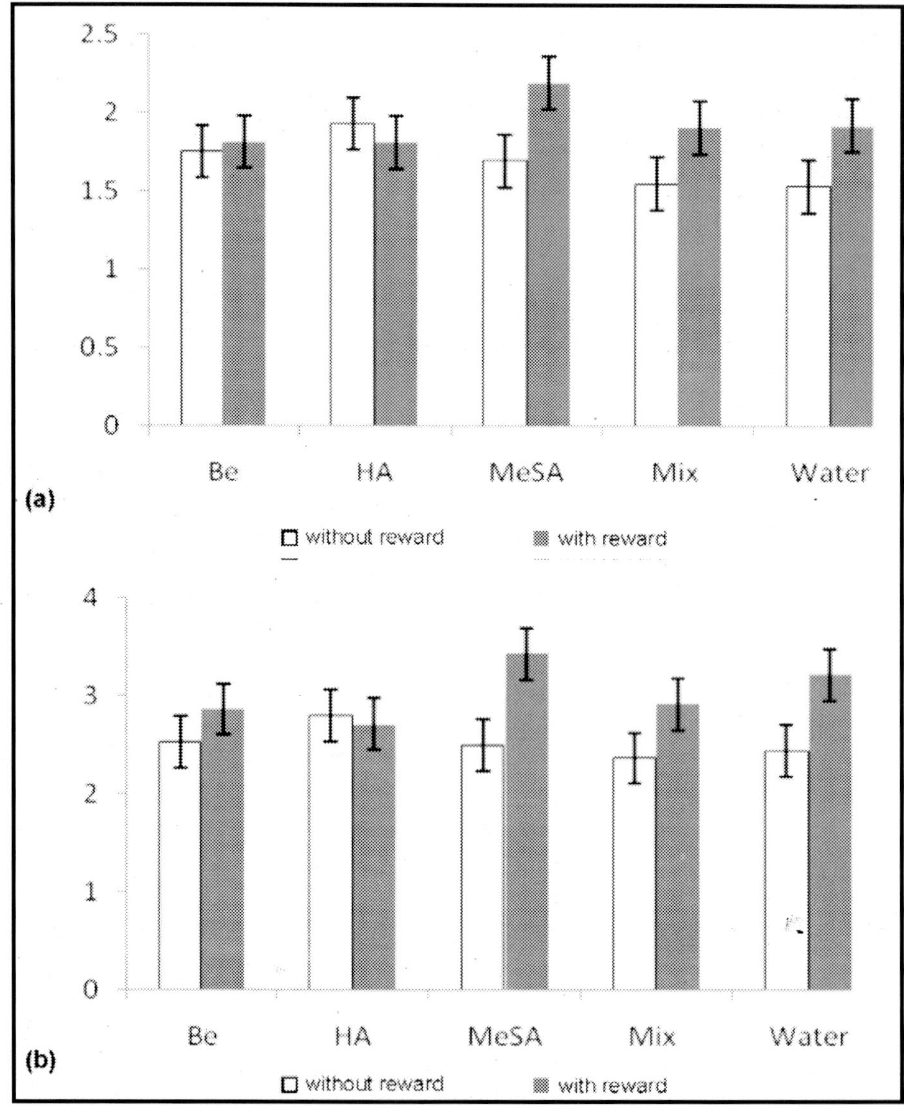

Fig. 2.2. Mean (√x+0.5) trap catches of parasitoids on non-attracting sticky traps adjacent to broccoli plants treated with synthetic herbivore-induced plant volatile compounds and with or without a reward of buckwheat flowers (all time periods pooled) (a) Encyrtidae, (b) all parasitic Hymenoptera

In the broccoli experiment, total catch of all parasitic Hymenoptera responded to the treatments with a significant (F=2.44, df=4, 150, P=0.049) HIPV* reward interaction effect. In the HIPV treatments without buckwheat reward plots, parasitoid catches were increased in HA treated plots (Fig 2.2a) but did not differ significantly from the water control and from other HIPV treatments. The presence of buckwheat led to a significant increase in parasitoid catch in the water control treatment, the MeSA and HIPV Mix treatments. For encyrtids, there was a significant (F = 2.46, df = 4, 150; P = 0.048) HIPV* reward interaction. Only HA led to a significant increase in catches compared with the control (Fig. 2.2b). For this taxon, the buckwheat treatment had a significant effect for MeSA, the HIPV Mix and water control treatments (Fig. 2.2b).

POTENTIAL FOR BIOLOGICAL CONTROL ENHANCEMENT

Work by Butler *et al.* (2007) addressed the tension between biodiversity conservation and the need for increased agricultural production. Conserving biodiversity on farmland is essential because agriculture is the dominant land use and wildlife may have nowhere else to survive. However, reconciling this need with the quest for sustainable development and adequate nutrition for the human population may not be entirely in conflict. Many species that might be conserved on farms can provide valuable ecosystem services. Biological control of crop and rangeland pests is one such service, and has been valued at US$ 123 × 10^9/year globally (Costanza, 2004). But pests continue to cause crop losses and drive the use of pesticides which are produced from non-renewable resources. A dependence on pesticides reduces biodiversity and also leads to external environmental and human health costs. We therefore, need good ecological science to manipulate farm biodiversity to increase the impact of biological control. Chemical ecology approaches that are based on natural products and the defence mechanisms of plants are an important dimension in conservation biological control. The present results add to the promising findings from the same project (Simpson *et al.*, 2011a,b) as well as other authors that have worked under field conditions (James, 2005; Yu *et al.*, 2008).

This study shows that synthetic HIPV spray applications to plant foliage and the incorporation of floral resources within the crop together increase certain parasitoids in sweet-corn and broccoli. The ability of small parasitoids to move towards synthetic compounds in the field has been demonstrated previously (Murai *et al.*, 2000; Krupke and Brunner, 2003; Fatouros *et al.*, 2008) but the present work is the one of the first to test the 'attract and reward' approach.

The most encouraging indication that this strategy offers benefits for boosting natural enemy density is from Eulophidae in the sweet-corn experiment. There parasitoids were approximately doubled in number comparing the water control with the addition of buckwheat flowers. Further, several of the HIPV treatments, most notably MeJA, resulted in parasitoid catches significantly above the water control level, particularly when the HIPV was combined with buckwheat flower rewards. The other parasitoid family that responded with significant HIPV reward effects was Encyrtidae but only the combination of MeJA and reward resulted in a catch greater than that in the water control. This suggests that the use of 'attract and reward' is far from a panacea. Just as choice of plant species used in conservation biological control is now recognised as important (Baggen and Gurr, 1998; Wäckers, 2004) and HIPV compounds are known to be complex in their effects on different taxa (Moraes *et al.*, 1998), the combined use of these approaches as an 'attract and reward' strategy going to require significant work to identify optimal compound and flower combinations suited to various natural enemy taxa. Certainly other work suggests that encyrtids can be made to respond to HIPVs. James and Grasswitz (2005) reported attraction of the encyrtid wasp *Metaphycus* sp. to MeJA applied to grapevines.

Encyrtids responded significantly in the present broccoli experiment. The use of either MeSA or the HIPV mix with and buckwheat reward gave a significantly greater numbers of insects than did the water control. Interestingly, the HA treatment without reward also elevated encyrtid numbers. Since encytrids were one of the more common parasitoid families captured, the same general trend applied to parasitic Hymenoptera as a whole but with the notable addition of Be and reward now giving a significant response.

Opportunities for Biological Control

For the development of effective conservation biological control (CBC) strategies it is crucial to achieve not only increased abundance of beneficial insects but also cascading effects on lower trophic levels: reduction in pest numbers and pest damage. More detailed reports from the same project that the present case study is drawn from (Simpson *et al.*, 2011b) provide that tantalising information: in the sweet-corn experiment there was reduced abundance of *Helicoverpa* spp. larvae and reduced levels of plant damage. Overall, much remains to be resolved before 'attract and reward' can be developed into a workable pest management strategy but results from the present paper and those referred to within offer encouraging support for this research direction.

CONCLUSIONS

Laboratory work over two decades has shown that plants under attack by arthropod herbivores produce volatile chemicals that attract predators and parasitoid wasps. Many of the herbivore-induced plant volatiles (HIPVs) responsible for this effect have been identified, synthesised and used in slow-release dispensers. Under field conditions they result in elevated catches of natural enemies. Remarkably, application of compounds such as jasmonic acid to plants can also induce the production of a natural blend of HIPVs. Such findings suggest that applying synthetic HIPVs to crops may attract – both directly and indirectly – the natural enemies that could protect crops from pest damage. That effect was evident in results from the first year of an Australian study. The present paper integrates results from the second year of that Australian field study and shows that technologies based on this natural mechanism could constitute the foundation for a new approach to increasing the ecosystem service of biological pest control, especially when combined with the presence of nectar plants to 'reward' natural enemies once attracted.

ACKNOWLEDGEMENTS

A.T. Simmons, D.G. James, G. Leeson and H. Nicol are acknowledged for their input into the Australian Research Council-funded project from which the case study is drawn from. The authors also thank Jeff West, Donna Read, Karen Gogala and Greg Simpson for help with field work, the growers Jeff McSpedden and Bruce Armstrong for allowing us to conduct the field experiments and Linton Staples for providing methyl anthranilate. Donna Read is thanked also for assistance with manuscript preparation.

REFERENCES

Arimura, G.I., Kost, C. and Boland, W. (2005) Herbivore- induced indirect plant defence. *Biochim. Biophys. Acta,* **1734**, 91–111.

Adler, L.S. and Bronstein, J.L. (2004) Attracting antagonists: Does floral nectar increase leaf herbivory? *Ecology,* **85**, 1519–1526.

Baggen, L.R. and Gurr, G.M. (1998) The influence of food on *Copidosoma koehleri* (Hymenoptera : Encyrtidae), and the use of flowering plants as a habitat management tool to enhance biological control of potato moth, *Phthorimaea operculella* (Lepidoptera : Gelechiidae). *Biol. Control,* **11**, 9–17.

Baggen, L.R., Gurr, G.M. and Meats, A. (1999) Flowers in tri-trophic systems: mechanisms allowing selective exploitation by insect natural enemies for conservation biological control. *Entomol. Exp. Appl.*, **91**, 155–161.

Begum, M., Gurr, G.M., Wratten, S.D., Hedberg, P.R. and Nicol, H.I. (2006) Using selective food plants to maximize biological control of vineyard pests. *J. Appl. Ecol.*, **43**, 547–554.

Berndt, L.A., Wratten, S.D. and Hassan, P.G. (2002) Effects of buckwheat flowers on leafroller (Lepidoptera : Tortricidae) parasitoids in a New Zealand vineyard. *Agric. Forest Entomol.*, **4**, 39–45.

Bowie, M.H., Wratten, S.D. and White, A.J. (1995) Agronomy and phenology of "companion plants" of potential for enhancement of insect biological control. *New Zealand J. Crop Hort. Sci.*, **23**, 423–427.

Butler, S.J., Vickery, A. and Norris, K. (2007) Farmland biodiversity and the footprint of agriculture. *Science*, **315**, 381–384.

Costanza R. D'Arge, R., De Groot, R., Farber, S., Grasso, M., Hannon, B., Limburg, K., Naeem, S., O'Neill, R.V., Paruelo, J., Raskin, R.G., Sutton, P. and Van Den Belt, M. (1997) The value of the world's ecosystem services and natural capital. *Nature*, **387**, 253–260.

De Moraes, C.M., Lewis, W.J., Paré, P.W., Alborn, H.T. and Tumlinson, J.H. (1998) Herbivore-infested plants selectively attract parasitoids. *Nature*, **393**, 570–573.

Dicke, M. and Sabelis, M.W. (1998) How plants obtain predatory mites as bodyguards. *Neth. J. Zool.*, **38**, 148–165 .

Fatouros, N.E., Dicke, M., Mumm, R., Meiners, T. and Hilker, M. (2008) Foraging behavior of egg parasitoids exploiting chemical information. *Behav. Ecol.*, **19**, 677–689.

Greathead, D.J. and Greathead, A.H. (1992) Biological control of insect pests by insect parasitoids and predators : the BIOCAT database. *Biocontr. News Inf.* **13**, 61N–68N.

Irvin, N.A., Scarratt, S.L., Wratten, S.D., Frampton, C.M., Chapman, R.B. and Tylianakis, J.M. (2006) The effects of floral understoreys on parasitism of leafrollers (Lepidoptera : Tortricidae) on apples in New Zealand. *Agric. Forest Entomol.*, **8**, 25–34.

James, D.G. (2003) Field evaluation of herbivore-induced plant volatiles as attractants for beneficial insects: methyl salicylate and the green lacewing, *Chrysopa nigricornis*. *J. Chem. Ecol.*, **29**, 1601–1609.

James, D.G. (2005) Further field evaluation of synthetic herbivore-induced plant volatiles as attractants for beneficial insects. *J. Chem. Ecol.*, **31**, 481–495.

James, D.G. (2011) Grape and hop plants sprayed with botanical oil pesticides containing herbivore-induced plant volatiles attract insect predators and parasitoids. *Environ. Entomol.* (In Press).

James, D.G. and Price, T.S. (2004) Field-testing of methyl salicylate for recruitment and retention of beneficial insects in grapes and hops. *J. Chem. Ecol.* **30**, 1613–1628.

James, D.G. and Grasswitz, T.R. (2005) Synthetic herbivore-induced plant volatiles increase field captures of parasitic wasps. *BioControl,* **50**, 871–880.

Krupke C.H. and Brunner J.F. (2003) Parasitoids of the consperse stink bug (Hemiptera: Pentatomidae) in north central Washington and attractiveness of a host-produced pheromone component. *J. Entomol. Sci.,* **38**, 84–92.

Lavandero, B., Wratten, S., Shishehbor, P. and Worner, S. (2005) Enhancing the effectiveness of the parasitoid *Diadegma semiclausum* (Helen): Movement after use of nectar in the field. *Biol. Control,* **34**, 152–158.

Lee, J.C. and Heimpel, G.E. (2005) Impact of flowering buckwheat on lepidopteran cabbage pests and their parasitoids at two spatial scales. *Biol. Control,* **34**, 290–301.

Lou, Y-G., Du, M-H., Turlings, T.C.J., Cheng, J-A. and Shan, W-F. (2005) Exogenous application of jasmonic acid induces volatile emissions in rice and enhances parasitism of *Nilaparvata lugens* eggs by the parasitoid *Anagrus nilaparvatae. J. Chem. Ecol.,* **31**, 1985–2002.

Murai, T., Imai, T. and Maekawa, M. (2000) Methyl anthranilate as an attractant for two thrips species and the thrip parasitoid *Ceranisus menes. J. Chem. Ecol.,* **26**, 2557–2565.

Naumann, I.D. (1991) *The Insects of Australia: A Textbook for Students and Research Workers.* 2nd ed. Melbourne University Press, Melbourne.

Payne, R.W., Murray, D.A., Harding, S.A., Baird, D.B. and Soutar, D.M. (2009) *Genstat for Windows (12th Ed.) Introduction.* VSN International, Hemel Hempstead.

Polis, G. A. and Strong, D. R. (1996) Food web complexity and community dynamics. *Am. Nat.* **147**, 813–846.

Simpson, M.R., Gurr, G.M., Simmons, A.T., Wratten, S.D. James, D.G., Leeson, G. and Nicol, H.I. (2011a) Insect attraction to synthetic herbivore-induced plant volatiles in field crops. *J. Appl. Ecol.* (in press). doi: 10.1111/j.1365-2664.2010.01946.x.

Simpson, M., Gurr, G.M., Simmons, A.T., Wratten, S.D., James, D.G., Leeson, G. and Nicol, H.I. (2011b) Insect attraction to synthetic herbivore-induced plant volatile-treated field crops. *Agric. Forest Entomol.* (in press). doi: 10.1111/j.14619563.2010.00496.x.

Stevens, N.B., Stephens, C.J., Iqbal, M., Jennings, J.T., La Salle, J. and Austin, A.D. (2007) *What Wasp is That? An Interactive Identification Guide to the Australasian Families of Hymenoptera.* Australian

Biological Resources Study (ABRS) and Centre for Biological Information Technology (CBIT), Australia.

Turlings, T.C.J., Tumilson, J.H. and Lewis, W.J. (1990) Exploitation of herbivore-induced plant odors by host seeking parasitic wasps. *Science*, **250**, 1251–1253.

Thies, C. and Tscharntke, T. (1999) Landscape structure and biological control in agroecosystems. *Science*, **285**, 893–895.

Wäckers, F.L. (2004) Assessing the suitability of flowering herbs as parasitoid food sources: flower attractiveness and nectar accessibility. *Biol. Control*, **29**, 307–314.

Wäckers, F.L., Romeis, J. and van Rijn, P. (2007) Nectar and pollen feeding by insect herbivores and implications for multitrophic interactions. *Annu. Rev. Entomol.*, **52**, 301–323.

Witting-Bissinger, B.E., Orr, D.B. and Linker, H.M. (2008) Effects of floral resources on fitness of the parasitoids *Trichogramma exiguum* (Hymenoptera : Trichogrammatidae) and *Cotesia congregata* (Hymenoptera: Braconidae). *Biol. Control*, **47**, 180–186.

Wolcott, G.N. (1942) The requirements of parasites for more then hosts. *Science*, **96**, 317–318.

Yu, H.L., Zhang, Y.J., Wu, K.M., Gao, X.W. and Guo, Y.Y. (2008) Field-testing of synthetic herbivore-induced plant volatiles as attractants for beneficial insects. *Environ. Entomol.*, **37**, 1410–1415.

3

Insect Pheromone Technology: A Vigorous Tree of Science Blessed with a Variety of Evergreen Branches, Including Nanotechnology

HANS E. HUMMEL[1,2]* AND DETLEF F. HEIN[1]

[1]Organic Agriculture, Justus-Liebig-University of Giessen
Karl-Glöckner-Strasse 21C, D-35394 Giessen, Germany
[2]Illinois Natural History Survey, Institute of Natural Resource Sustainability, University of Illinois, 1816 S. Oak Street
Champaign, Il.61820, USA
*E-mail: Hans.E.Hummel@agrar.uni-giessen.de

INTRODUCTION

Insect pheromones, "invented" and first described fifty years ago, experienced a time of vigorous, exponential growth during the period from the 1970s to 1990s when in many entomological journals reports on newly identified pheromone structures were abundant. These reports covered a variety of functions such as attractants and kairomones, repellents, new concepts of insect management, and a bewildering number of applications. Over the last 50 years, the pheromone tree which is itself an offshoot in the larger context of chemical ecology thus produced many healthy branches. As a result, the literature documents a rich variety of reports describing the management of pest insects in agriculture, horticulture, silviculture, medical entomology and stored product insects research. Integrated pest management, once a dream propagated by some pioneers (Butenandt, 1939; Butenandt *et al.*, 1959; Hecker and Butenandt, 1984; Metcalf and Luckmann, 1994) now is on its way to reality (Koul *et al.*, 2002; Bajwa and Kogan, 2004; Vreysen *et al.*, 2007; Horowitz *et al.*, 2009). Today, "monitoring" the presence and abundance of specific pest insects, their "mass trapping", the "attract and trap", "lure and kill", "MSD", (mass trapping combined with shielding and deflection; Hummel *et al.*, 2006), and such well proven approaches as male confusion or "female-male communication (mating) disruption" listed (Fig. 3.1) are common terms. As standard procedures of chemical ecology, they are practiced by entomologist all over the world. These efforts cover an increasingly large variety of insect species on a large variety of crops (Hummel and Miller, 1984; Jutsum and Gordon, 1989; Ridgway *et al.*, 1990; Cardé and Minks, 1997; Howse *et al.*, 1998; Hardie and Minks, 1999; Wyatt, 2003; Witzgall *et al.*, 2010). Pheromones are characterized by their proverbial specificity and eminent efficiency. Only minute amounts in the submilligram range are necessary for trapping while some 40 to 100 g/ha are required for mating disruption. Pheromones and equally kairomones are environmentally benign, biodegradable, nontoxic, virtually free of inducing resistance, and compatible with all other IPM approaches. They are key tools in our search for technology suitable for a sustainable management of this planet's limited resources. Unfortunately, a majority of conventional dispensers (for reviews see Cardarelli, 1979; Leonhardt and Beroza, 1982; Hummel, 1983; Jutsum and Gordon, 1989; Kydonieus and Beroza, 1982; Howse *et al.*, 1998; Wennemann and Hummel, 2001) used for these purposes are still less than satisfactory, in spite of about four decades of technical developments. Recently, however, an avenue was discovered linking pheromone technology to organic nanofiber technology (Hellmann *et al.*, 2009; Hein *et al.*, 2009a,b). Thus, a new twig has been identified on the pheromone tree (Fig. 3.1) which may further develop into a vigorous, independent branch. First results will be presented to substantiate this unexpectedly successful link.

NANOFIBER TECHNOLOGY

Organic nanofibers (not to be mistaken as inorganic nanoparticles) have been known for several decades. Virtues of the electrospun variety of nanofibers have recently been broadly reviewed by Greiner and Wendorff (2007) who added very substantially to their development and to their first application towards solving problems in agriculture (Hellmann *et al.*, 2009). Hein *et al.* (2009a,b; 2010) described the techniques necessary for inclusion of grapevine moth pheromone (E,Z)-7,9-dodecadienyl acetate into such nanofibers. They also demonstrated the feasibility for mating disruption of the European grapevine moth, *Lobesia botrana* (Lepidoptera: Tortricidae) in South

German vineyards of Staatliches Weinbauinstitut Freiburg/Breisgau. The "semifield cage approach" of Doye (2006) of the pheromone research group at Kaiserslautern, Germany, paved the way for demonstrating first successful disruption results in vineyards. In our experiments of 2009; three replications were performed. The experimental protocol called for (i) a negative control (without pheromone), (ii) a positive control (with the well proven Isonet® plastic "spaghetti" dispenser of Shin Etsu) and (iii) the new nanofiber dispenser containing 33 weight% of pheromone in the electrospun Ecoflex® polymer matrix. The samples were tested simultaneously side by side within the same vineyard of Freiburg during the summer and fall season of 2009.

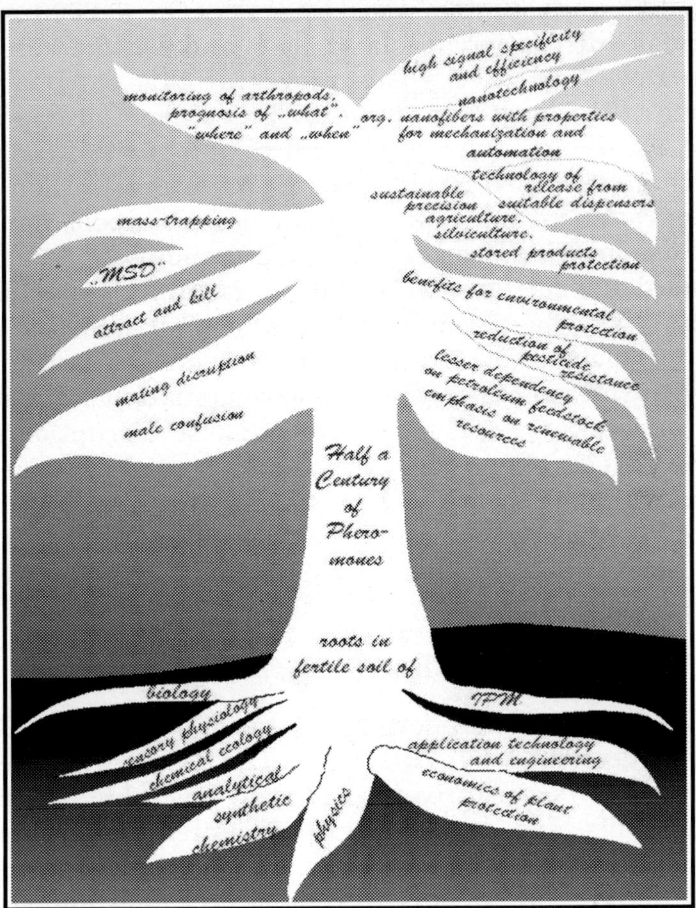

Fig. 3.1. Hypothetical pheromone tree. Branches on the left mark specific pheromone techniques. Branches on the right side list some typical advantages of pheromones. The tree is rooted in the biological, chemical and physical sciences. But IPM, application technology and economics also play their roles

Male disruption (Q) is defined as:

$$Q = R/Rk \times 100\ (\%)$$

Where, R = number of males in treated plot, Rk = number of males in untreated plot); When

Q = 100%, there is no difference between treated and untreated plot. The lower Q the better is the mating disruption.

The success in mating disruption or degree of efficiency (P) is defined as:

$$P = 100 - Q\ (\%)$$

First mating disruption results obtained during the course of 4 weeks in 2009 were plotted week by week. On a 2000 m^2 area of vineyard, male disruption levels of 90% and better were recorded within the Doye-field cages. This proves at this stage of technical development the longevity of the pheromone preparation for respectable 3.5 weeks. Twenty weeks duration of the (biodegradable) fibers are the ultimately desired goal of the future, without the need for another treatment and without removing the spent nanofibers at the end of the growing season. They will decompose by spontaneous environmental interactions. This will be generally beneficial for the environment, will save fuel costs, manpower, and the need for expensive supplementary treatments. In detail, disruption levels achieved with the nanofiber-pheromone-combination were 89, 89, 95, and 45%, for weeks 1 to 4, respectively. These disruption levels are significantly different from the untreated controls during weeks 1 to 3 while statistical significance was lacking with respect to the positive control. During week 4, the positive control was still active at the 92% disruption level, while the nanofiber-pheromone-combination became progressively depleted.

IMPLICATIONS FOR PEST MANAGEMENT

Ecoflex®, a copolymer of 1,4-butanediol, terephthalate, and adipate, was described by Witt *et al.* (2001). It was chosen as a polymer matrix for several reasons: First, it has a high affinity to the aliphatic alcohols and esters often encountered as characteristic components of pest insect pheromones. Secondly, Ecoflex® is biodegradable and decomposes rapidly in contact with soil microorganisms. There is no need to collect the spent fibers at the end of the growing season. If the need arises for various degrees of UV protection of the pheromone double bonds (which are vital to their function and are keys to their species specificity), a strongly UV absorbing sunscreen additive may be mixed into the polymer-pheromone solution. Thirdly, pheromone material is *inside* the fiber, not just adsorbed to the outside. Hellmann *et al.* (2009) provided proof for this notion by reacting the fibers with RuO_4 and finding electron dense inclusion islets *inside* the slim fibers which in themselves have the width of 1/100th of a human hair (Fig. 3.2).

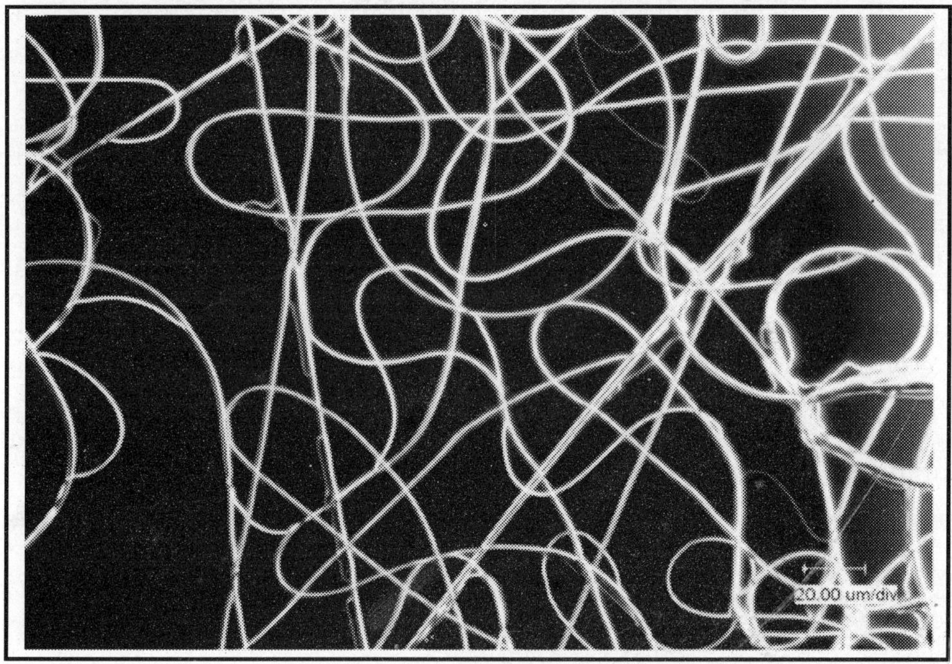

Fig. 3.2. Nanofibers derived from electrospinning of a solution of codlemone with Ecoflex® (courtesy A. Greiner, Philipps-University Marburg, with permission). A human hair would have a diameter of 100× that of the electrospun nanofibers

Choices for other suitable polymer materials are by no means exhausted. Experiments so far have been undertaken with only 3 classes of biodegradable polymers: polyamide (mimicking to some extent the natural spider silk), cellulose acetate, and polyesters. Dozens of other polymers and co-polymers with adaptable and optimized properties may be considered.

Not all pheromones are equally compatible with all polymers. The electrospinning process yielding a homogenous, extremely long fiber without humps and constrictions is in itself a complicated and mechanistically incompletely understood physical procedure (Fig. 3.2). Certain combinations do not blend and spin out well enough. Up to now, selecting compatible partners is a matter of empiricism. The best combination so far turned out to be Ecoflex® fiber with 33% (E,Z)-7,9-dodecadienyl acetate. Many more combinations must be tried in the future for obtaining a broader choice for selection. Of course, other pest insects with known pheromones will be tried as soon as stable nanofiber-pheromone combinations become available.

Best progress so far has been achieved with polymer fibers spun under controlled conditions in the laboratory and then transported to the vineyard to be employed there. However, *direct* electrospinning under field conditions will be the key to future applications on large acreages. Characteristic for large scale row crops are cotton, maize, wheat, soybeans, alfalfa, potatoes, rice, and tea, coffee, and tobacco, to name a few. Suitably engineered machinery for some of these purposes is at the beginning of its development (Fig. 3.3). In industrialized countries with their inherently high wage level, manpower saving strategies are best be met by partially or completely mechanized if not automated processes. Capital investments into these ventures are currently the rate limiting factor for further progress. Most investments of the past were spent for conventional droplet spraying technology with its obvious strength but also its environmental handicaps of wastefulness and unsustainability.

Fig. 3.3. Tractor with attached electrospinning platform (courtesy Prof. A. Greiner, Philipps-University Marburg, Germany, with permission).

THE SIGNIFICANCE OF PHEROMONES

How could we best prove the significance of pheromones? Simply by looking at the literature of the last fifty years where progress of the field is recorded better than by any historian of science. Because pheromones are active in such tiny amounts, specific techniques had to be created for their detection, quantification, and synthesis (Mori, 1992; Schulz, 2005). Physics, chemistry, and chemical ecology provided some of the roots but existing techniques had to be greatly refined. Justus Liebig's legendary analytical achievements of the mid 1850s one hundred years later were, also under the challenge of pheromone analytics, dwarfed by the advancement of chromatography and spectroscopy. Substance characterization and identification by establishing their C/H/O/N ratios, once crudely possible at the 10–100 mg level, suddenly was possible at the 10–100 ng level. This is a million fold gain in sensitivity. Some references quoted in Table 3.1 (part a), may highlight these accomplishments. Similarly, biological and hyphenated techniques (Table 3.1, part b) because of their built in specificity, were able to advance the field even further, as the senior author could witness during his own lifetime.

Through the investigations of Schneider (1957) and his associates, e.g. Kaissling (1971, 2009), we know how sensitive and species specific insect antennae are. In the case of *Bombyx mori*, 5000 molecules of bombykol will excite the male antenna and induce mating dances of males, themselves valuable and indispensable indicators in the hands and Butenandt and coworkers during their long search for the structure of bombykol (Fig. 3.4) and its structural proof by total synthesis (Butenandt and Hecker, 1961) (for a review see Hecker and Butenandt, 1984).

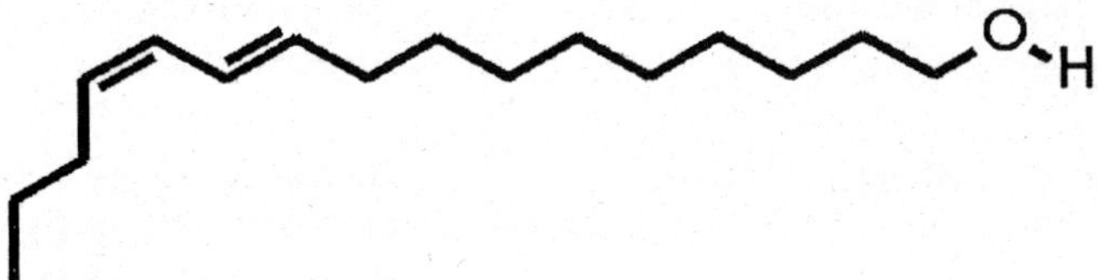

Fig. 3.4. Bombykol, (E,Z)-10,12-Hexadecadien-1-ol, the prototype of unsaturated straight chain aliphatic pheromones

Moreover, applications of pheromones (see left part of the pheromone tree in Fig. 3.1) are so abundant today that we cannot imagine pest management (IPM) without pheromones as an integral component.

Table 3.1. Significance of pheromones

Technique	Author(s)
***Part A:* Pheromone research advanced the development of the following *abiotic* techniques:**	
Microzonolysis, microhydrogenation, micro-derivatization and characterization of emerging products;	Hecker (1955), Huwyler (1973, 1976, 1979), Beroza (1975), Hecker & Butenandt (1984), Attygalle (1998)
Gas chromatography with flame ionization detector	Huwyler (1973, 1976, 1979), Weber & Weber (1982), Heath & Tumlinson (1984)
Hyphenated chromatographic-spectrometric techniques; GLC-MS, HPLC-MS	Webster *et al.* (1998)
Liquid chromatography	Heath & Tumlinson (1984), Millar & Haynes (1998)
Microspectrometric techniques, e.g. GLC-FT-IR, HPLC-UV	Beroza (1975), Heath & Tumlinson (1984), Silverstein *et al.* (1991), Leal (1998)
NMR	Heath & Tumlinson (1984), Webster & Kiemle (1998), Mori (1998)
***Part B:* Pheromone research advanced these *biotic* techniques:**	
EAG, GLC-EAD (insect antennae as detector)	Schneider (1957), Kaissling (1971, 2009), Arn *et al.* (1975), Cork *et al.* (1990), Bjostad (1998)
Single cell recordings (SCR), GLC-SCR	Wadhams (1984), Bjostad (1998)
EAG, SCR as parts of BioFET biosensors	Schuetz *et al.* (1996, 1999, 2000)
Insect flight activation chamber, t-GLC-behavior bioassay	Flaschenträger (1957), Bayer & Anders (1959), Gaston *et al.* (1966), Hummel (1984)
Flight tunnels for evaluating upwind orientation mechanisms of flying insects	Baker & Lynn (1984), Baker & Cardé (1984), Witzgall (1997), El-Sayed *et al.* (1999)
Orientation mechanisms of insects	Farkas & Shorey (1972), Kramer (1978, 1997), Kennedy *et al.* (1980)

Today and in hindsight, semiochemicals can be classified according to the following straightforward scheme (Fig. 3.5).

It is based on a body of about ten thousand references and is the result of worldwide cooperation. This in itself is a gigantic accomplishment, in spite of many presently existing deficiencies. A lot of fine print still will have to be added in the future because so far only the pheromones of roughly one thousand species received closer investigation whereas, the class of Insecta consists of one million and probably considerably more species with individual pheromone composition. In addition to the heuristic value of pheromones, kairomones, and allomones for the basic understanding of chemical communication (Meinwald and Eisner, 1995), pheromone applications must always be seen in the wider context of IPM (Hummel, 2007; Horowitz *et al.*, 2009; Pedigo and Rice, 2009).

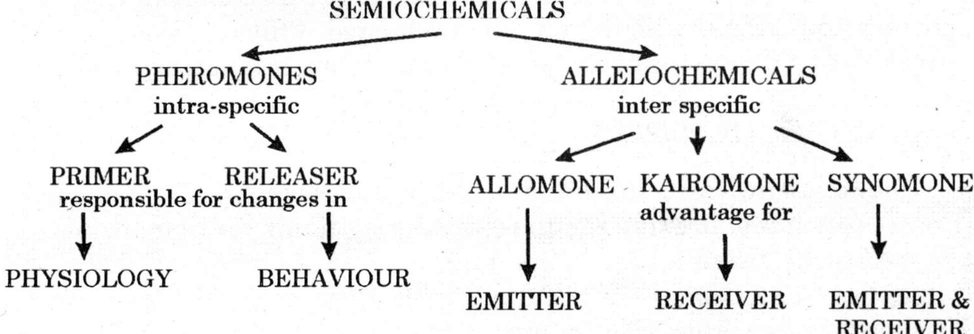

Fig. 3.5. Pheromones in the wider context of signal compounds known in chemical ecology (modified from references Karlson and Lüscher, 1959; Brown *et al.*, 1970; Nordlund, 1981).

Overall, pheromone technology once again proves to be a healthy tree blessed with a novel, unexpectedly productive branch whose fruits will facilitate new future applications in agricultural food production and environmental protection. Butenandt and his associates anticipated the usefulness of pheromones in agriculture. Their visions of 1939 and 1959 finally are finding their experimental proof.

CONCLUSIONS

The period of 2009-2011 is seeing the 50th anniversary of pheromones. Both the term and the first chemically identified member of this class of compounds originated from the same laboratory at Munich, Germany. Pheromones as a branch of the biological tree enjoyed extraordinary, exponential growth. Today, while reaching adulthood, this tree may be seen as an example for an interdisciplinary and truly international endeavour with far reaching consequences. Plant protection against pest arthropods is unthinkable without the input by pheromones at various levels. Far away from senescence, the pheromone tree is growing new branches, among them the challenging integration with nanotechnology. Organic nanofibers with insect pheromone dissolved in the nanofibers are an entirely novel technology whose applications could prove to be a milestone towards better, ideally the "intelligent" and fully adjustable dispensers of the future. These are expected to combine longevity of constant pheromone release, yet biodegradability of the polymer material, while protecting the pheromone against the UV radiation in sunlight, against rain, and dust. Some experimental parameters such as the electrospinning conditions, the polymer material, and the ratio of pheromone to polymer materials are open to future manipulation and optimization. This will result in prolonged efficacy and provide opportunities for savings through mechanization and automatization.

ACKNOWLEDGEMENTS

Dieter Schwarz Foundation (Heilbronn-Neckarsulm) and BLE (Bonn), Germany, provided high risk starting funds for demonstrating the technical feasibility of this approach. A number of dedicated colleagues (Greiner, Wendorff, Leithold, Breuer, Vilcinskas, Kratt, Kleeberg), associates (Hellmann, Deuker) and students (Lindner, Czyrt, Schmidt, Lahr) helped in translating the original ideas of 2004 into reality. Patent application advice of Drs. Stumpf and Lischka of TransMIT GmbH Giessen is gratefully acknowledged. Stimulating and critical discussions with Prof. U. Koch, Kaiserslautern, forced the first author's research idea into focus and challenged him to think about repeated modifications and refinements. Dipl.-Ing. agr. A. Deuker helped with formatting the figures and editing the text. The authors are grateful to Prof. Dr. A. Greiner, Marburg, for the permission to reprint Figs. 3.2 and 3.3.

REFERENCES

Attygalle, A.B. (1998) Microchemical techniques. In: J.G. Millar and K.F. Haynes (eds.), *Methods in Chemical Ecology*, Vol. 1, Kluwer Academic Publishers, Amsterdam, The Nethrlands, pp. 207–294.

Arn, H., Städler, E. and Rauscher, S. (1975) The electroantennographic detector - A selective and sensitive tool in the gas chromatographic analysis of insect pheromones. *Z. Naturforsch.*, **30C**, 722–725.

Bajwa, W.I. and Kogan, M. (2004) Cultural practices: Springboard to IPM. In: O. Koul, G.S. Dhaliwal and G.W. Cuperus (eds.), *Integrated Pest*

Management. Potential, Constraints and Challenges, CABI Publishing, Wallingford, pp. 21–38.

Baker, T.C. and Cardé, R.T. (1984) Techniques for behavioral bioassays. In: H.E. Hummel and T.A. Miller (eds.), *Techniques in Pheromone Research*, Springer, New York, pp. 45–73.

Baker, T.C. and Linn, C.E. Jr. (1984) Wind tunnels in pheromone research. In: H.E. Hummel and T.A. Miller (eds.), *Techniques in Pheromone Research*, Springer, New York, pp. 75–110.

Bayer, E. and Anders, E. (1959) Biologische objekte als detektoren für gaschromatographie. *Naturwissenschaften*, **46**, 380.

Beroza, M. (1975) Microanalytical methodology relating to the identification of insect sex pheromone and related behavior-control chemicals. *J. Chromatogr. Sci.*, **13**, 314–321.

Bjostad, L.B. (1998) Electrophysiological methods. In: J.G. Millar and K.F. Haynes (eds.), *Methods in Chemical Ecology*, Vol. 1, Kluwer Academic Publishers, Amsterdam, The Netherlands, pp. 339–375.

Brown, W.L., Eisner, T. and Whittaker, R.H (1970) Transspecific chemical messengers: Allomones and kairomones. *BioScience*, **20**, 21–22.

Butenandt, A. (1939) Zur Kenntnis der Sexual-Lockstoffe bei Insekten (Kurzfassung). In: *Jahrbuch der Preußischen Akademie der Wissens-chaften*, Mathematisch-Naturwissenschaftliche Klasse, Berlin, pp. 97–98.

Butenandt, A. and Hecker, E. (1961) Synthese des bombykols, des sexuallockstoffes des seidenspinners, und seiner geometrischen Isomeren. *Angew. Chem.*, **73**, 349–353.

Butenandt, A., Beckmann, R., Stamm, D. and Hecker, E. (1959) Über den sexuallockstoff des seidenspinners *Bombyx mori*. Reindarstellung und konstitution. *Z. Naturforsch.*, **14**, 283–284.

Cardarelli, N.F. (1979) The efficacy, environmental impact and mechanism of release and dispersal of pesticide materials emitted from a controlled-release dispenser. In: H. Geissbühler (ed.), *Advances in Pesticide Research*, part 3, IUPAC, Pergamon Press, Oxford, pp, 744–753.

Cardé, R.T. and Minks, A.K. (1997) *Insect Pheromone Research. New Directions.* Chapman & Hall, New York.

Cork, A., Beevor, P.S., Gough, A.J.E. and Hall, D.R. (1990) Gas chromatography linked to electroantennography: a versatile technique for identifying insect semiochemicals. In: A.R. McCaffery and I.D. Wilson (eds.), *Chromatography and Isolation of Insect Hormones and Pheromones*, Plenum, New York, pp. 271–274.

Doye, E. (2006) Entwicklung eines Freilandtests zur Überprüfung der Wirksamkeit von Pheromonanwendungen im Weinbau. Dissertation: TU Kaiserslautern, Biology. URN (NBN): de:hbz : 386-kluedo-19213.

El-Sayed, A., Gödde, J., Witzgall, P. and Arn, H. (1999) Characterization of pheromone blend for grapevine moth, *Lobesia botrana*, by using flight track recording. *J. Chem. Ecol.*, **25**, 389–400.

Farkas, S. and Shorey, H.H. (1972) Chemical trail-following by flying insects: a mechanism for orientation to a distant odor source. *Science*, **178**, 67–68.

Flaschenträger, B., Sayed Amin, El. and Jarczyk, H.J. (1957) Ein Lockstoff-analysator (Odouranalyzer) für Insekten. *Microchim. Acta*, **45**, 385–389.

Gaston, L.K., Fukuto, T.R. and Shorey, H.H. (1966) Sex pheromones of noctuid moths. IX. Isolation techniques and quantitative analysis for pheromones with special reference to that of *Trichoplusia ni* (Lepidoptera: Noctuidae). *Ann. Entomol. Soc. Am.*, **59**, 1062–1066.

Greiner, A. and Wendorff, J.H. (2007) Electrospinning: A fascinating method for the preparation of ultrathin fibers. *Angew. Chem. Int. Ed.*, **46**, 5670–5703.

Hardie, J. and Minks, A.K. (1999) *Pheromones of Non-Lepidopteran Insects Associated with Agricultural Plants.* CABI Publishing, Wallingford.

Heath, R.R. and Tumlinson, J.H. (1984) Techniques for purifying, analyzing, and identifying pheromones. In: H.E. Hummel and T.A. Miller (eds.), *Techniques in Pheromone Research*, Springer, New York, pp. 287–322.

Hecker, E. (1955) 4'-Nitro-azobenzol-carbonsäure-4-chlorid als reagenz auf alkohole. *Chem. Ber.*, **88**, 1666–1675.

Hecker, E. and Butenandt, A. (1984) Bombykol revisited. reflections on a pioneering period and on some of its consequences. In: H.E. Hummel and T.A. Miller (eds.), *Techniques in Pheromone Research*, Springer, New York, pp. 1–44.

Hein, D.F., Leithold, G., Hummel, H.E., Vilcinskas, A., Greiner, A., Wendorff, J.H., Dersch, R., Hellmann, Ch., Breuer, M., Beer, H., Schroer, S., Kratt, A., Kleeberg, H., Schulze, Ch. and Wahl, F. (2009a) Nanofasern als neuartige träger für flüchtige signalstoffe zur biotechnischen regulierung von schadinsekten im integrierten und ökologischen landbau. In: J. Mayer *et al.* (eds.), *Werte – Wege – Wirkungen: Biolandbau im Spannungsfeld zwischen Ernährungssicherung, Markt und Klimawandel.* 10. Wissenschaftstagung Ökologischer Landbau Zürich, 11-13 Februar 2009, Dr. Köster, Berlin, pp. 308–311.

Hein, D.F., Leithold, G., Hummel, H.E., Lindner, I., Vilcinskas, A., Greiner, A., Wendorff, J.H., Dersch, R., Hellmann, Ch., Breuer, M. and Kratt, A. (2009b) Nanofibers as novel carriers for semiochemicals in plant protection. *Mitt. Dtsch. Ges. Allg. Angew. Entomol.*, **17**, 105–108.

Hein, D.F., Breuer, M., Hummel, H.E., Greiner, A., Wendorff, J.H., Hellmann, Ch., Vilcinskas, A., Kratt, A., Kleeberg, H. and Leithold, G. (2010) Electrospun nanofibers as novel carriers of insect pheromones: communication disruption strategy against *Lobesia botrana*. In IOBC/

wprs Bulletin, Working Group "Integrated Protection in Viticulture", *Proceedings of the Working Group meeting at Staufen im Breisgau (Germany)*, 01 - 04 November, 2009.

Hellmann, Ch., Greiner, A. and Wendorff, J.H. (2009) Design of pheromone releasing nanofibers for plant protection. *Polymers for Advanced Technologies*, published online in Wiley InterScience, DOI: 10.1002/ pat.1532.

Horowitz, A.R., Ellsworth, P.C. and Ishaaya, I. (2009) Biorational pest control – An overview. In: I. Ishaaya and A.R. Horowitz (eds.), *Biorational Control of Arthropod Pests. Application and Resistance Management*, Springer, Dordrecht, pp. 1–20.

Howse, P., Stevens, I. and Jones, O. (1998) *Insect Pheromones and their Use in Pest Management.* Chapman & Hall, London.

Hummel, H.E. (1983) Insecticides and their design. *J. Nematol.*, **15**, 615–639.

Hummel, H.E. (1984) The tandem gas chromatography-Behavior bioassay. In: H.E. Hummel and T.A. Miller (eds.), *Techniques in Pheromone Research*, Springer, New York, pp. 191–216.

Hummel, H.E. (2007) *Diabrotica virgifera virgifera* LeConte: Inconspicuous beetle-formidable challenges to agriculture. *Commun. Appl. Biol. Sci. Ghent University*, **72**, 7–32.

Hummel, H.E. and Miller, T.A. (1984) *Techniques in Pheromone Research.* Springer, New York.

Hummel, H.E., Shaw, J.T. and Hein, D.F. (2006) A promising biotechnical approach to pest management of the western corn rootworm in Illinois maize fields shielded with a MCA kairomone baited trap line. *Mitt. Dtsch. Ges. Allg. Angew. Entomol.*, **15**, 131–135.

Huwyler, S. (1973) Ultramikromethoden. Katalytische hydrierung. *Experientia*, **29**, 1310–1311.

Huwyler, S. (1976) Ultramicromethods VI. Lithium aluminumhydride reduction od aldehydes and ketones. *J. Microchem.*, **21**, 135–139.

Huwyler, S. (1979) Ultramicromethods VII. Ozonolysis. *J. Microchem.*, **24**, 468–474.

Jutsum, A.R. and Gordon, R.F.S. (1989) *Insect Pheromones in Plant Protection.* Wiley, Chichester.

Kaissling, K.E. (1971) Insect olfaction. In: L.M. Beidler (ed.), *Handbook of Sensory Physiology,* Vol. 4, Springer, Berlin & New York, pp. 351–431.

Kaissling, K.E. (2009) Olfactory perireceptor- and receptor events in moths: a kinetic model revised. *J. Comp. Physiol. A*, **195**, 895–922.

Karlson, P. and Lüscher, M. (1959) "Pheromones", a new term for a class of biologically active substances. *Nature*, **183**, 55–56.

Kennedy, J.S., Ludlow, A.R. and Sanders, C.J. (1980) Guidance system used in moth sex attraction. *Nature*, **288**, 475–477.

Koul, O., Dhaliwal, G.S., Marwaha, S.S. and Arora, J.K. (eds.) (2002) *Biopesticides and Pest Management*. Vols. 1 and 2, Campus Books International, New Delhi.

Kramer, E. (1978) Insect Pheromones. In: G.L. Hazelbauer (ed.), *Taxis and Behaviour*, Chapman & Hall, London, pp. 207–229.

Kramer, E. (1997) A tentative intercausal nexus and its computer model on insect orientation in windborne pheromone plumes. In: R.T. Cardé and A.K. Minks (eds.), *Insect Pheromone Research. New Directions*, Chapman & Hall, New York, pp. 232–247.

Kydonieus, A.F. and Beroza, M. (1982) *Insect Suppression with Controlled Release Pheromone Systems*, Vols. 1 and 2, CRC Press, Boca Raton, Florida.

Leal, W.S. (1998) Infrared and ultraviolet spectroscopy techniques. In J.G. Millar and K.F. Haynes (eds.), *Methods in Chemical Ecology*, Vol. 1, Kluwer Academic Publishers, pp. 185-206.

Leonhardt, B.A. and Beroza, M. (eds.) (1982) *Insect Pheromone Technology: Chemistry and Applications*. ACS Symposium Series 190, American Chemical Society, Washington D.C.

Meinwald, J. and Eisner, T. (1995) The chemistry of phyletic dominance. In: T. Eisner, and J. Meinwald (eds.), *Chemical Ecology. The Chemistry of Biotic Interaction,* National Academy of Sciences, Washington DC, pp. 29–40.

Metcalf, R.L. and Luckmann, W.H. (1994) *Introduction to Insect Pest Management.* 3rd edition. Wiley, New York.

Millar, J.G. and Haynes, K.F. (1998) *Methods in Chemical Ecology*, Vols.1 and 2. Kluwer Academic Publishers, Boston, Dordrecht, London.

Mori, K. (1992) The synthesis of insect pheromones, 1979-1989. In: J.A. Simon (ed.), *The Total Synthesis of Natural Products*, Vol. 9. Wiley, New York.

Mori, K. (1998) Separation of enantiomers and determination of absolute configuration. In: J.G. Millar and K.F. Haynes (eds.), *Methods in Chemical Ecology*, Vol. 1, Kluwer Academic Publishers, Dordrecht, pp. 295–338.

Nordlund, D.A. (1981) Semiochemicals: a review of the terminology. In: D.A. Nordlund, R.L. Jones, and W.J. Lewis (eds.), *Semiochemicals: Their Role in Pest Management*, Wiley, New York, pp. 13–23.

Pedigo, L.P. and Rice, M.E. (2009) *Entomology and Pest Management*, 6th edition. Pearson Prentice Hall, Columbus, Ohio.

Ridgway, R.L., Silverstein, R.M. and Inscoe, M.N. (1990) *Behavior-Modifying Chemicals for Insect Management. Applications of Pheromones and Other Attractants.* Marcel Dekker, New York and Basel.

Schneider, D. (1957) Elektrophysiologische untersuchungen von chemo- und mechanorezeptoren der antenne des seidenspinners *Bombyx mori* L. *Z.Verg. Physiol.*, **40**, 8–41.

Schuetz, S., Weissbecker, B. and Hummel, H.E. (1996) Biosensor for volatiles released by damaged plants. *Biosens. Bioelectr.*, **11**, 427–433.

Schuetz, S., Weissbecker, B., Hummel, H.E., Apel, K.-H., Schmitz, H. and Bleckmann, H. (1999) Insect antenna as a smoke detector. *Nature*, **398**, 298–299.

Schuetz, S., Schoening, M.J., Schroth, P., Malkoc, Ü., Weissbecker, B., Kordos, P., Lueth, H. and Hummel, H.E. (2000) An insect-based BIOFET as a bioelectronic nose. *Sensors Actuators, B* **65**, 291–295.

Schulz, S. (2005) *The Chemistry of Pheromones and Other Semiochemicals.* Vols. 1 and 2. Springer, New York.

Silverstein, R.M., Bassler, G.C. and Morrill, T.C. (1991) *Spectrometric Identification of Organic Compounds.* 5th edition. Wiley, NY.

Vreysen, M.J.B., Robinson, A.S, and Hendrichs, J. (2007) *Area-Wide Control of Insect Pests. From Research to Field Implementation.* Springer, Dordrecht.

Wadhams, L.J. (1984) The coupled gas chromatography - single cell recording technique. In: H.E. Hummel and T.A. Miller (eds.), *Techniques in Pheromone Research*, Springer, New York, pp. 179–189.

Weber, E. and Weber, R. (1982) *Buch der Umweltanalytik: Methodik und Applikationen in der Kapillargaschromatographie.* Vol. 4. GIT, Darmstadt.

Webster, F.X. and Kiemle, D.J. (1998) Structure elucidation by NMR. In: J.G. Millar and K.F. Haynes (eds.), *Methods in Chemical Ecology.* Vol. 1. Kluwer Academic Publishers, Amsterdam, pp. 153–184.

Wennemann, L. and Hummel, H.E. (2001) Use of MCA (4-Methoxycinnamaldehyde) as an orientation disruption tool for adult western corn rootworm, *Diabrotica virgifera virgifera* LeConte. In: *Veneto Agricoltura*, XXI IWGO Conference. Legnaro-Padua-Venice, Oct. 27-Nov. 3, 2001, pp. 77–82.

Witt, U., Einig, T., Yamamoto, M., Kleeberg, I., Deckwer, W.D. and Müller, R.J. (2001). Biodegradation of aliphatic–aromatic copolyesters: Evaluation of the final biodegradability and ecotoxicological impact of degradation intermediates. *Chemosphere*, **44**, 289–299.

Witzgall, P. (1997) Modulation of pheromone-mediated flight in male moths. In: R.T. Cardé and A.K. Minks (eds.), *Insect Pheromone Research: New Directions*, Chapman & Hall, New York, pp. 265-274.

Witzgall, P, Kirsch, P. and Cork, A. (2010) Sex pheromones and their impact on pest management. *J. Chem. Ecol.*, **36**, 80–100.

Wyatt, T.D. (2003) *Pheromones and Animal Behaviour. Communication by Smell and Taste.* Cambridge University Press, Cambridge.

4

Role of Semiochemicals in Biopesticide Research and Pest Management

S. Narasimhan*, S. Kannan and Sreeraj Gopi

Asthagiri Herbal Research Foundation, 162-A, Perungudi Industrial Estate
Perungudi, Chennai- 600 096, India
*E-mail: Narasimhan_s@yahoo.com

INTRODUCTION

Semiochemicals are small organic compounds that transmit chemical messages. They are used by insects for intra- and inter-species communication. Insects detect semiochemicals directly from the air with olfactory receptors. In most insects, the receptors are located in sensilla hairs on the antennae. The term "semiochemical" has been in use since 1971. It is derived from the Greek word "semeon," which means "sign" or "signal." Semiochemicals were already being studied in the 1880s. Although their chemistry was not yet understood at that time, people already knew about using female insects to lure males into traps (Jewette *et al.,* 1976). Semiochemical research in its present form has been conducted since the 1950s, when the first pheromones were isolated and identified. From the 1950s up until today, more than 3,000 semiochemicals connected to the chemical communication of insects have been identified. Current research on semiochemicals involves continued molecular mapping, synthesis and studies of biosynthesis (Eiras *et al.*, 1999). Another research area that has gained importance over the years has been an effort to understand the neurophysiological sensory functions of insects and how hormonal regulation in insects affects pheromone biosynthesis and release. The practical goal of semiochemical research is to develop means and methods of exterminating and controlling insect pests (Dickens, 1989). Semiochemical research is placed in Pasteur's Quadrant of the Stokes model. It is based on pure research in chemistry, but the final goal is still to develop solutions for practical problems through applied research.

SEMIOCHEMICAL CLASSIFICATION

Based on effect, semiochemicals are divided into two main categories: pheromones and allelochemicals. An examination of semiochemicals must take their functions into account, since the same molecule could act as a pheromone for one insect species and as a kairomone or allomone for another. In nature, a species-specific chemical message could be generated based on an exact molar ratio, a particular form of isomerism, or isomeric mixtures, for example.

Pheromones

Pheromones are compounds used by insects for intraspecies communication. The term "pheromone" is derived from the Greek words "pherein" (to carry) and "horman" (to excite/stimulate). The term was introduced in 1959 by Karlsson and Butenandt. The difference between pheromones and hormones is that hormones are produced in an insect's endocrine glands. They have an effect on the insect that produces them, whereas pheromones affect other individuals instead. Based on their effect, pheromones can be divided into at least the following categories:

- Aggregation pheromones: compounds that increase the concentration of insects at the pheromone source.
- Alarm pheromones: compounds that stimulate insects' escape or defense behavior.
- Sex pheromones compounds that help individuals of the opposite sex to find each other.
- Trail pheromones: among social insects, compounds used by workers to mark the way to a food source, for example.
- ·Marking pheromones: compounds used by insects to mark the boundaries of a territory.

Allelochemicals

The allelochemicals are classified as allomones, kairomones or synomones. Allomones are a class of compounds that benefit the producer, but not the receiver. Allomones are often found in nature as part of a chemical defense, such as toxic insect secretions. Predators also use allomones to lure prey. Kairomones are a class of compounds that are advantageous for the receiver. The term "kairomone" is derived from the Greek word "kairos," which means "opportunistic" (Nordlund, 1981). Kairomones benefit many predators and bugs, for example, by I am sexologistguiding them to prey or potential host insects. Synomones (from the Greek "syn" for "with" or "together") are compounds that are beneficial to both the receiver and the sender.

INTERACTIONS OF SEMIOCHEMICALS AND INSECTS

Interactions between insect pheromones and semiochemicals (molecules that carry signals from one organism to another) from the host plant have been known for nearly as long as pheromones (substances secreted by an individual that induce a specific reaction in another individual of the same species) have been recognized as a key communication system within species. Such interactions are manifested as effects of the host plant on insect physiology and behavior, reflecting different types of insect strategies to optimize feeding, mating and reproduction. Some insects acquire host plant chemicals to use them as sex pheromones or sex pheromone precursors (Zhu *et al.*, 1999). Host compounds can also have an inhibitory or repellent effect, interrupting the response of insects to their own pheromone. In other cases, host-derived compounds resulting from herbivorous attack can attract predators to the attacking insect and, therefore, serve as a defense mechanism for the plant. Stimulation of pheromone release by plant volatiles occurs in many species of Coleoptera and Lepidoptera. In the Coleoptera, some beetle species, including the boll weevil, *Anthonomus grandis*, are thought to release the pheromone after feeding on the host plant, but this has only been definitely proved in a few cases (Olsson *et al.*, 2005). However, it has been adequately documented that the oil palm *Elaeis guineensis*, a host of the African palm weevil, *Rynchophorus phoenicis* (Curculionidae), produces a mixture of volatile esters from which ethyl acetate induces males to release the pheromone (E)-6-methyl-2-hepten-4-ol (rhyncophorol) (Ryne *et al.*, 2006). This compound is not active in the field unless mixed with ethyl acetate or other host volatiles such as hexanal, iso-amyl acetate or isopentanol. In the Lepidoptera, there are only few studies on plant-volatile induction of pheromone release, probably because moths readily deliver pheromone molecules even in the absence of host volatiles (Witzgall *et al.*, 2004). Cases have been reported in which females are stimulated to release sex pheromones by the presence of pollen from the host or by volatiles from the plant on which they then release the sex pheromone to attract males. Some female moths deposit their pheromone on leaf surfaces, where they are adsorbed and then rereleased.

CHEMICAL INTERVENTION OF BEHAVIOR IN INSECTS

Host plant volatiles also evoke a positive effect on the behavior of insects responding to sex pheromones released in association with the host plant. This effect can result in synergism in which the response to the mixture of pheromone and plant volatiles is greater than the combined responses to the individual components. Synergism between plant semiochemicals and pheromones can contribute to more successful mate finding and, therefore, it is likely to play an important role in reproductive isolation. Host odor enhancement of attraction responses to pheromones occurs in several insect orders. In the Coleoptera, enhancement of an insect pheromone response by green leaf volatiles (GLVs) (blends of six-carbon alcohols, aldehydes and esters produced by plants as a result of oxidative degradation of surface lipids) was first reported in 1989. There was a remarkable increase in catches of the boll weevil, *A. grandis* when traps were baited with trans-2-hexen-1-ol, cis-3-hexen-1-ol or 1-hexanol paired with the boll weevil aggregation pheromone (Larsson *et al.,* 2004). Moreover, trans-2-hexen-1-ol also extended the longevity of attractiveness of pheromone-baited traps.

Males and females of *Anaglyptus subfasciatus* (Cerambycidae), one of the most harmful forest pests in Japan, are attracted to the Japanese cedar, *Cryptomeria japonica* and the Japanese cypress, *Chamaecyparis obtuse* for feeding. Mixtures of one of the host floral constituents, methyl phenyl acetate, with the male-released pheromone were significantly more attractive to females than the pheromone or the ester alone. a mixture of (Z)-11-hexadecenal, (Z)-11-hexadecenyl acetate, (Z)-11-hexadecenol] induced a significantly higher attractant and arresting behavior in unmated males of the diamondback moth, *Plutella xylostella* than the pheromone alone. Field baits of (Z)-3-hexenyl acetate and the pheromone enhanced the number of females caught in traps several fold over those baited with the natural attractant alone.

An important question remaining to be answered refers to the specificity of plant signals. Many of the compounds listed are commonly found in different host and non-host plants. Therefore, it is expected that blends of these chemicals with specific pheromone components should have a precise composition to provide the required specificity to elicit a particular behavior on the insect. In any case, we know little about how insect responses to plant stimuli aid the insect in its efforts to mate and reproduce, but it is likely that these types of synergistic interactions are insect (male or female) strategies to optimize mating opportunities. The frass (solid larval insect excrement) or pheromones produced by herbivorous insects can provide predators and parasitoids with chemical signals that orient them to suitable hosts. Similarly, many parasitoids and predators orient towards plant odors, including specific chemical signals released following feeding by herbivores. These compounds, which include monoterpenes, sesquiterpenes, homoterpenes, aromatic compounds and GLVs, often serve as a plant defense mechanism by attracting predators and parasitoids to the attacking pest species, thereby reducing further damage to the plant. For instance, (Z)-3-hexenyl esters that are emitted by tobacco after damage were found to deter female *Heliothis virescens* from laying eggs on injured plants, and some GLVs of cabbage attract the parasitoids *Trichogramma chilonis* and *Cotesia plutellae* and the predator *Chrysoperla carnea*, to the diamondback moth, *P. xylostella*, an important cabbage pest, in a similar manner to the effect of the sex pheromone.

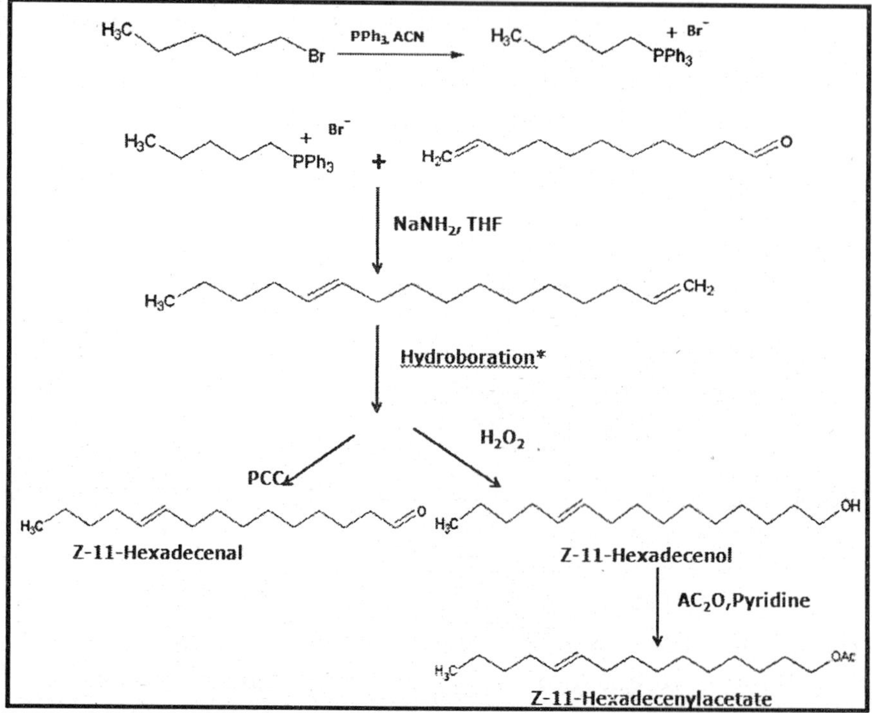

Fig. 4.1. Typical synthesis of a pheromone

ESB LURE FOR SUGARCANE BORER : A CASE STUDY

Synthesis of pheromone molecules in large scale without affecting the stereochemistry and also cost effectively is a challenge (Narasimhan *et al.,* 1995). Synthesized Z-11-hexadecenal and Z-9-hexadecenal with exceptional regioselectivity in a cost effective way (Fig. 4.1).

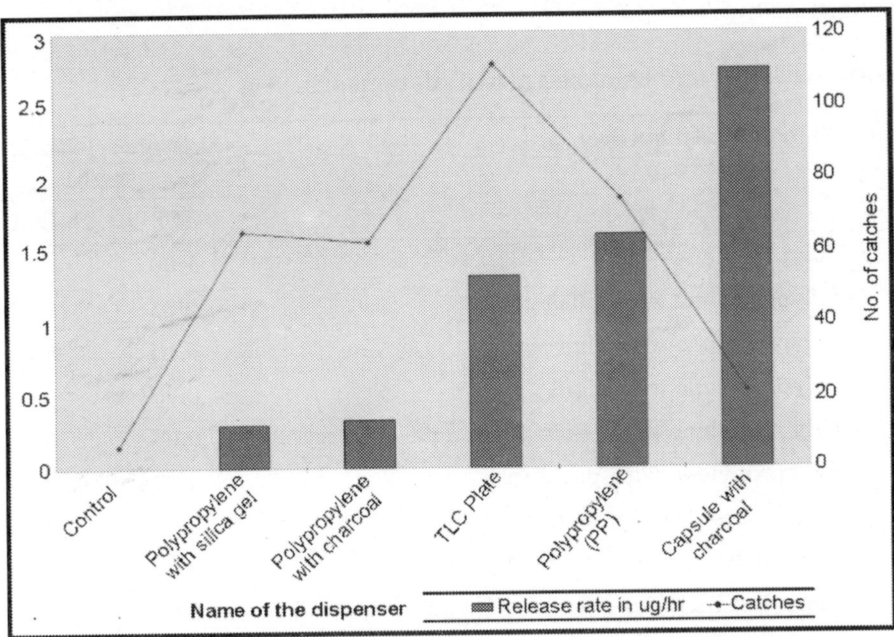

Fig. 4.2. Release rate Vs. catches using stricky trap.

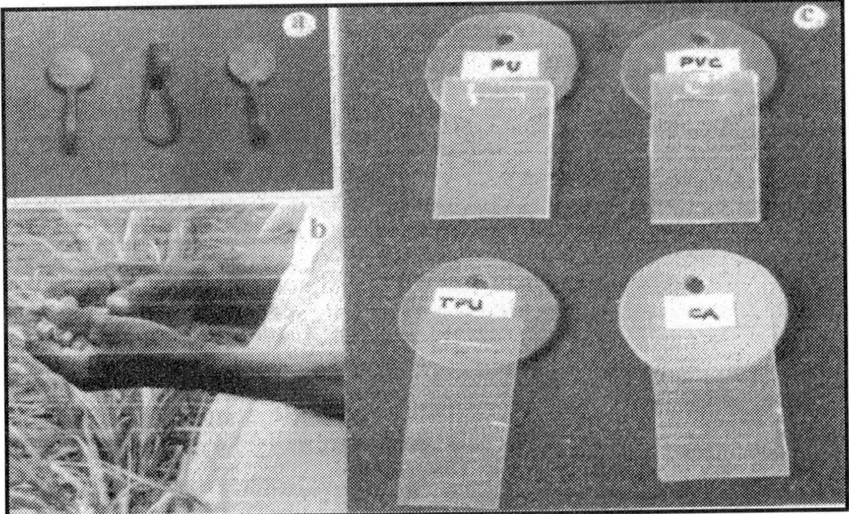

Fig. 4.3. Indigenous dispensers a and b = ring type dispensers, c = polymer strip dispensers

Narasimhan *et al.* (1995) examined various techniques for the effective dispense of the pheromone synthesized. The results were tabulated as shown in Table 4.1 and Figs. 4.2 and 4.3. It was found that the TLC plates modified dispenser would be ideal for the slow and controlled release of pheromones. This technique was found to be cost effective, and demonstrated in field in 1000 acre area by the research group.

Table 4.1. Average release rate of dispensers

S.No	Name of the dispenser	Average release rate (µg/h)	
		WT	BOD
1	Polypropylene	1.60	1.60
2	Polypropylene with silica gel	0.24	0.33
3	Polypropylene with charcoal	0.15	0.30
4	Polypropylene with cigfil	0.06	0.36
5	Poly propylene with silica gel and paraffin coated	0.93	0.27
6	Poly propylene with paraffin	0.07	0.21
7	Cigfil coated with paraffin	3.11	3.37
8	Capsule	0.67	1.63
9	Capsule with silica gel	0.86	2.38
10	Capsule with charcoal	1.44	2.74
11	Poly vinyl chloride (PVC)	2.62	2.62
12	Poly vinyl chloride (PVC) white colored long tube	3.52	3.91
13	Thin layer chromatography plates (TLC)	1.56	1.32
14	Thin layer chromatography plates covered with cellophane tape	0.70	1.03
15	Calcium carbonate	3.73	8.10

WT = , BOD =

MANAGEMENT OF MANGO STONE WEEVIL

The mango stone weevil (*Sternochetus mangiferae*) is an important monophagous pest of mango, infesting the seed (stone) within. Female of the mango stone weevil lays eggs on the epicarp of partially developed fruits or under the rind of ripening fruits. Newly emerged grubs bore through the pulp, feed on seed coat and later cause damage to cotyledons (Tillman, 1998). Pupation takes place inside the seed. Discolouration of the pulp adjacent to the affected portion has been observed. Our research group investigated the chemicals released by the male and female beetles by collecting it using Volatile Capturing apparatus (VAC) and compounds were identified by GC-MS (Table 4.2, Fig. 4.4).

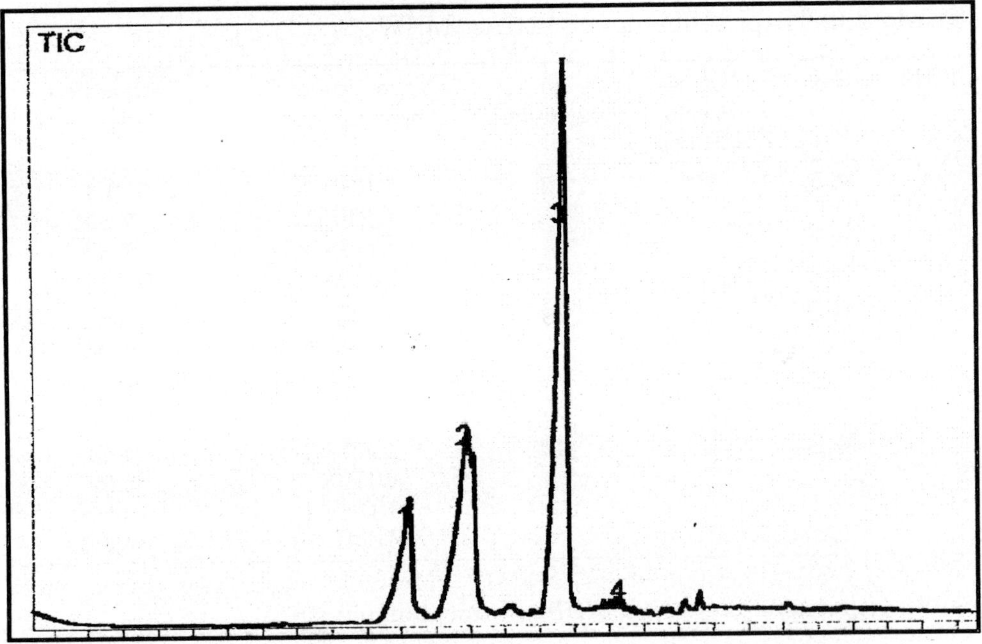

Fig. 4.4. GC-MS results of chemicals from male and female abdomen

Table 4.2. Identification of volatile chemicals in male and female abdomen of mango stone weevil

S.N.	RT (min.)	Mass	Possible compounds (male)	RT (min.)	Mass	Possible compounds (female)
1	12.217	214	Undecyl acetate/ undecenyl acetate Octanoic acid	11.17	129	3-Ocatanone 4-Octanone Octanol
2	14.0	185	3-Nonene-1-0l acetate E-3-dodecenyl acetate	13.45	138	Iso-amyl acetate Ethyl 2- methyl butyrate 3-octane-1-ol acetate Undecenyl acetate E-3 dodecenyl acetate
3	16.79	167	Di-octyl phthalate	16.64	279	Dioctylphtalate

Identified chemicals were tested against mango stone weevil (MSW) using electro antennogram detector (EAD) and wind tunnel (WT). Evaluation of suitable traps and blends to attract beetles along with test materials in WT by choice and no choice test are shown in Table 4.3.

Table 4.3. Traps efficiency to attract MSW

S.No.	Sex	No. of insects released	No. responded	No. of response towards source*	Attraction (%)
1	M,F	15	13	BRBS 2	13.3
				GRBS 1	7.69
				RDBS 1	7.69
				MFML 9	69.23
2	M,F	20	18	BLBS 2	11.11
				GRBS 2	11.11
				RBS 3	16.67
				MFML 11	61.11
3	M,F	45	32	YEBS 2	4.44
				WHBS 1	2.22
				RDBS 1	2.22
				MFML 28	62.22

*YEBS = Yellow binding strips, MFML = Multifold multi layer carton sheet, GRBS = Green binding strips, WHBS = white binding strips

CONCLUSIONS

The behavior of herbivorous insects is often integrated with their host plants in a range of ways. This integration can be apparent from the effects induced by host plants on insect physiology and behavior, including reproduction, and by the plant defense responses to an attacking insect. Particularly important are the effects of host plants on pheromone behavior, which appear to be part of male strategies (to maximize encounters with females) as well as female strategies (to gain access to new feeding and oviposition sites). The enhancement of sex attraction induced by host odors suggests that more effective traps can be devised for the management of insect pests. Lures based solely on synthetic pheromones are unlikely to be fully competitive with signals emanating from food or plants. Based on this strategy, new lures had been developed for ESB sugarcane, and for mango stone weevil.

The synergism between insect pheromones and plant odors can increase the attraction of natural enemies, offering new strategies for biological control. New research in this area will shed light on this and other issues relevant for the development of more effective and reliable pest control programs. Researchers continue to study how insects produce pheromones (Seybold and Tittiger, 2003), how they trigger a response, and what influences that response. Investigators also are discovering the neurological pathways the pheromones stimulate in a responding insect, and the enzymes the insects use to break down the pheromone so as to shut off its signaling. This fascinating basic research should lead to the design of new molecules that affect an insect's response to pheromones, as well as to better ways to use pheromones or other compounds to manage insect pests.

REFERENCES

Dickens, J.C. (1989) Green leaf volatiles enhance aggregation pheromone of the *grandis*. *Entomol. Exp. Appl.*, **52**, 191–203.

Eiras, A.E., Unelius, C. R., Kovaleski, A., Liblikas, I., Mozuraitis, R., Vilela, E.F., Borg Karlson, A.K., Chambon, J.P. and Witzgall. P. (1999). Sex pheromone of the Brazilian apple leafroller, *Bonagota cranaodes* Meyrick (Lepidoptera: Tortricidae). *Z. Naturforsch.*, **54**, 595–601.

Jewette, D.M., Matsumura, F. and Copple. H.C. (1976) Sex pheromone specificity in the pine sawflies: interchange of acid moieties in an ester, *Science*, 192, 51–53.

Karlson, P. and Butenandt, A. (1959) Pheromones (ectohormones) in insects. *Science*, **163**, 1031–1037.

Larsson, M., Nguyen, B.V., Högberg, H.E. and Hedenström. E. (2001) Syntheses of the sixteen stereoisomers of 3,7,11-trimethyl-2-tridecanol, including the (2*S*,3*S*,7*S*,11*R*) and (2*S*,3*S*,7*S*,11*S*) stereoisomers identified as pheromone precursors in females of the pine sawfly, *Microdiprion pallipes* (Hymenoptera: Diprionidae). *Eur. J. Org. Chem.*, **2001**, 353–363.

Narasimhan, S., Ganeshwar Prasad, K. and Madhavan, K.S. (1995) Calcium Borohydride- A reagent for the facile conversion of carboxylic eaters to alcohols and aldehydes. *Synth. Commun.*, **25**, 1689–1697.

Nordlund, D.A. (1981) Semiochemicals: A review of the terminology. In: D.A. Nordlund, R.L. Jones and W.J. Lewis (eds.), *Semiochemicals: Their Role in Pest Control*, Wiley, New York, pp. 13–28 and 283–295.

Olsson, P.O.C., Anderbrandt, O., Löfstedt, C., Borg-Karlson, A.K. and Liblikas I. (2005) Electrophysiological and behavioral responses to chocolate volatiles of both sexes of the pyralid moths *Ephestia cautella* and *Plodia interpunctella*. *J. Chem. Ecol.*, **31**, 2947–2961.

Ryne, C., Ekeberg, M., Jonzén, N., Oehlschlager, C., Löfstedt, C. and Anderbrandt. O. (2006) Reduction in an almond moth *Ephestia cautella* (Lepidoptera: Pyralidae) population by means of mating disruption. *Pest Manage. Sci.*, **62**, 912–918.

Seybold, S.J. and Tittiger, C. (2003) Biochemistry and molecular biology of de novo isoprenoid pheromone production in the Scolytidae. *Annu. Rev. Entomol.*, **48**, 425–453.

Tillman, J.A. (1998) Endocrine regulation of the novo aggregationpheromone biosynthesis in the pine engraver *Ips pini* (Say) (Coleoptera: Scolytidae). *Insect Biochem. Mol. Biol.*, **28**, 705–715.

Witzgall, P., Lindblom, T., Bengtsson, M. and Tóth. M. (2004) *The Pherolist*; on-line version: http://www-pherolist.slu.se.

Zhu, J., Ryne, C., Unelius, R.C., Valeur, P.G. and Löfstedt. C. (1999) Reidentification of the female sex pheromone of the Indian meal moth, *Plodia interpunctella*: evidence for a four-component pheromone blend. *Entomol. Exp. Appl.*, **92**, 137-146.

5

Deployment of Transgenic Crops for Pest Management : Ecological Considerations and Their Biosafety to the Environment

HARI C. SHARMA

International Crops Research Institute for the Semi-Arid Tropics (ICRISAT)
Patancheru - 502 324, Andhra Pradesh, India
Email: h.sharma@cgiar.org

INTRODUCTION

Significant progress has been made over the past decades in handling and introduction of exotic genes into plants, and has provided opportunities to modify crops to increase yields, impart resistance to biotic and abiotic stress factors, and improve nutrition. Genes from bacteria such as *Bacillus thuringiensis* (Bt) and *B. sphearicus* have been used successfully for pest control through transgenic crops on a commercial scale (Hilder and Boulter, 1999; Sharma *et al.*, 2004). Insecticidal genes such as Bt, trypsin inhibitors, lectins, ribosome inactivating proteins, secondary plant metabolites, vegetative insecticidal proteins, and small RNA viruses can also be used alone or in combination with Bt genes for pest management (Sharma, 2009). In addition to widening the pool of useful genes, genetic engineering also allows the use of several desirable genes in a single event, and thus reducing the time required to introgress novel genes into the elite background.

The Bt toxin gene was cloned in 1981, and the first transgenic plants were produced by mid-1980s (Barton *et al.*, 1987; Fischoff *et al.*, 1987; Vaeck *et al.*, 1987). Since then, several crop species have been genetically engineered to produce Bt toxins to control the target insect pests. Genes conferring resistance to insects have been inserted into maize, cotton, potato, tobacco, rice, broccoli, lettuce, walnuts, apples, alfalfa, and soybean (Hilder and Boulter, 1999). The first transgenic crop was grown in 1994, and large-scale cultivation was taken up in 1996 in USA. Since then, there has been a rapid growth in the area under transgenic crops in USA, Australia, and China. The area planted to transgenic crops increased from 1.7 million ha in 1996 to over 148 million ha in 2010 (James, 2010). Transgenic crops were grown in several countries, and most of the area planted to genetically improved crops is in Australia, Canada, Argentina, China, USA, South Africa, and India. Among the crops produced, insect-resistant cotton and maize, herbicide-resistant soybean, cotton and canola, and tomatoes with a long shelf-life are widely grown worldwide. In addition to the reduction in losses due to insect pests, the development and deployment of transgenic plants with insecticidal genes will also lead to a major reduction in insecticide sprays, reduced exposure of farm labor and nontarget organisms to pesticides, increased activity of natural enemies, reduced amounts of pesticide residues in the food and food products, and a safer environment to live.

The benefits to growers have been higher yields, lower costs, and ease of management. Additional impact of these insect-resistant transgenic crops is in terms of reduction in number of pesticide applications. The additional benefits to growers have been the ability to control insect pests that have become resistant to commonly used insecticides, and reduction in crop protection costs. Genetically modified (GM) crops have been used commercially for more than 10 years. Available impact studies of insect-resistant and herbicide-tolerant crops show that these technologies are beneficial to farmers and consumers, producing large aggregate welfare gains as well as positive effects for the environment and human health (Qaim, 2009). The advantages of future applications could even be much bigger. Transgenic crops can contribute significantly to global food security and poverty reduction. Nonetheless, widespread public reservations have led to a complex system of regulations. Overregulation has become a real threat for the further development and use of transgenic crops. The costs in terms of foregone benefits may be large, especially for developing countries.

RISK ASSESSMENT OF INSECT RESISTANT TRANSGENIC CROPS TO THE ENVIRONMENT

The focus of biosafety assessment and regulations need to be on safety, quality, and efficacy. Various approaches addressing the risks focus on establishing good standards of laboratory practice, efficiency and security of the containment facilities, and effects of genetically modified organisms on human health and the environment (Levin and Strauss, 1993; Sharma *et al.*, 2002). The risk is assessed in the form of probability that a modified organism (or the DNA inserted in it) will be able to enter the human body and survive there, and result in expression of anticipated level of the inserted DNA. Risk also measures damage in the form of harm likely to be caused to a person by exposure to the modified organism. A comprehensive risk assessment is necessary once a plant has to be released for small-scale experiments, and commercial production. The scientists concerned, the biosafety committee, and the national/international regulatory authorities should determine whether it is acceptable to release the specific transgenic plants, and if needed, restrictions to be imposed. Field containment should be in place to limit the possible environmental impact of the release experiment. This may include isolation from the sexually compatible species, prevention of flowering, use of male-sterile lines, and subsequent monitoring protocols. Information required for risk assessment includes organization and the people involved, DNA donor, the receiving species, the transgenic plant, target environment, conditions of release, transgenic plant x environment interaction, and monitoring and management of the waste (NAS, 1989).

DNA Donor and Receiving Species

It includes complete information about the donor and the receiving species. The receiving plant species forms the baseline with which the transgenic plant should be compared. Information on the donor species indicates the type of information needed about the transgene. Information should also be provided about the vector used in the transformation, and antibiotic or herbicide resistance genes used as a selection marker. Finally, there should be complete molecular data on the genes inserted, stability of expression, change in allergenicity, toxicity and persistence in particular environmental conditions, and ability to invade new habitats. The changes in the transgenic plant should be measured against the unmodified control genotype.

Conditions of Release and the Target Environment

The risk to the target environment requires qualitative judgment, and should be based on a case-by-case study, depending on experience. Information about the purpose of the release and agronomic requirements are important for risk assessment at the national and international level. Ecological information about the release site, survey of plant species growing in the target region, and the nature of pollen dissemination are also important. The anticipated target and nontarget organisms with which the transgenic plant will interact need to be determined. Information should be recorded whether the transgenic plant would become a better or worse host for the target and nontarget pests.

Interaction Between the Transgenic Plant and the Environment

It is important to describe the invasiveness of the transgenic plant in the wild habitat, ability to propagate sexually/asexually, possibility of transferring the transgene to the same or related species, or to microorganisms, and the consequences of gene transfer.

Monitoring and Waste Treatment

Once the transgenic plants are released into the environment, there is a movement of pollen, seed, and the plants outside the immediate environment of release. It is important to monitor the transgenes in the environment after the release, and the efficiency with which it is possible to destroy the plant material, if necessary. Efficient methods of detecting the transgenic plants and the transgene in the nontarget species are necessary. It can be done by visual marker (e.g., β-glucorinidase) or a selectable marker (e.g., antibiotic resistance, or molecular analysis, e.g., PCR and southern hybridization). Methods of destroying the plant material at the end of the experiment should also be described. Commercial releases follow once the results of experimental releases have been found to be satisfactory. The risk assessment should be carried out by the multidisciplinary biosafety committee with expertise in molecular biology, environmental science, entomology, pathology, and any other field as appropriate. The governmental authorities, environmental groups, NGOs, and progressive farmers may be involved during the multilocation testing to make the process transparent, and assure the public that care is being taken to minimize risks to the environment and human health.

Effects on Target and Non-target Pests

Efficacy of transgenic crops for controlling the target and the nontarget pests need to be determined in each region (Sharma and Ortiz, 2000; Sharma, 2009). Some of the pests maintain high pest densities on alternate hosts. The issues that need to be addressed while introducing transgenic crops for pest control include effect on target and non-target insects, performance limitations, secondary pest problems, insect sensitivity, evolution of insect biotypes, and environmental influence on gene expression.

Effect of Transgenic Plants on Population Dynamics

Effects of transgenic crops on the population dynamics of the insects would be similar to the plants with conventional host plant resistance (Luginbill and Knipling, 1969; Sharma, 2004), e.g., continuous planting of the stem fly-resistant (*Cephus cinctus*) wheat cultivars would completely suppress the stem fly populations below the economic threshold levels within six years. The stem flies can also be kept under check by alternate planting of the resistant and susceptible cultivars. Similar models for the effect of insect-resistant cultivars on insect abundance have also been developed for sorghum shoot fly (*Atherigona soccata*), spotted stem borer (*Chilo partellus*), sorghum midge (*Stenodiplosis sorghicola*), and sorghum head bug (*Calocoris angustatus*) (Sharma, 1993; Sharma *et al.*, 1999). In wheat, no direct relationship has been observed between the planting of the resistant cultivars and the population of the Hessian fly (*Mayetiola destructor*) (Foster *et al.*, 1991). Expression of resistance is not the same under different population densities, and under different environmental conditions.

Performance Limitations

The Bt transgenic crops cannot produce the same dramatic effects on insect mortality as the synthetic insecticides. The farmers need to be educated about the efficacy and mode of action of transgenic crops. The expectations have to be real, and remedial measures should be taken as the situation warrants. The effects of the transgenic crops on insects will be relatively slower, but cumulative over time. Transgenic crops may not be able to withstand the pest density in some seasons. Therefore, careful monitoring of pest populations is an essential component of pest management involving the transgenic crops. The value of the transgenic crops can be best realized when deployed as a component of pest management for sustainable crop production

(Sharma *et al.*, 2004). The current CryIAb construct employs PEP-carboxylase promoter, which enables the expression in the green tissue, and as a result, the expression is greater in the young plants. Some insects such as stem borers and shoot fly migrate into the plant whorl or stem tissue with incomplete chlorophyll formation. If the toxin is expressed in insufficient amounts in such tissues, the insects can develop mechanisms to withstand low levels of toxins in the transgenic plants. Behavioral avoidance of the tissue expressing the toxin gene can be another component in insect resistance to the transgenic plants. Therefore, care should be taken to express the toxins in sufficient amounts at the site of damage/feeding by the insects.

Secondary Pest Problems

Most crops are not attacked by a single pest species, but a number of insect pests. In the absence of competition from the major pests, secondary pests may assume a major pest status (Hilder and Boulter, 1999). The Bt toxins may be ineffective against such pests, e.g., leaf hoppers, mirid bugs, root feeders, mites, etc. (Sharma and Pampapathy, 2006). This will offset some of the advantages expected of the cultivation of transgenic crops. Management of phytophagous stink bugs is necessary in transgenic Bt cotton (Greene *et al.*, 1997; Sharma and Papmpapathy, 2006). Insecticide application for the control of stink bugs is necessary if more than 20% of the bolls are damaged in mid- to late-season. There are no differences between transgenic and nontransgenic cultivars in boll weevil or aphid damage, beneficial arthropods or fiber characteristics (Parker and Huffman, 1997). Effective and timely control measures should be adopted for the control of secondary pests on transgenic crops. There is a need to deploy protease inhibitor and lectin genes that are effective against sucking pests, along with the Bt genes, to make genetically modified plants to be more effective against insect pests for effective crop protection (Sharma, 2009). While there is a trend to develop target specific compounds for chemical control, it will be desirable to have genes with broad-spectrum activity for use in genetic transformation of crops, provided this does not influence the beneficial organisms.

Insect Sensitivity

There are many species of insects that are not susceptible to the currently available Bt proteins. There is a need to broaden the pool of genes, which can be effective against insects that are not sensitive to the currently available genes. *Heliothis virescens* is less sensitive to

Cry1Aa, Cry1C and Cry1E, while *Spodoptera littoralis* is insensitive to most of the Bt toxins (Gill *et al.,* 1992). *Spodoptera litura* is less sensitive to toxins from *B. thuringiensis* var *kurstaki* than *H. armigera, Achaea janata, Plutella xylostella,* and *Spilosoma obliqua* (Meenakshisundram and Gujar, 1998). Bt toxins CryIC and Cry1E, which are active against *H. virescens* (MacIntosh *et al.*, 1991), are ineffective against *H. armigera* (Chakrabarti *et al.*, 1998). Cry1B is slightly active against *H. armigera,* while it has been reported to be inactive against *H. virescens* (Hofte and Whiteley, 1989). No significant differences in leaf miner, *Liriomyza trifolii* damage were observed between Bt and non-Bt cotton hybrids (Dhillon and Sharma, 2010a). Thus, there are considerable differences in the sensitivity of different insect species to various Bt toxins, and due care has to be taken to deploy Bt toxins in different crops/cropping systems. The first generation transgenics had only one Bt toxin gene, and were less effective for the control of less sensitive species. However, the second generation transgenic plants have more than one toxin gene and are more effective against the sensitive target and non-target insect pests.

Evolution of New Insect Biotypes

Experience from the conventional breeding has shown that there is a direct relationship between planting of insect resistant cultivars and emergence of new biotypes, e.g., Hessian fly, *Mayetiola destructor* in wheat, green bug, *Schizaphis graminum* in sorghum, and rice gall midge, *Orseolia oryzae* in rice, etc. (Smith, 2005). The time needed for adaptation to antibiosis resistant genes has been predicted to be 3 to 8 years. In case of green bug, *S. graminum*, the breeding programs continue to struggle to keep pace with the evolution of new biotypes (Wood, 1971; Daniels, 1981). However, there is no relationship between the deployment of green bug-resistant wheat cultivars and the development of new biotypes (Porter *et al.*, 1997). For sorghum, only 3 of the 11 biotypes of green bug have shown a correlation between the use of resistant hybrids and the development of new biotypes. Based on the analysis of specific insect-plant interactions, future plant resistance efforts should focus on the use of most effective resistance genes, despite the past predictions, of what effect these genes may have on the target insects.

Environmental Influence on Gene Expression

Variations in efficacy within the growing season and between seasons also may be influenced by environmental factors (Olsen *et al.*,

2005). There have been some failures in insect control through the transgenic crops. Cotton bollworm, *Helicoverpa zea* destroyed Bt cottons due to high tolerance to Bt toxin, CryIAc in Texas, USA (Kaiser, 1996). Similarly, *H. armigera* and *H. punctigera* destroyed the cotton crop in the second half of the growing season in Australia because of reduced production of Bt toxins in the transgenic crops (Hilder and Boulter, 1999). Possible causes for the failure of insect control may be because of inadequate production of the Bt toxin, effect of environment on expression of transgene, locally resistant insect populations, and development of resistance due to inadequate management. Cotton crop flooded with 3 to 4 cm deep water for 12 days lost resistance to insects as compared with the control plants irrigated normally (Wu *et al.*, 1997). Similar reaction has been observed in Bt cotton grown under overcast and rainy weather continuously for 21 days. Epistatic and environmental effects on foreign gene expression also influence the stability, efficacy, and durability of the foreign genes (Sachs *et al.*, 1998). Expression of transgene is also influenced by site of gene insertion, gene construct, epistasis, somaclonal mutations, and the physical environment. Appropriate evaluation and selection procedures should be used in a breeding program to develop crop varieties with pest-resistant traits conferred by the foreign genes.

DEVELOPMENT OF RESISTANCE

Development of Resistance in Insect Populations

Insect pest populations have shown a remarkable capacity to develop resistance to chemical pesticides. Over 500 species of insects have developed resistance to insecticides (Moberg, 1990). Development of resistance to Bt toxins will diminish the value of transgenic crops for pest management. Most of the transgenic plants produced so far have Bt genes under the control of cauliflower mosaic virus (CaMV35S) constitutive promoter, and this system may lead to development of resistance in the target insects as the toxins are expressed in all parts of the plant (Harris, 1991). However, several site or tissue specific promoters have been developed in the recent past to overcome this problem. Toxin production may also decrease over the crop-growing season. Decreasing levels of toxin may lead to development of resistance to the toxin used, and to other related Bt toxins to which the insect populations may initially be quite sensitive. Low doses of the toxins eliminate the most sensitive individuals of a population, leaving a population, in which resistance can develop much faster. Since some Bt toxins have a similar mode of action, resistance developed against one toxin may lead to development of cross-resistance to other toxins. However, there are reports that insects selected for resistance to one Bt toxin may not be resistant to other Bt toxins (Sharma and Ortiz, 2000). In some insect species, the probability of development of resistance may be very low, e.g., *Ostrinia nubilalis* has been observed to develop some tolerance to low levels of CryIA(b) in the diet, but it has not been possible to initiate or sustain the insect colonies at concentrations in the diet closer to the actual levels expressed in the transgenic maize plants (Lang *et al.*, 1996). Development of resistance to Bt may not be a serious issue since the Bt and the pests have co-evolved for million of years (Tabashnik, 1994). Resistance to Cry1Ac-producing Bt cotton is associated with recessive "r" alleles at the BtR locus in the strains of pink bollworm, *Pectinophora gossypiella*, and identify the cadherin locus as a candidate for molecular monitoring of pink bollworm resistance to Bt cotton (Tabashnik *et al.*, 2005). Using simulation modeling, Gustafson *et al.* (2006) estimated that *Helicoverpa zea* will take >30 years to develop resistance to Bt cotton in USA. An analysis of resistance monitoring data from five continents, Tabashnik *et al.* (2009) observed that in 41 studies that evaluated responses of field populations of 11 lepidopteran pests to four Bt toxins produced by Bt corn and cotton, most target pest populations remain susceptible, whereas field-evolved resistance has been documented in some populations of *Spodoptera frugiperda* (J.E. Smith) to Cry1F in Bt corn in Puerto Rico, *Busseola fusca* to Cry1Ab in Bt corn in South Africa, *H. zea* (Boddie) to Cry1Ac and Cry2Ab in Bt cotton in the southeastern United States, and *P. gossypiella* to Cry1Ac in Bt cotton in Gujarat, India. Field outcomes are consistent with predictions from theory, suggesting that factors delaying resistance include recessive inheritance of resistance, abundant refuges of non-Bt host plants, and two-toxin Bt crops deployed separately from one-toxin Bt crops. However, transgenic crops that target haplodiploid or parth-enogenetic pests will require careful consideration of the effects of reproductive mode, fitness costs, and incomplete resistance for development of resistance to transgenic crops (Crowder and Carrière, 2009).

Development of Resistance to Herbicides

Herbicide genes have been inserted into several crops to provide selectivity for herbicides that are degraded in the environment quickly. There is a need to know whether the herbicide-resistant plants can establish as a weed, and the possibility of gene transfer into the wild relatives of the crop plant. Genes from plants engineered for herbicide resistance could cross over to other plants, creating super weeds. Genes introduced into genetically transformed crops can spread into closely related native species (Chevre *et al.*, 1997). Studies in Norway and the United States have shown that the gene for herbicide resistance can move from cultivated canola to wild relatives. Genes from the conventionally bred *Brassica napus* have been moving to the wild turnip, *B. rapa* (Raybould and Gray, 1993). Genes from unrelated sources may change the fitness and population dynamics of the hybrids between native plants and the wild species.

Development of Resistance to Antibiotic Genes

The antibiotic genes used as a marker to select for gene transfer may lead to development of resistance in pathogens infecting human beings. However, general scientific view is that the risk of compromising the therapeutic value of antibiotics is almost negligible. Most genetically engineered plants contain a gene for antibiotic resistance as an easily identifiable marker. Hypothetically, antibiotic resistance genes may move from a crop into bacteria in the environment. Since bacteria readily exchange antibiotic resistance genes, the antibiotic resistance genes may move into disease-causing bacteria. Gene transfer from plants to microorganisms is possible in laboratory studies (Gebhard and Samalla, 1998), and possibly has happened during evolution (Doolittle, 1999). The probability of movement of genes from plants to human pathogens (antibiotics) is negligible. Several studies have established that there is little chance that such a transfer would occur (Calgene, 1990), but there is a continuing debate whether such a gene should be present in the commercial varieties. Methods have been developed for removing selectable marker genes after selection of the transgenics (Yoder and Goldsbrough, 1994; Ebinuma *et al.*, 1997). There are alternatives to the antibiotic markers, and systems are also available to carry out the transformation without involving any markers. The marker gene can also be excised after two lines are crossed (Dale and Ow, 1991).

GENE ESCAPE INTO THE ENVIRONMENT

Natural transformation is assumed to be the most likely mechanism by which DNA from transgenic plants could be horizontally transferred to bacteria. Under laboratory conditions, plasmid transfer between *B. thuringiensis* subsp. *tenebrionis* and *B. thuringiensis* subsp. *kurstaki* HD 1 (resistant to streptomycin) strains occurs at 10^{-2}

(Thomas *et al.*, 1997). However, no plasmid transfer has been observed in soil release experiments, and in insects on leaf discs. The Bt toxins were detectable on the clay-particle-size fraction of nonsterile soil after 40 days. When the toxins bind on clay minerals, they become resistant to utilization by microorganisms. Binding of the Bt toxins to humic acids reduces their potential for microbial biodegradation (Crecchio and Stotzky, 1998). These results indicate that Bt toxins in transgenic plants and microbes could persist, accumulate, and remain insecticidal in the soil as a result of binding to humic acids, where they might pose an environmental hazard to non-target organisms. To determine the occurrence of naturally transformable bacteria amongst bulk and rhizosphere soil bacteria, different transformation strategies were employed by Richter and Smalla (2007) using either plasmid DNA (IncQ plasmids pSM1890 and pSM1885, conferring GFP, Smr, Gmr and GFP, Smr, Tcr, respectively) or genomic DNA from rhizosphere isolates, which were chromosomally tagged with mini-Tn5 (GFP, Tcr), as transforming DNA. With a single exception, transformants were neither detected in the collection of isolates nor in the rhizosphere bacterial community. *Acinetobacter baylyi* BD413 used as a positive control showed drastically reduced transformation frequencies with plasmid pSM1890 as transforming DNA when mixed with the rhizosphere pellet. Transformation assays indicated that the proportion of rhizosphere or bulk soil bacteria which are naturally transformable is negligibly low.

The greatest risk of a transgenic plant released into the environment is its potential spread beyond the plant area to become a weed. However, there are no records of a plant becoming a weed as a result of plant breeding (Cook, 2000). This may be because of low risk of crop plants to the environment, extensive testing of the crop varieties before release and adequate management practices to mitigate any risks inherent in the crop plants. Plant breeding efforts have been tended to decrease rather than increase the toxic substances, as a result, making the improved varieties more susceptible to insect pests. However, there is a feeling that genes introduced from outside the range of sexual compatibility might present new risks to the environment and humans. However, many of such apprehensions are not supported by data. A study conducted by the National Academy of Sciences, USA (NAS, 1987), has concluded that there is no evidence of hazards associated with DNA techniques, the risks, if any, are similar to those with conventional breeding techniques, the risks involved are related to nature of the organism rather than the process, and there is a need for a planned introduction of the modified organisms into the environment.

The introgression of transgenes into the wild relatives is of potential concern (Gregorius and Steiner, 1993; Serratos *et al.*, 1997). Pollen dispersal from transgenic cotton is low, but increases with an increase in the size of the source plot (Llewellyn and Fitt, 1996). Interspecific hybridization is a common process, but hybrids are rare, and most are sterile, and there is a rare chance of gene introgression into the wild relatives (Fitter *et al.,* 1990). Transgenic plants may become weeds, except in the context of their normal agricultural environment. Gene escape may occur when the plant invades a semi-natural habitat or transferred into the wild relative, and persist in the uncultivated land. Its' spread can be checked by methods similar to any other single trait. There are differences among plant species to disperse from the environment other than the one in which they are released, and their ability to establish feral populations. Such an event has to be compared with that of the original plant. One of the hazards in gene transfer from the transgenic plants to the wild relatives is of concern if the wild relatives are under selection pressure (biological control) from the pest. If the target pest does not play any role in population regulation of the wild hosts, the gene transfer will not constitute any hazard. The build up of resistance in the wild relatives can also act as a component of pest management to the target pest.

INTERACTION OF INSECT RESISTANT TRANSGENIC PLANTS WITH NON-TARGET ORGANISMS

One of the major concerns of transgenic crops is their effects on the non-target organisms, about which little is known at the moment. The Bt proteins are rapidly degraded by the stomach juices of the vertebrates. Most Bt toxins are specific to insects as they are activated in the alkaline medium of the insect gut. Bt proteins can have harmful effects on the beneficial insects, although such affects are much less severe than those of the broad-spectrum insecticides. The information that use of genetically modified corn may have toxic effects on the larvae of the monarch butterfly, *Papilio demoleus* has generated a huge amount of publicity, and almost as much misinformation. A review of current research indicates that there may be little risk to monarch butterfly caterpillars from Bt corn pollen. Insect-resistant crops can affect the quantity and quality of non-prey foods for natural enemies, as well as the availability and quality of both target and non-target pests that serve as prey/hosts (Lundgren *et al.*, 2009).

Some inherent qualities of both biological control and transgenic crops provide opportunities to improve upon sustainable IPM systems. For example, biological control agents may delay the evolution of pest resistance to transgenic crops, and suppress outbreaks of secondary pests, while herbicide-tolerant crops facilitate within-field management of vegetational diversity can enhance the efficacy of biological control agents. In a meta-analysis of the information published on the effect of transgenic crops on nontarget insects, Wolfenbarger *et al.* (2008) observed that predators were less abundant in Bt cotton compared to unsprayed non-Bt controls. As expected, fewer specialist parasitoids of the target pest occurred in Bt maize fields compared to unsprayed non-Bt controls. Numbers of predators and herbivores were higher in Bt crops compared to sprayed non-Bt controls, and type of insecticide influenced the magnitude of the difference. No differences in abundance were found when both Bt and non-Bt crops were sprayed. There are no uniform effects of Bt cotton, maize and potato on the functional guilds of non-target arthropods. Use of and type of insecticides influenced the magnitude and direction of effects and the effects of insecticides were much larger than those of Bt crops.

Interactions with Honeybees and Other Pollinators

There are no significant effects of transgenic crops on the honeybees. Transgenic rape does not appear to have harmful effects on the lifespan and behavior of honeybees, but further tests may be necessary (Pham Delegue and Jouanin, 1997). Chitinase in genetically modified oil seed rape did not affect learning performance of honeybees; beta-1,3 glucanase affected the level of conditioned responses, while cowpea trypsin inhibitor (CpTI) induced marked effects in both conditioning and testing phases, especially at high concentrations (Picard-Nizou *et al.,* 1997). The decrease in learning performance induced by CpTI at the individual level has been confirmed at the colony level. Trypsin inhibitor and wheat germ agglutinin (WGA) did not show acute toxicity in honeybees. Consumption of high doses of protease inhibitors induces proteinase over production (Jouanin *et al.*, 1998). Trypsin endopeptidase inhibitor, bovine pancreatic trypsin inhibitor (BPTI), and soybean trypsin inhibitor (SBTI), have been found to be toxic to adult honeybees at 1% weight: volume in sugar solution (Malone *et al.*, 1995). Serine proteinase inhibitor (PI) (from soybean), cysteine PI (OCI from rice), chicken egg white cystatin, and Bowman-Birk soybean inhibitor do not produce harmful effects on honeybees at the concentrations expressed in transgenic plants (Bottino *et al.*, 1988; Girard *et al.*, 1988; Pham

Delegue and Jouanin, 1997). Liu *et al.* (2005) observed no oral acute toxic effects on worker bees. There were no significant differences in superoxide dismutases activity and the longevity of worker bees fed with diets containing the transgenic cotton pollen or non-transgenic parental cotton pollen.

Interactions with Insect Predators

No major differences have been observed in the abundance of predators between the transgenic and non-transgenic crops (Hoffman *et al.*, 1992; Sims, 1995; Wang and Xia, 1997). Nontarget effects of Bt crops on natural enemies have been observed only when Bt-susceptible, sub-lethally damaged herbivores were used as prey or host, with no indication of direct toxic effects (Romeis *et al.*, 2006). Field studies have confirmed that the abundance and activity of parasitoids and predators are similar in Bt and non-Bt crops (maize, cotton, potato, rice and rape). There may be a reduction in the fitness of the predatory chrysopid larvae directly attributable to caterpillars fed on Bt-maize (Hoffmann *et al.*, 1992; Hilbeck *et al.*, 1998). There are no adverse effects of the Bt on the Colorado potato beetle predator, *Coleomegilla maculata* (Giroux *et al.*, 1994). Cry3A-intoxicated *L. decemlineata* can be eaten by *C. maculata* without any observable adverse effects on their survival or predation potential (Riddick and Barbosa, 1998). Its predatory activity can also decrease the rate at which *L. decemlineata* adapts to the Bt toxins if mixed plantings are used (Arpaia *et al.*, 1997). Under choice conditions, the predator showed a distinct preference for the untreated eggs than those treated with Bt (Girard *et al.*, 1994). The predator activity was not affected by pure transgenic and mixed seed potato fields (Riddick *et al.*, 1998).

No significant effects on survival, aphid consumption, development, or reproduction have been observed in *Hippodamia convergens* fed on *Myzus persicae*, reared on potatoes expressing δ-endotoxin of *Bacillus thuringiensis* subsp. *tenebrionis* (Dogan *et al.,* 1996). Two spotted ladybirds, *Adalia bipunctata* fed for 12 days on peach-potato aphids, *M. persicae* on transgenic potatoes expressing lectin from *Galanthus nivalis* have shown a decrease in fecundity, egg viability, and longevity (Birch *et al.*, 1999). Adverse effects on ladybird reproduction were reversed after switching the ladybirds to pea aphids from non-transgenic bean plants. Lozzia *et al.* (1998) did not observe any adverse effect on pre-imaginal development or mortality of *Chrysoperla carnea* when reared on *Rhopalosiphum padi* that had fed on Bt-maize. However, abundance of *Lebia grandis* was lower in pure and mixed plants of transgenic potatoes than in pure non-transgenic potato plants (Riddick *et al.*, 1998). There was no apparent effect of transgenic cotton on the relative abundance of predatory spiders (*Clubiona* sp. and *Neoscona* sp.), coccinellid (*Cheilomenes sexmaculatus*), and the chrysopid (*Chrysoperla carnea*) (Sharma *et al.*, 2007). However, the abundance of spiders, coccinellids, and chrysopids was quite low in insecticide protected plots towards end of the cropping season. Direct exposure of the coccinelllid predator, *C. sexmaculatus* larvae to Bt toxins resulted in reduced larval survival and adult emergence as compared to the controls (Dhillon and Sharma, 2009). However, there were no adverse effects of the Bt toxins on *C. sexmaculatus* when the larvae were reared on *Aphis craccivora* fed on different concentrations of Cry1Ab or Cry1Ac in the artificial diet.

Thus, the effects of transgenic plants on the activity of predators vary across crops and the insect species involved. Because Bt-transgenic varieties can lead to substantial reductions in insecticide use in some crops, they can contribute to integrated pest management systems with a strong biological control component.

Interactions with Insect Parasitoids

Nontarget effects of Bt crops on natural enemies have been observed only when Bt-susceptible, sub-lethally damaged herbivores were used as prey or host, with no indication of direct toxic effects (Romeis *et al.*, 2006). Increased levels of parasitism by *Campoletis sonorensis* have been observed on transgenic plants compared to the nontransgenic plants, which may be due to fewer larvae on the transgenic plants. *C. sonorensis* and transgenic plants act synergistically, decreasing the larval survival beyond the level expected for an additive interaction (Johnson and Gould, 1992). The parasitoid, *Cardiochiles nigriceps* does not reduce the survival of the host larvae significantly, and its' activity is not influenced by the transgenic plants (Johnson, 1997; Johnson *et al.*, 1997). Egg parasitism of third-generation noctuids on Bt-transgenic cotton has been observed to be lower than in the conventional cottons (Wang and Xia, 1997). Percentage of parasitism by *Diadegma insulare* was not significantly different between the mixed and non-mixed plots of transgenic crop (Riggin Bucci and Gould, 1997). There is no effect of transgenic corn on the parasitization of *O. nubilalis* by *Eriborus terebrans* and *Macrocentrus grandii* (Orr and Landis, 1997).

Intra-field mixtures could serve to decrease density of a target pest such as the diamondback moth, while not adversely affecting the activity of natural enemies. *Diadegma insulare* was not harmed by exposure to Cry1C, while similar studies involving insecticides significantly reduced parasitism rates on strains of *P. xylostella* resistant to these insecticides (Chen *et al.*, 2008). In another study, Cry1Ac was not detected in newly emerged parasitoid, *Cotesia vestalis*, but detected in *Chrysoperla carnea* larvae fed on Bt-resistant *P. xylostella* larvae reared on Bt transgenic oilseed rape (Wei *et al.*, 2008). However, no Cry1Ac could be detected in *C. carnea* larvae when the lacewings were transferred to *P. xylostella* larvae reared on conventional oilseed rape.

There was a significant reduction in cocoon formation and adult emergence of the ichneumonid parasitoid, *Campoletis chlorideae* reared on *H. armigera* larvae fed on the leaves of transgenic cottons before and after parasitization (Sharma *et al.*, 2007). However, no Bt toxins were detected in *H. armigera* larvae and the parasitoid cocoons with enzyme linked immunosorbent assay. Reduction in cocoon formation was because of early mortality of the *H. armigera* larvae due to Bt toxins in the leaves of transgenic cotton. Survival and development of *C. chlorideae* was also poor when *H. armigera* larvae were fed on the leaves of cotton hybrid Mech 184. The adverse effects of transgenic cotton on survival and development of *C. chlorideae* were largely due to early mortality and possibly poor nutritional quality of *H. armigera* larvae due to toxic effects of the transgene (Sharma *et al.*, 2007, 2008). Bt sprays on chickpea prolonged the larval period, and reduced pupation and adult emergence of the *H. armigera* parasitoid, *C. chlorideae* (Dhillon and Sharma, 2010b). The Bt-intoxicated *H. armigera* larvae also resulted in reduced weight of the cocoons and adults of *C. chlorideae*. Bt toxins were detected in *H. armigera* larvae fed on Bt-sprayed chickpeas, but not in *C. chlorideae*. The adverse effects of Bt on the parasitoid were largely through early mortality of *H. armigera* larvae or poor quality of the insect host.

The effects of transgenic crops on the natural enemies vary across crops and the cropping systems. Some of the variation may be due to differences in pest abundance between the transgenic and the non-transgenic crops. Wherever the transgenic crops have shown adverse effects on the natural enemies, these effects may still be far lower than those of the broad-spectrum pesticides.

Interactions with Microflora in the Rhizosphere

Under field conditions, the microflora of Bt transgenic potato plants has been observed to be minimally different from that of chemically and microbially treated commercial potato plants (Donegan *et al.*, 1996). It is unlikely that expression of Bt and any other genes in transgenic plants would have an adverse effect on the soil microflora. Decomposition dynamics and bacterial and fungal communities associated with decomposition were strongly affected by surface and incorporated placements, and by temporal factors (Lu *et al.*, 2010). However, no significant differences were observed between Bt and non-Bt rice variety in either decomposition dynamics or in the soil microbial communities associated with residue decay. Mocali *et al.* (2009) observed a significant difference in microbial respiration and diversity among with Bt and control egg plants, but no such effects were observed after 6 and 12 months, suggesting a strong correlation between plants and microorganisms, as well as a short-term impact. The antimicrobial peptide magainin II has activity against a range of micro-organisms. Tubers harvested from potatoes genetically modified (GM) to express a synthetic magainin gene show improved resistance to the bacterial pathogen, *Erwinia carotovora*. There is little likelihood of any major sustained non-target effect of genetic modification using a magainin II transgene on plant-associated and soil microflora and function (O'Callaghan *et al.*, 2008). Transplastomic modification of tobacco with antibiotic resistance marker-gene *aadA* caused a relative decline of a specific *Flavobacterium* population, but not of other bacteria (O'Callaghan *et al.*, 2008). Transgenic cotton leaves have no significant acute toxicity on the earthworm, *Eisenia fetida* from oral exposure to the transgenic cotton line, GK19 (Liu *et al.*, 2009). The average weight, numbers of cocoons and new offsprings of *E. fetida* in the GK19 was lower than in the Simian 3, but the differences were not significant.

CONCLUSIONS

The use of crop protection traits through transgenics will continue to expand in future, and gene pyramiding will become very common. This approach of controlling insects would offer the advantage of allowing some degree of selection for specificity effects, so that pests, but not the beneficial organisms are targeted. Despite numerous future promises, there is a multitude of concerns about the impact of transgenic crops on the environment. The major issues in the environmental assessment of transgenic crops are putative invasiveness, vertical or horizontal gene flow, other ecological impacts, and effects on biodiversity. These are all highly interdisciplinary and complex issues. A crucial component for a proper assessment is defining the appropriate baseline for comparison and decision. The most appropriate reference point is the impact of plants developed by traditional breeding as the latter is an integral and accepted part of agriculture. In general, there are no major adverse effects of genetically modified insect-resistant crops on the generalist predators, while some adverse effects have been observed on the host specific parasitoids, which are largely due to early mortality of the host larvae or poor nutritional quality of the inset host, rather than direct toxicity of the insecticidal proteins. Such effects are common for all pest control interventions, including synthetic insecticides, and are not regarded as a risk.

It is equally important to ensure the safety of food derived from transgenic crops based on the principle of nutritional equivalence. Insect-resistant transgenic crops will play a significant role in integrated pest management in the future, reducing the number of insecticide sprays and pesticide residues in food. Concerted efforts are required involving international and advanced research institutes, and the national research organizations to harmonize the regulatory requirements to assess the bio-safety of the food derived from genetically engineered crops and their effects on the non-target organisms for sustainable crop production and food security.

ACKNOWLEDGEMENTS

I thank Dr. M.K. Dhillon, Division of Entomology, Indian Agricultural Research Institute (IARI), New Delhi, for his constructive comments on the manuscript.

REFERENCES

Arpaia, S, Gould, F. and Kennedy, G. (1997) Potential impact of *Coleomegilla maculate* predation on adaptation of *Leptinotarsa decemlineata* to Bt-transgenic potatoes. *Entomol. Exp. Appl.*, **82**, 91–100.

Barton, K., Whiteley, H. and Yang, N.S. (1987) *Bacillus thuringiensis* δ-endotoxin in transgenic *Nicotiana tabacum* provides resistance to lepidopteran insects. *Plant Physiol.*, **85**, 1103–1109.

Birch, A.N.E., Geoghegan, I.E., Majerus, M.E.N., McNicol, J.W., Hackett, C.A., Gatehouse, A.M.R. and Gatehouse J.A. (1999) Tri-trophic interactions involving pest aphids, predatory 2-spot ladybirds and transgenic potatoes expressing snowdrop lectin for aphid resistance. *Mol. Breed.*, **5**, 75–83.

Bottino, M.B., Girard, C., Jouanin, L., Metayer, M. le, Picard Nizou, A.L., Sandoz, G., Pham Delegue, M.H., Lerin, J., Le Metayer, M. and Thomas, G. (1998) Effects of transgenic oilseed rape expressing proteinase inhibitors on pest and beneficial insects. *Acta Hort.*, **459**, 235–239.

Boulter, D. (1995) Plant biotechnology. Facts and public perception. *Phytochemistry*, **40**, 1–9.

Brinkmann, N. and Tebbe, C.C. (2007) Differences in the rhizosphere bacterial community of a transplastomic tobacco plant compared to its non-engineered counterpart. *Environ. Biosaf. Res.*, **6**, 113–119.

Calgene (1990) *Kan^r gene: Safety and Use in Genetically Engineered Plants.* Request for Advisory Opinion. Calgene Inc., CA.

Chakrabarti, S.K., Mandaokar, A.D., Ananda Kumar, P. and Sharma, R.P. (1998) Synergistic effect of Cry1Ac and Cry1F δ-endotoxin of *Bacillus thuringiensis* on cotton bollworm, *Helicoverpa armigera. Curr. Sci.*, **75**, 663–664.

Chen, M., Zhao, J.Z., Collins, H.L., Earle, E.D., Cao, J. and Shelton, A.M. (2008) A critical assessment of the effects of Bt transgenic plants on parasitoids. *PLoS ONE*, **3**, e2284. doi:10.1371/journal.pone.0002284.

Chevre, A.M., Eberm F., Baranger, A. and Renard, M. (1997) Gene flow from transgenic crops. *Nature*, 389, 924.

Cook, R.J. (2000) Science-based risk assessment for the approval and use of plants in agricultural and other environments. In: G.J. Persley and M.M. Lnatin (eds.), *Agricultural Biotechnology and the Poor*, Consultative Group on International Agricultural Research, pp. 123–130.

Crecchio, C. and Stotzky, G. (1998) Insecticidal activity and biodegradation of the toxin from *Bacillus thuringiensis* subsp. *kurstaki* bound to humic acids from soil. *Soil Biol. Biochem.*, **30**, 463–470.

Crowder, D.W. and Carrière, Y. (2009) Comparing the refuge strategy for managing the evolution of insect resistance under different reproductive strategies. *J. Theor. Biol.*, **261**, 423–430.

Dale, E.C., and Ow, D.W. (1991) Gene transfer with subsequent removal of the selection gene from the host genome. *Proc. Nat. Acad. Sci. USA*, **88**, 10558–10562.

Daniels, N.E. (1981) *Migration of Greenbugs in the Texas Panhandle in Relation to their Biotypes.* Texas Agriculture Experiment Station Miscellaneous Publication, MP-1487, Texas.

Dhillon, M.K. and Sharma, H.C. (2009) Effects of *Bacillus thuringiensis* δ-endotoxins Cry1Ab and Cry1Ac on the coccinellid beetle, *Cheilomenes sexmaculatus* (Coleoptera, Coccinellidae) under direct and indirect exposure conditions. *Biocontr. Sci. Technol.*, **19**, 407–420.

Dhillon, M.K. and Sharma, H.C. (2010a) Influence of seed treatment and abiotic factors on damage to Bt and non-Bt cotton genotypes by the serpentine leaf miner, *Liriomyza trifolii* (Diptera: Agromyzidae). *Int. J. Trop. Insect Sci.*, **30**, 127–131

Dhillon, M.K. and Sharma, H.C. (2010b) Chickpea-mediated effects of *Bacillus thuringiensis* on *Helicoverpa armigera* and its larval parasitoid, *Campoletis chlorideae. J. Appl. Entomol.*, **134**, 682–693.

Dogan, E.B., Berry, R.E., Reed, G.L. and Rossignol, P.A. (1996) Biological parameters of convergent lady beetle (Coleoptera: Coccinellidae) feeding on aphids (Homoptera: Aphididae) on transgenic potato. *J. Econ. Entomol.*, **89**, 1105–1108.

Donegan, K.K., Schaller, D.L., Stone, J.K., Ganio, L.M., Reed, G., Hamm, P.B. and Seidler, R.J (1996) Microbial populations, fungal species diversity and plant pathogen levels in field plots of potato plants expressing the *Bacillus thuringiensis* var. *tenebrionis* endotoxin. *Transgenic Res.*, **5**, 25–35.

Doolittle, W.F. (1999) Phytogenic classification and the universal tree. *Science*, **284**, 2124–2127.

Ebinuma, H., Sugita, K., Matsunaga, E. and Yamakado, M. (1997) Selection of marker-free transgenic plants using the isopentenyl transferase gene. *Proc. Natl. Acad. Sci. USA*, **94**, 2117–2121.

Fearing, P.L., Brown, D., Vlachos, D., Meghji, M. and Privalle, L. (1997) Quantitative analysis of CryIA(b) expression in Bt maize plants, tissues, and silage and stability of expression over successive generations. *Mol. Breed.*, **3**, 169–176.

Fischhoff, D.A., Bowdish, K.S., Perlak, F.J., Marrone, P.G., McCormick, S.M., Niedermeyer, J.G., Dean, D.A., Kusano-Kretzmer, K., Mayer, E.J., Rochester, D.E., Rogers, S.G. and Fraley, R.T. (1987) Insect tolerant tomato plants. *BioTechnology*, **5**, 807–812.

Fitt, G., Mares, C.L., and Llewellyn, D.J. (1994) Field evaluation and potential ecological impact of transgenic cottons (*Gossypium hirsutum*) in Australia. *Biocontr. Sci. Technol.*, **4**, 535–548.

Fitter, A., Perrins, J. and Williamson, M. (1990) Weed probability challenged. *Bio/Technology*, **8**, 473.

FAO (1986) *International Code of Conduct on the Distribution and Use of Pesticides*. Food and Agriculture Organization, Rome, Italy.

Foster, J.E., Ohim, H.W., Patterson, F.L. and Taylor, P.L. (1991) Effectiveness of deploying single gene resistances in wheat for controlling the damage by Hessian fly (Diptera: Cecidomyiidae). *Environ. Entomol.*, **20**, 964–969.

Gebhard, F. and Smalla, K. (1998) Transformation of *Acinetobacter* sp. strain BD413 by transgenic sugarbeet DNA. *Appl. Environ. Microbiol.*, **64**, 1550–1554.

Gill, S.S., Cowles, E.A. and Pietrantonio, F.V. (1992) The mode of action of *Bacillus thuringiensis* endotoxins. *Annu. Rev. Entomol.*, **37**, 615–636.

Girard, C., Picard Nizou, A.L., Grallien, E., Zaccomer, B., Jouanin, L. and Pham Delegue, M.H. (1998) Effects of proteinase inhibitor ingestion on survival, learning abilities and digestive proteinases of the honeybees. *Transgenic Res.*, **7**, 239–246.

Giroux, S., Cot, J.C., Vincent, C., Martel, P. and Coderre, D. (1994) Bacteriological insecticide M-one effects on predation efficiency and mortality of adult *Coleomegilla maculata lengi* (Coleoptera: Coccinellidae). *J. Econ. Entomol.*, **87**, 39–43.

Gould, F. (1998) Sustainability of transgenic insecticidal cultivars: integrating pest genetics and ecology. *Annu. Rev. Entomol.*, **43**, 701–726.

Greene, J.K., Turnipseed, S.G. and Sullivan, M.J. 1997. Treatment thresholds for stink bugs in transgenic Bt cotton. *Proceedings, Beltwide Cotton Conference,* 6-10 January 1997, Vol 2. National Cotton Council, New Orleans, LA, pp. 895–898.

Gregorius, H.R. and Steiner, W. (1993) Gene transfer in plants as a potential agent of introgression. In: K. Workman and J. Tommie (eds.), *Transgenic Organisms*, Birkhauser Verlag, Basel, Switzerland, pp. 83–107.

Gustafson, D.I., Head, G.P. and Caprio, M.A. (2006) Modeling the impact of alternative hosts on *Helicoverpa zea* adaptation to bollgard cotton. *J. Econ. Entomol.*, **99**, 2116–2124.

Harris, A. (1991) Comparison of costs and returns associated with *Heliothis* resistant Bt cotton to non-resistant varieties. *Proceedings, Beltwide Cotton Conference, 1991.* National Cotton Council, Memphis, Tannesse, pp. 249–297.

Hilbeck, A., Baumgartner, M., Fried, P.M. and Bigler, F. (1998) Effects of transgenic *Bacillus thuringiensis* corn-fed prey on mortality and development time of immature *Chrysoperla carnea* (Neuroptera: Chrysopidae). *Environ. Entomol.*, **27**, 480–487.

Hilder, V.A. and Boulter, D. (1999) Genetic engineering of crop plants for insect resistance – A critical review. *Crop Prot.*, **18**, 199–191.

Hoffmann, M.P., Zalom, F.G., Wilson, L.T., Smilanick, J.M., Malyj, L.D., Kiser, J., Hilder, V.A. and Barnes, W.M. (1992) Field evaluation of transgenic tobacco containing genes encoding *Bacillus thuringiensis* delta-endotoxin or cowpea trypsin inhibitor: efficacy against *Helicoverpa zea* (Lepidoptera: Noctuidae). *J. Econ. Entomol.*, **85**, 2516–2522.

Hofte, H, and Whiteley, H.R. (1989) Insecticidal crystal proteins of *Bacillus thuringiensis. Microbiol. Rev.*, **53**, 242–255.

James C. (2008) *Global Status of Commercialized Biotech/GM Crops: 2008.* ISAAA Briefs No. 39. International Service for Acquisition of Agri-Biotech Applications (ISAAA), Ithaca, New York.

Johnson, M.T. (1997) Interaction of resistant plants and wasp parasitoids of tobacco budworm (Lepidoptera: Noctuidae). *Environ. Entomol.*, **26**, 207–214.

Johnson, M.T. and Gould, R. (1992) Interaction of genetically engineered host plant resistance and natural enemies of *Heliothis virescens* (Lepidoptera: Noctuidae) in tobacco. *Environ. Entomol.*, **21**, 586–597.

Johnson, M.T., Gould, F. and Kennedy, G.G. (1997) Effects of natural enemies on relative fitness of *Heliothis virescens* genotypes adapted and not adapted to resistant host plants. *Entomol. Exp. Appl.*, **82**, 219–230.

Jouanin, L., Girard, C., Bonade Bottino, M., Metayer, M. le, Nizou A.L.P., Lerin, J., Delegue, M.H.P. le and Metayer, M. (1998) Impact of transgenic oilseed rape expressing proteinase inhibitors on coleopteran pests and honey bees. *Cahiers Agric.*, **7**, 531–536.

Kaiser, J. (1996) Pests overwhelm Bt cotton crop. *Nature*, 273, 423.

Lang, B.A., Moellenbeck, D.J., Isenhour, D.J. and Wall, S.J. (1996) Evaluating resistance to CryIA(b) in European corn borer (Lepidoptera: Pyralidae) with artificial diet. *Resist. Pest Manage.*, **8**, 29–31.

Levin, M. and Strauss, H.S. (1993) Overview of risk assessment and regulation of environmental biotechnology. In: M. Levin and H.S. Strauss (eds.), *Risk Assessment in Genetic Engineering*, McGraw Hill Inc., New York, pp. 1–17.

Liu, B, Xu, C.R., Yan, F.M. and Gong, R.Z. (2005) The impacts of the pollen of insect-resistant transgenic cotton on honeybees. *Biodiver. Conser.*, **14**, 3487–3496.

Liu B., Wang, L.A., Zeng, Q., Meng, J., Hu, W.J., Li, X.G., Zhou, K.X., Xue, K., Liu, D.D. and Zheng, Y.P. (2009) Assessing effects of transgenic Cry1Ac cotton on the earthworm *Eisenia fetida*. *Soil Biol. Biochem.*, **41**, 1841–1846.

Llewellyn, D. and Fitt, G. (1996) Pollen dispersal from two field trials of transgenic cotton in the Namoi Valley, Australia. *Mol. Breed.*, **2**, 157–166.

Lozzia, G.C., Furlanis, C., Manachini, B. and Rigamonti, I.E. (1998) Effects of Bt corn on *Rhopalosiphum padi* L. (Rhynchota: Aphididae) and on its predator *Chrysoperla carnea* Stephen (Neuroptera: Chrysopidae). *Bollettino Zool. Agrar. Bachicoltura*, **30**, 153–164.

Lu, H.H, Wu, Wu, W.X., Chen, Y.X., Zhang, X.J., Devare, M. and Thies, J.E. (2010) Soil microbial community responses to Bt transgenic rice residue decomposition in a paddy field. *Plant Soil*, **336**, 279–290.

Luginbill, P. Jr. and Knipling, E.F. (1969) Suppression of wheat stem fly with resistant wheat. *USDA/ARS Prod. Res. Report*, **107**, 1–9.

Lundgren, J.G., Gassmann, A.J., Bernal, J., Duan, J.J. and Ruberson, J. (2009) Ecological compatibility of GM crops and biological control. *Crop Prot.*, **28**, 1017–1030.

MacIntosh, S.C., Stone, T.B., Jokerst, R.S. and Fuchs, R.L. (1991) Binding of *Bacillus thuringiensis* proteins to a laboratory selected strain of *Heliothis virescens*. *Proc. Natl. Acad. Sci. USA*, **188**, 8930–8933.

Malone, L.A., Giacon, H.A., Burgess, E.P.J., Maxwell, J.Z., Christeller, J.T. and Laing, W.A. (1995) Toxicity of trypsin endopeptidase inhibitors to honey bees (Hymenoptera: Apidae). *J. Econ. Entomol.*, **88**, 46–50.

Meenakshisundaram, K.S. and Gujar, G.T. (1998) Proteolysis of *Bacillus thuringiensis* subspecies *kurstaki* endotoxin with midgut proteases of some important lepidopterous species. *Indian J. Exp. Biol.*, **36**, 593–598.

Moberg, W.K. (1990) Understanding and combating agrochemical resistance. In: M.B. Green, H.M. LeBaron and W.K. Moberg (eds.), *Managing Resistance to Agriochemicals*, American Chemical Society, Washington DC, pp. 3–16.

Mocali, S., Dentice, A., Marcucci, A. and Benedetti, A. (2009) The impact of post-harvest treatments of transgenic eggplant residues on soil quality and microbial diversity. *Agrochimica*, **53**, 296–307.

NAS (1987) *Introduction of Recombinant DNA-Engineered Organisms into the Environment: Key Issues.* National Academy of Sciences. Washington DC.

NAS (1989) *Field Testing of Genetically Modified Organisms.* National Academy of Sciences, Washington DC.

O'Callaghan, M., Gerard, E.M., Bell, N.L., Waipara, N.W., Aalders, L.T., Baird, D.B. Conner, A.J. (2008) Microbial and nematode communities associated with potatoes genetically modified to express the antimicrobial peptide magainin and unmodified potato cultivars. *Soil Biol. Biochem.*, **40**, 1446–1459.

Olsen, K.M., Daly, J.C., Holt, H.E. and Finnegan, E.J. (2005) Season-long variation in expression of Cry1Ac gene and efficacy of *Bacillus thuringiensis* toxin in transgenic cotton against *Helicoverpa armigera* (Lepidoptera: Noctuidae). *J. Econ. Entomol.*, **98**, 1007–1017.

Orr, D.B. and Landis, D.A. (1997) Oviposition of European corn borer (Lepidoptera: Pyralidae) and impact of natural enemy populations in transgenic versus isogenic corn. *J. Econ. Entomol.*, **90**, 905–909.

Parker, R.D. and Huffman, R.L. (1997) Evaluation of insecticides for boll weevil control and impact on non-target arthropods on non-transgenic and transgenic Bt cotton cultivars. In: *Proceedings, Beltwide Cotton Conference,* 6-10 January 1997, Vol. 2, National Cotton Council, Memphis, USA, pp. 11216–1221.

Pham Delegue, M.H. and Jouanin, L. (1997) The honey bee and transgenic plants. *Rev. Francaise d'Apic.*, **574**, 250–251.

Picard Nizou, A.L., Grison, R., Olsen, L., Pioche, C., Arnold, G. and Pham-Delegue, M.H. (1997) Impact on proteins used in plant genetic engineering toxicity and behavioral study in the honeybee. *J. Econ. Entomol.*, **90**, 1710–1716.

Porter, D.H., Burd, J.D., Shufran, K.A., Webster, J.A. and Teetes, G.L. (1997) Greenbug (Homoptera: Aphididae) biotypes: selected by resistant cultivars or preadapted opportunists. *J. Econ. Entomol.*, **90**, 1055–1065.

Potrykus, I. (1990) Gene transfer to plants: Assessment and perspectives. *Physiol. Plantarum*, **79**, 125–134.

Qaim, M. (2009) The economics of genetically modified crops. *Annu. Rev. Resource Econom.*, **1**, 665–694.

Raybould, A.F. and Gray, A.J.C. (1993) Genetically modified crops and hybridization with wild relatives: a UK perspective. *J. Appl. Ecol.*, **30**, 119–219.

Riddick, E.W. and Barbosa, P. (1998) Impact of Cry3A-intoxicated *Leptinotarsa decemlineata* (Coleoptera: Chrysomelidae) and pollen on consumption, development, and fecundity *of Coleomegilla maculata (*Coleoptera: Coccinellidae*). Ann. Entomol. Soc. Am.*, **91**, 303–307.

Riddick, E.W., Dively, G. and Barbosa, P. (1998) Effect of a seed-mix deployment of Cry3A-transgenic and nontransgenic potato on the abundance of *Lebia grandis (*Coleoptera: Carabidae) *and Coleomegilla maculata (*Coleoptera: Coccinellidae). *Ann. Entomol. Soc. Am.*, **91**, 647–653.

Riggin Bucci, T.M. and Gould, F. (1997) Impact of intraplot mixtures of toxic and nontoxic plants on population dynamics of diamondback moth (Lepidoptera: Plutellidae) and its natural enemies. *J. Econ. Entomol.*, **90**, 241–251.

Richter, B. and Smalla, K. (2007) Screening of rhizosphere and soil bacteria for transformability. *Environ. Biosaf. Res.*, **6**, 91–99.

Romeis, J., Meissle, M. and Bigler, F. (2006) Transgenic crops expressing *Bacillus thuringiensis* toxins and biological control. *Nat. Biotechnol.*, **24**, 63–71.

Sachs, E.S., Benedict, J.H., Stelly, D.M., Taylor, J.F., Altman, D.W., Berberich, S.A. and Davis, S.K. 1998. Expression and segregation of genes encoding CryIA insecticidal proteins in cotton. *Crop Sci.*, 38, 1–11.

Serageldin, I. (1999) Biotechnology and food security in the 21st Century. *Science*, **286**, 387–389.

Serratos, J.A., Willcox, M.C. Castillo-Gonzalez, F. (1997) *Gene Flow Among Maize Landraces, Improved Maize Varieties, and Teosinte: Implications for Transgenic Maize.* International Wheat and Maize Research Institute, Mexico, DF, Mexico.

Sharma, H.C. (1993) Host plant resistance to insects in sorghum and its role in integrated pest management. *Crop Prot.*, **12**, 11–34.

Sharma, H.C. (2009) Biotechnological Approaches for Pest Management and Ecological Sustainability. CRC Press/Taylor and Francis, New York.

Sharma, H.C., and Ortiz, R. (2000) Transgenics, pest management, and the environment. *Curr. Sci.*, **79**, 421–437.

Sharma, H.C., Ananda Kumar, P., Seetharama, N., Hari Prasad, K.V. and Singh, B.U. (1999) Role of transgenic plants in pest management in sorghum. In: *Symposium on Tissue Culture and Genetic Transformation of Sorghum,* 23-28 Feb 1999. ICRISAT, Patancheru, Andhra Pradesh, India.

Sharma, H.C., Pampapathy, G. and Arora, R. (2007) Influence of transgenic cottons with *Bacillus thuringiensis* cry1Ac gene on the natural enemies of *Helicoverpa armigera*. *BioControl*, **52**, 469–489.

Sharma, H.C., Crouch, J.H., Sharma, K.K, Seetharama, N. and Hash, C.T. (2002) Applications of biotechnology for crop improvement: prospects and constraints. *Plant Sci.*, **163**, 381–395.

Sharma, H.C., Dhillon, M.K. and Arora, R. (2008) Effects of *Bacillus thuringiensis* δ-endotoxin fed *Helicoverpa armigera* on the survival and development of the parasitoid *Campoletis chlorideae*. *Entomol. Exp. Appl.*, **126**, 1–8

Sharma, H.C., Sharma, K.K, Seetharama, N. and Crouch, J.H. (2004) Genetic transformation of crops for insect resistance: potential and limitations. *Crit. Rev. Plant Sci.*, **23**, 47–72.

Sharma, H.C., Sharma, K.K., Seetharama, N. and Ortiz, R. (2000) Prospects for using transgenic resistance to insects in crop improvement. *Electronic J. Biotechnol.*, **3**, http://www.ejb.org/content/vol 3/issue 2/full/20.

Sims, S.R. (1995) *Bacillus thuringiensis* var. *kurstaki* (CryIA (c) protein expressed in transgenic cotton: Effects on beneficial and other non-target insects. *Southwest. Entomol.*, **20**, 493–500.

Smith, C.M. (2005) *Plant Resistance to Arthropods – Molecular and Conventional Approaches.* Springer Verlag, Dordrecht, The Netherlands.

Tabashnik, B.E. (1994) Evolution of resistance to *Bacillus thuringiensis*. *Annu. Rev. Entomol.*, **39**, 47–79.

Tabashnik, B.E., Rensburg, J.B.J. van and Carrière, Y. (2009) Field-evolved insect resistance to Bt crops: Definition, theory, and data. *J. Econ. Entomol.*, **102**, 2011–2025.

Tabashnik, B.E., Biggs, R.W., Higginson, D.M., Henderson, S., Unnithan, D.C., Unnithan, G.C., Ellers-Kirk, C., Sisterson, M.S., Dennehy, T.J., Carrière, Y. and Morin, S. (2005) Association between resistance to Bt cotton and cadherin genotype in pink bollworm. *J. Econ. Entomol.*, **98**, 635–644.

Thomas, D.J.I., Morgan, J.A.W., Whipps, J.M. and Saunders, J.R. (1997) Plasmid transfer by the insect pathogen *Bacillus thuringiensis* in the environment. In: *Microbial Insecticides: Novelty or Necessity? Proceedings of a Symposium held at the University of Warwick,* 16-18 April 1997. British Crop Protection Council, Coventry, Farnham, UK, pp. 261–265.

Vaeck, M., Reynaerts, A., Hofte, H., Jansens, S., DeBeuckleer, M., Dean, C., Zabean, M., Van Montagu, M. and Leemans, J. (1987) Transgenic plants protected from insect attack. *Nature*, **327**, 33–37.

Wang, C.Y. and Xia, J.Y. (1997) Differences of population dynamics of bollworms and of population dynamics of major natural enemies between Bt transgenic cotton and conventional cotton. *China Cottons*, **24**, 13–15.

Wei, W., Schuler, T.H., Clark, S.J., Stewart, C.N.Jr. and Poppy, G.M. (2008) Movement of transgenic plant-expressed Bt Cry1Ac proteins through high trophic levels. *J. Appl. Entomol.*, **132**, 1–11.

Wolfenbarger, L.L., Naranjo, S.E., Lundgren, J.G., Bitzer, R.J. and Watrud, L.S. (2008) Bt crop effects on functional guilds of non-target arthropods: A meta-analysis. *PLoS ONE* **3**, e2118. doi:10.1371/journal.pone.0002118.

Wood, E.A. Jr. (1971) Designation and reaction of three biotypes of the greenbug cultured on resistant and susceptible cultivars of sorghum. *J. Econ. Entomol.*, **64**, 183–185.

Wu, J.Y., He, X.L., Shu, .C, Chen, S., Fu, C.X. and Huang, J.Q. (1997) Influence of waterlogging on the bollworm resistance of Bt cotton. *Jiangsu J. Agric. Sci.*, **13**, 231–233.

Yoder, J.I. and Goldsbrough, A.P. (1994) Transformation systems for generating marker-free transgenic plants. *Biotechnology*, **12**, 263–267.

6

Bacterial Toxins and their Application in Insect Pest Management

M. NAGESH*, SEEMA WAHAB[1], SALEEM JAVEED AND R. THIPPESWAMY

National Bureau of Agriculturally Important Insects, PB No. 2491
HA Farm Post, Bellary Road, Hebbal, Bangalore, 560 024
[1]Department of Biotechnology, Government of India, CGO Complex
Lodi Road, New Delhi 110 012
*E-mail: nagesh55@yahoo.com

INTRODUCTION

With the ever increasing demand for safe food and fodder for global nutritional and food security, with a mandatory commitment to environmental safety, natural resources and biodiversity security "Biopesticides" have become mainstay of Integrated Pest Management strategies in order to limit the use of synthetic molecules in bulk in agriculture. With the spectacular technological breakthroughs in Molecular Biology and Biotechnology, new vistas have opened up as offshoots of biological control and biopesticides. The important commercial breakthrough was the discovery of insecticidal toxin proteins (cry series) from bacterial pathogenesis of lepidopteran larvae, and its utilization in imparting host plant resistance to insect pests and commercial transgenic crops. The underlying principle was the utilization of information on bacterial pathogenesis and related toxins. Therefore, the microbial pathogenesis, more specifically, the bacterial toxins are emerging as potential sources of application in pest management. In the present contribution the authors attempted to focus on certain basic features of bacterial toxins and their possible applications.

Bacteria in general can be categorized in to gram positive or gram negative (Fig. 6.1), pathogenic or non-pathogenic. Most pathogenic bacteria have the ability to produce a variety of toxins which is also termed as toxigenesis. These toxin factors are considered as the offensive virulence factors which actively contribute to a successful infection. Therefore, these toxins can also be termed as virulence factors. Chemically they are of two types, lipopolysaccharides, associated with the cell wall of Gram-negative bacteria, and proteins, which are released from bacterial cells and may act at sites away from the site of bacterial growth. The cell-associated toxins are referred to as endotoxins and the extracellular diffusible toxins are referred to as exotoxins. Exotoxins are usually proteins or polypeptides, that act enzymatically or through direct action with host cells and stimulate a variety of host responses. Endotoxins are cell-associated structural components and most of them are located in the cell envelope. Endotoxins generally act in the vicinity of bacterial growth or presence.

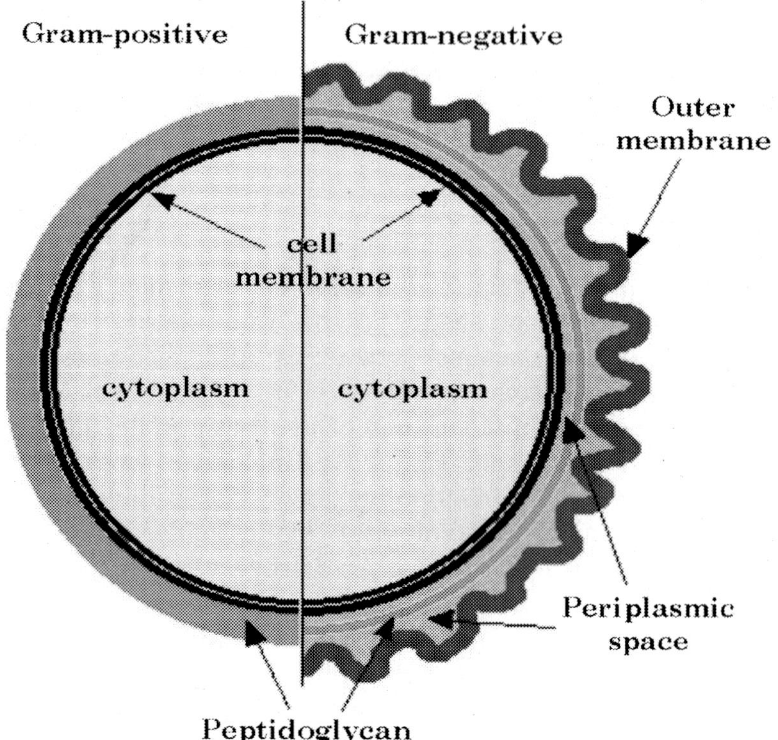

Fig. 6.1. Differentiation between gram positive and gram negative bacteria

The success of utilization of bacterial toxins in most cases of pathogenesis has been based on the successful insight in to the type toxin, mode of action, genes coding for the toxin, gene regulation and interactions of such toxins on one-to-one/many basis in an *in vivo* system. It has been shown that the information at the site of infection and pathogenesis is the central dogma for its utilization. Therefore, it is important to consider the nature and types of toxins at the basic level and for convenience, general and human pathogenic bacteria are taken as examples.

Generally the bacterial toxins act intracellularly and or on the cell surface. Like *Bt* crystal toxins, pore (channel) formation is also observed in toxins that are produced by other gram-positive bacteria like *B. anthracis, Clostridium botulinum, C. tetani, C. perfringens, Staphylococcus aureus* and in gram-negative bacteria like *Escherichia coli, Vibrio cholera* etc. *Bt* crystal toxins belong to the latter group of cell surface acting toxins (Lacy and Stevens, 1998).

Living bacteria usually secrete exotoxins during exponential growth. The production of the toxin is generally specific to a particular bacterial species that produces the disease associated with the toxin (e.g., only *Clostridium tetani* produces tetanus toxin; only *Corynebacterium diphtheriae* produces the diphtheria toxin). Usually, the toxin is the major determinant of virulence (e.g., tetanus and diphtheria).

Based on site of activity of the exotoxins, they can be enterotoxin, neurotoxin, leukocidin or hemolysin that indicate the target site of some well-defined protein toxins. A few bacterial toxins that obviously bring about the death of an animal are known simply as lethal toxins, and even though the tissues affected and the target site or substrate may be known, the precise mechanism by which death occurs is not clear (e.g. anthrax LF).

TYPES OF BACTERIAL TOXINS

Generally, three types of toxins from pathogenic bacteria are described, which include, Type I toxins that act from the cell surface; Type II that are cell membrane damaging toxins and Type III that are intracellular toxins. The Type I toxins include superantigens, while Type II toxins encompass pore forming toxins and toxins that enzymatically damage the membrane.

Pore-forming toxins, as the name suggests, insert a transmembranous pore into a host cell membrane, thereby disrupting the selective influx and efflux of ions across the membrane. This group of toxins includes the RTX toxins of Gram-negative bacteria, streptolysin O produced by *S. pyogenes*, and *S. aureus* alpha toxin. Generally, these toxins are produced as subunits that self-assemble as a pore on the eucaryotic membrane (Table 6.1). Type III toxins include a variety of groups of molecules such as Botulin neurotoxin, tetanus toxin, anthrax lethal toxin, metalloproteases, shiga toxin, *Bt* toxin etc.

Table 6.1. Some pore-forming bacterial toxins

Toxin	Bacterial source	Target	Disease
Perfringiolysin O	*Clostridium perfringens*	Cholesterol	Gas gangrene
Hemolysin	*Escherichia coli*	Cell membrane	UTI
Listeriolysin	*Listeria monocytogenes*	Cholesterol	Systemic; meningitis
Anthrax EF	*Bacillus anthracis*	Cell membrane	Anthrax (edema)
Alpha toxin	*Staphylococcus aureus*	Cell membrane	Abcesses
Pneumolysin	*Streptococcus pneumonia*	Cholesterol	Pneumonia; otitis media
Streptolysin O	*Streptococcus pyogenes*	Cholesterol	Strep throat
Leukocidin	*Staphylococcus aureus*	Phagocyte membrane	Pyogenic infections

Other than these, toxins that damage the extracellular matrix such as Hyaluronidases/ Collagenases also exist. Some bacterial toxins are utilized as invasins because they act locally to promote bacterial invasion. Examples are extracellular enzymes that degrade tissue matrices or fibrin, allowing the bacteria to spread. This includes collagenase, hyaluronidase and streptokinase. Other toxins, also considered invasins, degrade membrane components, such as phospholipases and lecithinases. The pore-forming toxins that insert a pore into eucaryotic membranes are considered as invasins.

Some protein toxins have very specific cytotoxic activity (i.e., they attack specific types of cells). For example, tetanus and botulinum toxins attack only neurons. But some toxins (as produced by staphylococci, streptococci, clostridia, etc.) have fairly broad cytotoxic activity and cause nonspecific death of various types of cells or damage to tissues, eventually resulting in necrosis. Toxins that are phospholipases act in this way. This is also true of pore-forming hemolysins and leukocidins.

Bacterial protein toxins are strongly antigenic. *In vivo*, specific antibody neutralizes the toxicity of these bacterial exotoxins (antitoxin). As proteins, many bacterial toxins resemble enzymes in a number of ways. Like enzymes, they are denatured by heat, acid and proteolytic enzymes, they act catalytically, and they exhibit specificity of action. The substrate (in the host) may be a component of tissue cells, organs or body fluid.

GENETICS AND OCCURRENCE

It has been shown that many of the toxin genes are located on the extra chromosomal DNA called "plasmids" in bacteria. The plasmids usually move among different strains of the same species and also rarely between different species and kingdom, with certain specificity and frequency (Davidson, 1999). In many instances insertion sequences (IS) and conjugative transposable elements (TE) play a vital role in the horizontal gene transfer. In *Bt*, majority of the toxin genes are borne on large plasmids and some of them are conjugative. Hence it is of interest to explore whether *Bt* would acquire any toxin genes by horizontal gene transfer (F+, Hfr, IS and TE). Interestingly the other *Bt* related pathogens like *B. popilliae* and *B. sphaericus*, to name a few, have different mode of action though they produce crystalline inclusions (Charles *et al.*, 1996).

The genetic ability to produce a toxin, including regulatory genes, may be found on the bacterial chromosome, plasmids and lysogenic bacteriophages. Sometimes they occur within pathogenicity islands. In any case, the processes of genetic exchange in bacteria, notably conjugation and transduction, can mobilize genetic elements between strains and species of bacteria. Horizontal gene transfer (HGT) of genes that encode virulence is known to occur between species of bacteria.

CONTROL OF SYNTHESIS AND RELEASE OF PROTEIN TOXINS

The regulation of synthesis and secretion of many bacterial toxins is tightly controlled by regulatory elements that are sensitive to environmental signals. For example, the production of diphtheria toxin is totally repressed by the availability of adequate amounts of iron in the medium for bacterial growth. Only under conditions of limiting amounts of iron in the growth medium does toxin production become derepressed. The expression of cholera toxin and related virulence factors (adhesins) is controlled by environmental osmolarity and temperature.

The processes by which protein toxins are assembled and secreted by bacterial cells are also variable. Many of the classic exotoxins are synthesized with an NH terminal leader (signal) sequence consisting of a few (1-3) charged amino acids and a stretch of (14-20) hydrophobic amino acids. The signal sequence may bind and insert into the cytoplasmic membrane during translation such that the polypeptide is secreted while being synthesized. The signal peptide is cleaved as the toxin (protein) is released into the periplasm. Alternatively, the toxin may be synthesized intracytoplasmically, then bound to a leader sequence for passage across the membrane. Frequently, chaperone proteins are required to guide this process. Some multicomponent toxins, such as the cholera toxin, have their subunits synthesized and secreted separately into the periplasm where they are assembled. In Gram-negative bacteria, the outer membrane poses an additional permeability barrier that a protein toxin usually has to mediate if it is to be released in a soluble form. It has been proposed that some Gram-negative exotoxins (e.g. *E. coli* ST enterotoxin) might be released in membrane vesicles composed of outer membrane components. Since these vesicles possibly possess outer membrane-associated attachment factors, they could act as "smart bombs" capable of specifically interacting with and possibly entering target cells to release their contents of toxin.

RELEVANCE TO PEST MANAGEMENT

Among several bacterial sources for novel toxin genes for utilization in modern agriculture, *Bacillus thurengiensis* and bacterial symbionts associated with entomopathogenic nematodes (EPN) form very important examples. Generally the bacterial toxins act on the susceptible cells in two ways, *viz.* intracellular and on the cell surface. Like *Bt* crystal toxins, pore (channel) formation is also observed in toxins that are produced by other gram-positive bacteria like *B. anthracis*, *Clostridium botulinum*, *C. tetani*, *C. perfringens*, *Staphylococcus aureus* and in gram-negative bacteria like *Escherichia coli*, *Vibrio cholera* etc. *Bt* crystal toxins belong to the latter group of cell surface acting toxins (Lacy and Stevens, 1998). The present article aims to discuss the possible utilization of promising candidate toxins and their genes from diverse bacterial sources with the success of *B. thuringiensis* and newer sources of bacterial gene pool from EPN as the back ground.

It has been shown that the many of the toxin genes are located on the extra chromosomal DNA called “plasmids” in bacteria. The plasmids usually move among different strains of the same species and also rarely between different species and kingdom, with certain specificity and frequency (Davidson, 1999). In many instances insertion sequences (IS) and conjugative transposable elements (TE) play a vital role in the horizontal gene transfer. In *Bt*, majority of the toxin genes are borne on large plasmids and some of them are conjugative. Hence it is of interest to explore whether *Bt* would acquire any toxin genes by horizontal gene transfer (F+, Hfr, IS and TE). Interestingly the other *Bt* related pathogens like *B. popilliae* and *B. sphaericus*, to name a few, have different mode of action though they produce crystalline inclusions (Charles *et al.*, 1996).

TOXINS OF *BACILLUS THURINGIENSIS* ORIGIN

There are various toxins having origin from *Bacillus thuringiensis*. These toxins are α exotoxins, β exotoxins, δ endotoxins, Vips and Cyt toxins. Except the δ endotoxins (crystal) the other toxins are produced during the exponential growth and the delta endotoxins are produced during the late exponential and stationary phase of growth. After the cell lysis the spore and crystal are released.

α-Exotoxin

Toumanoff (1954) was the first person to report the presence of this heat labile toxin from the subspecies *alseti* and it was found to be toxic to *Galleria mellonella* larvae. Heimpel (1967) coined the term alpha exotoxin for the heat labile exotoxin from *Bt*.

β-Exotoxin

The heat-tolerant exotoxin was discovered by McConnell and Richards in 1959 and has been known as the fly-factor, heat-stable toxin and thuringiensin. Levinson *et al.* (1990) chemically confirmed the existence of second heat tolerable toxin called type-II toxin and it was found to be more specific than the type-I toxin, which is very active against the potato beetle, *Leptinotarsa decemlineata*. The mode of action of the type-I toxin is inhibition of DNA-dependant RNA polymerase. Two β exotoxin based products area used in the former USSR. Thuringin 1 and 2 (2 and 10 per cent exotoxin) and Bitoxibacillin (0.6 to 0.8 per cent β exotoxin). The products are used effectively against the several species of red mites, against the larvae of houseflies and blowflies. Abbott laboratories had developed the product known as Di-beta (Beegle and Yamamoto, 1992).

Crystal Toxins (δ-Endotoxins)

At present there are about 133 different cry genes have been sequenced and the information on these genes are available in the public domain on the net (http://www.biols.susx.ac.uk/Home/Neil_Crickmore/Bt/index.html). Most of the *cry* genes are transcribed by the sporulation specific promoters and are produced predominantly in the late exponential and stationary phase, while cry3 genes are expressed during the vegetative phase of growth. In a cell generally only one crystal is seen but rarely two. The size of various cry genes are given in the table no. The cry1 group has extensive C-terminal, which has more of cysteine amino acid, which stabilizes the crystalline inclusion inside the cell. In a single crystal there are approximately one million protoxin crystals are packed. The crystal also includes other regulatory proteins like *npr* proteins and metal ions (Schnepf *et al.*, 1998). In some cases they also contain naturally truncated toxin molecule.

Vegetative Insecticidal Proteins (Vips)

The Vips do not form any parasporal crystal proteins and are apparently secreted form the cell. The Vips are presently excluded from the Cry protein nomenclature because they are not crystal forming proteins. The vip1A gene encodes a 100 kDa protein that is apparently processed from its N terminus to yield an 80 kDa protein upon secretion. This toxin is toxic to the western corn rootworm larvae in conjunction with the Vip2A protein, whose coding region is located just upstream of the of vip1A gene (Petosactal, 1997). Vip1A shows some sequence similarity to the protective antigen of *B. anthracis* toxin. The Vip3A gene encodes for an 88 kDa protein that is produced during the vegetative growth but is not processed upon secretion. Genes encoding Vip3A-type protein appear to be common among strains of Bt and *B. cereus*. Vip3A intoxication resembles the mode of action of Cry proteins (Yu *et al.*1997).

Cyt Toxins

Even before the structure of *cry3A* toxin was resolved by X-ray crystallography, comparison of the sequences of the large family of Cry toxins allowed some informed speculation about the toxin structure and the functional domains within the proteins. The same process was more difficult for the *CytA* toxin because only two genes had been sequenced, and these showed only a single amino acid difference

(Shimizu and Morikawa, 1996). The second member of the *Cyt* family is *CytB*. The structural similarity of these two toxins showed that there are five short beta strands and six longer alpha helices. In this toxin the beta sheets form the pore lipid bilayer. There is a weak immunological cross reactivity between Cyt and Cry4D and a weak similarity was apparent at the sequence level. Cyt toxins unlike Cry toxins are able to lyse wide variety of cell types *in vitro*. Cyt2A consists of a single domain in which two outer layers of alpha helices wrap around a mixed beta sheet. Cyt1A is believed to have similar structure.

Other Compounds Produced by Bt

Proteases, phospholipases, chitinases are a few of the enzymes that are involved in pathogenesis of insect host tissue be the bacteria.

TOXINS FROM BACTERIAL SYMBIONTS OF ENTOMOPATHOGENIC NEMATODES

Entomopathogenic nematodes are beneficial nematodes that invade the insects and cause mortality in 48-72 hours depending on the insect host and inoculum level. Entomopathogenic nematodes belong to the families Steinernematidae and Heterorhabditidae. These nematodes harbour and co-exist symbiotically with insect pathogenic bacteria *Xenorhabdus* and *Photorhabdus* species belonging to *Enterobacteriaceae* (Forst *et al.*, 1997). These bacteria reside in the guts of nematodes that invade insects and are released directly in to the open blood system of the insect.

A wide range of studies including genome analysis of these bacteria have confirmed that they contain a wide range of putative virulence facors, including toxins, exoenzymes, and hemolysins (Ffrench-Constant *et al.*, 2003). Genes coding for many of the virulence factors are known to occur in groups or in series termed as pathogenic islands or toxin complexes. The insecticidal genes are organized in the *tcd* island harbouring nine *tcd*- and *tcc*-like genes and several tc-like genes, while further nine *tcc*-like genes are scattered over the chromosome (Duchaud *et al.*, 2003). Four high molecular weight toxin complexes are termed as tca, tcb, tcc and tcd in case of *P. luminescens* or *X. nematophila*. Tca-like and Tcd-like proteins contribute predominantly to the oral toxicity of bacterial supernatants (Ffrench-Constant *et al.*, 2003).

All the three types of protein toxins (Type I, II and III) have been found to be associated with *Photorhabdus* and *Xenorhabdus* which are responsible for insecticidal activity and have similarity with that in *Yersinia enterocolitica,* a human pathogenic bacterium. Insecticidal toxins similar to cry, Vips, enzyme inhibitors and toxic enzymes are also found in these two bacterial species.

Extra-cellular toxins such as hemolysin/hemagglutinin-related proteins are associated with these bacteria. The mode of action in these bacteria is complex and multi-directional. Studies indicated a series of complex cell to cell signalling and stimuli that effect rapid mortality to infected insects. Transgenic plants with TccC gene complex from *Xenorhabdus* have already been created with resistance to lepidopteran larvae, which are in different stages of validation. This indicated that besides, toxins from *B. thuringiensis,* the gene pool on toxins from EPN-associated bacteria offer a range of new set of potential insecticidal molecules.

CONCLUSIONS

It is clear from the above discussion that even though *Bt* toxins and EPN-associated bacterial toxins share many similarities with the other toxins that are produced by gram positive and gram negative bacteria, yet they are functionally different. Molecular biology of the above toxins genes reveal that many of them are plasmid borne. Some of the plasmids are conjugative and readily transferred among closely related bacteria and also rarely among distantly related ones. The *Bt* plasmids specially the larger plasmids are conjugative and are found to shuttle between *Bt* and *B. cereus.* The plasmid exchange system that occurs in bacteria could enable bacteria to acquire a new phenotype based on the selection pressure that it often encounters. Hence *B. cereus* could have got new plasmids from its environment via the following methods, *viz.* transformation, transduction, and conjugation. For reasons not known, different strains of *Bt* are antigenically and toxin composition wise different and stable. Therefore the future line of investigation should be on the cataloguing of bacteria involved in insect pathogenesis and their toxins/virulence factors, genetic and molecular analysis of virulence factors and the mode of action governing the virulence and eventually identifying the potential toxin molecules for their possible use in pest management.

REFERENCES

Beegle, C.C and Yamamoto, T. (1992) History of *Bacillus thuringiensis* Berliner: Research and Development. *Can. Entomol.*, **124**, 587–616.

Carozzi, N and Koziel, M. (1998) *Advances in Insect Control: The Role of Transgenic Plants*. Taylor & Francis, London.

Charles, J.F., Le-Roux, C.N., and Delecluse, A. (1996) *Bacillus sphaericus* toxins: Molecular biology and mode of action. *Annu. Rev. Entomol.*. **41**, 451–472.

Christeller, J.T and Shaw, B.D. (1989) The interaction of a range of serine proteinase inhibitors with bovine trypsin and *Costelytra zealandica* trypsin. *Insect Biochem.*, **19**, 233–241.

Duchaud, E., Rusniok, C., Frangeul, L., Buchrieser, C., Givaudan, A., Taourit, S., Bocs, S., Boursaux-Eude, C., Chandler, M., Charles, J., Dassa, E., Derose, R., Derzelle, S., Freyssinet, G., Gaudriault, S., Médigue, C., Lanois, A., Powell, K., Siguier, P., Vincent, R., Wingate, V., Zouine, M., Glaser, P., Boemare, N., Danchin, A. and Kunst, F.(2003). The genome sequence of the entomopathogenic bacterium, *Photorhabdus luminescens*. *Nat. Biotechnol.*, **21**, 1307–1313.

Edmonds, H.S., Gatehouse, L.N., Hilder, V.A and Gatehouse, J.A. (1996) The antimetabolic effects of oryzacystatin on larvae of the Southern corn root worm (*Diabrotica undecimpunctata howrdiI*): Use of a bacterial expression system for oryzacystatin. *Entomol. Exp. Appl.*, **78**, 83–94.

Ffrench-Constant, R., Waterfield, N., Daborn, P., Joyce, S., Bennett, H., Au, C., Dowling, A., Boundy, S., Reynolds, S. and Clarke, D. (2003) Photorhabdus towards a functional genomic analysis of a symbiont and pathogen. *FEMS Microbiol. Rev.*, **26**, 433–456.

Forst, S., Dowds, B., Boemare, N. and Stackebrandt, E. (1997) *Xenorhabdus* and *Photorhabdus* spp.: bugs that kill bugs. *Annu. Rev. Microbiol.*, **51**, 47–72.

Gatehouse, A.M.R., Hilder, R.V and Boulter, D. (1992) *Plant Genetic Manipulation for Crop Protection*. CAB International, Wallingford, U.K

Heimpell, A.M. (1967) A critical review of *Bacillus thuringiensis* Berliner and other crystalliferous bacteria. *Annu. Rev. Entomol.*, **12**, 287–322.

Lacy, D.B and Stevens, R.C. (1998) Unraveling the structures and modes of action of bacterial toxins. *Curr. Opin. Struc. Biol.*, **8**, 778–784.

Levinson, B.L., Kasyan, K.J., Chiu, S.S, Currier, T.C and Gonzales, J.M. (1990) Identification of ß-exotoxin production plasmid encoding ß-exotoxin and a new exotoxin in *Bacillus thuringiensis* using high performance liquid chromatography. *J. Bacteriol.*, **172**, 3172–3179.

McGurl, B., Orozoco-Cardenas, M., Pearce, G and Ryan, C. (1994) Over expression of prosystemin gene in transgenic tomato plants generates a systemic signal that constitutively expresses protease inhibitor synthesis. *PNAS*, **91**, 9799–9802.

Petosa, C., Coiller,R.J., Klimpel, K.R., Leppal, S.H and Liddington, R.C. (1997) Crystal structure of the anthrax toxin protective antigen. *Nature*, 385, 833–838.

Schnepf, E., Crickmore, N., Van Rie, J., Lereclus, D., Baum, J., Feitelson, J., Zeigler, D.R and Dean, D.H. (1998) *Bacillus thuringiensis* and its pesticidal crystal proteins. *Microbiol. Mol. Biol. Rev.*, **62**, 775–806.

Shimizu, T and Morikawa, K. (1996) The beta prism: A new folding motif. *Trends Biochem. Sci.*, **21**, 3–6.

Terra, W.R. (1999) Evolution of digestive systems of insects. *Annu. Rev. Entomol.*, **85**, 181–200.

Thomas, K. and Nation, J. (1984) Protease, lipase and amylase activities in the mid gut and the hind gut of the cricket, *Gryllus rubens* and mole cricket, *Rhynchosiara* and its physiological significance. *Comp. Biochem. Physiol.*, **79A**, 297–304.

Todar's Online Textbook of Bacteriology (2004) The Good, the Bad, and the Deadly. *Science*, **304**, 1421.

Toumanoff, C. (1954) L' action de *Bacillus cereus* var. *alesti*. *Annal. Inst. Pasteur Paris*, **86**, 570–597.

Yu, C-G., Mullins, M.A., Warren, G.W., Koziel, M.G and Estruch, J.J. (1997) The *Bacillus thuringiensis* vegetative insecticidal protein Vip 3A lyses midgut epithelial cells of susceptible insects. *Appl. Environ. Microbiol.*, **63**, 532–536.

7

Genetic Improvement of Bioagents for Pest Management in Agriculture

MUKESH K. DHILLON

Division of Entomology, Indian Agricultural Research Institute
New Delhi- 110 012, India
E-mail: mukeshdhillon@rediffmail.com

INTRODUCTION

Bioagents are known to be effective, economically viable, and ecologically safe tool of insect pest management, however, their potential has not been fully exploited for the management of insect pests in agriculture. Several constraints such as widespread use of pesticides, difficulties involved in mass rearing, restricted host range, slow action, poor quality, and inability to withstand adverse conditions have been some of the notable constraints which make deployment of the bioagents practically difficult for the control of insect pests. Development of bioagents that are easy in quality mass production, with increased rate of multiplication, wide host range, longer persistence in the field, good host-finding ability, adaption to extremes of climatic conditions, and capable of tolerating pesticide effects applied for the control of other pests, could help successful deployment of bioagents to their full potential as pest management tool. Research efforts should also be continued to combine and develop new and innovative approaches for genetic manipulation of bioagents so that their mass production becomes economical. Genetic improvement has been proved to be practical and cost-effective for identifying the traits that limit the efficacy and retains the fitness of the improved strain (Headley and Hoy, 1987). Similarly, development and release of pesticide-tolerant natural enemies to maintain their parasitizing/predatory activity under pesticide application conditions for the management of other pests is another important approach of genetic improvement of bioagents.

Generally, the genetic improvement of bioagents have included screening and isolation of populations exhibiting the desired traits, artificial selection, cross hybridization, and more recently genetic manipulation such as mutagenesis and genetic engineering. However, other than bacteria and viruses, genetic improvement schemes have hardly been applied for the improvement in bioefficacy of the bioagents. Many of these bioagents are highly sensitive to change in climatic components like temperature, humidity, elevated atmospheric CO_2, UV radiation, etc. and there is need to develop the abiotic stress-tolerant strains of entomopathogens for their effective use in insect pest management (Dhillon and Gujar, 2010; Abbaszadeh *et al.*, 2011). The development of recombinant DNA techniques has made it possible to significantly improve the insecticidal efficacy of viruses, bacteria and fungi (Inceoglu *et al.*, 2006; Wang and St. Leger, 2007). More recently, molecular biology methods have elucidated pathogenic processes in several fungal biocontrol agents including two of the most commonly applied entomopathogenic fungi, *Metarhizium anisopliae* and *Beauveria bassiana*. The combination of molecular techniques have helped in identifying and characterizing genes involved in infection, manipulation of the genes of the pathogen to improve biocontrol performance, and allowed expression of a neurotoxin from the scorpion *Androctonus australis* (AaIT) in *M. anisopliae* strain ARSEF 549 (St. Leger and Wang, 2010). These advances have been achieved by combining new knowledge derived from basic studies of the molecular biology and genomics of these pathogens with technical developments that enable increases in gene expression and the use of genes from other organisms that encode insecticidal proteins to improve efficacy.

The recombinant DNA technologies have also made it possible to transfer genes specific to beneficial traits in natural enemies, and can be used to understand genetics and physiology of reproduction and control of sex ratio, broaden host range, and this information can be used to improve rearing of the natural enemies for biological control (Sharma, 2009). The use of genetic engineering methods for the improvement of natural enemies is advantageous over the artificial selection in the sense that the genetic improvement can be achieved rapidly without the generations of rearing and selecting from the available gene pool of the natural enemies, and where the gene from any species can be used for the genetic improvement. Although, genetic improvement of bioagents through selection, hybridization, or recombinant DNA technology, and recent advances in molecular biology have genetically improved some of the bioagents for a variety of traits, environmental issues related to release of genetically engineered bioagents such as potential risk of low level of successful establishment, speed of evolution and movement of genetic material from one species to another are equally important to address before their deployment (IAEA, 2006).

GENETIC IMPROVEMENT OF NATURAL ENEMIES

Conventional Approaches

Maintaining quality in laboratory-reared arthropod natural enemies is difficult due to possible genetic changes caused by inadvertent selection, inbreeding, genetic drift, and founder effects (Stouthamer *et al.*, 1992; Hopper *et al.*, 1993), however, if high quality and inexpensive artificial diets for predators and parasitoids are available, biocontrol programs would no longer be restricted by inefficient mass production methods. It is also likely that increase in temperature as a result of climate change will have a great bearing on the effectiveness of natural enemies for pest management, which in turn will affect insect host-natural enemy associations, crop production, and food security (Dhillon and Sharma, 2009). Therefore, altering longevity, research for mechanisms of ageing, shortening developmental rate, enhancing progeny production, altering sex ratio, extending temperature and relative humidity tolerances, and altering host or habitat preferences could enhance the effectiveness of biological control agents (Hoy, 1976; Dhillon and Sharma, 2009; Dhillon and Gujar, 2010). Another approach is to develop and release pesticide tolerant natural enemies so that parasitizing/predatory activity of the resistant natural enemy is not disrupted even when pesticide application is continued for management of other pests. Some of the significant achievements made in genetic improvement of predators and parasitoids through conventional approaches (Table 7.1) have been discussed hereunder:

Predators. Field efficacy of an organophosphate resistant strain of the predatory mite, *Metaseiulus occidentalis* has been demonstrated in mite pest infested grapevine orchards in Russia, which were also subjected to chemical insecticide spray applications (Petrushov, 1987).

Pesticide-resistant predatory mites have been identified and evaluated in the field (Hoy, 1990), and if the natural populations of predatory mites could be replaced with pesticide-resistant ones, there will be considerable increase in predatory mite population and decrease in pesticide application. Various parameters for optimum mass production of phytoseiid mite predator for biological control of spider mites have been standardized at University of Agricultural Sciences, Bangalore, India, while biological control strategy involving the conservation and augmentation of the two major predators, viz. *Dipha aphidivora* and *Microuns igorotus* has been developed for the management of sugarcane woolly aphid, *Ceratovacuna lanigera* Zehntner at National Bureau of Agriculturally Important Insects, Bangalore, India (Wahab, 2009).

Table 7.1. Genetically improved natural enemies through conventional approaches

Natural enemy	Improved trait	Reference(s)
Predators		
Predatory mite, *Metaseiulus occidentalis*	Organophosphate resistance	Petrushov (1987)
Chrysoperla spp.	Insecticide and high temperature tolerant	Wahab (2010)
Phytoseiid mite	Mass production standardized	Wahab (2010)
Dipha aphidivora & *Microuns igorotus*	Conservation and augmentation technique for the control of sugarcane woolly aphid	Wahab (2010)
Parasitoids		
Trichogramma japonicum	Resistance to fenvalerate and decamethrin	Xiong *et al.* (1988)
Aphytis lingnanensis	Tolerance to azinphosmethyl	Havron *et al.* (1991)
Anisopteromalus calandrae	Resistance to malathion	Baker and Thorne (1995)
Trichogramma chilonis	Tolerance to endosulfan, fenvalerate, monocrotophos, and high temperature (38°C)	Jalali *et al.* (2006); Wahab (2009)

Parasitoids. Several insect parasitoids resistant/tolerant to insecticides such as fenvalerate and decamethrin tolerant *Trichogramma japonicum* Ashmead (Xiong *et al.*, 1988), azinphosmethyl tolerant *Aphytis lingnanensis* Compere (Havron *et al.*, 1991), malathion resistant *Anisopteromalus calandrae* (Howard) (Baker and Thorne, 1995), and endosulfan tolerant *Trichogramma chilonis* Ishii (Jalali *et al.*, 2006) identified either from field populations or developed through laboratory selections, have been deployed for pest management in agriculture. However, certain biological parameters of insecticide tolerant strains could be affected in the process of selection, thus reducing their efficacy in the field (Caprio and Hoy, 1995). Project Directorate of Biological Control, Bangalore, has developed genetically improved strain of egg parasitoid *T. chilonis* having combined tolerance to three insecticides (endosulfan at 700 ppm, fenvalerate at 20 ppm, monocrotophos at 540 ppm), and high temperature (38°C), while temperature and insecticide tolerant strain of *T. chilonis* has also been developed at G.B. Pant University of Agriculture and Technology, Pantnagar, India (Wahab, 2010). Furthermore, Dhillon and Sharma (2009) reported that the ichneumonid parasitoid, *Campoletis chlorideae* stored at 12°C increased the adult longevity by about two month and exhibited higher fecundity than at 27°C (Fig. 7.1), however, the efficiency of parasitism and adult emergence were quite low (Table 7.2). However, the parasitoid showed better survival and lived about one month longer at 18°C than at 27°C, suggesting that the parasitoid adults stored at 18°C could be used for parasitism, while the immature stages should be reared at 27°C for mass production of the parasitoid for biological control of *H. armigera*.

Table 7.2. Effect of storage temperature on parasitoid, *Campoletis chlorideae*

Temperature	Fecundity	Parasitization (%)	Adult emergence (%)	Sex ratio (M:F)
12 ± 2°C	262.8	56.0	39.2	0.34
27 ± 2°C	217.6	92.0	81.3	1.38
LSD (P ≤ 0.05)	28.51	16.17	12.69	0.19

Source: Dhillon and Sharma (2009).

Biotechnological Approaches

Recombinant DNA techniques offer the potential of genetically modified more robust natural enemies with resistance to pesticides, or better adaptation to extremes of climatic conditions, increased host range, and efficient rearing systems (Atkinson *et al.*, 2001). The genetic improvement of natural enemies involves several components which are required to ultimately improve, identify and deploy them for pest management under field conditions (Fig. 7.2). Genes for use in genetic engineering of arthropods can be isolated from either closely or distantly related organisms for insertion into the target insect species. It may also be possible to isolate a gene from the species being manipulated, alter it, and reinsert it into a germ line. Although, the release of genetically improved arthropods for suppression of pest populations has been undertaken in the past (Braig and Yan, 2002), there has been a limited success in developing genetically engineered arthropods, except *Drosophila* (Kramer, 2004).

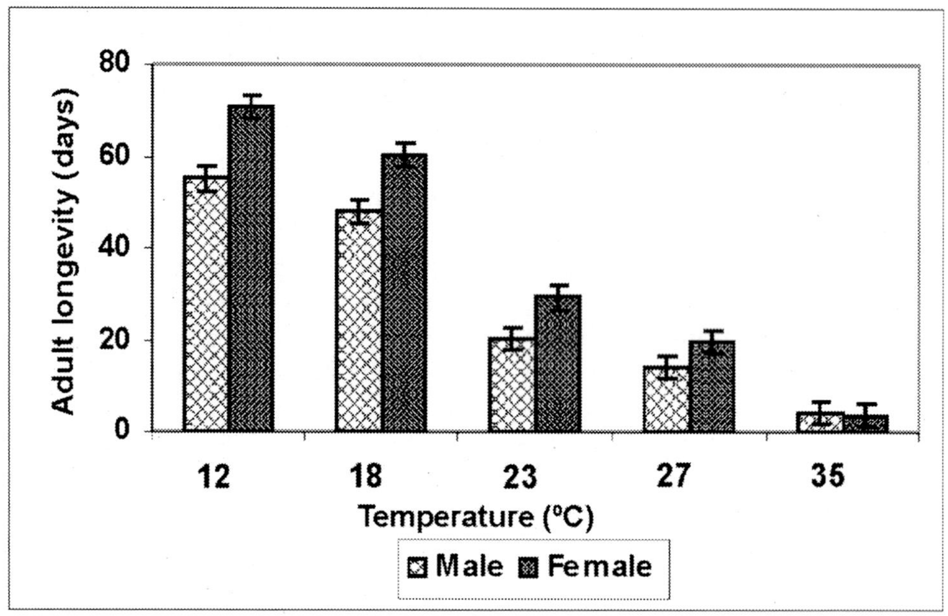

Fig. 7.1. Effect of temperature on longevity of *Campoletis chlorideae* adults (After: Dhillon and Sharma, 2009)

Transgenic natural enemies of crop pests with some desirable characteristics such as pathogen resistance, environmental hardiness, increased fecundity, and improved host-seeking ability have also been developed (O'Brochta and Atkinson, 1998), but additional work is required to make subtropical or tropical species of arthropod natural enemies to become adapted to a much broader range of climates. Several useful insecticide resistance genes have been cloned and expressed in insects, such as parathion hydrolase gene (*opd*) from *Pseudomonas diminuta*, cyclodiene resistance gene (*γ-aminobutryic acid A*) from *Drosophila melanogaster,* benomyl resistance gene β–*tubulin* from *Neurospora crassa* and *Septoria nodorum*, acetylcholinesterase gene (*AChe*) from *D. melanogaster* and *Anopheles stephensi,* glutathione-S-transferase gene (*GST1*) from *Musca domestica,* a cytochrome P450-B1 gene (*CYP6A2*) associated with DDT resistance in *D. melanogaster,* and amplification core and esterase *B1* gene from *Culex* responsible for organophosphate resistance (Atkinson *et al.*, 2001), and organophosphate degrading enzymes that confer resistance to paraoxon and parathion in *D. melanogaster* (Benedict *et al.*, 1994, 1995) and *Cardiochiles diaphaniae* (Presnail and Hoy, 1996), which can be deployed for the development of arthropod natural enemies for resistance to insecticides (Table 7.3).

Table 7.3. Future insecticide resistance genes cloned from insect and non-insect sources for genetic improvement of natural enemies

Gene	Trait	Source
Parathion hydrolase gene (opd)	Parathion resistance	*Pseudomonas diminuta*
γ-aminobutryic acid A	Cyclodiene resistance	*Drosophila melanogaster*
β-tubulin	Benomyl resistance	*Neurospora crassa*
Acetylcholinesterase gene (AChe)	Organophosphate resistance	*D. melanogaster* & *Anopheles stephensi*
glutathione-S- transferase gene (GST1)	Organophosphate resistance	*Musca domestica*
cytochrome P450-B1 gene (CYP6A2)	DDT resistance	*Drosophila melanogaster*
Amplification core and esterase B1 gene	Organophosphate resistance	*Culex* spp.
Organophosphate degrading enzymes	Paraoxon and Parathion	*D. melanogaster* & *Cardiochiles diaphaniae*

Transgenic predatory mite, *Metaseiulus occidentalis* has been used for the control of spider mite, *Tetranychus urticae* (Presnail *et al.*, 1997). Increased tolerance to cold in frost-susceptible arthropods or for tolerance to heat and dry conditions, e.g. in *Trichogramma* spp., can be achieved by genetic transformation (Almeida *et al.*, 1998). Most of the research on genetic engineering of insects has been aimed at gaining an understanding of gene regulation or developmental processes. Cloned genes could also be modified by *in vitro* mutation to achieve a desired phenotype. Several techniques have been employed for genetic engineering of insects such as DNA injection into eggs, e.g., in phytoseiid mite, *M. occidentalis* (Miller *et al.*, 1987) and maternal micro-injection, e.g., in adult parasitoid, *C. diaphaniae* (Presnail and Hoy, 1996), which could be used to deliver injected DNA to produce transformed insects including natural enemies of crop pests.

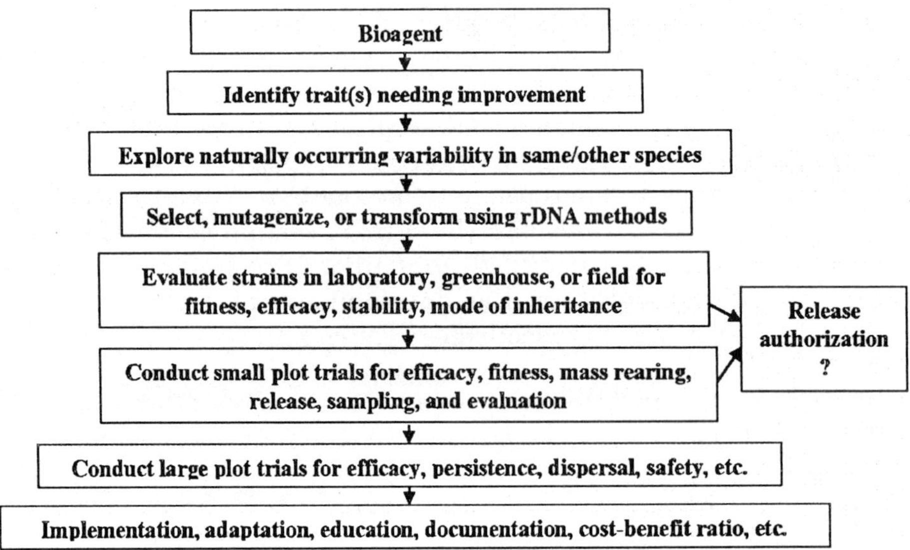

Fig. 7.2. Components of genetic improvement of biocontrol agents.

ENTOMOPATHOGENIC MICROORGANISMS

Growing awareness about the hazardous effects of synthetic pesticides has led to increasing interest in use of environmental friendly bioagents *viz.*, entomopathogenic fungi, bacteria, viruses, nematodes and microsporidia. However, many of these bioagents are highly sensitive to the environmental conditions, and the inconsistent performance of microbial pesticides has been attributed mainly to problems with formulation of entomopathogens, UV sensitivity, temperature, and humidity (Dhillon and Gujar, 2010). In recent review by Abbaszadeh *et al.* (2011) on effects of climatic conditions on the bioefficacy of biopesticides have elaborated that change in climatic components like temperature, humidity, elevated atmospheric CO_2, UV radiation, etc., from its normal result in alteration in efficacy of biopesticides, which could be positive or negative, depending on the prevailing climatic conditions, and environmental conditions have been emphasized as a criteria for their selection. Recent advances in molecular biology have also shown promise in improving the bioefficacy related traits of insect pathogenic microorganisms, which can also be deployed in developing the abiotic stress-tolerant strains of entomopathogens (Abbaszadeh *et al.*, 2011). Some notable advances in genetic improvement of bioagents, *viz.* bacteria, fungi, viruses, protozoa, and nematodes, for their effective use in insect pest management, are discussed as under.

Entomopathogenic Bacteria

Bacillus thuringiensis (Bt) is the most successfully commercialized biocontrol agent used across the world for the control of lepidopteran and coleopteran pests of vegetable and field crops. However, several factors like narrow range of biological activity in comparison to insecticides and varying degree of sensitivity in multi-species pest complex of the same order of insects, limit the use of Bt in plant protection. Efforts have been made to improve Bt for increasing its host range and stability (van Frankenhuyzen, 1993). Genetic manipulation of *cry* genes in Bt promises scope in increasing efficacy. For example, combinations of Cry proteins (Bradfisch *et al.*, 1996; Lee *et al.*, 1996), and mixing of spores with the Cry proteins (Moar *et al.*, 1995; Dubois and Dean, 1995), have been found to exhibit synergistic toxicity towards lepidopteran and dipteran insect pests. Plasmid curing and plasmid transfer from one Bt strain to another through conjugation-like process have also been used to improve the insecticidal properties of different strains of Bt (Gonzalez *et al.*, 1982; Baum *et al.*, 1998). However, curing and transfer of native *cry* plasmids has its limitations like most *cry* genes are not readily transformed by conjugate transfer. The use of recombinant DNA technologies can circumvent these problems associated with Bt strain improvement. Although, conjugation and transduction methods can also be used to transfer recombinant plasmid from donor to recipient Bt strain, the most preferred method of such plasmid transfer is electroporation (Gawron-Burke and Baum, 1991). Site-specific recombination system has also been deployed to selectively delete ancillary or foreign DNA elements such as antibiotic resistance genes from recombinant plasmids after their introduction into Bt host (Baum *et al.*, 1996). With the use of such techniques, several genetically modified Bt products (Table 7.4) have been registered for agricultural use (Navon, 2000).

Bioencapsulation of Bt endotoxin gene, and delivery of Bt toxin through *Pseudomonas* bacterium has been proved to be a successful technology for enhancement of field persistence and efficacy of Bt toxins against lepidopteran pests (Baum *et al.*, 1998). Epiphytosis and endophytosis approaches of transfer of truncated or full length sequences of insecticidal crystal protein genes into other bacteria associated with crop plants have also been successfully deployed to increase the efficacy of Bt biocide in the field (Obukowicz *et al.*, 1987; Oeda *et al.*, 1987; Skot *et al.*, 1990; Herrera *et al.*, 1994). Mosquito larvicidal genes from *Bacillus sphaericus* have been cloned from cellular DNA, and expressed in protease-deficient strain of *B. sphaericus*, which has enhanced insecticidal efficacy against aquatic dipterans (Rajamohan and Dean, 1996). A chimeric gene, *cry* 2Ax1 of Bt has been constructed by engineering C-terminal region of domain-III in Cry2Aa to improve the toxicity of its protein against *Helicoverpa armigera* (Hubner). The constructed chimeric *cry2Ax1* gene expressed in transformant of acrystalliferous Bt strain analyzed for proteins toxicity against larvae of *H. armigera* showed about 22-fold higher activity in the chimeric Cry2Ax1 protein than in the Cry2Ab (Wahab, 2010).

Table 7.4. Genetically modified *Bacillus thuringiensis* products for pest management in agriculture

Bt strain	Product	Target insect pest
aizawai recipient *kurstaki* recipient	Agree, Design (transconjugant)	Lepidoptera
kurstaki recipient	Condor, Cutlass (transconjugant)	Lepidoptera
	Raven (recombinant)	Lepidoptera, Coleoptera
kurstaki	Leptinox (recombinant)	Lepidoptera
δ-endotoxin encapsulated in *Pseudomonas fluorescens*	MVP, MATTCH, MTRACK	Lepidoptera
	CellCap®	Lepidoptera, Coleoptera

Entomopathogenic Fungi

Deployment of entomopathogenic fungi for the control of insect pests offers an environmentally safe strategy of pest management. However, their effectiveness is variable and depends on the genetic variability to cause mortality and spread within populations, environmental parameters (temperature, humidity), and development of suitable formulation techniques. Pathogenicity is the result of the interaction of numerous individual traits determining germination and penetration, followed by successful growth within the target host, which is prerequisite for induction of mortality. Despite of having a wide array of potential pesticidal proteins produced by entomopathogenic fungi, fungal genes have not played significant role in agricultural biotechnology. Entomopathogenic fungi, *B. bassiana* and *M. anisopliae* have excellent characteristics for insect pest control; however, their effectiveness is relative humidity-dependent in the field after application of fungal spores. The insecticidal efficacy of these entomopathogenic fungi in terms of mass production, stability, and application, has been optimized by conventional means. However, little attention has been paid for their genetic improvement due to lack of information on molecular and biochemical bases of fungal pathogenesis. Identification of genes responsible for insect pathogenesis is necessary for the improvement of insecticidal efficacy by genetic manipulation. The molecular and biochemical information on pathogenesis is only available for few fungal species, for example, the molecular and biochemical studies on pathogenesis of *M. anisopliae* have shown that this fungus causes green muscardine disease in insects, and various genes responsible for penetration in insect cuticle by the fungus, virulence, and nutritional stress have been cloned from this fungus (St. Leger, 1995).

Pathogenicity of entomopathogenic fungi, *Verticillium lecanii* against glasshouse whitefly, *Trialeurodes vaporariorum* and the aphid, *Macrosiphoniella sanborni*; and *M. anisopliae* against rice brown planthopper, *Nilaparvata lugens* have been genetically improved via manipulation of the parasexual cycle (Heale *et al.*, 1989). Incorporation of benomyl resistance into the entomopathogenic fungus, *M. anisopliae* have also been tried to make this fungus more useful in integrated pest management (Goettel *et al.*, 1989). Benomyl-resistant transformants of *B. bassiana* have also obtained by Sandhu *et al.* (2001). The enzymes secreted by the fungus facilitate penetration through insect cuticle and provide nutrition for fungal growth, that include subtilisin- and trypsin-like proteases, metalloproteases, and exopeptidases, and the specificity of the proteases confers some degree of host specificity to the fungus (St. Leger, 1995; Joshi *et al.*, 1997). Further research on fungal pathogenesis has suggested that the physical force and enzymes triggered by the fungus in conjunction with pH modify insect cuticle (St. Leger *et al.*, 1998).

With the help of genetic engineering additional copies of *PrI* gene have been inserted in the genome of *M. anisopliae*, which encodes subtilisin-like protease involved in host cuticle penetration, to enhance its insecticidal efficacy (St. Leger *et al.*, 1996). The *PrI* gene is normally expressed during growth on the cuticle, produced in the hemocoel of lepidopteran larvae infected with the recombinant fungi, resulting in partial hydrolysis of haemolymph proteins and extensive melanization of the larvae, leading to reduced feeding and early death of the infected larvae as compared to non-recombinant *M. anisopliae*. Manipulation of protease genes and site-directed mutagenesis of enzymes involved in pathogenesis could also be used to understand the catalytic mechanism of the enzymes with the potential for enhancement through genetic manipulation. Two genetically improved strains of the entomopathogenic fungus, *B. bassiana* have been obtained by crosses with the insecticidal toxin-producing strain, *Beauveria sulfurescens* using protoplast fusion technique for more effective control of *Ostrinia nubilalis* and *Leptinotarsa decemlineata*. The stability of the virulence after passing through insect-host and stability of molecular structure of the fusion products suggested that asexual genetic recombination by protoplast fusion may provide an attractive method for the genetic improvement of entomopathogenic fungi (Couteaudier *et al.*, 1996).

Recently, molecular biology methods have elucidated pathogenic processes in several fungal biocontrol agents. For example, the modified fungus *M. anisopliae* induced the same mortality rates in tobacco hornworm, *Manduca sexta* at 22-fold lower spore doses than the wild type, and survival time reduced by 40% (Wang and St. Leger, 2007). Similar results were obtained with mosquitoes and coffee berry borer, *Hypothenemus hampei* (Pava-Ripoll *et al.*, 2008). Fan *et al.* (2007a) constructed several *B. bassiana* hybrid chitinases where the chitinase was fused to chitin-binding domains derived from plant, bacterial, or insect sources. A hybrid chitinase containing the chitin binding domain from the silkworm *Bombyx mori* chitinase fused to the *B. bassiana* chitinase showed the greatest ability to bind to chitin compared to other hybrid chitinases. Constitutive expression of this hybrid chitinase gene by *B. bassiana* reduced time to death of insect hosts by 23% compared to the wild-type fungus showing that genetic components of the host insect incorporated into the fungal pathogen can increase virulence, and synthetic chitinases with improved enzyme activities through DNA shuffling have also been produced (Fan *et al.,* 2007b).

Signal transduction pathways including a protein kinase A (MaPka1) that are master regulators of insect infection processes have also been identified (Fang *et al.*, 2009), which could be helpful in expressing the hybrid chitinase gene in fungal pathogens. Furthermore, the *Pr1* gene and an esterase gene (*AaIT*) from *M. anisopliae* has also been used to increase the virulence of *B. bassiana*, however, coexpression of *Pr1* and *AaIT* genes in *B. bassiana* failed to achieve any synergistic virulence, since the expressed *AaIT* was digested by *Pr1* produced in the hemolymph (Lu *et al.*, 2008). The complete sequencing of pathogenic fungi *viz.*, *B. bassiana*, *M. anisopliae* var. *acridium*, *M.*

anisopliae var. *anisopliae*, and *Cordyceps militaris*, is also under progress, wherein combining transcriptome analysis with an insertional mutagenesis strategy to tag virulence genes will significantly increase the value of genome data.

Entomopathogenic Baculoviruses

In order to enhance the efficiency of nucleopolyhedrosis viruses (NPVs) either by reducing the time to kill host insects or to reduce their feeding capacity, different recombinant baculoviruses have been constructed expressing insect-selective toxins, hormones, enzymes involved in insect metabolism, the maize mitochondrial protein URF13, and by deletion of the *egt* gene which significantly increased the speed of host incapacitation (O'Reilly and Miller, 1991; Kamita *et al.*, 2005). Efforts have been made in *baculoviruses* to incorporate genes that produce the proteins, which kill the insects at a faster rate (Bonning and Hammock, 1992). Corsaro *et al.* (1989) optimized transfection conditions for cloning *UND-K* derivative of the IPLB-HZ 1075 cell line of the noctuid, *Helicoverpa zea* using a calcium-phosphate co-precipitation technique and the DNA genome of the *H. zea* S-type nuclear polyhedrosis virus. The technique permitted relatively efficient *in vitro* manipulation of the virus genome. Human 5-lipoxygenase (EC 1.13.11.34), the key enzyme involved in the transformation of arachidonic acid to the potent biologically active leukotrienes, has been over-expressed in cells of *Spodoptera frugiperda* using *Autographa californica* nuclear polyhedrosis virus strain E2 expression system (Funk *et al.*, 1989). Although infected cells were able to express mutant 5-lipoxygenase protein, enzyme activity was not substantially altered, suggesting the nonessential nature of certain histidines in binding iron at the putative ferric catalytic site.

The major antibacterial proteins, attacin, cecropin, and lysozyme are secreted into the haemolymph of the saturniid, *Hyalophora cecropia* upon challenge with bacteria. Attacin has been isolated and the corresponding amino acid sequence indicated that the basic attacin is synthesized as a 233-residue pre-proprotein, which was confirmed by cloning the cDNA fragment encoding the basic attacin in the *A. californica* nuclear polyhedrosis virus downstream of the polyhedrin promoter and expressing the protein in *S. frugiperda* cells. Attacin has also been produced in last instar larvae of *Trichoplusia ni* after injection of recombinant virus (Gunne *et al.*, 1990). Based on protein processing pattern, it has been suggested that a protease produced by the *S. frugiperda* cells cleaves *Arg45-Arg46,* producing mature attacin. The instability of the attacin proteins is rationalized in terms of their random-coil structure, which was deduced from circular dichroism measurements.

Baculovirus express-ing insect-selective toxins that affect the ion-channels of excitable cells of insects have proved to be highly promising in increasing the viral efficiency in insect pest control (Moscardi, 1999). Regev *et al.* (2003) have generated a recombinant AcMNPV expressing both excitatory (*LqhIT1*) and depressant (*LqhIT2*) genes, which resulted in synergistic effects with marked reduction in time to kill 50% host larvae. A lepidopteran-selective toxin, *Buthus tamulus* insect-selective toxin (ButaIT), from the venom of the South Indian red scorpion, *Mesobuthus tamulus* has also been engineered into a polyhedrin positive *Autographa californica* nuclear polyhedrosis virus (AcMNPV) genome, which showed enhanced insecticidal activity, reduced feeding and time to kill the *Heliothis virescens* larvae (Wudayagiri *et al.*, 2001; Rajendra *et al.*, 2006).

Entomopathogenic Protozoa

The protozoans belonging to Phylum Microspora have found with potential to control insect pests. Practically, genetic engineering could be applied to the microsporidia to enhance host range and pathogenicity. Although, phylogenetic and evolutionary relationships have been analyzed using ribosomal RNA analysis (Baker *et al.*, 1998), the basic knowledge and genetic engineering for *in vitro* culture, formulation, and to enhance pathogenesis by protozoan's are lacking, and need clear understanding before being deployed for insect pest management. A new species of microsporidium isolated from the larvae of *Spodoptera litura* has been reported to be highly pathogenic to these larvae, and specifically to lepidopteran insects (Johny *et al.*, 2006). This new isolate has been from the genus *Nosema*.

Entomopathogenic Nematodes

The nematode families Steinernematidae and Heterorhabditidae have a symbiotic association with *Xenorhabdus* (Enterobacteroaceae) bacteria, which are responsible for insect pathogenicity in more than 250 insect species. They are highly virulent, safe for nontarget organisms, and amenable to genetic selection. However, their susceptibility to environmental stress, temperature extremes, solar radiation, and desiccation are some of the major limitations of entomopathogenic nematodes for pest control. Natural variation is available among nematode strains for virulence, desiccation tolerance, host-finding ability, and activity at low temperatures, and hence natural selection can be used to optimize these parameters (Gaugler and Hashmi, 1996).

Heterorhabditis bacteriophora has been engineered to express *Caenorhabditis elegans hsp70* to enhance tolerance to high temperatures, where the transformed nematodes exhibited normal growth and development, and survived on exposure to 40°C (Hashmi *et al.*, 1998). The transgene was inherited extrachromosomally, and the transformation was stable for 15 generations. This research furthermore, demonstrates that the use of *C. elegans* could be used as a model for further research on transformation for enhancement of pathogenicity traits of the Steinernematidae and Heterorhabditidae. Heat tolerance gene from the natural heterorhabditid isolate IS-5 of entomopathogenic nematode, *H. bacteriophora* has been transferred in commercial HP88 strain to improve the pathogenicity of this nematode stain under high temperature conditions (Segal and Glazer, 2000).

Genetically engineered bacteria associated with entomopathogenic nematodes, e.g. *Xenorhabdus nematophilus*, have also been examined for nematode specificity genes, virulence, and insecticidal genes (Hurlbert, 1994). Genes involved in sex determination in *H. bacteriophora* to increase the proportion of hermaphrodites in the population in order to increase the reproductive potential of this nematode have also been identified (Kahel-Raifer and Glazer, 2000). This could increase the reproductive potential of these nematodes when mass reared in fermentors, since the yield will not depend on male-female encounters. In recent past, Ciche and Sternberg (2007), demonstrated that RNAi can be used effectively in *H. bacteriophora*, which can also be applied for analyses of nematode genes involved in symbiosis and parasitism. Genetically improved strains of *Heterorhabditis indica* and *Steinernema riobrave* for tolerance to environment and enhanced efficacy against *H. armigera* have also been developed at Central Institute for Cotton Research, Nagpur, India (Wahab, 2010).

VECTORS, MARKERS AND PROMOTERS

The use of the polymerase chain reaction (PCR) to identify DNA markers, offers a highly efficient method for detecting genetic changes in arthropod natural enemy populations. For example, RAPD-PCR method has a potential value for monitoring, establishment, and dispersal of specific biotypes of arthropod natural enemies (Chapco *et al.,* 1992; Edwards and Hoy, 1993). Pesticide resistance genes have been used in arthropods with the goal to develop insecticide-resistant strains of natural enemies of insect pests. The β-galactosidase gene (*lacZ*) isolated from *E. coli* and regulated by the *Drosophila hsp70* promoter, has been expressed in phytoseiid predator, *M. occidentalis* (Presnail and Hoy, 1992). For a stable integration of the transgene, two separate plasmids, viz. donor plasmid, carrying the transgene of interest and a visible transformation marker are used. Four different transposable elements from four separate families of eukaryotic transposable elements can be used to transform nondrosophilid insects (Atkinson and O'Brochta, 1999). In addition, viral and bacterial endosymbiont gene delivery systems have also been developed for some insect species.

Many genes have been cloned and inserted into *Drosophila* by P-element-mediated transformation, but very few cloned genes are of value for genetic manipulation of beneficial arthropods. Transposable elements, called P-elements (mobile pieces of DNA that have the ability to move from one site on a chromosome to another) can be used to transform insects by macro-injecting the embryos with two circular DNA plasmids (Liao *et al.*, 2000). They move vertically to the progeny, but they can also move between individuals of different species horizontally (Kidwell, 1992). The P-elements are effective only in case of *Drosophila*, but type II transposable elements, unrelated to P-elements such as *Minos* (a member of *Tc1* family) in case of *Ciratitis capitata* (Loukeris *et al.*, 1995) and mosquito, *Anopheles stephensi* (Catteruccia *et al.*, 2000); *Mariner* in *Lucilia cuprina* (Coates *et al.*, 1997), *hobo* in *Musca domestica* and *H. armigera* (O'Brochta *et al.*, 1994; Pinkerton *et al.*, 1996), and *PiggyBack* in several insects belonging to Diptera, Coleoptera and Lepidoptera (Horn *et al.*, 2002; Kramer, 2004; Uchino *et al.*, 2007; Ferguson *et al.*, 2011), have been used to transform the insects. Furthermore, the vectors like *PiggyBac* with promoter *3xP3-ECGFP* play an important role in present day insect genetic transformations, and could be tested for their usefulness in transformation of natural enemies of arthropods.

ENVIRONMENTAL ISSUES AND POTENTIAL RISKS

Release of insects modified through conventional approaches has occurred for many years, however, concerns about the safety of biological control agents in terms of preservation of native flora and fauna have been raised by the environmentalists (Howarth, 1991). The era of accepting classic biological control as environmentally risk-free appears to have passed (Hoy, 1992). Protocols for evaluating the risks associated with releasing parasitoids and predators that have been manipulated with rDNA techniques do not currently exist (Tiedje *et al.*, 1989; Hoy, 1995), their deployment might have a potential risk to the environment, and need careful attention before their commercialization. Stability of the introduced gene is necessary to ensure that the transgenic population will carry the gene with adequate expression, and that the released population will not pose a hazard to the environment. Transposable elements have the ability to move within the insect genome, and this may influence the stability of the introduced gene if the transposable element is used from the same family as the vector (O'Brochta and Atkinson, 1996). For example, *Hermes* and *hobo* transposable elements belong to the same *hAT* family, although their transpose elements are only 55% identical (Sundararajan *et al.*, 1999). Mating between transgenic and wild population would expose the transgenic insects to a range of genetic backgrounds that would allow for broad non-additive effects. These interactions may alter vector competence, host range, and influence the life history parameters by affecting survival and reproduction (Gubler, 1993).

The risks associated with accidental release of a transgene are largely associated with the nature of transgene, the marker gene, and their expression patterns. Therefore, release of transgenic arthropods will have to be evaluated on the basis of probable benefits and potential risks, and following assays need to be considered for risk assessment of genetically modified organisms:

- Attributes of the species like potential for gene flow to other species, role in ecosystem, mobility, longevity, and fecundity.
- Function and stability of transgene.
- Alternate hosts, presence of species in the environment, and possibilities for indirect dissemination.
- Comparison of the transgenic with non-transgenic phenotype.

There are no established procedures in place to conduct risk assessment for transgenic insects, as the goal is to spread them as far as possible (Hoy, 2000). Therefore, each transgenic beneficial arthropod species should be evaluated on a case-by-case basis (IAEA, 2006). Therefore, there is a need to evolve suitable guidelines, standardize protocols, and develop science-led global database on regulatory guidelines for the assessment of nontarget effects of genetically manipulated bioagents, to address the environmental issues and potential risk of release of genetically modified bioagents.

CONCLUSIONS

A complex of naturally occurring bioagents, *viz.* predators, parasitoids, or pathogens (entomopathogenic viruses, bacteria, fungi, protozoa, and nematodes), have been found to be effective in suppression of many insect pests, and considered as potential, economically viable, and ecologically safe tool of insect pest management. However, widespread pesticide use, difficulties involved in mass production, restricted host range, slow action, poor quality, limited persistence in the field, and inability to withstand adverse environmental conditions have been experienced as the practical difficulties in commercial deployment of these bioagents for the control of several insect pests. Biocontrol agents that are genetically improved for a variety of traits, such as environmental hardiness, easy mass production and quality control, increased multiplication rate, broad host range and host-finding ability, persistence, adaption to extremes of climatic conditions, and capable of tolerating sub-lethal doses of pesticides can play an important role in pest management.

Maintaining quality in laboratory-reared arthropod natural enemies is difficult due to possible genetic changes caused by inadvertent selection, inbreeding, genetic drift, and founder effects. If high quality and inexpensive artificial diets for predators and parasitoids are available, biological control programs would no longer be restricted by inefficient mass production methods. Development and release of pesticide-tolerant natural enemies to maintain their parasitizing/ predatory activity under pesticide application conditions for the management of other pests is another important approach of genetic improvement of bioagents. Selection, hybridization, or recombinant DNA technology, and recent advances in molecular biology have genetically improved some of the bioagents for a variety of traits, such as, pesticide-resistant *Trichogramma* spp., improvement of Bt for increasing its host range and stability, baculoviruses with more toxic proteins, and incorporation of benomyl resistance into the entomopathogenic fungus, *M. anisopliae*, which kill the insects at a faster rate. The recombinant DNA technologies have also made it possible to transfer genes specific to beneficial traits in natural enemies and manipulated genes involved in insect infection by entomopathogens to improve pathogenicity. Environmental issues related to release of genetically engineered bioagents such as potential risk of low level of successful establishment, speed of evolution and movement of genetic material from one species to another, and deployment of mechanisms that drive the gene of interest through natural populations, testing, deployment, and risk assessment have also been discussed.

Molecular approaches will play an important role in exploring diversity for beneficial traits, while genetic engineering will enable the production of bioagents with known functions. Biotechnological tools offer the potential of genetically modifying more robust natural enemies with resistance to pesticides, or better adaptation to extremes of climatic conditions, increased host range, and efficient rearing systems; and gene manipulation in entomopathogens to improve pathogenicity. Although considerable progress has been made in the laboratory, these advances have not been accompanied by product development, delivery, and risk assessment. Experience with plant and animal biotechnology may serve as a guideline, and also as a warning, if the public does not understand completely the benefits of the development of genetically modified bioagents. There is a need to develop mechanisms to drive the gene of interest through natural populations, testing, deployment, and risk assessment.

REFERENCES

Abbaszadeh, G., Dhillon, M.K., Srivastava, C. and Gautam, R.D. (2011) Effect of climatic factors on bioefficacy of biopesticides in insect pest management. *Biopestic. Int.*, **7**, 1–14.

Almeida de, R.P., Silva da, C.A.D. and Medeiros de, M.B. (1998) Biotechnology for mass production and management of *Trichogramma* for use in biological control. *Doc. Centro Nac. Pesq. Algodao*, **60**, 1–61.

Atkinson, P.W. and O' Brochta, D.A. (1999) Genetic transformation of non-drosophilid insects by transposable elements. *Ann. Entomol. Soc. Am.*, **92,** 930–936.

Atkinson, P.W., Pinkerton, A.C. and O'Brochta, D.A. (2001) Genetic transformation systems in insects. *Annu. Rev. Entomol.*, **46,** 317–346.

Baker, J.F. and Thorne, J.E. (1995) Evaluation of a resistant parasitoid for biological control of weevils in insecticide treated wheat. *J. Econ. Entomol.*, **88,** 1570–1579.

Baker, M.D., Vossbrinck, C.R., Becnel, J.J. and Andreadis, T.G. (1998) Phylogeny of Amblyospora (Microsporida: Amblyosporidae) and related genera based on small subunit ribosomal DNA data: A possible explanation of host parasite cospeciation. *J. Invertebr. Pathol.*, **71**, 199–206.

Baum, J.A., Johnson, T.B. and Carlton, B.C. (1998) *Bacillus thringiensis*: natural and recombinant bio insecticide products. In: F.R. Hall and J.J. Menn (eds.), *Biopesticides: Use and Delivery*, Humana Press, Totowa, pp. 189–209.

Baum, J.A., Kakefuda, M. and Gawron-Burke, C. (1996) Engineering *Bacillus thuringiensis* bio-pesticides with an Indigenous site-specific recombination system. *Appl. Environ. Microbiol.*, **62**, 4367–4373.

Benedict, M.Q., Salazar, C.E. and Collins, F.H. (1995) A new dominant selectable marker for genetic transformation; *Hsp70-opd. Insect Biochem. Mol. Biol.*, **25**, 1061–1065.

Benedict, M.Q., Scott, M.J. and Cockburn, A.F. (1994) High-level expression of the bacterial *opd* gene in *Drosophila melanogaster*: improved inducible insecticide resistance. *Insect Mol. Biol.*, **3**, 247–252.

Bonning, B.C. and Hammock, B.D. (1996) Development of recombinant baculoviruses for insect control. *Annu. Rev. Entomol.*, **41**, 191–210.

Bradfisch, G.A., Thompson, M. and Schwab, G.E. (1996) Pesticidal compositions. US Patent 5508264.

Braig, H.R. and Yan, G. (2002) The spread of genetic constructs in natural insect populations. In: D.K. Letourneau and B.E. Burrows (eds.), *Genetically Engineered Organisms: Assessing Environmental and Human Health Effects*, CRC Press, Boca Raton, Florida, pp. 251–314.

Caprio, M.A. and Hoy, M.A. (1995) Pre-mating isolation in a simulation model generates frequency dependent selection and alters establishment rates of resistant natural enemies. *J. Econ. Entomol.*, **88**, 205–212.

Catteruccia, F., Nolan, T., Loukeris, T.G., Blass, C., Savakis, C., Kafatos, F.C. and Crisanti, A. (2000) Stable germline transformation of the malaria mosquito *Anopheles stephensi. Nature,* **405**, 959–962.

Chapco, W., Ashton, N.W., Martel, R.K.B. and Antonishyn, N. (1992) A feasibility study of the use of random amplified polymorphic DNA in the population genetics and systematics of grasshoppers. *Genome*, **35**, 569–574.

Ciche, T.E. and Sternberg, P.W. (2007) Postembryonic RNAi in *Heterorhabditis bacteriophora*: A nematode insect parasite and host for insect pathogenic symbionts. *BMC Develop. Biol.*, **7, 101**, 1–11.

Coates, C.J., Turney, C.L., Frommer, M., O'Brochta, D.A. and Atkinson, P.W. (1997) Interplasmid transposition of the *mariner* transposable element in non-drosophilid insects. *Mol. Gen. Genet.*, **253**, 728–733.

Corsaro, B.G., DiRenzo, J. and Fraser, M.J. (1989) Transfection of cloned *Heliothis zea* cell lines with the DNA genome of the *Heliothis zea* nuclear polyhedrosis virus. *J. Virol. Methods*, **25**, 283–292.

Couteaudier, Y., Viaud, M. and Riba, G. (1996) Genetic nature, stability, and improved virulence of hybrids from protoplast fusion in *Beauveria. Microb. Ecol.*, **32**, 1–10.

Dhillon, M.K. and Gujar, G.T. (2010). Impact of climate change on arthropod biodiversity and plant protection. *Int. J. Environ. Consum.*, **6**, 41–45.

Dhillon, M.K. and Sharma, H.C. (2009) Temperature influences the performance and effectiveness of field and laboratory strains of the ichneumonid parasitoid, *Campoletis chlorideae*. *BioControl*, **54**, 743–750.

Dubois, N. and Dean, D.H. (1995) Synergism between Cry1A insecticidal crystal proteins and spores of *Bacillus thuringiensis*, other bacterial spores, and vegetative cells against *Lymantria dispar* (Lepidoptera: Lymantriidae) larvae. *Environ. Entomol.*, **24**, 1741–1747.

Edwards, O.R. and Hoy, M.A. (1993) Polymorphism in two parasitoids detected using random amplified polymorphic DNA (RAPD) PCR. *Biol. Control: Theory Applic. Pest Manage.*, **3**, 243–257.

Fan, Y., Fang, W., Guo, S., Pei, X., Zhang, Y., Xiao, Y., Li, D., Jin, K., Bidochka, M.J. and Pei, Y. (2007a) Increased insect virulence in *Beauveria bassiana* strains overexpressing an engineered chitinase. *Appl. Environ. Microbiol.*, **73**, 295–302.

Fan, Y., Fang, W., Xiao, Y., Yang, X., Zhang, Y., Bidochka, M.J. and Pei, Y. (2007b) Directed evolution for increased chitinase activity. *Appl. Microbiol. Biotechnol.*, **76**, 135–139.

Fang, W., Pava-Ripoll, M., Wang, S.B. and St. Leger, R.J. (2009) Protein kinase A regulates production of virulence determinants by the entomopathogenic fungus, *Metarhizium anisopliae*. *Fungal Genet. Biol.*, **46**, 277–285.

Funk, C.D., Gunne, H., Steinar, H., Izumi, T. and Samuelsson, B. (1989) Native and mutant 5-lipoxygenase expression in a baculovirus/insect cell system. *Proc. Nat. Acad. Sci. USA*, **86**, 2592–2596.

Ferguson, H.J., Neven, L.G., Thibault, S.T., Mohammed, A. and Fraser, M. (2011) Genetic transformation of the codling moth, *Cydia pomonella* L., with *PiggyBac* EGFP. *Transgenic Res.*, **20**, 201–214.

Gaugler, R. and Hashmi, S. (1996) Genetic engineering of an insect parasite. In: J.K. Setlow (ed.), *Genetic Engineering: Principles and Methods*, Plenum, New York, pp. 135–157.

Gawron-Burke, C. and Baum, J. (1991) Genetic manipulation *Bacillus thuringiensis* insecticidal crystal protein genes in bacteria. In: J.K. Setlow (ed.), *Genetic Engineering: Principles and Methods*, Plenum Press, New York, pp. 237–263.

Goettel, M.S., St. leger, R.J., Bhairi, S., Jung, M.K., Oakley, B.R. and Staples, R.C. (1989) Transformation of the entomopathogenic fungus *Metarhizium anisopliae* using the *Aspergillus nidulans benA3* gene. *Curr. Genet.*, **17**, 129–132.

Gonzalez Jr., J.M., Brown, B.J. and Carlton, B.C. (1982) Transfer of *Bacillus thuringiensis* plasmids coding for δ-endotoxin among strains of *B. thringiensis* and *B. cereus*. *Proc. Natl. Acad. Sci. USA*, **79**, 6951–6955.

Gubler, D.J. (1993) Release of exotic genomes. *J. Am. Mosq. Control Assoc.*, **9**, 78–83.

Gunne, H., Hellers, M. and Steiner, H. (1990) Structure of preproattacin and its processing in insect cells infected with a recombinant baculovirus. *Eur. J. Biochem.*, **187**, 699–703.

Harrison, R.L. and Bonning, B.C. (1999) Genetic engineering of biocontrol agents for insects. In: J.E. Rechcigl and N.A. Rechcigl (eds.), *Biological and Biotechnological Control of Insect Pests*, Lewis Publishers, Boca Raton, Florida, pp. 243–280.

Hashmi, S., Hashmi, G., Glazer, I. and Gaugler, R.R. (1998) Thermal response of *Heterorhabditis bacteriophora* transformed with the *Caenorhabditis elegans hsp70* encoding gene. *J. Exp. Zool.*, **281**, 164–170.

Havron, A., Kenan, G. and Rosen, D. (1991) Selection for pesticide resistance in *Aphytis*. *Entomol. Exp. Appl.*, **61**, 229–235.

Headley, J.C. and Hoy, M.A. (1987) Benefit/cost analysis of an integrated mite management program for almonds. *J. Econ. Entomol.*, **80**, 555–559.

Heale, J.B., Isaac, J.E. and Chandler, D. (1989) Prospects for strain improvement in entomopathogenic fungi. *Pestic. Sci.*, **26**, 79–92.

Herrera, G., Snyman, S. J. and Thomson, J.A. (1994) Construction of a bioinsecticidal strain of *Pseudomonas fluorescens* active against the sugarcane borer, *Eldana saccharina*. *Appl. Environ. Microbiol.*, **60**, 682–690.

Hopper, K.R., Roush, R.T. and Powell, W. (1993) Management of genetics of biological control introductions. *Annu. Rev. Entomol.*, **38**, 27–51.

Howarth, F.G. (1991) Environmental effects of classical biological. *Annu. Rev. Entomol.*, **36**, 485–509.

Hoy, M.A. (1976) Genetic improvement of insects: fact or fantasy. *Environ. Entomol.*, **5**, 833–839.

Hoy, M.A. (1988) Biological control of arthropod pests: traditional and emerging technologies. *Am. J. Alter. Agric.*, **3**, 63–68.

Hoy, M.A. (1990) Pesticide resistance in arthropod natural enemies: variability and selection response. In: R.T. Roush and B.E. Tabashnik (eds.), *Pesticide Resistance in Arthropods,* Chapman and Hall, New York, pp. 203–236.

Hoy, M.A. (1992) Criteria for release of genetically-improved phytoseiids: an examination of the risks associated with release of biological control agents. *Environ. Appl. Acarol.*, **14**, 393–416.

Hoy, M.A. (1995) Impact of risk analyses on pest-management programs employing transgenic arthropods. *Parasitol. Today*, **11**, 229–232.

Hoy, M.A. (2000) Transgenic arthropods for pest management programs: risks and realities. *Exp. Appl. Acarol.*, **24**, 463–495.

Hurlbert, R.E. (1994) Investigations into the pathogenic mechanisms of the bacterium-nematode complex. *ASM News*, **60**, 473–478.

IAEA (International Atomic Energy Agency) (2006) Status and risk assessment of the use of transgenic arthropods in plant protection. In: *Proceedings of a Technical Meeting Organized by the Joint FAO/IAEA Programme of Nuclear Techniques in Food and Agriculture and the Secretariat of the International Plant Protection Convention*, held in Rome, 8 to 12 April 2002. FAO/IAEA Agriculture and Biotechnology Laboratory, International Atomic Energy Agency, Vienna, Austria.

Inceoglu, A.B., Kamita, S.G. and Hammock, B.D. (2006) Genetically modified baculoviruses: a historical overview and future outlook. *Adv. Virus Res.*, **68**, 323–360.

Jalali, S.K., Singh, S.P., Venkatesan, T., Murthy, K.S. and Lalitha, Y. (2006) Development of endosulfan tolerant strain of an egg parasitoid *Trichogramma chilonis* Ishii (Hymenoptera: Trichogrammatidae). *Indian J. Exp. Biol.*, **44**, 584–590.

Johny, S., Kanginakudru, S., Muralirangan, M.C. and Nagaraju, J. (2006) Morphological and molecular characterization of a new microsporidian (Protozoa: Microsporidia) isolated from *Spodoptera litura* (Fabricius) (Lepidoptera: Noctuidae). *Parasitology*, **132**, 803–814.

Joshi, L., St. Leger, R.J. and Roberts, D.W. (1997) Isolation of a cDNA encoding a novel subtilisin-like protease (PrIB) from the entomopathogenic fungus *Metarhizium anisopliae* using differential display-RT-PCR. *Gene*, **197**, 1–8.

Kahel-Raifer, H. and Glazer, I. (2000) Environmental factors affecting sexual differentiation in the entomopathoganic nematode *Heterorhabditis bacteriophora*. *J. Exp. Zool.*, **287**, 158–166.

Kamita, S.G., Kang, K.D., Hammock, B.D. and Inceoglu, A.B. (2005) Genetically modified baculoviruses for pest insect control. In: L.I. Gilbert, K. Iatrou and S.S. Gill (eds.), *Comprehensive Molecular Insect Science*, Vol. **6,** Elsevier, Pergamon, Oxford, pp. 271–322,.

Kidwell, M.G. (1992) Horizontal transfer of P-elements and other short inverted repeat transposons. *Genetica*, **86**, 275–286.

Kramer, M.G. (2004) Recent advances in transgenic arthropod technology. *Bull. Entomol. Res.*, **94**, 95–110.

Lee, M.K., Curtiss, A., Alcantara, E. and Dean, D.H. (1996) Synergistic effect of the *Bacillus thuringiensis* toxins Cry1Aa and Cry1Ac on the gypsy moth, *Lymantria dispar*. *Appl. Environ. Microbiol.*, **62**, 583–586.

Liao, G.C., Rehm, E.J. and Rubin, G.M. (2000) Insertion site preferences of the P-transposable element in *Drosophila melanogaster*. *Proc. Natl. Acad. Sci. USA*, **97**, 3347–3351.

Loukeris, T.G., Livadaras, I., Arca, B., Zabalkou, S. and Savakis, C. (1995) Gene transfer into the Medfly, *Ceratitis capitata,* with a *Drosophila hydei* transposable element. *Science*, **270**, 2002–2005.

Lu, D.D., Pava-Ripoll, M., Li, Z.Z. and Wang, C.S. (2008) Insecticidal evaluation of *Beauveria bassiana* engineered to express a scorpion neurotoxin and a cuticle degrading protease. *Appl. Microbiol. Biotechnol.*, **81**, 515–522.

Miller, L.H., Sakai, R.K., Romans, P., Gwadz, W., Kantoff, P. and Coon, H.G. (1987) Stable integration and expression of a bacterial gene in the mosquito *Anopheles gambiae*. *Science*, **237**, 779–781.

Moar, W.J., Carey, M.P., Faassen, H.V., Bosch, D., Furtos, R., Rang, C., Luo, K. and Adang, J.M. (1995) Development of *Bacillus thuringiensis* CryIC resistance by *Spodoptera exigua* (Lepidoptera: Noctuidae). *Appl. Environ. Microbiol.*, **61**, 2086–2092.

Moscardi, F. (1999) Assessment of the application of baculovirus for control of Lepidoptera. *Annu. Rev. Entomol.*, **44**, 257–289.

Navon, A. (2000) *Bacillus thuringiensis* insecticides in crop protection-reality and prospects. *Crop Prot.*, **19**, 669–676.

O'Brochta, D.A. and Atkinson, P.W. (1996) Transposable elements and gene transformation in non-drosophilid insects. *Insect Biochem. Mol. Biol.*, **26**, 739–753.

O'Brochta, D.A. and Atkinson, P.W. (1998) Building a better bug - inserting new genes into a few specific insect species could stop some infectious diseases, benefit agriculture and produce innovative materials. *Sci. Am.*, **279**, 60–65.

O'Brochta, D.A., Warren, W.D., Saville, K.J. and Atkinson, P.W. (1994) Interplasmid transposition of *Drosophila hobo* elements in non-drosophilid insects. *Mol. Gen. Genet.*, **244**, 9–14.

O'Reilly, D.R. and Miller L.K. (1991) Improvement of a baculovirus pesticide by deletion of the EGT gene. *Biotechnology*, **9**, 1086–1089.

Obukowicz, M.G., Perlak, F.J., Bolton, S.L., Kusano-Kretzmer, K., Mayer, E.J. and Watrud, L.S. (1987) IS50L as a non-self transposable vector used to integrate the *Bacillus thuringiensis* δ-endotoxin gene into the chromosome of root colonizing pseudonomads. *Gene*, **51**, 291–296.

Oeda, K., Oshie, K., Shimizu, M., Nakamura, K., Yamomoto, H., Nakayama, I. and Ohkawa, H. (1987) Nucleotide sequence of the insecticidal protein gene of *Bacillus thuringiensis* strain *aizawai* IPL7 and its high-level expression in *Escherchia coli*. *Gene*, **53**, 113–119.

Pava-Ripoll, M., Posada, F.J., Momen, B., Wang, C.S. and St. Leger, R.S. (2008) Increased pathogenicity against coffee berry borer, *Hypothenemus hampei* (Coleoptera: Curculionidae) by *Metarhizium anisopliae* expressing the scorpion toxin (AaIT) gene. *J. Invertebr. Pathol.*, **99**, 220–226.

Petrushov, A.Z. (1987) Results of the introduction into USSR of insecticide resistant populations of the predatory mite *Metaseiulus occidentalis* (Acarina: Phytoseiidae). *Zool. Zhurnal*, **66**, 674–680.

Pinkerton, A.C., O'Brochta, D.A. and Atkinson, P.W. (1996) Mobility of *hAT* transposable elements in the old world bollworm, *Helicoverpa armigera*. *Insect Mol. Biol.*, **5**, 223–227.

Presanil, J.K. and Hoy, M.A. (1992) Stable genetic transformation of a beneficial arthropod by microinjection. *Proc. Natl. Acad. Sci. USA*, **89**, 7732–7736.

Presnail, J.K. and Hoy, M.A. (1996) Maternal microinjection of the endoparasitoid *Cardiochiles diaphaniae* (Hymenoptera: Braconidae). *Ann. Entomol. Soc. Am.*, **89**, 576–580.

Presnail, J.K., Jeyaprakash, A., Li, J. and Hoy, M.A. (1997) Genetic analysis of four lines of *Metaseiulus occidentalis* (Acari : Phytoseiidae) transformed by maternal microinjection. *Ann. Entomol. Soc. Am.*, **90**, 237–245.

Rajamohan, F. and Dean, D.H. (1996) Molecular biology of bacteria for the biological control of insects. In: M. Gunasekaran and D.J. Weber (eds.), *Molecular Biology of the Biological Control of Pests and Diseases of Plants*, CRC Press, Boca Raton, Florida, pp. 105–122.

Rajendra, W., Hackett, K.J., Buckley, E. and Hammock, B.D. (2006) Functional expression of lepidopteran-selective neurotoxin in baculovirus: Potential for effective pest management. *Biochim. Biophys. Acta*, **1760**, 158–163.

Regev, A., Rivkin, H., Inceoglu, B., Gershburg, E., Hammock, B.D., Gurevitz, M. and Chejanovsky, N. (2003) Further enhancement of Baculovirus insecticidal efficacy with scorpion toxins that interact cooperatively. *FEBS Lett.*, **537**, 106–110.

Sandhu, S.S., Unkles, S.E., Rajak, R.C. and Kinghorn, J.R. (2001) Generation of benomyl resistant *Beauveria bassiana* strains and their infectivity against *Helicoverpa armigera*. *Biocontr. Sci. Technol.*, 11, 250–254.

Segal, D. and Glazer, I. (2000) Genetics for improving biological control agents: the case of entomopathogenic nematodes. *Crop Prot.*, **19**, 685–689.

Sharma, H.C. (2009) *Biotechnological Approaches for Pest Management and Ecological Sustainability*. CRC Press, Boca Raton, Florida.

Skot, L., Harrison, S.P., Nath, A., Mytton, L.R. and Clifford, B.C. (1990) Expression of insecticidal activity in rhizobium containing δ-endotoxin gene cloned from *Bacillus thuringiensis* subsp. *tenebrionis*. *Plant Soil*, **127**, 285–295.

St. Leger, R.J. (1995) The role of cuticle-degrading proteases in fungal pathogenesis of insects. *Can. J. Bot.* (Supplement 1), **73**, S1119–S1125.

St. Leger, R.J. and Wang, C. (2010) Genetic engineering of fungal biocontrol agents to achieve greater efficacy against insect pests. *Appl. Microbiol. Biotechnol.*, **85**, 901–907.

St. Leger, R.J., Joshi, L. and Roberts, D. (1998) Ambient pH is a major determinant in the expression of cuticle-degrading enzymes and hydrophobin by *Metarhizium anisopliae*. *Appl. Environ. Microbiol.*, **64**, 709–713.

St. Leger, R.J., Joshi, L., Bidochka, M.J. and Roberts, D.W. (1996) Construction of an improved mycoinsecticide overexpressing a toxic protease. *Proc. Natl. Acad. Sci. USA*, **93**, 6349–6354.

Stouthamer, R., Luck, R.F. and Werren, J.H. (1992) Genetics of sex determination and the improvement of biological control using parasi-toids. *Environ. Entomol.*, **21**, 427–535.

Sundararajan, P., Atkinson, P.W. and O'Brochta, D.A. (1999) Transposable element interactions in insects: crossmobilization of *hobo* and *Hermes*. *Insect Mol. Biol.*, **8**, 359–368.

Tiedje, J.M., Clowell, R.K., Grossman, Y.L., Hodson, R.E., Lenski, R.E., Mack, R.M. and Regal, P.J. (1989) The planned introduction of genetically engineered organisms: Ecological considerations and recommendations. *Ecology*, **70**, 298–315.

van Frankenhuyzen, K. (1993) The challenge of *Bacillus thuringiensis*. In: P.F. Entwistle, J.S. Cory, M.J. Bailey and S. Higgs (eds.), *Bacillus thuringiensis, An Environmental Biopesticide: Theory and Practice*, Wiley and Sons, London, pp. 171–191.

Wahab, S. (2010) Biotechnological approaches in the management of plant pests, diseases and weeds for sustainable agriculture. In: O. Koul, G.S. Dhaliwal, S. Khokhar and R. Singh (eds.), *Biopesticides in Food and Environment Security* Souvenir Third Conference on Biopesticides: Emerging Trends (BET 2010), October 20–22, 2010, Hisar, India. Society of Biopesticide Sciences India, Jalandhar, India, pp. 15–40.

Wang C.S. and St. Leger, R.J. (2007) A scorpion neurotoxin increases the potency of a fungal insecticide. *Nat. Biotechnol.*, **25**, 1455–1456.

Wudayagiri, R., Inceoglu, A.B., Herrmann, R., Derbel, M., Choudary, P.V. and Hammock, B.D. (2001) Isolation and characterization of a novel lepidopteran selective toxin from the venom of South Indian red scorpion, *Mesobuthus tamulus*. *BMC Biochem.*, **2**, 16–23.

Xiong, H., Huang, L.K., Fen, L.Y., Zi, M.Q., Ying, L.L. and Chu, Z.L. (1988) Preliminary study on the selection for insecticide resistant strain of *Trichogramma japonicum* Ashmead. In: J. Voegele, J.K. Waage and J.C. van Lenteren (eds.), *Trichogramma and other Egg Parasitoids,* II International Symposium, Guangzou (China), 10-15 November, 1986, Antibes INRA, Paris, No. 43, pp. 411–418.

8

Developments in the Genetic Improvement of Entomopathogenic Nematodes

VIKAS JINDAL*, HARIT K. BAL, RUISHENG AN AND PARWINDER S. GREWAL**

Department of Entomology
The Ohio State University, OARDC, Wooster, OH-44691, USA
E mail: *vikas_ento@yahoo.co.in, **grewal.4@osu.edu

INTRODUCTION

During the era of green revolution use of chemical pesticides has dominated pest management practices around the world. Overreliance on chemical pesticides has resulted in harm to non-target organisms, development of resistance in target pests and pathogens, and degradation of the environment. As an alternative, a diverse array of biological pest control agents including microorganisms, predators and parasitoids is being developed for pest management. Entomopathogenic nematodes (EPNs) are one such alternative pest control agents in which interest has increased steadily over the past three decades and they are now used commercially to control insect pests in many parts of the world (Grewal *et al.*, 2005). Two genera, *Steinernema* (Panagrolaimomorpha: Steinernematidae) and *Heterorhabditis* (Rhabditida: Heterorhabditidae), are the most successful which depend on symbiotic bacteria (*Photorhabdus* for *Heterorhabditis* and *Xenorhabdous* for *Steinernema*) for insect pathogenictiy and their own growth and reproduction. The use of EPNs as biological control agents have been found effective against white grubs, sciarid flies, fungus gnats, European crane fly, citrus root weevil, black vine weevil, pecan weevil, grape root borer, cutworms, codling moth, fruit fly etc. (Grewal *et al.*, 2005). However, their susceptibility to environmental stresses including temperature extremes, solar radiation, and desiccation, limited storage stability, and high price restrict their wide spread use (Kaya and Gaugler, 1993; Grewal *et al.*, 2005). Consequently, much recent research has been focused on EPN strain discovery and improvement. Classical techniques such as artificial selection, hybridization, mutagenesis and molecular techniques including genetic engineering have been applied for strain improvement (Burnell, 2002; Brey and Hashmi, 2003). In this chapter, we briefly describe the progress in the discovery and identification of new EPN species and strains, and the genetic improvement of strains for enhanced biological pest control potential. Future trends in this field are also envisioned.

LIFE CYCLE OF EPNs

The third stage juvenile, known as infective juvenile (IJ), is the only free living stage in the life cycle of the EPNs which can survive outside the host without feeding for months (Grewal *et al.*, 2011). These IJs are associated with their symbiotic bacteria and carry them in their intestinal lumen (Ciche, 2007). The IJs locate their host and penetrate through intestine, exoskeleton, buccal cavity, spiracles, etc to reach the haemocoel to resume their development. Inside the haemocoel, IJs release cells of the bacterial symbiont from their intestine. The bacteria start multiplying in the haemocoel and secrete hydrolytic enzymes and toxins (Bowen *et al.*, 1998; An *et al.*, 2009). The host insect normally dies due to septicemia within 24–72 h after nematode penetration. The IJs feed on the multiplying bacteria and host tissues of the cadaver to develop into adults and complete 2-3 generations inside the cadaver. The host larvae turn reddish brown when infected with *Heterorhabditis bacteriophora* and pale yellow when infected with *Steinernema carpocapsae* (Fig. 8.1). However, the color of the cadaver is dependent upon the species/ strains of EPNs. In *Heterorhabditis*, the IJs develop into hermaphrodites, but in *Steinernema*, they develop into amphimictic males and females. As food resources are depleted, new IJs develop and reassociate with the symbiotic bacteria, and subsequently emerge out of the cadaver to initiate new host search. EPNs after hatching from the egg pass through four juvenile stages to enter into the adult stage (Fig. 8.2). Some eggs are laid whereas others hatch inside the female or hermaphrodites resulting in a condition known as *endotokia matricida* (an intra uterine development of juveniles) (Johnigk and Ehlers, 1999).

Fig. 8.1. *Galleria mellonella* cadaver infected with *Steinernema carpocapsae* (left) and *Heterorhabditis bacteriophora* (right)

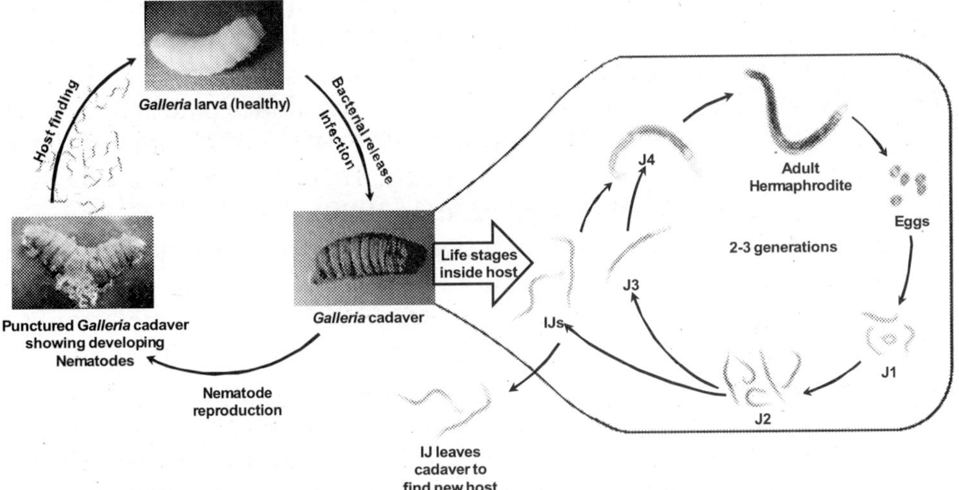

Fig. 8.2. Generalized infection and reproduction cycle of heterorhabditid nematodes. Enlarged view shows nematode reproduction cycle in the insect cadaver. J1-J4 are juvenile stages and IJs are infective juveniles

GENETIC IMPROVEMENT

Strain Discovery and Evaluation

The natural biodiversity of EPNs is a rich source of strains with desired traits. The local populations of EPNs often show genetic heterogeneity as well as variability in biological traits (Blouin *et al.*, 1999; Hominick *et al.*, 1999; Somasekhar *et al.*, 2002; Grewal, *et al.*, 2002b). Till date, in the family Steinernematidae only two genera,

Steinernema Travassos (>30 species) and *Neosteinernema* Nguyen and Smart (1 species), and in Heterorhabditidae, one genus, *Heterorhabditis* (9 species) have been reported (Stock and Hunt, 2005). Therefore, continuing efforts to explore the naturally occurring species/strains of EPNs with superior traits is a fruitful exercise as seen in recent studies (Grewal *et al.*, 2002b, 2004; Koppenhöfer and Fuzy, 2003; Shapiro *et al.*, 2005).

The differences in various traits such as host finding and virulence in different strains of a species have been documented (Salem *et al.*, 2007; Grewal *et al.*, 2002b; Bedding *et al.*, 1983). Bedding *et al.* (1983) started screening the local EPN species for inter- and intra-specific differences in their infectivity to insect pests and found *Heterorhabditis* spp D1 strain highly virulent against *Lucilia cuprina.* Further, Gaugler *et al.* (1989b) reported higher phenotypic variation in host finding ability among strains of *S. carpocapsae* (=*S. feltiae*) and reported Florida strain exhibiting better host finding ability towards *Galleria mellonella* as compared to other strains. This was followed by studies that determined greater potential of *S. glaseri* strains NJ-43 and strain S1-12 for their virulence against *Popillia japonica* (Selvan *et al.*, 1994) and characterization of *H. bacteriophora* IS-5 for high heat tolerance (Glazer *et al.*, 1996) and *S. feltiae* IS-6 for desiccation tolerance (Solomon *et al.*, 1999). These IS-5 and IS-6 strains are being used to establish the role of gene/proteins involved in their superior traits. The locally discovered strains, namely *H. bacteriophora* HP88 (Poinar and Georgis, 1990) and *H. bacteriophora* GPS11 (Grewal *et al.*, 2002b) possessing high virulence against important insect pests have been further used for experiments on genetic improvement and for release as commercial products. The New Jersey (NJ) strains of *H. bacteriophora* (NJ-2) and *S. glaseri* (NJ-43) were also found to possess higher virulence against Japanese beetle than the strains cultured in laboratory (Selvan *et al.*, 1993).

Another example worth mentioning is of the mixed strain of *S. riobrave*, obtained by pooling all new strains, exhibited high virulence against citrus root weevil, *Diaprepes abbreviatus* in comparison with the commercial strain SR55 (Stuart *et al.*, 2004). In more recent studies, differences in higher virulence against insect pests in strains of *S. carpocapsae* (Laznik *et al.*, 2010a), *S. riobrave* (Cabanillas, 2003; Stuart *et al.* 2004; Jenkins *et al.* 2007), *S. feltiae* (Campos-Herrera and Gutierrez, 2009; Laznik *et al.*, 2010b), *S. glaseri* (Vasconcelos *et al.*, 2004), *H. bacteriophora* (Toepfer *et al.*, 2005; Shu-Sen *et al.*, 2009), *H. indica* (Ebssa *et al.*, 2004; Jenkins *et al.*, 2007; Maneesakorn *et al.*,

2010) and *Heterorhabditis* sp. (Lee *et al.*, 2002; El-Lakmah and Azazy, 2010) have been reported. Improved host finding has also been reported in populations of *S. riborave* and *S. glaseri* against boll weevil and white grub, respectively (Cabanillas, 2003; Koppenhoffer and Fuzy, 2008). The *S. carpocapsae* (strains, Agriotos, ALL and Sal) and *H. bacteriophora* (strains Beeshaani and DE6) have been reported to be higher in desiccation tolerance than the other contemporary strains (Shapiro *et al.*, 2003; Mukuka *et al.*, 2008; Mukuka *et al.*, 2010a).

Mixed strains often possess greater genetic diversity that provides greater scope for adaptations of the nematodes to locate and penetrate host insects and overcome their immune defenses (Wang *et al.*, 1995). This suggests the role of genetic diversity in maintaining higher virulence in nematode strains for inundative biological control.

Earlier, Grewal *et al.* (2002a) showed variations in IJs longevity and their stress tolerance in several populations of *H. bacteriophora*. The genetic variation among these isolates of *Heterorhabdistis* spp. was also confirmed using isozyme analysis (Jagdale *et al.*, 2006). In a recent study, the EPN populations, isolated from diverse sites across Israel, were found to vary in virulence, infection rates, and heat and desiccation tolerance (Salame *et al.*, 2010b). This natural genetic variation is important for the genetic improvement of EPNs for enhanced pest management potential.

Genetic Selection and Breeding

Trait improvement. The genetic selection and selective breeding are the most widely used techniques for improving beneficial traits that are encoded by large number of genes. In EPNs, the use of genetic selection (domestication) for improvement of beneficial traits was first suggested by Gaugler (1986). Glazer *et al.* (1991) reported genetic variation in *H. bacteriophora* HP88 strain for tolerance to heat, desiccation, ultraviolet radiation, and host finding ability. The heritability values were high, indicating that artificial selection and breeding would be an appropriate method for improving these traits in *Heterorhabditis* strains. Studies on the application of selection and breeding to improve different traits of EPNs are discussed below.

(i) *Host finding ability.* One of the main targets for genetic improvement of EPNs is increased efficacy which greatly depends on host finding. Improved nematode host-finding ability is important as it would enhance motility in nematodes, enabling them to locate hosts from greater distances. The host finding behavior of EPNs based upon their response to host cues has been classified along a continuum, with one extreme being a cruiser and the other being an ambusher (Grewal *et al.*, 1994b). Cruisers have been characterized by active mobility (Lewis *et al.*, 1992; Campbell and Gaugler, 1993), ability to orientate to volatile long-range host cues (Lewis *et al.*, 1993) and ability to find sedentary hosts (Alatorre-Rosas and Kaya, 1990). The searching ability of a cruiser, *S. glaseri,* varies from a wide to a localized search in response to hosts and host-associated cues, thus limiting their search area and rate of movement which is mediated by the presence of host or host cues (Lewis *et al.*, 1992). On the other hand, ambushing nematodes are known for low motility (Lewis *et al.*, 1992; Campbell and Gaugler, 1993) and lack of response to long range volatile cues (Gaugler *et al.*, 1989b; Lewis *et al.*, 1993). They respond to host volatile cues either after contact with the host cuticle (Lewis *et al.*, 1995) or during bouts of standing on their tails (Campbell and Kaya, 2002), which was earlier termed as nictation (Campbell and Gaugler, 1993), but recently proposed to be called as 'tail standing' (Kruitbos and Wilson, 2010). Another component of the continuum is the intermediate foragers such as *S. feltiae,* which neither nictate (Campbell and Gaugler, 1993), nor respond to host volatile cues in a manner similar to other cruise foragers (Grewal *et al.*, 1994b; Lewis *et al.*, 1995).

Genetic selection has been used as a classic approach to enhance host finding and host invasion abilities of EPNs (Gaugler, 1993). For effective selection of a desired trait, enough variability should occur in the traits in the population. For instance, such variability was observed in the host finding ability among different strains of *S. carpocapsae* by Gaugler *et al.* (1989b). This genetic variability was further maximized by creating a foundation population by intra-specific hybridization from a series of round-robin matings among 10 strains. This hybridized Foundation strain selected for enhanced host finding produced a 20- to 27-fold increase for host finding ability of *S. carpocapsae* towards the wax moth *G. mellonella* after 13 rounds of selection (Gaugler *et al.*, 1989a). A 25 per cent increase in the proportion of IJs initiating host seeking behavior after 6 rounds of selection was also reported (Gaugler *et al.*, 1989a). Improved host finding in strain G13 of *S. carpocapsae,* achieved by selective breeding, did not impair other fitness traits, viz. vigor, pathogenicity, sex ratio, morphology, etc. (Gaugler *et al.*, 1990). Gaugler *et al.* (1991) reported significant host-finding response of *S. carpocapsae* G13 towards *G. mellonella*, as compared to wild-type strains. Since *S. carpocapsae* is a generalist that kills several insect species, there will be enhanced host finding capability not only against one host but for other insect species as well. For example, house cricket, *Acheta domesticus* received similar host finding response of *S. carpocapsae* as *G. mellonella.*

Further studies by Gaugler and Campbell (1991), where G13 was subjected to 33 rounds (13 against *G. melonella* and 20 against scarab larvae) of selection, resulted in improved host finding against Japanese beetle, *P. japonica.* A 153- and 72-fold increase in host finding ability was achieved as compared to strains Foundation and All, respectively and subsequently this new strain was designated as S20. However, the selected strain failed to provide enhanced parasitism against Japanese beetle larvae in field trials. Poor field efficacy was attributed to the tendency of IJs to respond to CO_2 released by the host, followed by aggregation of IJs near the host spiracle that remain protected with sieve plates in Scarab larvae, thereby preventing nematode entry (Gaugler *et al.,* 1991; Gaugler *et al.,* 1994). Additionally, poor field results were also found with S17 and S20 strains (genetically selected over 17 and 20 generation for enhanced host finding abililty) against sweet potato weevils, *Cylas* sp. (Jansson *et al.*, 1993), despite positive results in laboratory trials with parent G13 strain (Jansson *et al.*, 1990). In commercial mushroom farm, to control mushroom sciarid fly, *Lycoriella mali* the genetically selected strain of *S. feltiae*, ScP did not outperform the wild type strain (Grewal *et al.,* 1993). These studies suggest that successful laboratory selection does not necessarily lead to improved field performance in selected strains. Therefore, only nematode species already known to effectively parasitize target insect should be chosen, while making selection for enhanced field efficacy.

Nevertheless, poor field efficacy of genetically selected strains should not be taken as an indication of unsuitability of genetic selection approach for nematode trait enhancement. For instance, Tomalak (1994a) showed significant (92.5%) mortality of mushroom fly, *Lycoriella solani* by *S. feltiae* strain selected after 34 rounds of selection than unselected strain along with good IJ persistence in the mushroom casing due to nematode recycling in the host insect. Similarly, Selvan *et al.* (1994) reported a significant increase in field efficacy of a strain of *S. glaseri* selected against *P. japonica* larvae, over the parental NC strain. The *H. zealandica* strain X1 and *H. bacteriophora* strain GPS11, identified as the most virulent in laboratory, were found to be most effective against *P. japonica* and *Cyclocephala borealis* Arrow as compared to *H. bacteriophora* strain HP88, *S. glaseri* strains NJ and MB; and *S. kraussei* strain UK in field condition (Grewal *et al.*, 2004). Their consistency in the control is equal to or better than the curative application of chemical insecticide, trichlorfon. These findings encourage the need of a holistic approach to utilize genetic selection for efficacy of enhanced traits, whereby it is important to make a good choice of the beneficial characteristic for selection and select an appropriate strain and appropriate bioassay design with which to measure the trait under selection.

Artificial selection of EPNs for high responsiveness to another foraging cue has been shown to increase their host finding ability. Recently, a compound (*E*)-β-caryophyllene released by maize roots was reported to be attractive to *H. bacteriophora* (Hiltpold *et al.*, 2010a). The strain, artificially selected for six generations, moved twice as rapidly towards (*E*)- β-caryophyllene source than the original strain. In field trials, the selected strain was significantly more effective than the original strain in reducing western corn rootworm populations in plots with a maize variety that releases (*E*)- β-caryophyllene, but not in the plots with a maize variety that does not emit this root signal. Selected strains had slightly less infectiousness but no negative effect was reported in their establishment and persistence in the field (Hiltpold *et al.*, 2010b). These results illustrate the great potential of improving the host finding ability of EPNs.

(ii) *High temperature tolerance.* High heritability (h^2= 0.98) in *H. bacteriophora* strain HP88 for heat tolerance, suggested that the genetic selection could be successful for this trait in EPNs (Glazer *et al.*, 1991). Grewal *et al.* (1994a) determined the thermal niche breadth of entomopathogenic nematodes collected from different climatic regimes. They concluded that the conspecific populations isolated from different localities did not differ in their thermal niche breadth, however, different species isolated from the same locality differ in their thermal niche breadth. Grewal *et al.* (1996a) reported that thermal reproductive niche breadth of *H. bacteriophora* was extended from 15–30°C to 12–32°C by genetic selection. Virulence of *S. anomali* was enhanced across the entire thermal niche breadth, but *H. bacteriophora* improved in one portion of the thermal niche and declined in another portion (Grewal *et al.*, 1996a). Shapiro *et al.* (1997) transferred the heat tolerance trait from *H. bacteriophora* strain IS5 to the commercial strain HP88 though hybridization. As hybridization in *Heterorhabditis* is challenging due to the difficulty of discerning hermaphrodites from females, they used a genetic marker mutant (dumpy mutant *Hdpy-2*) to verify that mating had occurred. The progeny from the cross, exhibited enhanced heat tolerance at 40°C for

2 h. The fitness of hybrid with respect to virulence was more, reproductive potential almost similar and storage ability lesser than HP88. Hussaini *et al.* (2004) studied the thermotolerance in native isolates of *Heterorhabditis* spp. and *Steinernema* spp. in India. The pre-conditioning of EPNs at 37°C enhanced survival rate and viability of IJs on subsequent exposure at 40°C. The F1 population had superior tolerance to high temperature compared to wild population; however, the pathogenicity was only marginally affected after exposure to 42°C as compared to wild population.

In another study, after four selection steps of *H. bacteriophora*, the mean tolerated temperature increased from 38.5 to 39.2°C (Ehlers *et al.*, 2005). Mukuka *et al.* (2010d) screened 36 natural populations and 18 hybrid or inbred strains for their response to high temperature. A high variability in heat tolerance and the relatively high heritability was recorded in the adapted (35°C for 3 h followed 25°C for 1 h) strains of *H. bacteriophora*. However, the mean tolerated temperature ranged from 33.3 to 40.1°C for nonadapted and from 34.8 to 39.2°C for adapted strain populations.

Recently, an overall increase in mean heat tolerance of 5.5°C was achieved in *H. bacteriophora* by cross breeding and successive genetic selection (Mukuka *et al.*, 2010b). The heat-tolerant strains were superior or similar in fitness to the commercial strain EN 01 with respect to virulence, host penetration and reproductive potential (Mukuka *et al.*, 2010c). In case of *S. carpocapsae,* hybrid strain (Italian strain-highly virulent x DD-136 strain-highly tolerant to heat and desiccation) showed enhanced mortality of pecan weevil, *Curculio caryae* (Shapiro *et al.*, 2005).

(iii) *Cold tolerance.* For selective breeding at low temperature, Griffin and Downes (1994) assessed the so called realized heritability (h^2= 0.33) of a hybrid strain (HF85) of *Heterorhabditis megidis* calculated by the difference between the basic hybrid strain and a population obtained by selective breeding. The mean temperature at which *H. bacteriophora* (strain HF85) kills 50% of *Tenebrio molitor* larvae was reduced from 10 to 5°C. However, Griffin and Downes (1994) did not provide evidence that the breeding success could be maintained in the selected population as no further selection steps were performed after they had reached 5°C. It was also reported that cold selection enhanced *S. feltiae* virulence (insect mortality) by 5.3- and 6.6-fold after six and twelve passages at 8°C, respectively (Grewal *et al.*, 1996b). Ehlers *et al.* (2005) estimated the heritability (h^2 = 0.38) for activity at low temperature for *H. bacteriophora* hybrid strain PS7.

The mean temperature at which the IJs of *H. bacteriophora* were active, could be reduced from 7.3 to 6.1°C during the first 5 selection steps, however, it increased during the following 5 steps to 7.1°C (Ehlers *et al.*, 2005). It is known from other studies that storage at low temperatures can also increase the pathogenicity of EPNs (Surrey, 1996; Ogura and Nakashima, 1997).

If the increased pathogenicity would be mainly a result of physiological adaptation, the improvement is lost once the selection pressure is removed from the population. The natural capacity of EPNs for freezing tolerance may also prove to be the first step towards developing a method for their long-term storage. Ali and Wharton (2009) reported *S. feltiae* to be freeze-tolerant if frozen at relatively high sub-zero temperature. However, slow freezing at high sub-zero temperature (-1°C) enhanced the nematodes' survival at lower temperatures (down to -12°C). Brown and Gaugler (1996) worked on the freezing tolerance of *S. anomali, S. feltiae,* and *H. bacteriophora* and showed that all the species exhibit some degree of freezing tolerance. They also illustrated that *S. feltiae* acclimated better to freezing temperature after sudden temperature decrease from 25 to 10°C under a direct acclimation regime, whereas freezing tolerance of *H. bacteriophora* increased under a gradual stepwise acclimation regime (25 to 15°C, 10°C and 5°C for 3 days at each temperature).

(iv) *Desiccation tolerance.* Selective breeding is also a feasible approach to increase tolerance of *H. bacteriophora* to desiccation stress only if IJs are stress-adapted prior to exposure to desiccation conditions. The heritability (h^2= 0.48) for desiccation tolerance has been reported by Strauch *et al.* (2004). The mean of the tolerated a_w-value (tolerated water activity) continuously dropped from 0.89 to 0.81 when the adaptation of IJs was done before selection. Mukuka *et al.* (2010b) could also decrease the a_w value from 0.67 to 0.65 after adaptation and from 0.9 to 0.7 without prior adaptation in *H. bacteriophora*. However, further selection did not enhance the tolerance in both cases. The strains with increased desiccation tolerance were usually less fit, which may be due to possible trade-off effect of selection for desiccation tolerance (Mukuka *et al.*, 2010c). Salame *et al.* (2010a) bred heterogenous population of *S. feltiae* to enhance desiccation tolerance. A survival rate of 80-90% for tolerance of rapid desiccation and >85% for tolerance to slow desiccation was observed after 10 and 20 selection cycles, respectively.

(v) *Nematicide tolerance.* The only report of genetic selection for enhanced tolerance to the nematicides is that of Glazer *et al.*

(1997). After 11 rounds of selection of the HP88 strains of *H. bacteriophora,* a 8 to 9-fold increase in resistance to fenamiphos and avermectin, and a 70 -fold increase in resistance to oxamyl was reported. No further efforts were made in this aspect, however, this study provides a platform to select the strains tolerant to other insecticides/ fungicides/herbicides so that they can become an integral part of integrated pest management (IPM) in any crop.

(vi) *Enhanced propagation.* The genetic improvement of traits related to the propagation potential in liquid culture is also important for commercial production of EPNs. Johnigk *et al.* (2002) produced inbred lines of *H. bacteriophora* which on propagation in liquid culture resulted in higher IJ recovery and yield.

Trait stability. Before development of an EPN strain into a commercial product, nematodes go through many generations for evaluations. During the process of evaluation in the laboratory, inadvertent adaptations can cause dramatic changes in the biological attributes of EPNs which may result in deterioration of the desirable traits (Wang and Grewal, 2002). Reductions in heat, UV and desiccation tolerance, storage stability, and reproductive potential have been reported in EPNs during laboratory maintenance (Shapiro *et al.,* 1996; Wang and Grewal, 2002; Bilgrami *et al.,* 2006). Bai *et al.* (2005) established that inbreeding can be a useful technique to fix a desired trait in the population during laboratory rearing. The inbred lines of *H. bacteriophora* remained stable as compared to serially cultured foundation population (16 passages) for virulence, reproductive capacity, heat tolerance and host seeking ability (Bai *et al.,* 2005).

Sandhu (2005) established 35 inbred lines in the GPS11 strain of *H. bacteriophora* and identified lines with superior IJ longevity, heat and ultra violet (UV) tolerance, and virulence against wax moth *G. mellonella* larvae. Out of these, seven lines were tested for differences in virulence against two white grub species, *P. japonica* and *C. borealis*. The inbred line A2 was found to be significantly more virulent towards *P. japonica* than the parent strain GPS11 and inbred lines A7, A8, A12 and A21. In case of *C. borealis,* significantly higher mortality was observed in lines A2, A8, A21 and A12 and the parent strain GPS11 (Fig. 8.3) (Grewal *et al.*, unpublished data). Further, the total number of IJs of different inbred lines, penetrated in *P. japonica* and *C. borealis* grubs was not significantly different, however, the higher proportion of living nematodes for line A2 than A12 were recorded (Fig. 8.4), which showed that the A2 line has more ability to survive in the grubs (Grewal *et al.*, unpublished data). This study describes that inbreeding not only would maintain the stability of the traits but can also result in the identification of inbred lines with traits superior to the parent.

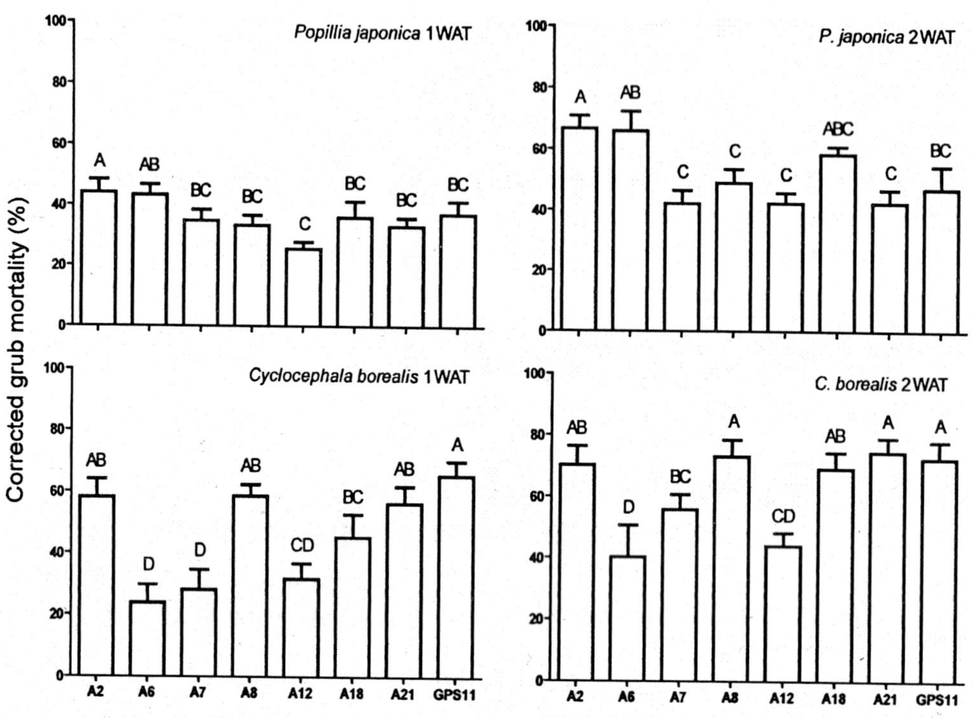

Fig. 8.3. Mean percentage mortality of *Popillia japonica* and *Cyclocephala borealis* grubs caused by inbred lines and parent of *H. bacteriophora* GPS11 strain 1 and 2 weeks after treatment (WAT). Bars (mean ± SEM) with same letter(s) are not significantly different according to $P < 0.05$ (Tukey's HSD test). (Grewal *et al.*, unpublished data)

Mutagenesis

Mutagenesis may also provide an alternative means of strain improvement. Zioni *et al.* (1992) induced mutations in a homozygous inbred line in *H. bacteriophora* strain HP88 by exposing young hermaphrodites to 0.05 M ethyl methanesulfonate (EMS). In F2 generation, a dumpy mutant (designated *Hdpy-l*) was isolated from mutagenized population. In another attempt, a dumpy mutant *RHdpy* was produced through EMS in *H. bacteriophora* strain HP88 (Rahimi *et al.*, 1993). Koltai *et al.* (1994) produced two dumpy mutants of *H. bacteriophora* (*Hbdpy-2* and *Hbdpy*-3) using the same technique. In *S.*

feltiae, the first morphological and behavioural mutants were described by Tomalak (1994b). The mutation occurred in a single gene locus designated as *Sfdpy-1* and the resulting morphology has been classified as dumpy. The individuals isolated from dumpy populations had the ability of fast movement and dispersal, and were also more effective in penetration into the hemocoel of *L. solani* larvae. Three desiccation tolerant mutants of *H. megidis* strain UK211, from rapid desiccation screen (57% RH, 210 min), have also been isolated following EMS mutagenesis by O'Leary and Burnell (1997). The mutants were more tolerant to dehydration at low relative humidity but not at high relative humidity. The temperature tolerance and host finding ability of the IJs were not impaired in the desiccation tolerant mutants which demonstrated the importance of use of EMS mutagenesis for making desiccation tolerant mutants (O'Leary and Burnell, 1997). Till date, the potential of this technique for strain improvement of EPNs remains unexplored.

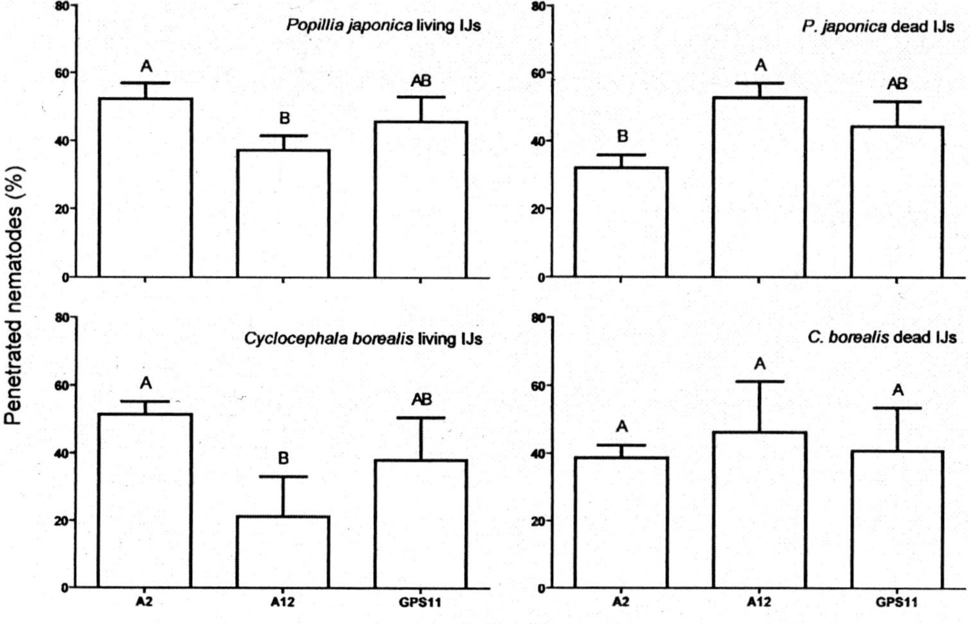

Fig. 8.4. Mean percentage of live and dead infective juveniles of inbred lines, A2 and A12 and parent GPS11 strain of *Heterorhabditis bacteriophora* found in *Popillia japonica* and *Cyclocephala borealis* grubs 24 h after incubation. Bars (mean ± SEM) with same letter(s) are not significantly different according to $P < 0.05$ (Tukey's HSD test) (Grewal *et al.,* unpublished data)

Bacterial Switching

Improvement of EPN strains may also be achieved through novel combinations of compatible nematodes and bacterial partners (Gerritsen *et al.*, 1998; Shapiro *et al.*, 2005). It was also described that the bacterial partner plays an important role in trait stability (Bilgrami *et al.*, 2006). Grewal *et al.* (1997) evaluated the effect of bacterial switch in *Steinernema scapterisci* and found the IJs containing non-symbiont bacteria (*X. nematophilus* and *Xenorhabdus* sp. 'R') to be more virulent to *G. mellonella* than those containing natural symbiont bacteria (*Xenorhabdus* sp. 'S'). Through novel combin-ation with bacterial partners, it has been observed that the virulence of *H. megidis* strains against *Tipula oleracea* can be improved (Gerritsen *et al.*, 1998). Clear evidence of improvement in virulence has been also detected in the modified strain of *S. carpocapsae* created through the transfer of its bacterial partner *Xenorhabdus nematophila* from the Italian strain, which shows high virulence against a key pest of pecan, *Curculio caryae,* to the DD-136 strain, which exhibits high levels of heat and desiccation tolerance but low virulence to the weevil. The modified strain causes *C. caryae* mortality similar to the Italian strain and possesses high levels of heat and desiccation tolerance similar to the DD-136 strain, indicating that bacterial transfer could be a valuable tool in improving the biocontrol efficacy of the EPNs (Shapiro *et al.*, 2005).

Bacterial switching has some advantages over other methods of EPN strain improvement since it is simpler than transgenic appro-aches, and probably more rapid than either transgenic or selection methods (Gaugler, 1987; Gaugler *et al.*, 1989a, 1990). However, the success of bacterial switching depends on the genetic makeup of the trait that is targeted (e.g. dominance/recessive factors), and it is always possible that some or all the hybrids will be inferior to the parent strain. Therefore, it is required to determine the level of varia-tion in outcomes when switching bacterial partners among strains.

Use of Molecular Techniques

The recent advances in molecular biology techniques have led to increased efforts to develop genetically engineered EPNs. Hashmi *et al.* (1995) made the first successful transformed *H. bacteriophora* HP88 by microinjection of plasmid vectors carrying *hsp-16* gene coding for 16-kDa heat shock protein as well as *rol-6* gene coding for roller phenotype. In F1 generation, 6 to 7% were transformed, as shown by β-galactosidase activity in the muscle and hypodermis after exposure to heat shock. The frequency of transformation declined after three or four generations, indicating the introduced DNAs were not incorporated into the genome. The roller phenotype was not observed in *H. bacteriophora*, although *rol6* gene was present in the progeny.

Genetic enhancement for thermotolerance in *H. bacteriophora* was achieved by transformation with the gene encoding a 70kDa heat-shock protein (hsp70) from *C. elegans* (Hashmi *et al.*, 1998). Successful transformation of the gene was confirmed, transcripts were increased several-fold in transgenic nematodes upon heat shock, and survival of the IJs dramatically increased in the transgenic nematodes. This strain was released in turfgrass microplots in the field in the spring, summer and fall of 1996. Although transgenic nematodes did not differ in their ability to persist, the risk assessment study supported the regulatory view that the transgenic nematode strain is unlikely an environmental threat (Gaugler *et al.*, 1997; Wilson *et al.*, 1999). Another study generated *S. feltiae* transgenic nematodes carrying a cloned stress-resistance gene *trehalose-phosphate synthase1* from yeast (Vellai *et al.*, 1999). A dramatic increase in the tolerance of transgenic adult nematodes was reported but no further developments have occurred.

Recent advances in decoding of sequence of genomes and progress to look at many genes, proteins and genetic pathways have lead to great revolution in biological sciences (Koltai, 2009). The first cDNA-sequencing project of the EPNs was reported by Sandhu *et al.* (2006). Out of 1246 ESTs (expressed sequence tags) generated in *H. bacteriophora* strain GPS11, 1072 ESTs were categorized into functional categories and 613 ESTs did not find matches in any of the searched databases suggesting potentially novel genes in the EPNs. The ESTs were compared to animal-parasitic, human-parasitic, plant-parasitic, and free-living nematodes and 36 ESTs had been identified similar to ESTs of only parasitic nematodes, suggesting that they are involved in parasitic nematode-specific functions (Bai *et al.*, 2007).

In case of dehydration of *S. feltiae* strain IS-6, 10-fold increase in concentration of a unique heat stable, water stress related protein DESC47 encoded by the *desc-47* gene was reported by Solomon *et al.* (2000). Downregulation of another gene glycogen synthase (*Sf-gys-1*), a rate limiting enzyme in the synthesis of glycogen, which is supposed to play a role during desiccation survival of EPN was observed by Gal *et al.* (2001). Further, eighty-one unique ESTs were identified in IS-6 strain of *S. feltiae* during desiccation (Gal *et al.*, 2003). Out of these, genes *Sf-ALDH, HSP40, ubq-2* and *glycerol kinase* were upregulated only during 8 to 24 h of desiccation, however *Sf-LEA-1* of LEA (late embryogenesis abundant) was highly upregulated even after 24 h of desiccation. The gene expression was correlated with the ability of EPNs to withstand desiccation stress by Somvanshi *et al.* (2008). They found inverse relationship between the ability of the nematodes to survive desiccation-tolerance and the expression profiles of selected desiccation-tolerance-related genes. The desiccation tolerant strains of *S. carpocapsae* and *S. riobrave* have lowest expression of genes, viz. *aldehyde dehydrogenase, nucleosome assembly protein*1, *glutathione peroxidase,* and *heat-shock protein40.* However, *H. bacteriophora* TTO1 strain, most susceptible to desiccation, showed the highest expression of these genes. Hao *et al.* (2010) analysed cDNA transcripts expressed in parasitic phase of *S. carpocapsae* to find genes involved in parasitism. About 32% of total ESTs generated had no homology with any known gene in public databases, which showed the presence of novel genes involved in parasitism of EPN.

Recently, genome of the TTO1 strain of *H. bacteriophora* has been sequenced at the Washington University Genome Sequencing Center in St. Louis (Grewal, P.S., Personal communication). A draft genome of about 68 Mbp assembled in 4,652 contigs, which ranges from 2,506 bp to 201,449 bp has been obtained (Bai *et al.,* unpublished data). In the most comprehensive analysis of 31,485 high quality ESTs of the TTO1 strain, 554 parasitic nematode-specific ESTs were identified (Bai *et al.*, 2009). The completion of the genome sequence of *H. bacteriophora* establishes a solid foundation for the much needed functional genomics studies on the genes that are involved in critical biological processes such as dauer formation, stress resistance, sex determination, etc. The functions of these genes in *H. bacteriophora* need to be elucidated using RNA interference technology and other means. Ciche and Sternberg (2007) have already indicated the feasibility of postembryonic RNAi in *H. bacteriophora* through soaking, which may be a potent approach for studying gene function in EPNs. The success of postembryonic RNAi was evidenced by sterility, defective gonads and germline proliferation in adult animals. Using these genomic resources, we may be able to generate genetically modified *H. bacteriophora* with improved biological control potential. These approaches may be extended to develop more genetically improved strains of the other EPN species.

In addition, the feasibility of improving EPN efficacy through bacterial switching is attributed to the fact that the virulence of the nematode–bacterium combination depends on the virulence of the nematode, the bacterium, and the interaction between the two (Gerritsen *et al.*, 1998). Thus, understanding mechanisms of the virulence of the two symbiotic partners and how they interact with each other is particularly important for long-term goal to improve the efficacy of the EPNs in biological control. Currently the genomes of *P. luminescens* subsp. *luminescens* (the bacterial symbiont of *H. bacteriophora* strain TTO1) and the dual human–insect pathogen *P. asymbiotica* ATCC43939 have been published (Duchaud *et al.*, 2003; Wilkinson *et al.*, 2009). The genomes of *X. nematophila* and *X. bovienii* (the symbionts of *S. carpocapsae* and *S. feltiae*, respectively) have recently been completed (www.genoscope.cns.fr/agc/mage). Several pathogenicity islands have been identified in the genome of *P. luminescens* (Duchaud *et al.*, 2003; Waterfield *et al.*, 2002). These pathogenicity islands are predicted to encode several toxins, some of which have been studied in detail, for example the Tc (toxin complex) and Mcf (makes caterpillars floppy) toxins (Waterfield *et al.*, 2001; Daborn *et al.*, 2002), which are also encoded in the *X. nematophila* and *X. bovienii* genomes (Morgan *et al.*, 2001; Sergeant *et al.*, 2006). Both *P. luminescens* Tc and Mcf induce apoptosis in eukayotic cells, suggesting their role in crippling the insect immune response to facilitate bacterial growth (Dowling *et al.*, 2004; Waterfield *et al.*, 2005). Using a differential fluorescence induction approach, Munch *et al.* (2008) identified 29 promoters induced by *P. luminescens* in the presence of isolated *G. mellonella* homogenate. An *et al.* (2009) compared the *in-vivo* gene expression of *Photorhabdus* and *Xenorhabdus* during infection to the white grub using a selective capture of transcribed sequences approach (SCOTS). Forty genes in *P. temperata* and thirty-nine by *X. koppenhoeferi* were identified to be differentially expressed at 24 h post bacterial infection of the white grub *Rhizotrogus majalis*. The two bacteria share more genes in stress response category than any other functional group. More than 60% of the identified genes were uniquely induced in either bacterium suggesting vastly different molecular mechanisms of pathogenicity to the same insect host. These results provide valuable information regarding the set of genes induced in *Photorhabdus* and *Xenorhabdus* bacteria upon infection of insect host and also highlight differences in molecular features used by these two closely related bacteria to promote their pathogenicity in the same insect host. For future research to facilitate the strategy of improving EPN efficiency through bacterial switching, particular virulence genes which are identified to be unique in one bacterial strain can be trans-engineered. Additional summaries of advances in understanding *Xenorhabdus* and *Photorhabdus* virulence are available from the recent review articles (Richards and Goodrich-Blair, 2009; Waterfield *et al.*, 2009).

Further, fully understanding mechanisms of symbiotic interactions between the nematode and the bacteria is also critical. Incompatibility among strains of *Photorhabdus* and *Heterorhabditis* species has led to unsuccessful combinations (Gerritsen and Smits, 1993; Gerritsen *et al.*, 1998; Han *et al.*, 1990). In contrast to *Heterorhabditis*, *Steinernema* nematodes are more amenable to bacterial transfer because they are considered less narrow in association with the symbiotic bacteria and thus more capable of growing successfully on bacteria of other strains or species (Grewal *et al.,* 1997). In addition, for *Steinernema* nematodes, they can be reared axenically, whereas *Heterorhabditis* nematodes cannot (Lunau *et al.*, 1993). Nonetheless, more research on combinations among various *Heterorhabditis* and *Steinernema* species and strains, and their symbionts will be necessary before any solid conclusions can be drawn regarding whether one group is more amenable to bacterial transfer than the other. In any case, dissecting the partnership between the nematode and the bacteria benefits, not only to the symbiosis studies through serving as model system, but also for improving the nematode biocontrol efficacy via novel combinations of the two partners. Through direct use of the genetic sequences, a recent study has investigated the nematode-bacteria partnership from a co-evolutionary perspective (Maneesakorn *et al.,* 2011). They observed that while the nematode strains with distant geographic distributions showed no detectable phylogenetic divergence within *H. bacteriophora* or *Heterorhabditis georgiana* monophyletic groups, their respective symbiotic bacteria speciated into two *Photorhabdus* species. Although such evolutionary process which is termed as duplication reduces the phylogenetic congruence between *Heterorhabditis* nematodes and *Photorhabdus* bacteria, global cophylogenetic tests detected a highly significant correlation between the two phylogenies, and the associations between *Heterorhabditis zealandica, H. indica* and *H. megidis* and their symbiotic bacteria contribute to the overall cophylogenetic structure significantly (Maneesakorn *et al.,* 2011). The effort has been also put into identification of the symbiotic factors to unravel the intricate mechanisms shaping the specific associations between the nematode and bacteria. Previous studies have characterized the essential molecular determinants of *Xenorhabdus* bacteria for vesicle colonization of the *Steinernema* IJs using a signature-tagged mutagenesis (STM) approach (Heungens *et al.*,

2002). Characterizing the genes encoding these determinants suggests that a fitness cost is incurred by the nematode (Mitani *et al.*, 2004), and the fitness cost may result from the provision of nutrition to the bacterial partner (Martens *et al.* 2005; Flores-Lara *et al.*, 2007).

Using SCOTS, a recent study has further investigated gene expression of *X. koppenhoeferi* residing in the enduring IJs of *S. scarabaei* to predict the potential suite of molecular mechanisms employed by the bacteria to persist in the nematode's intestinal vesicle (An and Grewal, unpublished data). Studies from *Photorhabdus-Heterorhabditis* interactions suggest that *Photorhabdus* bacteria undergo major transcriptional reshaping in the nematode intestine (An and Grewal, 2010), and metabolic switches are involved in lifestyle decision in these bacteria (An and Grewal, 2010; Lango and Clark, 2010). Besides general starvation mechanisms, the bacteria induce cellular acidification to slow down growth, switch to pentose phosphate pathway to overcome oxidative stress and nutrition limitation, and shed motility but develop biofilm to persist in the nematode intestine until being released into the insect hemolymph (An and Grewal, 2010). In future, the ability to genetically manipulate the bacterial partners based on the understanding of their molecular basis in pathogenecity and symbiosis will ultimately lead to improvement in the efficiency of these nematode-bacterial complexes.

CONCLUSIONS

The major improvement in EPNs has been achieved using classical genetic tools. For the genetic improvement of EPNs, the natural diversity of EPNs must be explored continuously for new beneficial traits. Till date, attempts have been made to improve the traits like heat, cold, desiccation tolerance, host finding ability, but there are some other traits on which studies must be initiated for the success of EPNs as biological control agents. These may include mass multiplication in liquid culture, pesticide tolerance, UV tolerance, etc. The new vistas for the genetic improvement of EPNs have opened with the use of molecular biology techniques. With the advancement of genome sequence of *H. bacteriophora* and identification of various ESTs, the genetically engineered EPNs with the desired trait can be produced. Only one example of genetically transformed EPNs for heat tolerance trait is through the microinjection technique (Hashmi *et al.*, 1995), however, no other successful report of microinjection is available. The strandardization of microinjection technology for developing genetically engineered EPNs is a need of the hour. The identification of the role of gene encoding for beneficial traits in EPNs will also be helpful to develop molecular markers for marker assisted selection of new isolates found in nature. Grewal *et al.* (2006) described that the understanding of physiological, genetic and molecular mechanisms associated with the beneficial traits is also important for the development of genetically engineered EPNs. The classical genetic techniques like selection, breeding and mutagenesis can not be ignored even in the modern era of molecular biology, e.g. if a particular trait is being regulated by number of genes, selective breeding will be the best approach to transfer that trait in the desired species. As the EPNs exhibit strong symbiosis with the bacteria, which are ultimately responsible for the death of host insect, enhanced understanding the mechanism of their symbiosis and pathogenesis at the molecular level will lead to new dimensions in use of EPNs as biological control agents. The bacteria with improved virulence and the EPNs with improved stress tolerance and host finding ability may improve their potential as biological control agents against insect pests. The use of genetically improved EPNs in pest management programme will help not only to produce pesticide free food but also safe environment.

REFERENCES

Alatorre-Rosas, R. and Kaya, H. K. (1990) Interspecific competition between entomopathogenic nematodes in the genera *Heterorhabditis* and *Steinernema* for an insect host in sand. *J. Invertebr. Pathol.,* **55,** 179–188.

Ali, F. and Wharton, D.A. (2009) Cold tolerance of entomopathogenic nematodes. *New Zealand Pl. Prot.*, **62**, 396.

An, R. and Grewal, P.S. (2010) Molecular mechanisms of persistence of mutualistic bacteria *Photorhabdus* in the entomopathogenic nematode host. *PLoS ONE*, **5**, e13154.

An, R., Sreevatsan, S. and Grewal, P.S. (2009) Comparative *in vivo* gene expression of the closely related bacteria *Photorhabdus temperata* and *Xenorhabdus koppenhoeferi* upon infection of the same insect host, *Rhizotrogus majalis*. *BMC Genomics*, **10**, 433.

Bai, C., Shapiro-Ilan, D.I., Gaugler, R. and Hopper, K.R. (2005) Stabilization of beneficial traits in *Heterorhabditis bacteriophora* through creation of inbred lines. *Biol. Contr.,* **32**, 220–227.

Bai, X., Grewal, P.S., Hogenhout, S.A., Adams. B.J., Ciche, T.A., Gaugler, R. and Sternberg, P.W. (2007) Expressed sequence tag analysis of gene representation in insect parasitic nematode *Heterorhabditis bacteriophora*. *J. Parasitol.,* **93,** 1343-1349.

Bai, X., Adams, B.J., Ciche, T.A., Clifton, S., Gaugler, R., Hogenhout, S.A., Spieth, J., Sternberg, P.W., Wilson, R.K. and Grewal, P.S. (2009) Transcriptomic analysis of the entomopathogenic nematode *Heterorhabditis bacteriophora* TTO1. *BMC Genomics*, **10,** 205-218.

Bedding, R.A., Molyneux, A.S. and Akhurst, R.J. (1983) *Heterorhabditis* spp., *Neoaplectana* spp., and *Steinernema kraussei*: Interspecific and Intraspecific differences in infectivity for insects. *Exp. Parasitol.*, **55**, 249–257.

Bilgrami, A.L., Gaugler, R., Shapiro-Ilan, D.I. and Adams, B.J. (2006) Source of trait deterioration in entomopathogenic nematodes *Heterorhabditis bacteriophora* and *Steinernema carpocapsae* during *in vivo* culture. *Nematology,* **8,** 397–409.

Blouin, M.S., Liu, J. and Berry, R.E. (1999) Life cycle variation and genetic structure of nematode populations. *Heredity*, **83**, 253–259.

Bowen, D., Rocheleau, T.A., Blackburn, M., Andreev, O., Golubeva, E., Bhartia, R. and ffrench-Constant, R.H. (1998) Insecticidal toxins from the bacterium *Photorhabdus luminescens*. *Science*, **280**, 2129–2132.

Brey, C.W. and Hashmi, S. (2003) Genetic improvement of entomopathogenic nematodes for insect control. In: R.K. Upadhyaya (ed.) *Advances in Microbial Control of Insect Pests*, Kluwer Academic/Plenum Publishers, New York, USA, pp. 297–312.

Brown, I.M. and Gaugler, R. (1996) Cold tolerance of steinernematid and heterorhabditid nematodes. *J. Therm. Biol.,* **21**, 115–121.

Burnell, A. (2002) Genetics and genetic improvement. In: R. Gaugler (ed.), *Entomopathogenic Nematology*, CAB International, Wallingford, UK, pp. 241–264.

Cabanillas, H.E. (2003) Susceptibility of the boll weevil to *Steinernema riobrave* and other entomopathogenic nematodes. *J. Invertebr. Pathol.*, **82**, 188–197.

Campbell, J.F. and Gaugler, R. (1993) Nictation behavior and its ecological implications in host search strategies of entomopathogenic nematodes. *Behavior,* **126,** 155–169.

Campbell, J.F. and Kaya, H.K. (2002) Variation in entomopathogenic nematode (Steinernematidae and Heterorhabditidae) infective-stage jumping behavior. *Nematology,* **4**, 471–482.

Campos-Herrera, R. and Gutiérrez, C. (2009) Screening Spanish isolates of steinernematid nematodes for use as biological control agents through laboratory and greenhouse microcosm studies. *J. Invertebr. Pathol.*, **100**, 100–105.

Ciche, T.A. (2007) *The Biology and Genome of Heterorhabditis bacteriophora*. The *C. elegans* Research Community, Wormbook. http://www.wormbook.org.

Ciche, T.A. and Sternberg, P.W. (2007) Postembryonic RNAi in *Heterorhabditis bacteriophora*: A nematode insect parasite and host for insect pathogenic symbionts. *BMC Develop. Biol.,* 7, 101.

Daborn, P.J., Waterfield, N., Silva, C.P., Au, C.P., Sharma, S. and ffrench-Constant, R.H. (2002) A single *Photorhabdus* gene, makes caterpillars floppy (mcf), allows *Escherichia coli* to persist within and kill insects. *Proc. Natl. Acad. Sci. USA*, **99**, 10742–10747.

Dowling, A.J., Daborn, P.J., Waterfield, N.R., Wang, P., Streuli, C.H. and ffrench-Constant, R.H. (2004) The insecticidal toxin makes caterpillars floppy (mcf) promotes apoptosis in mammalian cells. *Cell. Microbiol.*, **6**, 345–353.

Duchaud, E., Rusniok, C., Frangeul, L., Buchrieser, C., Givaudan, A., Taourit, S., Bocs, S., Boursaux-Eude, C., Chandler, M., Charles, J.F., Dassa, E., Derose, R., Derzelle, S., Freyssinet, G., Gaudriault, S., Medigue, C., Lanois, A., Powell, K., Siguier, P., Vincent, R., Wingate, V., Zouine, M., Glaser, P., Boemare, N., Danchin, A. and Kunst, F. (2003) The genome sequence of the entomopathogenic bacterium *Photorhabdus luminescens. Nat. Biotechnol.*, **21**, 1307–1313.

Ebssa, L., Borgemeister, C. and Poehling, H.M. (2004) Effectiveness of different species/strains of entomopathogenic nematodes for control of western flower thrips (*Frankliniella occidentalis*) at various concentrations, host densities, and temperatures. *Biol. Contr.*, **29**,145–154.

Ehlers, R-U., Oestergaard, J., Hollmer, S., Wingen, M. and Strauch, O. (2005) Genetic selection for heat tolerance and low temperature activity of the entomopathogenic nematode–bacterium complex *Heterorhabditis bacteriophora–Photorhabdus luminescens. BioControl*, **50,** 699–716

El-Lakwah, S.F. and Azazy, A.M. (2010) Physiological and biological studies of some entomopathogenic nematode species of families (steinernematidae and heterorabditidae) *Egypt. Acad. J. Biol. Sci.*, **2**, 45–50.

Flores-Lara, Y., Renneckar, D., Forst, S., Goodrich-Blair, H. and Stock, P. (2007) Influence of nematode age and culture conditions on morphological and physiological parameters in the bacterial vesicle of *Steinernema carpocapsae* (Nematoda : Steinernematidae). *J. Invertebr. Pathol.*, **95**, 110–118.

Gal, T.Z., Glazer, I. and Koltai, H. (2003) Differential gene expression during desiccation stress in the insect-killing nematode *Steinernema feltiae* IS-6. *J. Parasitol.,* 89, 761-766.

Gal, T.Z., Solomon, A., Glazer, I. and Koltai, H. (2001) Alterations in the levels of glycogen and glycogen synthase transcripts during desiccation in the insect-killing nematode *Steinernema feltiae* IS-6. *J. Parasitol.,* 87, 725-732.

Gaugler, R. (1986) Feasibility of genetically improving steinernematid and heterorhabditid nematodes. In: R.A. Samson, J.M. Vlak and D. Peters (eds.), *Fundamental and Applied Aspects of Invertebrate Pathology,*

Proceedings of the Fourth International Colloquium of Invertebrate Pathology, Veldhoven, August 18-22, The Netherlands, pp. 300-303.

Gaugler, R. (1987) Entomogenous nematodes and their prospects for genetic improvement. In: K. Maramorosch (ed.), *Biotechnology in Invertebrate Pathology and Cell Culture*, Academic Press, San Diego, CA, pp. 457-484.

Gaugler, R. (1993) Ecological genetics of entomopathogenic nematodes. In: R. Bedding, R. Akhurst and H. Kaya, H (eds.), *Nematodes and the Biological Control of Insect Pests*. CSIRO Publishing, East Melbourne, Australia, pp. 89–95.

Gaugler, R. and Campbell, J.F. (1991) Selection of enhanced host-finding of scarab larvae (Coleoptera, Sacarabaeidae) in an entomopathogenic nematodes. *Environ. Entomol.,* **20,** 700–706.

Gaugler, R., Campbell, J.F. and Gupta, P. (1991) Characterization and basis of enhanced host-finding in a genetically improved strain of *Steinernema carpocapsae. J. Invertebr. Pathol.,* **57**, 234–241.

Gaugler, R., Campbell, J.F. and Mc Guire, T.R. (1989a). Selection for host-finding in *Steinernema feltiae. J. Invertebr. Pathol.,* **54,** 363–372.

Gaugler, R., McGuire, T.R. and Campbell, J.F. (1989b) Genetic variability among strains of the entomopathogenic nematode *Steinernema feltiae. J. Nematol.,* **21,** 247–253.

Gaugler, R., Campbell, J.F. and McGuire, T.R. (1990) Fitness of a genetically improved entomopathogenic nematode. *J. Invertebr. Pathol.,* **56,** 106–116.

Gaugler, R., Wilson, M. and Shearer, P. (1997) Field release and environmental fate of a transgenic entomopathogenic nematode. *Biol. Contr.,* **9,** 75–80.

Gaugler, R., Glazer, I., Campbell, J.F. and Liran, N. (1994) Laboratory and field- evaluation of an entomopathogenic nematode genetically selected for improved host-finding. *J. Invertebr. Pathol.,* **63,** 68–73.

Gerritsen, L.J.M. and Smits, P.H. (1993) Variation in pathogenicity of recombinations of *Heterorhabditis* and *Xenorhabdus luminescens* strains. *Fundam. Appl. Nematol.*, **16**, 367–373.

Gerritsen, L.J.M., Wiegers, G.L. and Smits, P.H. (1998) Pathogenicity of new combinations of *Heterorhabditis* spp. and *Photorhabdus luminescens* against *Galleria mellonella* and *Tipula oleracea*. *Biol. Contr.*, **13**, 9–15.

Glazer, I., Gaugler, R. and Segal, D. (1991) Genetics of the nematode *Heterorhabditis bacteriophora* strain HP88: the diversity of beneficial traits. *J. Nematol.*, **42,** 342–333.

Glazer, I., Salame, L. and Segal, D. (1997) Genetic enhancement of nematicide resistance in entomopathogenic nematodes. *Bio. Sci. Technol.,* **7,** 499–512.

Glazer, I., Kozodoi, E., Hashmi, G. and Gaugler, R. (1996) Biological characteristics of the entomopathogenic nematode *Heterorhabditis bacterio-*

phora IS-5: a heat tolerant nematode from Israel. *Nematologica*, **42,** 481–492.

Grewal, P.S., Bai, X. and Jagdale, G.B. (2011) Longevity and stress tolerance of entomopathogenic nematodes. In: R.N. Perry and D. Wharton (eds), *Molecular and Physiological Basis of Nematodes Survival,* CAB International Publishing, Wallingford, UK, pp. 157–181.

Grewal, P.S., Ehlers, R.U. and Shapiro-Ilan, D.I. (2005) Critical issues and research needs for expanding the use of nematodes in biocontrol. In: P.S. Grewal, R.U. Ehlers and D.I. Shapiro-Ilan (eds), *Nematodes as Biocontrol Agents,* CAB International, Wallingford, UK, pp. 479–489.

Grewal, P.S., Gaugler, R. and Shupe, C. (1996a) Rapid evolution of thermal sensitivity of entomopathogenic nematodes in response to selection at temperature extremes. *J. Invertebr. Pathol.,* **68,** 65–73.

Grewal, P.S., Gaugler, R. and Wang, Y.I. (1996b) Enhanced cold tolerance of the entomopathogenic nematode *Steinernema feltiae* through genetic selection. *Ann. Appl. Biol.,* **129,** 335–341.

Grewal, P.S., Matsuura, M. and Converse, V. (1997) Mechanisms of specificity of association between the nematode *Steinernema scapterisci* and its symbiotic bacterium. *Parasitology,* **114,** 483–488.

Grewal, P.S., Selvan, S. and Gaugler, R. (1994a) Thermal adaptation of entomopathogenic nematodes: niche breadth for infection, establishment, and reproduction. *J. Therm. Biol.,* **19**, 245–253.

Grewal, P.S., Wang, X. and Taylor, R.A.J. (2002a) Dauer juvenile longevity and stress tolerance in natural populations of entomopathogenic nematodes: Is there a relationship? *Int. J. Parasitol.*, **32**, 717–725.

Grewal, P.S., Grewal, S.K., Malik, V.S. and Klein, M.G. (2002b) Differences in susceptibility of introduced and native white grub species to entomopathogenic nematodes from various geographic localities. *Biol. Contr.*, **24**, 230–237.

Grewal, P.S., Lewis, E.E., Gaugler, R. and Campbell, J.F. (1994b) Host finding behavior as a predictor of foraging strategy in entomopathogenic nematodes. *Parasitology,* **108,** 207–215.

Grewal, P.S., Tomalak, M., Keil, C.B.O. and Gaugler, R. (1993) Evaluation of a genetically selected strain of *Steinernema feltiae* against the mushroom sciarid *Lycoriella mali. Ann. Appl. Biol.,* **123**, 695–702.

Grewal, P.S., Power, K. T., Grewal, S.K., Suggars, A. and Haupricht, S. (2004) Enhanced consistency in biological control of white grubs (Coleoptera: Scarabaeidae) with new strains of entomopathogenic nematodes. *Biol. Contr.,* **30,** 73–82.

Grewal, P.S., Bornstein-Forst, S., Burnell, A.M., Glazer, I. and Jagdale, G.B. (2006) Physiological, genetic, and molecular mechanisms of chemoreception, thermobiosis, and anhydrobiosis in entomopathogenic nematodes. *Biol. Contr.,* **38,** 54–65.

Griffin, C.T. and Downes, M.J. (1994) Selection of *Heterorhabditis* sp. for improved infectivity at low temperatures. In: A.M. Burnell, R.U. Ehlers and J.P. Masson (eds), *Genetics of Entomopathogenic Nematode–Bacterium Complexes*. Office for Official Publications of the European Communities, Luxembourg, Belgium, pp. 120–128.

Hao, Y.J., Motiel, R., Abubucker, S., Mitreva, M. and Simoes, N. (2010) Transcripts analysis of the entomopathogenic nematodes *Steinernema carpocapsae* induced *in vitro* with insect haemolymph. *Mol. Biochem. Parasitol.,* **169**, 79–86.

Han, R., Wouts, W.M. and Li, L. (1990) Development of *Heterorhabditis* spp. strains as characteristics of possible *Xenorhabdus luminescens* subspecies. *Rev. Nematol.*, **13**, 411–415.

Hashmi, S., Hashmi, G., and Gaugler, R. (1995) Genetic transformation of an entomopathogenic nematode by microinjection. *J. Invertebr. Pathol.,* **66,** 293–296.

Hashmi, S., Hashmi, G., Glazer, I. and Gaugler, R. (1998) Thermal response of *Heterorhabditis bacteriophora* transformed with *Caenorhabditis elegans hsp70* encoding gene. *J. Exp. Zool.,* **281,** 164–170.

Hiltpold, I., Baroni, M., Toepfer, S., Kuhlmann, U. and Turlings, T.C.J. (2010a) Selection of entomopathogenic nematodes for enhanced responsiveness to a volatile root signal helps to control a major root pest. *J. Exp. Biol.,* **213,** 2417–2423.

Hiltpold, I., Baroni, M., Toepfer, S., Kuhlmann, U. and Turlings, T.C.J. (2010b) Selective breeding of entomopathogenic nematodes for enhanced attraction to a root signal did not reduce their establishment or persistence after field release. *Pl. Signal. Behav.,* **5**, 1450–1452.

Hominick, W.M., Hunt, D.J., Reid, A., Briscoe, B.R. and Bohan, D.A. (1999) Biosystematics, phylogeny, and population genetics of entomopathogenic nematodes. In: N. Boemare, P. Richardson and F. Coudert (eds), *Taxonomy, Phylogeny and Gnotobiological Studies of Entomopathogenic Nematodes Bacterium Complexes*, European Commission Publication, COST819, Brussels, pp. 45–53.

Heungens, K., Cowles, C.E. and Goodrich-Blair, H. (2002) Identification of *Xenorhabdus nematophila* genes required for mutualistic colonization of *Steinernema carpocapsae* nematodes. *Mol. Microbiol.*, **45**, 1337–1353.

Hussaini, S.S., Nagesh, M. and Shakeela, V. (2004) Selection for high temperature tolerance in some native *Steinernema* and *Heterorhabditis* species. *Indian J. Nematol.*, **34,** 185–192.

Jagdale, G.B., Saeb, A.T.M., Somasekhar, N. and Grewal, P.S. (2006) Genetic variation and relationships between isolates and species of the entomopathogenic nematode genus *Heterorhabditis* deciphered through isozyme profiles. *J. Parasitol.*, **92**, 509–516.

Jansson, R.K., Lecrone, S.H. and Gaugler, R. (1993) Field efficacy and persistence of entomopathogenic nematodes (Nematoda: Steinernematidae, Heterorhabditidae) for control of sweetpotato weevil (Coleoptera: Apionidae) in southern Florida. *J. Econ. Entomol.,* **86,** 1055–1063.

Jansson, R.K., Lecrone, S.H., Gaugler, R. and Smart, G.C. Jr. (1990) Potential of entomopathogenic nematodes as biological control agents of sweetpotato weevil (Coleoptera: Curculionidae). *J. Econ. Entomol.,* **83,** 1818–1826.

Jenkins, D.A., Shapiro-Ilan, D.I. and Goenaga, R. (2007) Virulence of entomopathogenic nematodes against *Diaprepes abbreviatus* in an oxisol. *Fla. Entomol.*, **90**, 401–403.

Johnigk, S.A. and Ehlers, R-U. (1999) *Endotokia matricida* in hermaphrodites of *Heterorhabditis* spp. and the effect of the food supply. *Nematology,* **1,** 717–726.

Johnigk, S.A., Hollmer, S., Strauch, O., Wyss, U. and Ehlers , R-U. (2002) Heritability of the liquid culture mass production potential of the entomopathogenic nematode *Heterorhabditis bacteriophora. Biocontr. Sci. Technol.,* **12,** 267–276.

Kaya, H.K. and Gaugler, R. (1993) Entomopathogenic nematodes. *Annu. Rev. Entomol.,* **38**, 181–206.

Koltai, H. (2009) Genomics and genetic improvement of entomopathogenic nematodes. In: S.P. Stock, J. Vandenberg, I. Glazer and N. Boemare (eds), *Insect Pathogens: Molecular Approaches and Techniques,* CAB International, Wallingford, UK, pp. 346–364.

Koltai, H., Glazer, I. and Segal, D. (1994) Phenotypic and genetic characterization of two new mutants of *Heterorhabditis bacteriophora. J. Nematol.,* **26,** 32–39.

Koppenhofer, A.M. and Fuzy, E.M. (2003) Ecological characterization of *Steinernema scarabaei*, a scarab-adapted entomopathogenic nematode from New Jersey. *J. Invertebr. Pathol.*, **83**, 139–148.

Koppenhofer, A.M. and Fuzy, E.M. (2008) Attraction of four entomopathogenic nematodes to four white grub species. *J. Invertebr. Pathol.*, **99**, 227–234.

Kruitbos, L.M. and Wilson, M.J. (2010) Is it time to 'wave' goodbye to 'nictating' nematodes? *Nematology,* **12,** 309–310.

Lango, L. and Clarke, D.J. (2010) A metabolic switch is involved in lifestyle decisions in *Photorhabdus luminescens*. *Mol. Microbiol.*, **77**, 1394–1405.

Laznik, Z., Strukelj, M. and Trdan, S. (2010a) Activity of entomopathogenic nematodes (Rhabditida) against cereal leaf beetle (*Oulema melanopus* [L.], Coleoptera, Chrysomelidae) adults under laboratory conditions. *Acta Agric. Slov.*, **95**, 141–148.

Laznik, Z., Toth, T., Lakatos, T., Vidrih, M. and Trdan, S. (2010b) The activity of three new strains of *Steinernema feltiae* against adults of

Sitophilus oryzae under laboratory conditions. *J. Food Agric. Environ.*, **8**, 150–154.

Lee, D.W., Choo, Y.H., Kaya, H.K., Lee, S.M., Smitley, D.R., Shin, H.K. and Park, C.G. (2002) Laboratory and field evaluation of korean entomopathogenic nematode isolates against the oriental beetle *Exomala orientalis* (Coleoptera: Scarabaeidae). *J. Econ. Entomol.*, **95**, 918–926.

Lewis, E.E., Gaugler, R. and Harrison, R. (1992) Entomopathogenic nematode host finding: response to host contact cues by cruise and ambush foragers. *Parasitology,* **105,** 309–315.

Lewis, E.E., Gaugler, R. and Harrison, R. (1993) Response of cruiser and ambusher entomopathogenic nematodes (Steinernematidiae) to host volatile cues. *Can. J. Zool.,* **7,** 765–769.

Lewis, E.E., Selvan, S., Campbell, J.F. and Gaugler, R. (1995) Changes in foraging behavior during the infective stage of entomopathogenic nematodes. *Parasitology,* **110**, 583–590.

Lunau, S., Stoessel, S., Schmidtpeisker, A.J. and Ehlers, R.U. (1993) Establishment of monoxenic inocula for scaling-up in-vitro cultures of the entomopathogenic nematodes *Steinernema* spp and *Heterorhabditis* spp. *Nematologica*, **39**, 385–399.

Maneesakorn, P., An, R., Grewal, P.S. and Chandrapatya, A. (2010) Virulence of four new strains of entomopathogenic nematodes from Thailand against second instar larva of the japanese beetle, *Popillia japonica* (Coleoptera: Scarabaeidae). *Thai. J. Agric. Sci.*, **43** (2), 61–66.

Maneesakorn, P., An, R., Daneshvar, H., Taylor, K., Bai, X., Adams, B.J., Grewal, P.S. and Chandrapatya, A. (2011) Phylogenetic and co-phylogentic relationships between the entomopathogenic nematodes (*Heterorhabditis*: Rhabditida) and their symbiotic bacteria (*Photorhabdus*: Enterobacteriacae). *Mol. Phylog. Evol.,* **59** (2), 271–280.

Martens, E.C., Russell, F.M. and Goodrich-Blair, H. (2005) Analysis of *Xenorhabdus nematophila* metabolic mutants yields insight into stages of *Steinernema carpocapsae* nematode intestinal colonization. *Mol. Microbiol.*, **58**, 28–45.

Mitani, D.K., Kaya, H.K. and Goodrich-Blair, H. (2004) Comparative study of the entomopathogenic nematode, *Steinernema carpocapsae*, reared on mutant and wild-type *Xenorhabdus nematophila*. *Biol. Contr.*, **29**, 382–391.

Morgan, J.A., Sergeant, M., Ellis, D., Ousley, M. and Jarrett, P. (2001) Sequence analysis of insecticidal genes from *Xenorhabdus nematophilus* PMFI296. *Appl. Environ. Microbiol.*, **67**, 2062–2069.

Mukuka, J., Strauch, O. and Ehlers, R.U. (2008) Variability in desiccation tolerance among different strains of the entomopathogenic nematodes *Heterorhabditis bacteriophora*. *Commun. Agric. Appl. Biol. Sci.,* **73**, 669–672.

Mukuka, J., Strauch, O. and Ehlers, R.U. (2010a) Variability in desiccation tolerance among different strains of the entomopathogenic nematode *Heterorhabditis bacteriophora. Nematology* **12,** 711–720.

Mukuka, J., Strauch, O., Hoppe, C. and Ehlers, R.U. (2010b) Improvement of heat and desiccation tolerance in *Heterorhabtidis bacteriophora* through cross- breeding of tolerant strains and successive genetic selection. *BioControl,* **55,** 511–521.

Mukuka, J., Strauch, O., Hoppe, C. and Ehlers, R.U. (2010c) Fitness of heat and desiccation tolerant hybrid strains of *Heterorhabditis bacteriophora* (Rhabditidomorpha: Heterorhabditidae). *J. Pest Sci.,* **83,** 281–287

Mukuka, J., Strauch, O., Waeyenberge, L., Viaene, N., Moens, M. and Ehlers, R.U. (2010d) Heat tolerance among different strains of the entomopathogenic nematode *Heterorhabditis bacteriophora. BioControl,* **55,** 423-434.

Munch, A., Stingl, L., Jung, K. and Heermann, R. (2008) *Photorhabdus luminescens* genes induced upon insect infection. *BMC Genomics,* **9,** 229.

O'Leary, S.A. and Burnell, A.M. (1997) The isolation of mutants of *Heterorhabditis megidis* (Strain UK211) with increased desiccation tolerance. *Fundam. Appl. Nematol.,* **20,** 197–205.

Ogura, N. and Nakashima, T. (1997) Cold tolerance and preconditioning of infective juveniles of *Steinernema kushidai* (Nematoda: Steinernematidae). *Nematologica,* **43,** 107–115.

Poinar Jr., G.O. and Georgis, R. (1990) Characterization and field application of *Heterorhabditis bacteriophora* strain HP88 (Heterorhabditidae: Rhabditida). *Rev. Nematol.,* **13** (4), 387–393.

Rahimi, F.R., McGuire, T.R. and Gaugler, R. (1993) Morphological mutant in the entomopathogenic nematode, *Heterorhabditis bacteriophora. J. Hered.,* **84,** 475–478.

Richards, G.R. and Goodrich-Blair, H. (2009) Masters of conquest and pillage: *Xenorhabdus nematophila* global regulators control transitions from virulence to nutrient acquisition. *Cell. Microbiol.,* **11,** 1025–1033.

Salame, L., Glazer, I., Chubinishvilli, M.T. and Chkhubianishvilli, T. (2010a) Genetic improvement of the desiccation tolerance and host seeking ability of the entomopathogenic nematode *Steinernema feltiae. Phytoparasitica,* **38,** 359–368.

Salame, L., Glazer, I., Miqaia, N. and Chkhubianishvili, T. (2010b) Characterization of populations of entomopathogenic nematodes isolated at diverse sites across Israel. *Phytoparasitica,* **38,** 39–52.

Salem, S.A., Rahman, A., Zebitz, H.A., Saleh, C.P.W., Ali, M.M.E., Fawkia, I. and El-Kholy, M.Y. (2007) Evaluation of entomopathogenic nematodes in controlling some cabbage pests. *J. Appl. Sci. Res.,* **3,** 323–328.

Sandhu, S.K. (2005) Genetic and molecular analysis of infective juvenile longevity and stress tolerance in the entomopathogenic nematode *Heterorhabditis bacteriophora.* M.S. Thesis. The Ohio State University, Columbus, Ohio, USA.

Sandhu, S.K., Jagdale, G.B., Hogenhout, S.A. and Grewal, P.S. (2006) Comparative analysis of the expressed genome of the infective juvenile entomopathogenic nematode, *Heterorhabditis bacteriophora*. *Mole. Biochem. Parasitol.,* **145,** 239–244.

Selvan, S., Gaugler, R. and Campbell, J.F. (1993) Efficacy of entomopathogenic nematode strains against *Popillia japonica* (Coleoptera: Scarabaeidae) Larvae. *J. Econ. Entomol.,* 86, 353-359.

Selvan, S., Grewal, P., Gaugler, R. and Tomalak, M. (1994) Evaluation of steinernematid nematodes against *Popillia japonica* (Coleoptera: Scarabaeidae) larvae: species, strains, and post-application rinse. *J. Econ. Entomol.* **78,** 342–345.

Sergeant, M., Baxter, L., Jarrett, P., Shaw, E., Ousley, M., Winstanley, C. and Morgan, J.A.W. (2006) Identification, typing, and insecticidal activity of *Xenorhabdus* isolates from entomopathogenic nematodes in United Kingdom soil and characterization of the xpt toxin loci. *Appl. Environ. Microbiol.*, **72,** 5895–5907.

Shapiro-Ilan, D.I., Glazer, I., Segal, D. (1996) Trait stability and fitness of the heat tolerant entomopathogenic nematode *Heterorhabditis bacteriophora* IS5 strain. *Biol. Contr.,* **6,** 238–244.

Shapiro, Ilan, D.I., Galzer, I. and Segal, D. (1997) Genetic improvement of heat tolerance in *Heterorhabditis bacteriophora* through hybridization. *Biol. Contr.,* **8,** 153–159.

Shapiro-Ilan, D.I., Stuart, R. and McCoy, C.W. (2003) Comparison of beneficial traits among strains of the entomopathogenic nematode, *Steinernema carpocapsae*, for control of *Curculio caryae* (Coleoptera: Curculionidae). *Biol. Contr.*, **28,** 129–136.

Shapiro-Ilan, D.I., Stuart, R.J. and McCoy, C.W. (2005) Targeted improvement of *Steinernema carpocapsae* for control of the pecan weevil, *Curculio caryae* (Horn) (Coleoptera: Curculionidae) through hybridization and bacterial transfer. *Biol. Contr.,* **34,** 215–221.

Shu-Sen, L., Ke-Bin, L., Chun-Qin, L., Qing-Lei, W., Jiao, Y. and Ya-Zhong, C. (2009) Identification of a strain of *Heterorhabditis* (Nematoda: Heterorhabditidae) from Hebei and its virulence to white grubs. *Acta Entomol. Sin.*, **52,** 959–966.

Solomon, A., Paperna, I. and Glazer, I. (1999) Desiccation survival of the entomopathogenic nematode *Steinernema feltiae*: induction of anhydrobiosis. *Nematology*, **1,** 61–68.

Solomon, A., Solomon, R., Paperna, I. and Glazer, I. (2000) Desiccation stress of entomopathogenic nematodes induces the accumulation of a novel heat-stable protein. *Parasitology,* **121,** 409–416.

Somasekhar, N., Grewal, P.S. and Klein, M.G. (2002) Genetic variability in stress tolerance and fitness among natural populations of *Steinernema carpocapsae. Biol. Contr.,* **23,** 303–310.

Somvanshi, V.S., Koltai, H. and Glazer, I. (2008) Expression of different desiccation-tolerance related genes in various species of entomopathogenic nematodes. *Mol. Biochem. Parasitol.*, **158**, 65–71.

Stock, S.P. and Hunt, D.J. (2005) Morphology and systematic of nematodes used in biocontrol. In: P.S. Grewal, R.U. Ehlers and D.I. Shapiro-Ilan (eds.), *Nematodes as Biocontrol Agents,* CAB International, Wallingford, UK, pp. 3–43.

Strauch, O., Oestergaard, J., Hollmer, S. and Ehlers, R-U (2004) Genetic improvement of the desiccation tolerance of the entomopathogenic nematode *Heterorhabditis bacteriophora* through selective breeding. *Biol. Contr.,* **31**, 218–226.

Stuart, R.J., Shapiro-Ilan, D.I., James, R.R., Nguyen, K.B. and McCoy, C.W. (2004) Virulence of new and mixed strains of the entomopathogenic nematode *Steinernema riobrave* to larvae of the citrus root weevil *Diaprepes abbreviatus*. *Biol. Contr.,* **30,** 439–445.

Surrey, M.R. (1996) The effect of rearing method and cool temperature acclimation on the cold tolerance of *Heterorhabditis zealandica* infective juveniles (Nematoda: Heterorhabditidae). *Cryo-Letters,* **17,** 313–320.

Toepfer, S., Gueldenzoph, C., Ehlers, R.U. and Kuhlmann, U. (2005) Screening of entomopathogenic nematodes for virulence against the invasive western corn rootworm, *Diabrotica virgifera virgifera* (Coleoptera: Chrysomelidae) in Europe. *Bull. Entomol. Res.*, **95**, 473–482.

Tomalak, M. (1994a) Selective breeding of *Steinernema feltiae* (Filipjev) (Nematoda: Steinernematidae) for improved efficacy in control of a mushroom fly, *Lycoriella solani* Winnertz (Diptera: Sciaridae). *Biocontr. Sci. Technol.,* 4, 187–198.

Tomalak, M. (1994b) Phenotypic and genetic characterization of dumpy infective juvenile mutant in *Steinernema feltiae* (Rhabditida: Steinernematidae). *Fundam. Appl. Nematol.*, **17,** 485–495.

Vasconcelos, V.O., Furlong, J., Freitas, G.M., Dolinski, C., Aguillera, M.M., Rodrigues, R.C.D. and Prata, M. (2004) *Steinernema glaseri* Santa Rosa strain (Rhabditida: Steinernematidae) and *Heterorhabditis bacteriophora* CCA Strain (Rhabditida: Heterorhabditidae) as biological control agents of *Boophilus microplus* (Acari: Ixodidae). *Parasitol. Res.*, **94,** 201–206.

Vellai, T., Molnar, A., Lakatos, L., Ba'nfalvi, T., Fodor, A. and Saringer, G. (1999) Transgenic nematodes carrying a cloned stress resistance gene from yeast. In: I. Glazer, P. Richardson, N. Boemare and F. Coudert (eds.), *Entomopathogenic Nematodes-Survival of Entomopathogenic Nematodes,* Office for Official Publications of the EC, Luxemberg EUR, pp. 105–119.

Wang, X. and Grewal, P.S. (2002) Rapid genetic deterioration of environmental tolerance and reproductive potential of an entomopathogenic nematode during laboratory maintenance. *Biol. Contr.*, **23**, 71–78.

Wang, Y., Campbell, J.F. and Gaugler, R. (1995) Infection of entomopathogenic nematodes *Steinernema glaseri* and *Heterorhabditis bacteriophora* against *Popillia japonica* (Coleoptera: Scarabaeidae) larvae. *J. Invertebr. Pathol.*, **66**, 178–184.

Waterfield, N.R., Ciche, T. and Clarke, D. (2009) *Photorhabdus* and a host of hosts. *Ann. Rev. Microbiol.*, **63**, 557–574.

Waterfield, N.R., Daborn, P.J. and ffrench-Constant, R.H. (2002) Genomic islands in *Photorhabdus*. *Trends Microbiol.*, **10**, 541–545.

Waterfield, N.R., Bowen, D.J., Fetherston, J.D., Perry, R.D. and ffrench-Constant, R.H. (2001) The tc genes of *Photorhabdus*: a growing family. *Trends Microbiol.*, **9**, 185–191.

Waterfield, N., Hares, M., Yang, G., Dowling, A.J. and ffrench-Constant, R.H. (2005) Potentiation and cellular phenotypes of the insecticidal toxin complexes of *Photorhabdus* bacteria. *Cell. Microbiol.*, **7**, 373–382.

Wilkinson, P., Waterfield, N.R., Crossman, L., Corton, C., Sanchez-Contreras, M., Vlisidou, I., Barron, A., Bignell, A., Clark, L., Ormond, D., Mayho, M., Bason, N., Smith, F., Simmonds, M., Churcher, C., Harris, D., Thompson, N.R., Quail, M., Parkhill, J. and ffrench-Constant, R.H. (2009) Comparative genomics of the emerging human pathogen *Photorhabdus asymbiotica* with the insect pathogen *Photorhabdus luminescens*. *BMC Genomics*, **10**, 302.

Wilson, M., Xin, W., Hashmi, S. and Gaugler, R. (1999) Risk assessment and fitness of a transgenic entomopathogenic nematodes. *Biol. Contr.*, **15**, 81–87.

Zioni, C.N.S., Glazer, I. and Segal, D. (1992) Phenotypic and genetic analysis of a mutant of *Heterorhabditis bacteriophora* strain HP88. *J. Nematol.*, **24**, 359–364.

9

Protease Inhibitors as Biopesticides: Potential and Constraints

NEHA KHANDELWAL, RAKESH JOSHI, VIDYA S. GUPTA AND ASHOK P. GIRI*

Plant Molecular Biology Unit, Division of Biochemical Sciences
National Chemical Laboratory (CSIR), Dr. Homi Bhabha Road
Pune- 411008, India
*E-mail: ap.giri@ncl.res.in

INTRODUCTION

Agricultural production in India plays major role in country's economic development by accounting over 28% of largest global pulse production. India is the second largest producer of vegetables, cotton, wheat and rice. Agriculture in India has gone through immense transitions ranging from low productivity due to pest attack and high productivity by the use of chemical pesticides. About 3% of the total pesticides used in the world are utilized in India. The used pattern of pesticides reflects that cotton crop alone consumes 44.5% pesticides estimated US$ 250 million for the control of bollworms which is followed by rice, accounting 22.8% of pesticide consumption for the control of yellow stem borer. These two crops consume more than two thirds of the total quantity of the pesticides used in the country. Other key pests of similar importance are stem borers of sorghum and maize, fruit and shoot borer of brinjal, fruit and pod borer of tomato and chickpea, and diamondback moth of cruciferous crops, cabbage and cauliflower. These pests are perennial and persistently cause losses to these economically important crops (Reddy and Zehr, 2004).

An average estimation of the damage caused by insects throughout the year has been shown in Fig. 9.1. During the past fifty years, the usage of chemical pesticides is at their peak for pest control in various crop species. Around 50% of the total insecticide consumption is that of organophosphates, followed by the synthetic pyrethroids (19%), organochlorines (18%), carbamates (4%) and biopesticides (1%). Most of the chemical pesticides become detrimental to human health as well as raises environmental concern when used above threshold limit and results in eradication of beneficial insects. In addition, most of the agronomically important pests have developed resistance towards pesticides, which causes the use of such insecticides inefficient at moderate level and demands for their increased use on agricultural crops. Sustainable agricultural practices are thus of global concern today which take maximum utilization of environment without causing any harm to it.

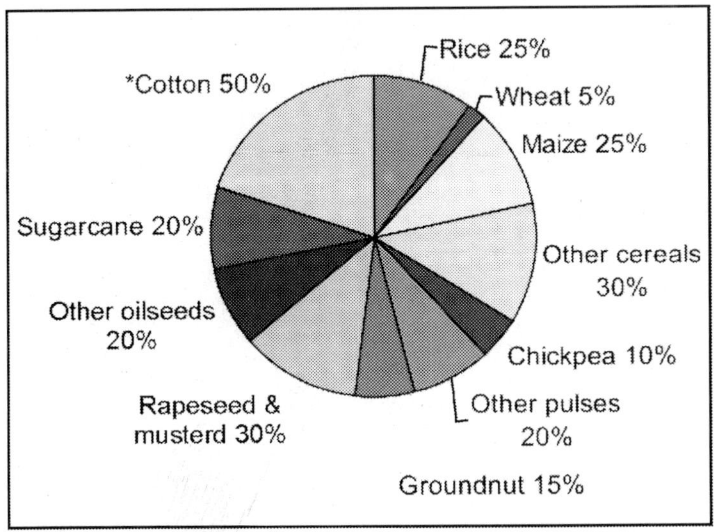

Fig. 9.1. The figure represents productivity loss of individual crop by insect pests in terms of percentage. Cotton (*) is highly affected by insect infestation leading to 50% loss of crop whereas wheat crop is least affected with total loss of 5%

Despite the use of various protective measures, pests have remained as a potential threat to agricultural crops. Insects that cause serious economic loss belong to lepidopteran, coleopteran, hemipteran, orthopteran and dipteran families. As pollinators of many plants, lepidopteran adult moths and butterflies are usually beneficial insects that feed on nectar. Caterpillars, however, consist of chewing mouth parts which allow them to feed on plant parts causing extensive damage. Unlike lepidopteran, coleopteran beetles have been known to cause damage usually in their adult stage, for instance *Caryedon serratus* and *Phyllotreta cruciferae* infect groundnut and seedling phase of cabbage (Ranga Rao *et al.*, 2010; Mayuri and Mikunthan, 2009). Much of the emphasis has been given on lepidopterans which represent the most damaging class of insects that feeds on cereals, pulses, vegetables, and oilseeds. Research in last few decades is much focused in studying about the interaction between plants and insects to develop plant protection strategy against them. Crop biotechnology, which broadly includes areas of development of transgenic crops, structural and functional genomics and marker-assisted breeding, could provide us with the vital breakthroughs to achieve improvements in both quality and quantity in a sustainable manner (Sharma *et al.*, 2003).

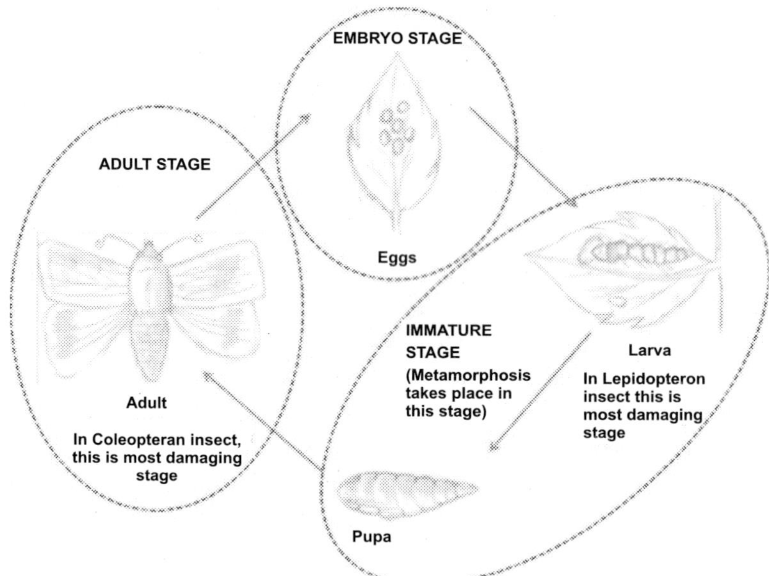

Fig. 9.2. Lifecycle of an insect includes four stages: from egg, larva and pupa to adult. Larva of the insect passes through various instars (six in lepidopteran and three in coleopteran) which allow it to attain complete maturity. Followed by pupation period, the adults (moth or butterfly in Lepidoptera and beetles in Coleoptera) emerge. Female adult lays hundreds of eggs on the leaves, which further hatch into 1st instar larvae.

PLANT-INSECT INTERACTIONS

In natural ecosystem, plants and insects are in constant interaction with each other. This complex and dynamic interaction between plants and insects is being studied from molecular, ecological and evolutionary perspective. Insects, as pollinators and predators of undesirable pests provide beneficial role, though many insects share same life cycle (Fig. 9.2), but many have emerged as harmful due to their feeding habit on enormous number of plant species. The number of plant attackers seemed to be almost limitless whose knowledge is essential in respect to develop effective strategy for the protection of crop plants. Insects that feed on plants are categorized into monophagous, oligophagous and polyphagous, depending upon the number of species or families they feed on.

A. **Monophagous or specialized pests.** Insects which are known to feed on single species of plant and during the absence of their main host plants they shift to some other related host species temporarily. Examples includes pink bollworm (*Pectinophora gossypiella*), rice stem borer (*Scirpophaga incertulas*), brinjal shoot and fruit borer (*Leucinodes orbonalis*) (Hanur, 2008).

B. **Oligophagous pests.** Insects that confine their feeding activity belonging to one family, a well known example is cabbage diamondback moth (*Plutella xylostella*; Plutellidae, Lepidoptera).

C. **Polyphagous or generalized pests.** These feed on various species of plants belonging to more than one taxonomic group or family, e.g. cotton bollworm or pod borer (*Helicoverpa armigera*), tobacco caterpillar (*Spodoptera litura*).

Polyphagous insects are considered as the most threatening among above categories due to their adaptation to various host plants posing problem in controlling them by employing same strategy with different crops. *Spodoptera exigua,* for example, feeds on approximately 50 plant species including pigweed, *Amaranthus hybridus*, and cotton, *Gossypium hirsutum* (Showler *et al.*, 2001). Similarly, *Heliothis virescens* mostly feeds on cotton, soybean, flax, alfalfa but found at high density on tobacco being its most preferable crop (Tilmann, 2006). Polyphagous pests possess the ability to adapt their digestive system in response to diverse type of host plant they feed upon, by switching on/off genes involved in the metabolism of the host plant components. The well known example is *Helicoverpa armigera* (Lepidoptera : Noctuidae) which has been widely studied for its adaptability on various host plants.

Voracious caterpillars of *H. armigera* feed on approximately 200 plant species including number of forest and fruit trees along with agronomically important crops such as cotton, chickpea, pigeon pea, tomato, maize, okra, etc. by metabolizing nutrients for their growth and development. In order to grow on varied type of host plants these insects bring about molecular changes in their gut complement (Patankar *et al.*, 2001). These molecular flexibilities enable insect's digestive system to react efficiently upon exposure to different crops by synthesizing more isoforms of proteases, if feeding on protein rich diet and amylases if feeding on starch rich diet. Thus, insects are capable of surviving on enormous number of plants due to which *H. armigera* has attained the status of notorious pest. Several studies have been attempted to control *H. armigera*, which alone is responsible for causing heavy losses of economically important crops ranging from minimum 14% to maximum loss of 100%. Currently, control of insect pests is mainly dependent on the use of chemical insecticides, which are commercially available in large numbers. Insect pest management by chemicals obviously has brought about considerable protection to crop yields over the past five decades. Pesticides such as lambda cyhalothrin, cypermethrin, and quinalphos are amongst few chemicals, which have been formulated and used in pest management. Though they have been found to be efficient in reducing insect proliferation, their detrimental effect on ecosystem has to be considered. In search of environmentally benign approach, transgenic plants have emerged as an efficient strategy for the pest management. Bt transgenics were successfully implemented for the control of major field pests, their implications and constrains are discussed partly in the following sections.

PLANT DEFENSE SYSTEM AGAINST INSECTS

Since years of interaction, plants and insects, both have coevolved in order to survive in their changing niches. Insects gradually adapt to take maximum benefit from the host while plants, on the other hand, have evolved to protect themselves. Being sedentary in nature, plants have developed the ability to react spontaneously against insect attack by synthesizing number of defensive molecules. These responses of plants have been extensively studied to exploit them for crop protective measures. Plants show different types of defence responses against pathogens and pests which are categorized into direct and indirect responses (Fig. 9.3). Direct mode includes production of toxic, repellent or anti digestive molecules like cyanogenic glucosides, glucosinolates, alkaloids, phenolics, proteinase inhibitors, etc., along with this they also pose some physical barriers (e.g. cuticle, trichomes, thorns, etc.) which play a role in primary protection against infestation (Bennett *et al.*, 1994). In indirect mode, plant emits some volatiles, which produce the extra floral nectar that attracts predators of insect herbivores to gain indirect means of protection (Kessler and Baldwin, 2001, 2002).

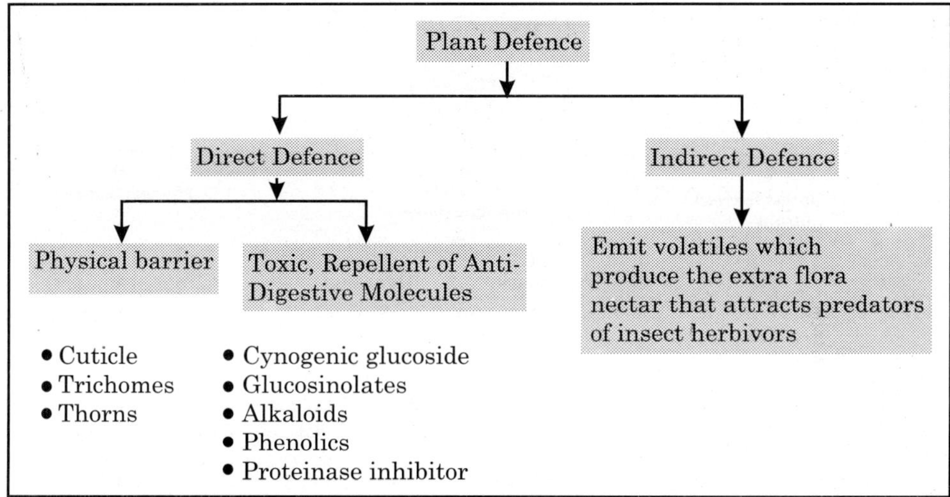

Fig. 9.3. Plant's response to insect attack is categorized into direct and indirect defense, which involves in preventing insect proliferation. Direct response comprises defensive chemicals which directly affect target pest whereas indirect defense involves attraction of herbivore predators by emission of volatiles.

Wounding by insect herbivory switches on different signalling reflexes in plants, and leads to induction of different defence mechanisms (Fig. 9.4). When an insect feeds on plant, some elicitor molecules are transferred through the saliva of insect. These elicitor molecules cause local secretion of some peptide hormones like prosystemin, threonin deaminase, etc. (Ferry *et al.*, 2004: Kang *et al.*, 2006). Prosystemin undergoes proteolytic cleavage to produce active systemin. Production of systemin switches the octadeconide (OD) pathway, which catalyzes the breakdown of linolenic acid and the formation of jasmonic acid (JA) (Ferry *et al.*, 2004). Jasmonic acid reacts with free isoleucine, which produces active signalling molecule. On cleavage, threonine gives isoleucine (reacts with jasmonic acid to gives JA-Ile) and α-keto butyrate (toxic to insect), this reaction is catalyzed by threonin deaminase (Kang *et al.*, 2006). JA-Ile switches on early genes in vascular bundle, which cause activation of NADPH oxidase. NADPH oxidase catalyzes the production of hydrogen peroxide (H_2O_2) (Orozco-Cardenas and Ryan, 1999). H_2O_2 is transferred to mesophyll cells, where they activate late gene responsible for production of defence molecule like proteinase inhibitor (PIs). In other way of defence, jasmonic acid induces the release of some volatiles that attract natural predators of the insect pest. Jasmonic acid on methylation produces methyl jasmonate. Methyl jasmonate acts as a mediator molecule to give plant-plant interaction (Orozco-Cardenas *et al.*, 2001; Rojo *et al.*, 2003).

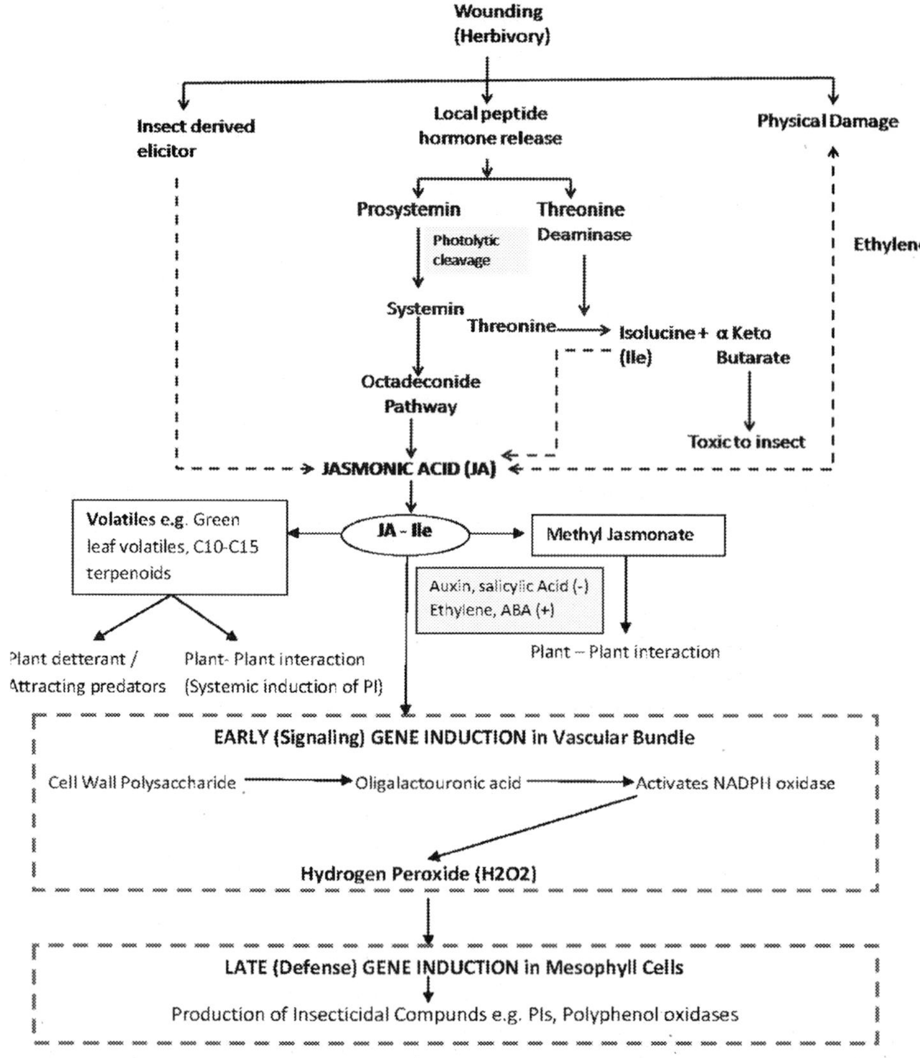

Fig. 9.4. Reflexes produced by a plant during insect infestation, which involves production of jasmonic acid (JA) by systemin through octadecanoid pathway. JA in the form of jasmonic isoleucine (JA-Ile) triggers expression of early and late defense genes

ROLE OF PROTEASE INHIBITORS IN PLANTS

Protease inhibitors (PIs) are of common occurrence in the plant kingdom and initially were found to be abundant in storage tissues such as tubers and seeds. They protect storage organs from predators by inhibiting hydrolytic activity of target proteases and play developmental role by the regulation of protein turnover. Later on, their presence has also been detected in the aerial parts of plants where their expression is induced in response to pest attack (Fig. 9.5). Protease inhibitors mostly have been characterized from Fabaceae (legumes), Poaceae (cereals) and Solanaceae families. Since last decade, insects' digestive proteolytic enzymes have come under intensive investigation thus major attention has been given in exploitation of defensive role of protease inhibitors, which are induced upon insect attack for the purpose of crop protection. Thorough understanding of protease inhibitors from its induction to its inhibitory effect on insects has been determined in various studies to characterize their potential role in controlling pest population. PIs are responsible in causing reduction in the availability of amino acids by blocking the hydrolytic activity of proteases and were first shown as plant defensive proteins in 1972 when their induction in potato and tomato was observed due to wounding and insect herbivory. Subsequently, Gatehouse and co-workers demonstrated the resistance of a cowpea variety to a bruchid beetle (Coleoptera) due to the elevated trypsin inhibitor (Boutler *et al.*, 1990).

PIs act on insects by interfering with the metabolic breakdown of the protein in the insect midgut thereby arresting their catabolism that leads to the accumulation of the protein causing physiological burden on insects. Most of the lepidopteran insects prefer protein rich diet hence employment of protease inhibitor has been proven to be the most promising approach for the development of effective strategy against insect pests. Natural mutants of trypsin inhibitors in wild tobacco plants recently provided the first true evidence that the inhibitors exert a certain level of control even against their natural 'adapted' pests (Jongsma and Beekwilder, 2008). In addition to resistance from insects, PIs are also known to confer resistance from abiotic stresses such as salt stress, change in pH and osmolarity (Tamura *et al.*, 2003). Strategically, use of protease inhibitors in integrated crop management requires the knowledge of the synthesis and role of PIs in plant system in response to insect attack.

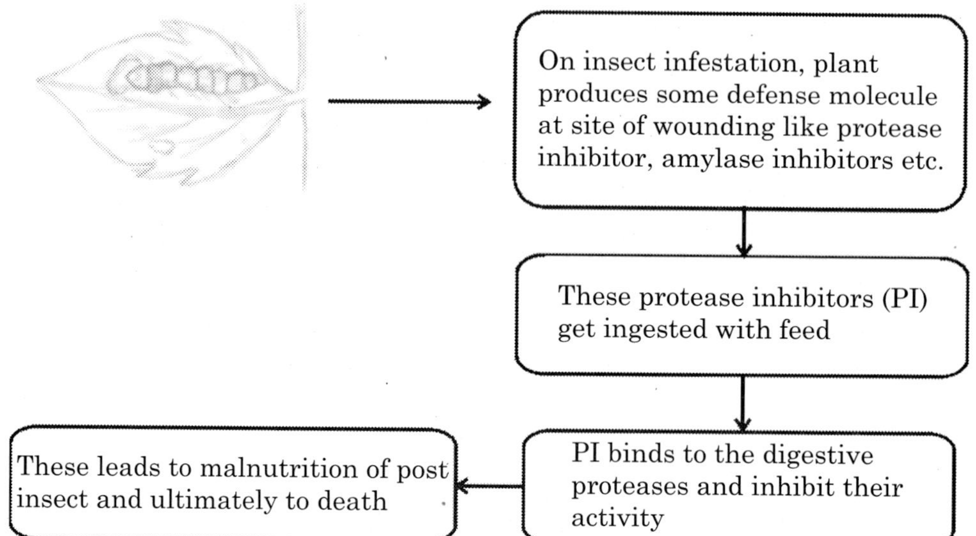

Fig. 9.5. PI exists as the plant's natural defense mechanism, which upon ingestion by insects interacts with the midgut proteases. PI acts by blocking active site of protease rendering it inactive for protein catabolism. Thus, accumulation of the protein in the gut creates a physiological burden on insect, ultimately leading to death of the insect.

MODE OF ACTION OF PROTEASE INHIBITORS

The different classes of inhibitors are distinguished by the structure of their polypeptide backbone. However, the mechanism of binding of the plant protease inhibitors to the insect proteases appears to be similar with all the four classes of inhibitors. Interaction between enzyme and inhibitor occurs in canonical (substrate like) manner. The inhibitor binds to the active site on the enzyme to form a complex with a very low dissociation constant (10^7 to 10^{14} M at neutral pH values), thus effectively blocking the active site. For serine proteinase inhibitors, most residues interacting with the proteases are located on a single loop, to which the P_1 residue is central. The P_1 residue of protease substrates is the one on the amino-side of the hydrolyzed bond, and is often crucial for substrate recognition. The surrounding residues (P_2, P_3, etc. at its N-terminus, and P_1', P_2', etc. at its C-terminus) play secondary roles in the interaction of proteases and their substrates. Both in inhibitors and in substrates, the P_1 residue fits into the S_1 substrate binding site of the proteinase. Unlike normal peptide substrates, the inhibitor residues around P_1 interact with the enzyme through complementary polar and hydrophobic interactions, and are held in position by strong bonds with the inhibitor scaffold. This prevents immediate dissociation of the complex, thus keeping the enzyme inactive. P_1 residue in reactive loop of inhibitor is primary determinant of inhibitor specificity. If P_1 residue is positively charged like Arg and Lys, they act as trypsin inhibitor. In case of chymotrypsin inhibitor, P_1 is hydrophobic in nature (Leu, Ile, Trp or Phe). The inhibitor residues around P_1 residue are from gene family which are more hypervariable in nature relative to most of protein.

DISTINCT CLASSES OF PROTEASE INHIBITORS

Protease inhibitors are categorized into four classes on the basis of the mechanistic classes of the proteases they inhibit: serine PIs, cysteine PIs, asparatate PIs and metallo proteases PIs (Fig. 9.6). Serine PIs have a widespread distribution in plant kingdom and are extensively studied as most of the lepidopteran insects utilize active serine proteases for their nutrient requirement by metabolic break-down of the protein. Cysteine PIs represent the second most well studied class of protease inhibitors whereas little is known about asparatate and metallo protease inhibitors.

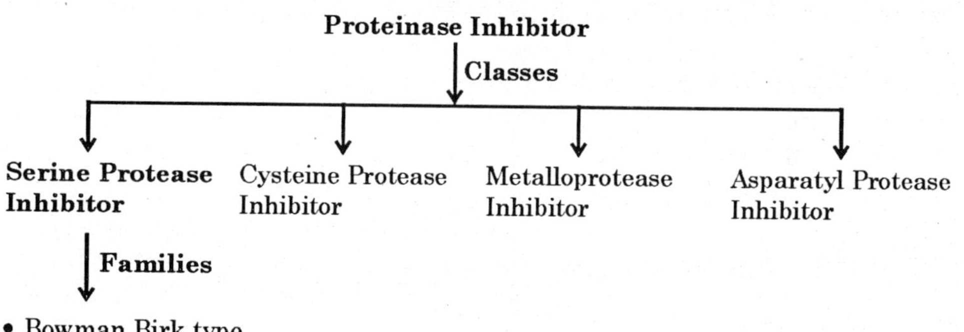

- Bowman Birk type
- Kunitz type
- Squash type
- Serpin type
- Mustard seed type
- Ragi seed type
- Wound Inducible Inhibitor
    1. Potato Inhibitor type 1
    2. Potato Inhibitor type 2

Fig. 9.6. Proteinase inhibitors are classified into four classes depending upon the proteases they inhibit. Serine protease inhibitors are further divided into various families based on their abundance, structure and function.

Serine Protease Inhibitors

Plant serine PIs act against proteases like trypsin, chymotrypsin, elastases and subtilisin (bacterial enzyme). Trypsin- and chymotrypsin-like enzymes are predominantly present in the midgut of lepidopteran insects. These enzymes are active in the alkaline pH from 9 to 11 of gut proteolytic environment. Protease function of cleaving proteins at specific sites, enables insects to utilize them for their nutrient enrichment. Trypsin cleaves at the C-terminal residue of lysine or arginine containing basic side chains whereas chymotrypsin cuts at the C-terminal residue of phenylalanine, trypsin and leucine, all containing hydrophobic side chain. Various researchers have identified the presence of serine protease inhibitors in plants for their defensive role and divided into various families of which four major types are: Kunitz, Bowman birk, squash and wound inducible type. Latter is further categorized into two families namely potato inhibitor type 1 (Pin I) and potato inhibitor type 2 (Pin II). Different families of serine PIs were recognized based on their distribution in plant kingdom, sequence and functional divergence.

As wounding induces high accumulation of the PIs in plant parts, one might ask about their consequential effect on native proteinases of the plants. Major digestion of protein in plants occurs by the proteases, which contain cysteine residues at their active site rather than serine residue thus the significant amount of serine PIs in plants does not affect plant metabolic function and cannot be used for regulating endogenous proteases. This clearly reflects the presence of different types of proteinase inhibitors for the regulation of protein turnover in plants. However, few reports have emerged describing endogenous function of serine proteinase inhibitors in plants (Hartl *et al.*, 2010). Moreover, many plants express a number of different PIs and it was unknown if these proteins work synergistically as defenses or if they also have other functions. In case of serine PIs, gene- and tissue-specific expression patterns suggest multiple functions in generative tissues, including a possible involvement in development. Role of PIs has been characterized from various plants towards their defensive function against pests. Recently, protease inhibitors from *Capsicum annuum* plant were also accessed for their expression in different parts of plants. Single and double inhibitory domain proteins were found to be strongly expressed in stems, whereas multi IRD protein expression was restricted to leaves, which clearly distinguish their role in defence from endogenous function (Tamhane *et al.*, 2009). Role of plant serine PIs in defence has been well documented as they were shown to inhibit insect digestive enzymes leading to retardation in their growth and development. Broadway and Duffey (1986) in their studies have demonstrated the antinutritional effect of serine protease inhibitors which showed inhibition of *Helicoverpa Zea* and *Spodoptera exigua* after incorporation in artificial diet. In another similar study, PIs of tobacco (*Nicotiana alata*) have shown to affect *H. punctigera* larval and black field cricket (*Teleogryllus commodus)* growth (Heath *et al.*, 1997). Following this, several serine protease inhibitors from non-host plants were investigated for their potential in conferring resistance to plants and attempts were made to develop transgenic plants.

Cysteine Protease Inhibitors

Cysteine protease inhibitors are also called plant cystatins or phytocystatins and are divided into two groups, one group containing single inhibitory domain whereas second group includes PIs carrying multiple domains. Cysteine PIs have been characterized from several plants, viz. cowpea, potato, cabbage, ragweed, carrot, papaya, apple fruit, avocado, chestnut and various crop plants like sunflower, rice, wheat, maize, soybean and sugarcane (Fan and Wu, 2005). Cysteine PIs act as antagonists for cysteine proteases, which are active in the insect midgut of pH ranges from 5 to 7. Cysteine proteases are usually found in coleopteran order of insects such as cowpea weevil, *Callosobruchus maculates*; bruchid, *Zabrates subfaceatus*; flour beetle, *Tribolium castaneum*; Mexican bean beetle, *Epilachna varivestis* and the bean weevil, *Acanthoscelides* spp.. Studies revealed that in few insects serine protease inhibitor in addition to cysteine PIs also inhibits cysteine proteases, e.g. cowpea and soybean trypsin inhibitor (serine PI) inhibits *C. maculatus* proteolytic enzymes (Lawrence, 2002). It could be possible if they share high sequence homology between cysteine and serine protease inhibitors, and thus cross interaction might occur. Oryzacystatin from rice has been well characterized among other cysteine PIs for its role in plant protection and attempts were also made to develop transgenic plants. Oryzacystatin contains highly conserved motifs glu-val-val-ala-gly (QVVAG) and pro-trp (PW) of which the former is the primary region, which involve in interaction with the enzyme and PW motif acts like a cofactor (Arai *et al.*, 1991). Other than inhibitory effect on herbivory, cysteine proteases and their inhibitors are known to participate in program cell death according to the study conducted on soybean cells. Endogenous cysteine protease inhibitors may function in modifying PCD that is activated during oxidative stress and pathogen attack.

Asparatate and Metallo Protease Inhibitors

Asparatyl and metallo protease inhibitors are the least studied class of protease inhibitors and have been described here together. Asparatyl protease inhibitors are active against aspartate proteases, which are usually found in hemipteran insects. These digestive enzymes display proteolytic activity at slightly acidic pH 3 to 5. Asparatyl protease inhibitors have been found in sunflower, barley, cardoon flowers and also in potato tubers. An asparatyl PI, cathepsin D has considerable homology with soybean trypsin inhibitor and inhibits trypsin and chymotrypsin along with asparatyl protease cathepsin D but does not inhibit other asparatyl proteases like rennin, cathepsin E and pepsin (Habib and Khalid, 2007). This suggests that specificity of protease inhibitors is sequence dependent. Small variation in sequence may result in changing the specificity and function of protease inhibitors. Such property can be exploited in order to design potent PI, which will be efficient against not only one class of proteases but other classes as well. Insects are also known to modulate their digestive protease complement in response to asparatate inhibitors to compensate for the loss of protease activity (Graham *et al.*, 1997). Metalloprotease inhibitors are divided into two families, that inhibit metallo carboxypeptidases called metallocarboxy-peptidase inhibitor which have been characterized from tomato and potato, and other is cathepsin D inhibitor family from potatoes. Metallocarboxypeptidase inhibitors have been found to play develop-mental as well as protective role as their existence has been observed in potato tubers along with proteinase inhibitor 1 and 2. On the other hand, their presence is also reported in potato leaf tissues as a response of wounding (Hollander-Czytko *et al.*, 1985; Liang and MacManus, 2002).

FAMILIES OF SERINE PROTEASE INHIBITORS (SPIs)

Serine protease inhibitors are well studied type of inhibitors. Such studies have provided a basic understanding of the mechanism of action that applies to most serine proteinase-inhibitor families. Depe-nding upon amino acid composition, reactive loop confirmation and their structure, serine protease inhibitors are categorized into diffe-rent families as follows.

Kunitz Type Protease Inhibitor (KPI) Family

Its occurrence is widespread in cereals, solanaceous species, and *Arabidopsis thaliana*, but they were first described in legumes (Liang and MacManus, 2002). They possess the ability to inhibit different proteases like trypsin, chymotrypsin, subtilisin, tissue plasminogen activator, etc. Kunitz type protease inhibitors are also produced under stress condition, as has been found in potato tubers (*S. tuberosum*) and inhibits Cathespin D (aspartic proteinase) and papain (cysteine proteinase) (Brzin *et al.*, 1988; Ritonja *et al.*, 1990). Their molecular weight is around 18 to 22 KD and its structure consists of two disulfide bonds, which reduces flexibility in reactive loop. The structure of Kunitz family inhibitor has been determined from several species. Soybean trypsin inhibitor is one of the well-studied candidates; roughly, it is spherical molecule with 12 antiparallel ß strands. Reactive site is on protruding loop and the P1 carbonyl group remains in plane and forms a nominal trigonal geometry (Song and Suh, 1998). The inhibitor is canonical inhibitor, as it forms a tight complex with the target proteinase which only slowly dissociate into mixture of cleaved (between its P1 and PÍ residue) and uncleaved forms (Iwanaga *et al.*, 1999). Mutation studies of *Erythrina variegata* seed kunitz inhibitor, by changing P1 and surrounding residue, have shown that there is 5-fold decreases in KI (Iwanaga *et al.*, 1998).

Bowman Birk Inhibitors (BBI-PI) Family

The Bowman Birk family is abundantly found in legumes and cereals, with reports from other plant families, which have not yet been substantiated. Presence of these inhibitors has been initially detected in seeds, their induction in leaves upon wounding is also reported. The family is named after D.E. Bowman and Y. Birk, who were the first to identify and characterize a member of this family from soybean (*Glycine max*). The soybean inhibitor is amongst the most studied members of this family and is often referred as classic BBI. Small cyclic inhibitors have also been identified in flowers of *Helianthus annus* called sunflower trypsin inhibitor 1 (SFTI-1) (Luckett *et al.*, 1999; Zablotna *et al.*, 2002). The inhibitor acts like a substrate for the enzyme and hence called as canonical (substrate like) inhibitors. The BBIs from dicotyledonous plant typically have a molecular mass of ~ 8 kD. They are double headed and can inhibit two serine proteinases at same time. The BBI from monocotyledonous plants are of two types, one group consists of single polypeptide chain with molecular mass of 8 kD and has single reactive site. Another group has molecular mass of 16 kD with two reactive sites. Dicot 8 kD double-headed with two homologous domain, carries separate reactive site for the cognate proteases. It has been suggested that larger inhibitors were derived from smaller inhibitors by an evolutionary mechanism of gene duplication.

The loss of the second domain in the small monocot inhibitor and the gene duplication event that formed the larger double domain inhibitor has been suggested to occur after divergence of the monocot and dicot. The first reactive site in these inhibitors is usually specific for trypsin, chymotrypsin and elastase. The active site configuration in these inhibitors are stabilized by the presence of seven conserved disulfide bonds (Chen *et al.*, 1992; Lin *et al.*, 1993). In case of double-headed BBIs, relative affinity of binding of protease is altered of which one site is already occupied as evident by the behaviour of peanut (*Arachis hypogea*) inhibitors. These inhibitors exhibit no activity against chymotrypsin, which is pre-occupied with trypsin and vice versa (Tur *et al.*, 1972). In same way, the activity of soybean BBIs decreases 100-fold when trypsin in bound at other site. The BBIs family of protease inhibitors contains a unique disulphide linked nine-residue loop that adopts a characteristic canonical confirmation, this loop is called protease-binding loop.

Squash Family

This is a unique inhibitor family, characterized only from the seeds of Curbitaceae family. Seven serine PIs belonging to this family have been isolated and characterized from seed of wild cucumber (*Cyclanthera pedata*). It is smallest inhibitor family known having only 28 to 30 amino acids and P_1 at ~5th aa from N terminal. They have 3 disulphide bridges and fold in novel knottin structure (Nguyen *et al.*, 1990; Heitz *et al.*, 2001) with similarity to potato carboxy peptidase inhibitor. Two different but inter convertible (*cis-trans* isomers) inhibitors have been isolated and characterized from seeds of wax gourd (*Benin Casa hispida* [Thumb] *cong*) (Atiwietin *et al.*, 2006). Because of small size of these inhibitors, studies have been carried out on native inhibitors, variants and chimera that have been prepared by chemical synthesis. The structure of squash inhibitors and inhibitor and proteinase complex has been determined by X-ray crystallography and NMR spectroscopy. These inhibitors have been shown to follow the standard mechanism for inhibition. X-Ray crystallography and NMR analysis shows that it has ellipsoidal shape; interpreted as lacking helices and β-sheets but consisting of turns and short interconnecting polypeptide structure or possessing very short helix [3 aa] and triple standard β-sheet (2,2 and 5 aa) (Bode *et al.*, 1989). Again the typical protrude reactive site loop is inserted into the proteinase active site and cleavage occurs residue adjacent, with H-bond patterning virtually identical to those of all serine proteinase-protein inhibitor complex. Changes at P_1 residue cause predictable changes in the inhibitors specificity, which are consistent with standard mechanism. Natural squash inhibitors of trypsin have arginine or lysine at P_1, whereas elastase inhibitors have leucine. The P_1 arginine of several inhibitors has been altered to alanine or valine to create elastase inhibitors and to phenylalanine to make chymotrypsin inhibitors (Favel *et al.*, 1989; Ling *et al.*, 1993).

Serpin Family

This family is widespread among the organisms, being the only proteinase inhibitor found in eukarya, bacteria, arechea and viruses. Whereas prokaryotes generally have single serpin gene, large multi-cellular eukaryotic organism, on the other hand, have several genes. Plant serpins have been purified and characterized from cereals' seed, pollens and from the phloem exudates of *Cucurbita maxima* (Ostergaard *et al.*, 2000).

Serpins of Arabidopsis have been shown to act on metacaspases like protein *in vivo* and play a role in the plant immune response. It has been suggested that rather than directly interacting with pathogen, plant serpins may have role in the complex pathway involved in up regulating the host immune response (Vercammen *et al.*, 2006). Plant serpins appears to be exclusively inhibitors of serine proteinases, including trypsin, elastase and chymotrypsin. However, serpins that inhibit casapases and papain like cysteine proteases have been also reported. Irreversible inhibition and the inhibitor - proteinase complex is stable to boiling in SDS, which is indicative of covalent acyl bond formation. The inhibitor cleavage position is within the same structural motifs as other. They have mixed proteinase specificities as Barley Serpin (BSZx) is potent inhibitor of trypsin and chymotrypsin at overlapping reactive site (Dahl *et al.*, 1996). Plant serpins show a typical irreversible inhibition with a second order association rate constant Ka of about 10^5/M/s indicative of inhibition. The cleavage of appropriate peptide bond occurs in reactive centre loops of inhibitor so that catalysis does not proceed beyond the formation if an acyl enzyme complex (Huntington *et al.*, 2000).

Ragi seed Trypsin/ α-Amylase Inhibitor Family

The trypsin/α-amylase inhibitor family comprises of single chain protein having molecular mass of about 13 kD and 5 disulfide bonds.

Family members have serine protease inhibitor activity, α-amylase inhibitory activity or both. A large number of inhibitors in this family have only α-amylase inhibitor activity. However, inhibitors from barley (*Hordeum vulgare*), rye (*Secale cereals*) and tall fescue (*Festuca arundinacea*) are active against trypsin (*Eleusine coracana*) inhibitor; show dual activities and can inhibit serine proteases as well as α-amylase (Mahoney *et al.*, 1984). The vast majority of characte-rized inhibitors have dual activities. The vast majority of characte-rized inhibitors from this family have only α-amylase inhibitory activity suggesting that serine proteinase inhibitory activity has evolved more recently. The structure of ragi inhibitor was initially solved by NMR and subsequently by X–ray analysis with the *Tenebrio molitor* α-amylase complex (Strobl *et al.*, 1995). Proteinase binding loop appeared to be adopted a canonical confirmation. Inhibitors consist of single polypeptide containing 5 disulfide bonds with a molecular mass of about 13 kD. From this structure, it is also shown that proteinase-binding loop is in the canonical confirmation for proteinase inhibitors, with the arginine at P_1 of the sessile bond fully exposed and projecting away from the body of the molecule. Both ragi and maize inhibitor have a novel structure containing four alpha helices and two short antiparallel β strands similar to that of the α-amylase inhibitor. This type of inhibitor has reversible inhibition mechanism with an equilibrium constant between 0.1 to 10 dependants on pH.

Mustard Seed Trypsin Inhibitor Family (Sinapis)

These types of inhibitors are mostly found in cruciferous species. These are small inhibitors with molecular weight around 7 kD. It forms tight binding complex with trypsin and being cleaved by trypsin (Ceciliani *et al.*, 1994). They are wound inducible inhibitors. They show divergent evolutionary relationship with Bowman Birk Inhibitor with 27% sequence similarity. The structure has been solved using NMR, the structure has an alpha helix and two beta sheet stabilized by 4 disulfide bonds (Zhao *et al.*, 2002).

Potato Type I PIs (Pin-I)

The inhibitors of this family are widespread in plants and have been described in many species, including potato tubers, tomato fruit, squash phloem exudates and in tomato leaves in response to wounding (Ryan, 1962). These inhibitors have the molecular mass of ~8 kD and are generally monomeric. While the inhibitors from cucurbit and potato tubers contain a single disulphide bond, the inhibitors in this family in general lack any disulphide bridges (Cai *et al.*, 1995). The inhibitory mechanism in this family is considered to fit the standard model.

X-ray crystallographic structure of potato type I inhibitor protein (Pin I), both free inhibitor and inhibitor fragment also have been solved (CI-2, CMTi-V). The inhibitor has constrained loop protruding from surface of molecule, which interacts directly with active site of proteinase. Comparison between free and bound inhibitor shows that there are only small conformation changes in the inhibitor upon binding, with increased stability of reactive loop (Maphalen *et al.*, 1988). Reactivity and specificity of reactive loop is varying according to P_1 residue, for example, presence of methionine or leucine at P_1 position inhibits chymotrypsin. In case of barley, CI-2 inhibitor which has specificity for subtilisin shows alanine at P_1 and in eglin C with anti-elastase activity containing leucine at P_1 site. The identity of these residues corresponds to the known cleavage specificities of proteinases and where changes to the P1 residue have been made (Longstaff *et al.*, 1990).

Potato Type II PIs (Pin-II)

These protease inhibitors are expressed as precursor protein containing multiple inhibitory repeat domains (IRDs), which is further processed to release individual inhibitory unit or domain. These inhibitory repeat domain show specificity with insect gut proteases, which signifies co-evolutionary response of plant against insect pests in due course of time (Mishra *et al.*, 2010). These precursor proteins are processed in the plant itself and in the insect gut upon ingestion by the action of proteases. The members of this group have been reported only from the members of Solanaceae family. Initially characterized from potato tubers, these inhibitors have been found in leaves, flowers, fruit and phloem of other solanaceaous species (Liang *et al.*, 2005). A low molecular-mass inhibitor of this family has been found to be constitutively present in jasmme tobacco (*Nicotiana alata*) flowers (Atkinson *et al.*, 1993). Six small wound-inducible proteinase inhibitors of this family have been reported from tobacco leaves (Barrette-Ng *et al.*, 2003). An analysis of these inhibitors and genes has shown that they are composed of multiple repeat units ranging from 1 to 8 IRDs. Inhibitors of this family have been reported to inhibit chymotrypsin, trypsin, elastase, oryzin, pronase E and subtilisin.

The best known representatives are the wound-induced proteinase inhibitors which contain up to eight sequence-repeats (the 'IP repeats') coded by the second exon of the gene. The architecture of the Pin-II genes is conserved: the first exon encoding the N-terminus of the signal peptide, and the second, major exon encoding the C-terminus of the signal peptide and a variable number of IP repeats, are always separated by a type I intron of 100–200 bp. However, the sequence of the IP repeats is variable in different proteins, presence of eight cysteine residues constituting four disulfide bridges and a single proline residue are conserved throughout the known repeat sequences.

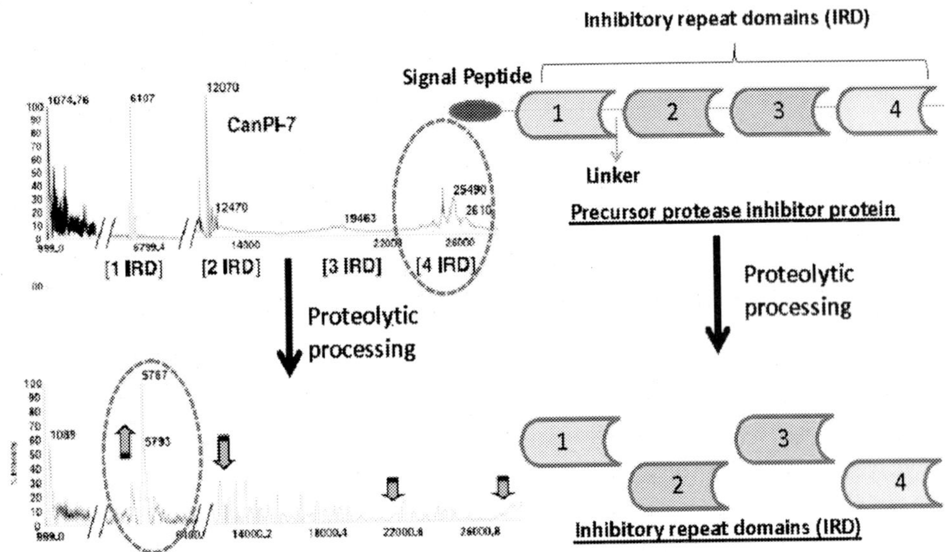

Fig. 9.7. This figure represents the processing of multi inhibitory repeat domains of PI protein (IRDs) to active single inhibitory domain. (a) A representative PI with 4 IRD (precursor form) upon cleavage at the linker region gets converted into single IRD (mature form). (b) Mass spectrometric analysis describes the eventual release of single IRD as shown by its increased peak intensity whereas the diminishing peak intensities of higher IRD (4IRD, 3IRD and 2 IRD) forms of a specific PinII type PI, CanPI-7 from *Capsicum annuum*

Precursor proteins of Pin-II type inhibitors in various plants consist of 1 to 8 IRDs, which upon cleavage by endogenous proteases release single inhibitor protein active against one or multiple serine proteinases (Fig. 9.7) (Heath *et al.*, 1995; Tamhane *et al.*, 2007). The ancestral Pin-II inhibitors of single inhibitory repeats are distributed in several mono- and dicotyledonous plants, whereas the multiple inhibitor repeat precursors arising from a series of gene duplication and/or domain duplication events are restricted to Solanaceae (Barta *et al.*, 2002). The Pin-II precursor PIs proteins have a conserved structure with putative endoplasmic reticulum signal peptides at the N-terminal followed by inhibitory repeat domains (IRDs) (1 to 8) and a C-terminal region, which probably functions as a vacuolar sorting signal. Pin-II type PIs have been isolated and characterized from different Solanaceous plants. NMR-based structures of series of 6KD *Nicotiana* sp. PIs have been determined, while the structure of the precursor PIs has also been reported in *N. alata* PIs and tomato PI–II, respectively (Nielsen *et al.*, 1995; Scanlon *et al.*, 1999). A circular conformation of 43KD inhibitory precursor of *N. alata* acquired by joining the partial repeats at the N- and C-terminals generates the sixth functional IRD. This type of 'clasped bracelet' fold has been adopted by the members of Pin-II family regardless of the number of inhibitory repeat.

In particular, *C. annuum* Pin-II type PI proteins and genes have been identified and characterized. A 6-KD protein PSI1.1 purified from seeds has shown the ability to inhibit trypsin and chymotrypsin (Moura and Ryan, 2001). In another study, a protein PSI1.2 from seeds has been reported to be a circularly permuted, ancestral member of Pin-II family. Seven small 6-KD PIs from leaves of *C. annuum* have been purified and the N-terminal sequenced to show homology to various repeat regions of previously reported *C. annuum* PI genes. Three repeat PI genes with the NCBI accession numbers AF221097 and AF039398 have also been isolated from cDNA libraries constructed from leaves and pericarp of *C. annuum* infected with the TMV. These CanPI genes are expressed constitutively as well as in significant amounts upon induction by various elicitors. The potential of these proteins for inhibiting insect growth and development has been demonstrated in several transgenic plants (Heath *et al.*, 1997; Sin *et al.*, 2004). Tobacco plants over-expressing potato Pin-II gene reduced the feeding and larval growth of *Manduca sexta* (Johnson *et al.*, 1989). Similarly, transgenic rice plants expre-ssing potato Pin-II gene were protected from *Sesmia inferens* infestation, a major lepidopteran pest of rice (Duan *et al.*, 1996).

STRATEGICAL USE OF PROTEASE INHIBITORS FOR INTEGRATED PEST MANAGEMENT

Biotechnological and molecular biology approaches have opened new avenues in the field of crop protection strategy. Genetic engineering has allowed integration of non-native gene of desired trait into crop plants resulting in the enhancement of productivity. Transgenic plants have gained much attraction over a period for developing resistance against insect attack. Studies carried on *Bacillus thuringiensis* transgenic plants have proved an effective strategy in integrated crop management. This strategy is applied to protect cotton and corn plants expressing *cry* genes from the damage of bollworm and corn borer as evident by their low rate of survival if compared to non-Bt plants (Kanglai *et al.*, 1996). Success with Bt further led to the development of transgenic chickpea plant expressing Cry1Ac or Cry2Aa or both proteins. Though Bt technology has emerged as a potential breakthrough in crop protection management, prevalence of Bt has raised concern regarding development of Bt resistant insects (Kanglai *et al.*, 1996). Thus, the deployment of integrated crop management strategies such as refugee and pyramiding of gene in transgenic plants will ensure durable insect resistance.

Protease inhibitors have attained considerable interest for their defensive role against insect pests and investigated as a viable alternative to Bt. These ubiquitous regulators of proteolytic enzymes have readily been identified as potential candidates for the development of transgenic plants. Many studies have revealed their effectiveness on insect's proteases by conducting *in vitro* assays and incorporation of PI in diet. Transgenic rice, cotton, tomato, tobacco, potato, and many more have been developed and their role is assessed in response to insect attack (Table 9.1). The first insect-resistant transgenic plants expressing a PI were produced in 1987 by expressing cowpea trypsin inhibitor (cysteine protease inhibitor) in tobacco plant (Hilder *et al.*, 1987). Several serine PIs have also been expressed in transgenic plants for resistance against insect pests of the order Lepidoptera while cysteine PIs have been expressed against the coleopteran pests. Inhibition in growth of *H. armigera* is observed when they are allowed to feed on transgenic cotton and peas (Julia *et al.*, 1999). Transgenic rice harbouring potato protease inhibitor type II has shown resistance from pink stem borer (*Sesamia inferens*) (Duan *et al.*, 1996). Large-scale field trials are currently being conducted from transgenic rice carrying serine protease inhibitor (Schluter *et al.*, 2010). However, yet no single transgenic crop plant expressing protease inhibitor gene(s) is available for farmers. This suggests we need further research and design of appropriate strategies for Pi-based insect resistance.

Table 9.1. Few of the transgenics harboring protease inhibitor genes for the control of target pest

Transgenic	Protease inhibitors	Class	Target pest	Reference
Tobacco	NA-P1	Serine	*Spodoptera litura*	Shrinivasan *et al.* (2009)
			Helicoverpa armigera	Charity *et al.* (1999)
Peas	Na-P1	Serine	*H. armigera*	Charity *et al.* (1999)
Oilseed rape	MTI-2	Serine	*Plutella xylostella*	De Leo *et al.* (2001)
Potato	Tomato Cathepsin d inhibitor	Cystain	*Leptinotarsa decemlineata*	Brunelle *et al.* (2004)
Barley	Cystain HvCPI-1 C68	Cystain	*L. decemlineata*	Alvarez-Alfageme *et al.* (2007)
Rice	Barley T1	Serine	*Sitophilus oryzae*	Alfonso-Rubi *et al.* (2003)
	Soybean T1	Serine	*Nilaparvata lugens*	Li *et al.* (2005)
			Cnaphalocrocis medinalis	Li *et al.* (2005)
	M-P1	Serine	*Chilo suppresssalis*	Vila *et al.* (2005)
	Cowpea T1	Serine	*C. medinalis*	Han *et al.* (2007)
	Brassica juncea T1	Serine	*Spodoptera litura*	Mandal *et al.* (2002)
	Mustard T1-2 De	Serine	*Spodoptera littoralis*	Leo and Gallerani (2002)
	Tobacco T1	Serine	*S. litura*	Srinivasan *et al.* (2009)
			H. armigera	Srinivasan *et al.* (2009)
	Potato Pin II	Serine	*Sesamia inferens*	Duan *et al.* (1996)
	Sporamin + Taro cystain	Serine + Cysteine	*H. armigera*	Senthilkumar *et al.* (2010)
	Buckwheat serine PI	Serine	*Trialeurodes vaporariorum*	Khadeeva *et al.* (2009)
Tomato	Taro cystatin	Cystatin	*Meloidogyne incognita*	Chan *et al.* (2010)
Cotton	Na-PI and St Pin1A	Serine	*H. armigera*	Dunse *et al.* (2010)

Gene Pyramiding

Application of integrated pest management strategies is essential in order to delay insect adaptation to insecticidal protein whether it is

Bt or PI. Researchers have developed transgenic plants carrying two or more insect resistant genes referred to as pyramiding of genes. The unique feature of gene pyramiding defines the long-term insect control. Thus, simultaneous expression of two or more resistant traits will prove an effective way to inhibit the insect adaptation. Complete resistance to diamondback moth (*Plutella xylostella*) is observed in transgenic brocolli lines expressing Cry1A and Cry1C gene (Cao *et al.*, 2002). In an interesting study performed by Dunse *et al.* (2010) where transgenic cotton plants were developed expressing Pin I and Pin II family genes in combination. *N. alata* protease inhibitor (NaPI) belong to Pin II family contains four trypsin inhibitory domains and two chymotrypsin domains. They have observed the induction of resistant chymotypsin in *Helicoverpa* species upon exposure to NaPI. Thus in order to provide complete protection from *H. armigera* and *H. punctigera* Pin I type PI from *Solanum tuberosum* (St-Pin1A) is used in combination with NaPI which ensured full protection (Dunse *et al.*, 2010). In the present chapter though much emphasis has been given on inhibitory role of protease inhibitor against insects, their effect on phytopathogenic microorganisms is also reported in various studies. Combined expression of sporamin (trypsin inhibitor) and cysteine PI from taro is expressed in transgenic tobacco plants exhibiting inhibition of *H. armigera* and other phytopathogens, *Erwinia carotovora* and *Pythium aphanidermatum*. Thus, the selection of two or more different traits is essentially critical to develop an efficient resistance mechanism in plants. It is evident from the above studies that gene pyramiding in transgenic plants is an effective tool to control long-term insect adaptation.

OTHER APPLICATIONS AS ALTERNATIVE TO TRANSGENICS

Development of transgenics has been proven the most promising approach in recent times to develop insect resistant crops. However, existence of other alternatives is reported especially when various regulatory issues have been posed for the employment of transgenics. Most insecticides are produced as emulsifiable concentrates, suspension concentrates, wettable granules or wettable powders. Previously chemical and botanical insecticidal formulations were prepared to control lepidopteran infestation, e.g. cyalothrin microemulsion formulation, Bt formulation, mixture of leaf extracts, etc. (Mahesh and Men, 2008; Dadang *et al.*, 2009; Zhao *et al.*, 2009). Since last five decades, Bt spray formulations are being used in field to control several insect pests. Bt sprays are commercially available with different brand names like Delfin, Dipel, etc. Applicability of similar formulations can be extended by using other bio-based defensive molecules such as protease inhibitors. The possible applications for these systems intended to form products with an extended shelf life and easy to use for the delivery of active ingredients. Attempts to develop insecticide formulations were centred on its non-toxicity to plants, maintaining active ingredient concentration and stability. These systems allow the flexibility in the use of various defensive molecules in the same system or incorporation of more than one defensive molecule for the efficient plant protection. Until now no reports on the use of protease inhibitor spray formulation are available. Development of PI based formulation will be fruitful in order to provide resistance against plant pests.

LIMITATIONS OF PROTEASE INHIBITORS

Upon single herbivore attack, several intricate activities occur in plants in order to protect themselves. Owing to the dynamics and complexity between plants and herbivores, it is complicated to study their interaction from single outlook. Plants accumulate defensive compounds not only in herbivore-damaged leaves but also in distal intact (systemic) leaves. Clearly, a signal travels to other parts of the plants and transmits an herbivory alert. Although this systemic response was found almost forty years ago, the identity of the mobile signal remains unknown (Dieter Hermsmeier and Baldwin, 2001). Thus in an attempt to generate a transgenic plant, effect of the insertion of gene may bring about various physiological changes in plants via cross signalling which is needed to be considered and regulated.

Insect Response to PIs

Darwin's theory of "survival of the fittest" defines the molecular, phenotypical, and physiological changes to enable an organism to compete with other species in the population in response to their changing environment. This theory is not only applied to higher organisms but lower species as well. In due course of time, plants and insects have coevolved in order to survive and to obtain maximum benefit of their habitat. Plants have developed various complicated mechanisms to protect themselves from genetical and molecular perspectives, of which synthesizing protease inhibitor is one such mechanism that is detailed above. Insects, however, remain undefeated, and have evolved to counter adapt plant responses. Despite the dramatically raised plant PI levels in defence against herbivorous insects, the effects on herbivore mortality or development are often relatively minor or even absent. Yet, even small effects on development may still be ecologically relevant, and affect the next generation with severely reduced fecundity (De Leo *et al.*, 2002). However, insects respond to the exposure by PIs by inducing over expression of those proteases that are of same class but normally are in less abundance or over expression of target proteases (Cloutier *et al.*, 2000). *H. armigera* when allowed to feed on giant taro PI (GtPI), decrease in total proteinase activity was measured but analysis of spectrum of digestive proteases in midgut revealed the increase in the chymotrypsin and elastase activity and decrease in the trypsin activity (Wu *et al.*, 1997). Insects have developed other strategies as well to combat the inhibitory effect of PIs such as use of presence of diet induced weakly sensitive proteases that are able to degrade PIs (Brunelle *et al.*, 2004; Gurden *et al.*, 2004; Koo *et al.*, 2008). Insects may also make use of complex proteases belonging to different functional classes that will compensate for the lost activity of proteases (Kiggundu *et al.*, 2010). Small sequence variation or mutation in proteases may lead to the inefficient binding of PIs with proteases contributing PI resistance (Dunse *et al.*, 2010). Such evolutionary mechanism allows insects to adapt on previously reported as potent PIs.

TRANSGENIC ISSUES

Taking into account the different modes of action and activity ranges of recombinant pesticidal proteins expressed in transgenic crops, the striking complexity of biotic interactions and food web relationships in agroecosystems, and the random insertion of transgene sequences in recipient plant genomes, three main factors generally determine the environmental impact of a given pest- (or pathogen) resistant transgenic crop. (i) the overall efficiency of the introduced resistance trait against the target herbivore (or pathogen), (ii) the activity range or functional specificity of the recombinant trait, and (iii) the (bio)chemical composition and physiological status of the host plant following transgene insertion and expression.

Overall Efficiency of the Recombinant Trait

Currently, most transgenic crops grown in agricultural fields worldwide are used for weed or herbivorous insect control (James *et al.*, 2009). By definition, any pest control measure adopted in the field, whether relying on transgenic crop lines or not, may exert direct and indirect effects on microbial, animal, and plant communities. The direct effects of insect-resistant (e.g. Bt toxin-expressing) plants are limited essentially to target pest populations, but the high pesticidal efficiency of these plants may indirectly impact the fitness of non-target organisms and the overall organization of non-target populations in the field. A well-known example of this is the negative impact of Bt toxin (Cry protein)-expressing plants on arthropod parasitoids and predators provided with 'poor quality' herbivore hosts or preys suffering recombinant toxin ingestion (Naranjo *et al.*, 2009). Another example is the adjustment of secondary herbivore, auxiliary carnivore, and soil detritivore arthropod populations in agricultural fields due to the efficient repression of primary herbivore pests with Bt toxin-expressing lines (Marvier *et al.*, 2007; Cloutier *et al.*, 2008).

Functional Specificity of the Recombinant Trait

Herbivore pest resistance traits introduced into crop plant genomes are usually selected for the control of a specific pest, but direct unintended effects on non-target organisms cannot be excluded (Groot *et al.*, 2002). The monarch butterfly controversy and the reported toxicity of Cry1A and Cry2A toxins against different lepidopteran insects are examples of the preferential, but non-exclusive action of pesticidal proteins against specific target pests (Sims *et al.*, 1995,1996; Van Frankenhuyzen *et al.*, 2002).

Alteration of the Host Plant's Characteristics

The heterologous expression of metabolic regulators such as protease inhibitors in transgenic crops also raises questions about their effects in *planta* and their resulting impact on the composition and physiology of the host plant, which represents a central element in the whole food chain (Vishal-Shah *et al.*, 2000; Khalf *et al.*, 2010). Unlike Cry toxins, acting on specific membrane protein receptors in the digestive tract of target herbivores (Pigott *et al.*, 2007), protease inhibitors may interact with plant endogenous protease targets structurally and functionally related to herbivore digestive proteases (Goulet *et al.*, 2008). Plants possess diverse type of proteases, which function endogenously in plant metabolism and growth. Thus, transgenic plants harbouring proteinase inhibitors of different functional classes though will be efficient in inhibiting insects, their effect on plant native proteases is of major concern. Further, it is also essential to assess their effect on humans and other organisms.

CONCLUSIONS

Employment of protease inhibitors highlights an effective strategy in insect crop management owing to its potential impact on agronomics. Several studies have reported the potency of protease inhibitors in retarding growth and development of insects but only few studies have demonstrated the generation of novel inhibitors by protein engineering for enhancing specificity and potency against mixture of gut proteases. Diverse type of isoforms or variants of proteases inhibitors have been reported which display sequence and functional differences that ultimately leads to the assorted interaction between the two. Selection of variants that exhibits increased potency against target proteases and less effectivity against non-target proteases can be done by identifying relevant target sites for site directed mutagenesis by performing rational designing based on 3-D models of protease inhibitors: proteases complexes (Duan *et al.*, 1996). Characterization of the molecular basis of resistance of an insect trypsin/ chymotrypsin to the plant proteinase inhibitor is required (Dunse *et al.*, 2010). Understanding how insects develop resistance to PIs is fundamental to the future rational design of PIs for the protection of agronomically important plants against insect pests. This knowledge will allow the development of a strategy to produce lepidopteran resistant transgenic plants that will incorporate multiple inhibitors with structural diversity that target different enzyme specificities. Constructing synthetic PIs with high specificity will also be beneficial in terms of delaying resistance by insects. *In vitro* molecular evolution schemes involving recombination or random mutagenesis in functionally relevant regions of the gene (protein) sequence, combined with high throughput screening approaches such as molecular phage display for the selection of improved inhibitor variants, probably represent an effective way to generate functionally diverse inhibitor populations. Insect response to plant PIs by inducing insensitive or different functional class of proteases can be understand by the knowledge of insect responsive promoters which will not only allow researchers to engineer insect resistance, but also to analyze the contribution of individual transgenes to the darwinian fitness of natural plant populations exposed to herbivory.

ACKNOWLEDGEMENTS

The Council of Scientific and Industrial Research (CSIR), Government of India, New Delhi supported this project under network project grants to National Chemical Laboratory, Pune (NWP0003).

Neha Khandelwal and Rakesh S. Joshi acknowledge CSIR for the research fellowships.

REFERENCES

Arai, S., Watanabe, H., Kondo, H., Emori, Y. and Abe, K. (1991) Papain inhibitory activity of oryzacystatin, a rice seed cysteine proteinase inhibitor, depends on the central Gln-Val-Val-Ala-Gly region conserved among cystatin superfamily members. *Biochemistry*, **109**, 294–298.

Atiwetin, P., Harada, S. and Kamei, K. (2006) Serine protease inhibitor from Wax Gourd (*Benincasa hispida*) seeds. *Biosci. Biotech. Biochem.*, **70,** 743–745.

Atkinson, A.H., Heath, R.L., Simpson, R.J., Clarke, A.E. and Anderson, M.A. (1993) Proteinase inhibitors in *Nicotiana alata* stigmas are derived from a precursor protein, which is processed into five homologous inhibitors. *The Plant Cell*, **5**, 203–213.

Barrette-Ng, I.H., Ng, K.K., Cherney, M.M. and Pearce, G. (2003) Structural basis of inhibition revealed by a 1:2 complex of the two-headed tomato inhibitor-II and subtilisin Carlsberg. *J. Biol. Chem.*, **278**, 24062–24071.

Barta, E., Pintar, A. and Pongor, S. (2002) Repeats with variations: accelerated evolution of the Pin2 family of proteinase inhibitors. *Trends Genet.*, **18**, 600–603.

Bennett, R.N. and Wallsgrove, R.M. (1994) Secondary metabolites in plant defense- mechanisms. *New Phytol.*, **127**, 617–633.

Bode, W., Greyling, H.J., Huber, R., Otlewski, J., and Wilusz T. (1989) The refined 2.0 A X-ray crystal structure of the complex formed between bovine beta-trypsin and CMTI-I, a trypsin inhibitor from squash seeds (Cucurbita maxima). Topological similarity of the squash seed inhibitors with the carboxypeptidase A inhibitor from potatoes. *FEBS Lett.*, **242**, 285–292.

Boulter, D., Gatehouse, J.A., Gatehouse, A.M.R. and Hilder, V.A. (1990) Genetic engineering of plants for insect resistance. *Endeavour*, **14**, 185–190.

Broadway, R.M. and Duffey, S.S. (1986) Plant proteinase inhibitors: Mechanism of action and effect on the growth and digestive physiology of larval *Heliothis zea* and *Spodoptera exigua*. *J. Insect Physiol.*, **32**, 827–833.

Brunelle F, Cloutier, C. and Michaud D. (2004) Colorado potato beetles compensate for tomato cathepsin D inhibitor expressed in transgenic potato. *Arch. Insect Physiol. Biochem.*, **55**, 103–113.

Brzin, J., Popovi, T., Drobnic-Kosorok, M., Kotnik, M. and Turk, V. (1988) Inhibitors of cysteine proteinases from potato. *Biol. Chem. Hoppe-Seyler*, **369**, 233–238.

Cai, M., Gong, Y., Kao, J. and Krishnamoorthi, R. (1995) Three-dimensional solution structure of *Cucurbita maxima* trypsin inhibitor-V determined by NMR spectroscopy. *Biochemistry*, **34**, 5201–5211.

Cao, J., Zhao, J.Z., Tang, J., Shelton, A. and Earle, E. (2002) Broccoli plants with pyramided cry1Ac and cry1C Bt genes control diamond back moths resistant to Cry1A and Cry1C proteins. *Theor. Appl. Genet.*, **105**, 258–264.

Ceciliani F, Bortolotti, F., Menegatti, E., Ronchi S, Ascenzi, P. and Pakmieri, S. (1994) Purification, inhibitory properties, amino acid sequence and identification of the reactive site of a new serine proteinase inhibitor from oil-rape (*Brassica napus*) seed. *FEBS Lett.*, **342**, 221–224.

Charity, J.A., Anderson, M.A., Bittisnich, D.J., Whitecross, M. and Higgins, T.J.V. (1999) Transgenic tobacco and peas expressing a proteinase inhibitor from *Nicotiana alata* have increased insect resistance. *Mol. Breed.*, **5**, 357–365.

Chen, P., Rose, J., Love, R., Wei, C.H. and Wang, B.C. (1992) Reactive sites of an anticarcinogenic Bowman-Birk proteinase inhibitor are similar to other trypsin inhibitors. *J. Biol. Chem.*, **267**, 1990–1994.

Christeller, J. and Liang, W. (2005) Plant serine protease inhibitors. *Protein Peptide Lett.*, **12**, 439–447.

Cloutier C., Jean, C., Fournier, M., Yelle, S. and Michaud, D. (2000) Adult Colorado potato beetles, *Leptinotarsa decemlineata* compensate for nutritional stress on oryzacystatin I-transgenic potato plants by hypertrophic behaviour and over-production of insensitive proteinases. *Arch. Insect Physiol. Biochem.*, **44**, 69–81.

Cloutier, C., Boudreault, S. and Michaud, D. (2008) Impact de pommes de terre re´ sistantes au doryphore sur les arthropodes non vise´ s: une me´ ta-analyse des facteurs possiblement en cause dans l'e´ chec d'une plante transgenique Bt. *Cahiers Agric.*, **17**, 388–394.

Dadang, Fitriasari, E.D. and Prijono, D. (2009) Effectiveness of two botanical insecticide formulations to two major cabbage insect pests on field application. *J. ISSAAS*, **15**, 42–51.

Dahl, S.W., Rasmussen, S. K. and Hejgaard, J. (1996) Heterologous expression of three plant serpins with distinct inhibitory specificities. *J. Biol. Chem.*, **271**, 25083–25088.

De Leo, F. and Gallerani, R. (2002) The mustard trypsin inhibitor 2 affects the fertility of *Spodoptera littoralis* larvae fed on transgenic plants. *Insect Biochem. Mol. Biol.*, **32**, 489–496.

Dieter Hermsmeier, U.S., and Baldwin, I.T. (2001) Molecular interactions between the specialist herbivore *Manduca sexta* (Lepidoptera, Sphingidae) and its natural host *Nicotiana attenuata*. I. Large-scale changes in the accumulation of growth- and defense-related plant mRNAs. *Plant Physiol.*, **125**, 683–700.

Duan, X., Li, X., Xue, Q., Abo-EI-Saad, M., Xu, D. and Wu, R. (1996) Transgenic rice plants harboring an introduced potato proteinase inhibitor II gene are insect resistant. *Nat. Biotechnol.*, **14**, 494–498.

Dunse, K.M., Kaas, Q., Guarino, R.F., Barton, P.A., Craik, D.J. and Anderson, M.A. (2010) Molecular basis for the resistance of an insect chymotrypsin to a potato type II proteinase inhibitor. *PNAS*, **107**, 15016–15021.

Fan, S.G., and Wu, G.J. (2005) Characteristics of plant proteinase inhibitors and their applications in combating phytophagous insects. *Bot. Bull. Acad. Sin.*, **46**, 273–292.

Favel, A., Nguyen, D.L., Coletti-Previero, M.A. and Castro, B. (1989) Active site chemical mutagenesis of *Ecballium elaterium* Trypsin Inhibitor II: New microproteins inhibiting elastase and chymotrypsin. *Biochem. Biophys. Res. Commun.*, **162**, 79–82.

Ferry, N., Edwards, M.G., Gatehouse, J.A. and Gatehouse, A.M. (2004) Plant-insect interactions: Molecular approaches to insect resistance. *Curr Opin Biotech.*, **15**, 155–161.

Goulet, M.C., Dallaire, C., Vaillancourt, L.P., Khalf, M., Badri, M.A., Preradov, A., Duceppe, M.O., Goulet, C., Cloutier, C. and Michaud, D. (2008) Tailoring the specificity of a plant cystatin toward herbivorous insect digestive cysteine proteases by single mutations at positively selected amino acid sites. *Plant Physiol.*, **146**, 1010–1019.

Graham, J.S. and Ryan, C.A. (1997) Accumulation of metallocarboxypeptidase inhibitor in leaves of wounded potato plants. *Biochem. Biophys. Res. Commun.*, **101**, 1164–1170.

Groot, A.T. and Dicke, M. (2002) Insect-resistant transgenic plants in a multi-trophic context. *Plant J.*, **31**, 387–406.

Gruden K, Kuipers, A.G., Guncar, G., Slapar, N., Strukelj, B. and Jongsma, M.A. (2004) Molecular basis of Colorado potato beetle adaptation to potato plant defence at the level of digestive cysteine proteinases. *Insect Biochem. Mol. Biol.*, **34**, 365–375.

Hanur, V.S. (2008) Bt resistance and monophagous pests: Handling with prudence. *Curr. Sci.*, **95**, 449–451.

He, K., Wang, Z., Bai, S., Zheng, L., Wang, Y. and Cui, H. (2006) Efficacy of transgenic Bt cotton for resistance to the Asian corn borer (Lepidoptera: Crambidae). *Crop Prot.*, **25**, 167–173.

Heath R.L., McDonald, G., Christeller, J.T., Lee, M., Bateman, K., West, J., Heeswijck, R.V. and Anderso, M.A. (1997) Proteinase inhibitors from *Nicotiana alata* enhance plant resistance to insect pests. *J. Insect Physiol.*, **43**, 833–842.

Heath, R.L., Barton, P.A., Simpson, R.J., Reid, G.E., Lim, G. and Anderson, M.A. (1995) Characterization of the protease processing sites in a multidomain proteinase inhibitor precursor from *Nicotiana alata*. *Eur. J. Biochem.*, **230**, 250–257.

Heath, R.L., McDonald, G., Christeller, J.T., Lee, M., Bateman, K., West, J., Heeswijck, R.H. and Anderson, M.A. (1997) Proteinase inhibitors from *Nicotiana alata* enhance plant resistance to insect pests. *J. Insect Physiol.*, **43**, 833–842.

Heitz, A., Hernadez J.F., Gagnon, J., Hong, T.T., Pham, T.T., Nguyen, T.M., Le-Nguyen, D. and Chiche L. (2001) Solution structure of the squash trypsin inhibitor MCoTI-II. A new family for cyclic knottins. *Biochemistry*, **40**, 7973–7983.

Hilder, V.A., Gatehouse, A.M.R., Sheerman, S.E., Barker, R.F. and Boulder, D. (1987) A novel mechanism of insect resistance engineered into tobacco. *Nature*, **300**, 160–163.

Hollander-Czytko, H. Andersen, J.L. and Ryan, C.A. (1985) Vacuolar localization of wound-induced carboxy peptidase inhibitor in potato leaves. *Plant Physiol.*, **78**, 76–79.

Huma, H. and Fazili, K.M. (2007) Plant protease inhibitors: a defense strategy in plants. *Biotechnol. Mol. Biol. Rev.*, **2**, 68–85.

Huntington, J.A., Read, R.J. and Carrell, R.W. (2000) Structure of serpin protease complex shows inhibition by deformation. *Nature*, **407**, 923–926.

Iwanaga, S., Nagata, R., Miyamoto, A., Kouzuma, Y., Yamasaki, N. and Kimura, M. (1999) Conformation of the primary binding loop folded through an intramolecular interaction contributes to the strong chymotrypsin inhibitory activity of the chymotrypsin inhibitor from *Erythrina variegata* seeds. *J. Biochem.*, **126**, 162–167.

Iwanaga, S., Yamasaki, N. and Kimura, M. (1998) Chymotrypsin inhibitor from *Erythrina variegata* seeds: involvement of amino acid residues within the primary binding loop in potent inhibitory activity toward chymotrypsin. *J. Biochem.*, **124**, 663–669.

James, C. (2009) *Global Status of Commercialized Biotech/GM Crops:* 2009. International Service for the Acquisition of Agri-biotech Applications (ISAAA), Ithaca, New York. ISAAA Brief No. 41.

Johnson, R., Narvaez, J., An, G. and Ryan, C.A. (1989) Expression of proteinase inhibitors I and II in transgenic plants effects on natural defense against *Manduca sexta* larvae. *Proc. Natl. Acad. Sci. USA*, **86**, 9871–9875.

Jongsma, M.A. and Beekwilder, J. (2008) Plant protease inhibitors: Functional evolution for defense. In: A. Schaller (ed.), *Induced Plant Resistance to Herbivory*, Springer, BV, pp. 235–251.

Kang, J.H., Wang, L., Giri, A.P. and Baldwin, I.T. (2006) Silencing threonine deaminase and JAR4 in *Nicotiana attenuata* impairs jasmonic acid-isoleucine-mediated defenses against *Manduca sexta*. *Plant Cell*, **18**, 3303–3320.

Kessler, A. and Baldwin, I.T. (2001) Defensive function of herbivore induced plant volatile emissions in nature. *Science*, **291**, 2141–2144.

Kessler, A. and Baldwin, I.T. (2002) Plant responses to insect herbivory: the emerging molecular analysis. *Annu. Rev. Plant Biol.*, **53**, 299–328.

Khalf, M., Goulet, C., Vorster, J., Brunelle, F., Anguenot, R., Fliss, I. and Michaud, D. (2010) Tubers from potato lines expressing a tomato Kunitz protease inhibitor are substantially equivalent to parental and transgenic controls. *Plant Biotechnol. J.*, **8**,155–169.

Kiggundu, A., Muchwezi, J., Van der Vyver, C., Viljoen, A., Vorster, J., Schluter, U., Kunert, K. and Michaud, D. (2010) Deleterious effects of plant cystatins against the banana weevil *Cosmopolites sordidus*. *Arch. Insect Biochem. Physiol.*, **73**, 87–105.

Koo, Y.D., Ahn, J.E., Salzman, R.A., Moon, J., Chi, Y.H., Yun, D.J., Lee, S.Y., Koiwa, H. and Zhu-Salzman, K. (2008) Functional expression of an insect cathepsin B-like counter-defence protein. *Insect Mol. Biol.*, **17**, 235–245.

Lawrence, P.K. (2002) Plant protease inhibitors in control of phytophagous insects. *Electron. J. Biotechnol.*, **5**, 93–109.

Liang, W.A., MacManus, M.T. and Allan, A.C. (2002) Protein-protein interaction in plant. *Annu. Plant Rev.*, **7**, 77–119.

Lin, G., Bode, W., Huber, R., Chi, C. and Engh, R.A. (1993) The 0.25 nm X-ray structure of the Bowman-Birk type inhibitor from mung bean in ternary complex with porcine trypsin. *Eur. J. Biochem.*, **212**, 549-555.

Ling, M.H., Qi, H.Y., and Chi, C.W. (1993) Protein, cDNA and genomic DNA sequences of the towel gourd trypsin inhibitor, A squash family inhibitor, *J. Biol. Chem.*, **268**, 810–814.

Longstaff, C., Campbell, A.F. and Fersht, A.R. (1990) Recombinant chymotrypsin inhibitor 2: expression, kinetic analysis of inhibition with α-chymotrypsin and wild-type and mutant subtilisin BPN', and protein engineering to investigate inhibitory specificity and mechanism. *Biochemistry*, **29**, 7339–7347.

Luckett, S., Garcia, R.S., Barker, J.J., Konare, A.V., Shewry, P.R., Clarke, A.R. and Brady, R.L. (1999) High-resolution structure of a potent, cyclic proteinase inhibitor from sunflower seeds. *J. Mol. Biol.*, **290**, 525–533.

Mahesh, P. and Men, U.B. (2008) Effect of Bt formulations on yield of brinjal by managing *Leucinodes orbonalis*. *Anal. Plant Prot. Sci.*, **16**, 485–547.

Mahoney, W.C., Hermodson, M.A., Jones, B., Powers, D.D., Corfman, R.S. and Reeck, G.R. (1984) Amino acid sequence and secondary structural analysis of the corn inhibitor of trypsin and activated Hageman Factor. *J, Biol. Chem.*, **259**, 8412–8416.

Markus, H., Giri, A.P., Kaur, H. and Baldwin, I.T. (2010) Serine protease inhibitors specifically defend *Solanum nigrum* against generalist herbivores but do not influence plant growth and development. *Plant Cell*, **22**, 4158–4175.

Marvier, M., McCreedy, C., Regetz, J. and Kareiva, P. (2007) A meta-analysis of effects of Bt cotton and maize on non-target invertebrates. *Science*, **316**, 1475– 1477.

Mayuri, K. and Mikunthan, G. (2009) Damage pattern of cabbage flea beetle, *Phyllotreta cruciferae* (Goeze) (Coleoptera: Chrysomelidae) and its associated hosts of crops and weeds. *Am-Eur. J. Agric. Environ. Sci.*, **6**, 303–307.

McPhalen, C.A. and James, M.N.G. (1988) Structural composition of two serine proteinase-protein inhibitor complexes: eglin-c-subtilisin Carlsberg and CI-2 subtilisin Novo. *Biochemistry*, **27**, 6582–6598.

Mishra, M., Tamhane, V.A., Khandelwal, N., Kulkarni, M.J., Gupta, V.S. and Giri, A.P. (2010) Interaction of recombinant CanPIs with *Helicoverpa armigera* gut proteases reveals their processing patterns, stability and efficiency. *Proteomics*, **10**, 2845–2857.

Moura, D.S. and Ryan, C.A. (2001) Wound-inducible proteinase inhibitors in pepper. Differential regulation upon wounding, systemin and methyl jasmonate. *Plant Physiol.*, **126**, 289–298.

Naranjo, S.E. (2009) Impacts of Bt crops on non-target invertebrates and insecticide use patterns. *CAB Rev.: Persp. Agric, Vet. Nutr. Nat. Resour.*, **4**, 1–23.

Nguyen, D.L., Heitz, A., Chiche, L., Castro, B., Boigegrain, R.A., Favel, A. and Coletti-Previero, M.A. (1990) Molecular recognition between serine proteases and new bioactive microproteins with a knotted structure. *Biochimie*, **72**, 431–435.

Nielsen, K.J., Heath, R.L., Anderson, M.A. and Craik, D.J. (1995) Structures of a series of 6-kDa trypsin inhibitors isolated from the stigma of *Nicotiana alata*. *Biochemistry*, **34**, 14304–14311.

Orozco-Cardenas, M. and Ryan, C.A. (1999) Hydrogen peroxide is generated systemically in plant leaves by wounding and systemin via the octadecanoid pathway. *Proc. Nat. Acad. Sci. USA*, **96**, 6553–6557.

Orozco-Cardenas, M.L., Narvaez-Vasquez, J. and Ryan, C.A. (2001) Hydrogen peroxide acts as a second messenger for the induction of defense genes in tomato plants in response to wounding, systemin, and methyl jasmonate. *Plant Cell*, **13**, 179–191.

Østergaard, H., Rasmussen, S.K., Roberst, T.H. and Hejgaard, J. (2000) Inhibitory serpins from wheat grain with R active centers resembling glutamine-rich repeats of prolamin storage proteins. *J. Biol. Chem.*, **275**, 33272–33279.

Patankar, A.G., Giri, A.P. Harsulkar, A.M., Sainani, M.N., Deshpande, V.V. Ranjekar, P. K. and Gupta, V.S. (2001) Complexity in specificities and expression of *Helicoverpa armigera* gut proteinases explains polyphagous nature of the insect pest. *Insect Biochem. Mol. Biol.*, **31,** 453–464.

Pigott, C.R. and Ellar, D.J. (2007) Role of receptors in *Bacillus thuringiensis* crystal toxin activity. *Microbiol. Mol. Biol. Rev.*, **71**, 255–281.

Rao, R.G.V., Rao, R.V. and Nigam, S.N. (2010) *Post-Harvest Insect Pests of Groundnut and their Management.* International Crops Research Institute for the Semi-Arid Tropics, Patancheru, Andhra Pradesh, India.

Reddy, K.V.S. and Zehr U.B. (2004) Novel strategies for overcoming pests and diseases in India. *Proc. 4th Int. Crop Sci. Congr.*, 1-8.

Ritonja, A., Krizaj, I., MeSko, P., Kopitar, M., Lucovnik, P., Strukelj, B., Pungercar, J., Buttle, D. J., Barrett, A. J. and Turk, V. (1990) The amino acid sequence of a novel inhibitor of cathepsin D from potato. *FEBS Lett.*, **267**, 13–15.

Rojo, E., Solano, R. and Sanchez-Serrano, J.J. (2003) Interactions between signaling compounds involved in plant defence. *J. Plant Growth Regul.*, **22,** 82–98.

Ryan, C.A. and Balls, A.K. (1962) An inhibitor of chymotrypsin from *Solanum tuberosm* and its behavior toward trypsin. *Proc. Natl. Acad. Sci.*, **48,** 1839–1844.

Scanlon, M.J., Lee, M.C., Anderson, M.A. and Craik, D.J. (1999) Structure of a putative ancestral protein encoded by a single sequence repeat from a multidomain proteinase inhibitor gene from *Nicotiana alata. Struct. Fold. Des.*, **7**, 793–802.

Schluter, U., Benchabane, M., Munger, A., Kiggundu, A., Vorster, J., Goulet, M.C., Cloutier, C. and Michaud, D. (2010) Recombinant protease inhibitors for herbivore pest control: A multitrophic perspective. *J. Exp. Bot.*, **61**, 4169–4183.

Sharma, M., Charak K.S. and Ramanaiah T.V. (2003) Agricultural biotechnology research in India: Status and policies. *Curr. Sci.*, **84**, 297–302.

Showler, A.T. (2001) *Spodoptera exigua* ovisposition and larval feeding preferences for pigweed, *Amaranthus hybridus*, over squaring cotton, *Gossypium hirsutum*, and a comparison of free amino acids in each host plant. *J. Chem. Ecol.*, **27**, 2013–2028.

Sims, S.R. (1995) *Bacillus thuringiensis* var. *kurstaki* [CryIA(c)] protein expressed in transgenic cotton: effects on beneficial and other nontarget insects. *Southwest. Entomol.*, **20**, 493–500.

Sims, S.R. (1997) Host activity spectrum of the CryIIA *Bacillus thuringiensis* subsp. *kurstaki* protein: effect on Lepidoptera, Diptera, and non-target arthropods. *Southwest. Entomol.*, **22**, 395–404.

Sin, S.F. and Chye, M.L. (2004) Expression of proteinase inhibitor II proteins during floral development in *Solanum americanum*. *Planta*, **219**, 1010–1022.

Song, H.K. and Suh, S.W. (1998) Kunitz-type soybean trypsin inhibitor revisited: refined structure of its complex with porcine trypsin reveals an insight into the interaction between a homologous inhibitor from *Erythrina caffra* and tissue-type plasminogen activator1. *J. Mol. Biol.*, **275**, 347–363.

Strobl, S., Muhlhahn P., Bernstein, R., Wiltscheck, R., Maskos, K., Wunderlich, M., Huber, R., Glockshuber, R. and Holak, T.A. (1995) Determination of the three-dimensional structure of the bifunctional alpha-amylase/trypsin inhibitor from ragi seeds by NMR spectroscopy. *Biochemistry*, **34,** 8281–8293.

Tamhane, V.A., Giri, A.P. Kumar, P. and Gupta, V.S. (2009), Spatial and temporal expression patterns of diverse Pin-II proteinase inhibitor genes in *Capsicum annuum* Linn. *Gene*, **442,** 88–98.

Tamhane, V.A., Giri, A.P., Sainani, M.N. and Gupta, V.S. (2007) Diverse forms of Pin-II family proteinase inhibitors from *Capsicum annuum* adversely affect the growth and development of *Helicoverpa armigera. Gene*, **403**, 29–38.

Tamura, T., Hara, K., Yamaguchi, Y., Koizumi, N. and Sano, H. (2003) Osmotic stress tolerance of transgenic tobacco expressing a gene encoding a membrane-located receptor-like protein from tobacco plants. *Plant Physiol.*, **131**, 454–462.

Tilmann, P.G. (2006) Tobacco as a trap crop for *Heliothis virescens. J. Entomol. Sci.*, **41**, 305–320.

Tur-sinal, A., Birk, Y., Gertler, A. and Rigbi, M. (1972) Basic trypsin and chymotrypsin inhibitor from groundnut (*Arachis hypogea*). *Biochim. Biophys. Acta.*, **263**, 666–672.

Van Frankenhuyzen, K. and Nystrom, C. (2002) The *Bacillus thuringiensis* toxin specificity database. Internet webpage: http:// www.glfc.cfs.nrcan. gc.ca/bacillus.

Vercammen D., Beatrice, B., van de Cotte, B., Beunens, T., Gavigan, J.A., De Rycke R.D., Brackenier, A., Inze, D., Harris, J.L. and Breusegem F.V. (2006) Serpin1 of *Arabidopsis thaliana* is a suicide inhibitor for metacaspase. *J. Mol. Biol.*, **364,** 625–636.

Visal-Shah, S., Brunelle, F. and Michaud, D. (2000) Multiple protease/ inhibitor interactions in plant-pest systems. In: D. Michaud (ed.), *Recombinant Protease Inhibitors in Plants*, Landes Bioscience, Georgetown, TX, pp. 107-113.

Wu, Y., Llewellyn, D., Mathews, A. and Dennis, E.S. (1997) Adaptation of *Helicoverpa armigera* (Lepidoptera : Noctuidae) to a proteinase inhibitor expressed in transgenic tobacco. *Mol. Breed.*, **3,** 371–380.

Zablotna, E., Kazmierczak, K., Jaskiewicz A., Stawikowski, M., Kupryszewski, G., and Rolka, K. (2002) Chemical synthesis and kinetic study of the smallest naturally occurring trypsin inhibitor SFTI-1 isolated from sun-flower seeds and its analogues. *Biochem. Biophys. Res. Commun.*, **292**, 855–859.

Zhao, F., Xia, H.-Y. and He, J.-L. (2009) Formulation design of cyhalothrin pesticide microemulsion. *Curr. Sci.*, **97**, 1458–1462.

Zhao, Q., Chae, Y.K., and Markley, J.L. (2002) NMR solution structure of ATTp, an *Arabidopsis thaliana* trypsin Inhibitor. *Biochemistry*, **41,** 12284–12296.

10

Botanical Biopesticides in Pest Management: Potential and Constraints

SURESH WALIA[1]*, DEVA KUMAR[2] AND OPENDER KOUL[3]

[1]Division of Agricultural Chemicals, Indian Agricultural Research Institute
New Delhi-110 012, India
[2]Indian Council of Agricultural Research, KAB II, New Delhi-110012, India
[3]Insect Biopesticide Research Centre, Parkash Nagar
Jalandhar-144 003, India
*E-mail: suresh_walia@yahoo.com

INTRODUCTION

Crop losses to insect pests, diseases and weeds in tropical countries like India are significantly high. In Southeast Asia estimated losses vary from 10 to 30% depending on crop, pest infestation and the environment. Such losses assume greater proportion when a new pest appears or an existing pest develops resistance to a pesticide. Since pest management in India rely mainly on synthetic pesticides, their overuse or misuse has led to health and environmental problems of various kinds. A safer alternative to chemical pesticides is therefore urgently required. According to the Environment Protection Agency (EPA) safer pesticide policy norms, safer pesticide is that which is less persistence in the environment, less toxic to non target organisms, less likely to contaminate groundwater, allow lesser exposure to humans and the environment, and can be rid off with more practical disposal technology. Natural pest management chemicals of botanical origin are the safer formulations of the plant extracts, their active ingredients or both (Koul and Dhaliwal, 2001; Koul and Walia, 2009). These include response elicitors such as insect antifeedants, repellents, semiochemicals, metamorphosis inhibitors, etc. Such plant based products offer higher level of safety to humans, animals, fish and other non-target organisms. In spite of the ever-increasing emphasis on natural pest control agents, they still account for a less than 1 per cent of total pesticide usage.

In fact, plants have developed natural protection mechanisms to defend themselves from insect attack, and as a result produce defensive secondary metabolites. Scientists throughout the world are surveying promising plants for such bioactive secondary metabolites and evaluate them for their pesticidal properties against major pests, and look for new classes of chemicals with novel modes of action. Plants with pesticide properties can be exploited either by using parts of plants in powder or as crude extracts in water or other solvents, or as purified extracts or active individual isolated compounds like nicotine, rotenone, pyrethrins, azadirachtin, etc. Lead botanicals can also be used as a key template to synthesize a chemical compound which then can be produced industrially. However, such synthetic analogues are not considered natural and have to go through the registration regulatory mechanism laid for synthetic pesticides. Today there is considerable interest among chemists to screen plant biodiversity for bioactive secondary metabolites and such phytochemicals have been comprehensively discussed in several reviews (Arnason *et al.*, 1989; Parmar and Walia, 2001; Walia *et al.*, 2001; Koul, 2004a; Walia, 2005; Walia and Parmar, 2006; Copping and Duke, 2007; Walia and Koul, 2008, Rosell *et al.*, 2008).

TRADITIONAL BOTANICAL BIOPESTICIDES

Nicotines, rotenones, pyrethrins and Sabadill alkaloids are the first generation of botanical pesticides that have been extensively used in the past to control household and agricultural pests. Nicotine extracted from *Nicotiana tabacum*, *N. rustica* and *N. glutinosa* is known for its insect control properties (Ujvary, 1999). Other nicotine analogues such as nor-nicotine and anabasine also exhibit significant insecticidal activity. It is used primarily against insects with piercing-sucking moth parts such as aphids, leafhoppers, whiteflies, thrips and mites. Because of its high mammalian toxicity and detrimental effect on human health, the use of nicotine as an insecticide has decreased tremendously. Rotenone is another botanical isolated from *Lonchocarpus*, *Tephrosia* and *Derris* species and is effective against aphids, borers and household pests. Its use had also been discontinued due to high fish and/or mammalian toxicity. Pyrethrum, the most widely used botanical insecticide is extracted from the flowers of *Tanacetum* (previously called *Chrysanthemum*) *cinerariaefolium* (pyrethrum). It is highly effective against houseflies, mosquitoes, fleas and lice. The toxins namely pyrethrins-I and II, cinerins-I and II and jasmolins-I and II have unusual insecticidal properties, most striking being the immediate knockdown or paralysis on contact which causes most flying insects to drop almost immediately upon exposure (Casida and Quistad, 1994). Because of its low toxicity to vertebrates, it finds wide acceptance worldwide. Like other natural pesticides, pyrethrins have limited stability and shelf life and under field conditions are rapidly degraded by sunlight and heat. Natural pyrethrins, among the lot exhibit high mammalian safety. Sabadilla alkaloid generally referred to as *Veratrum* alkaloid is another first generation insecticide that is extracted from (*Schoenocaulon officinale*) and other *Veratrum* species. The active ingredient veratrine with an oral LD_{50} of 4000 to 5,000 mg/kg is considered among the least toxic of botanical insecticides.

Under the paradigms of sustainable agriculture and self-reliance, efforts are being made to develop ecologically sound crop protectants from natural resources. Such plant derived second generation crop protectants alone or in combination with other botanicals, synergists, stabilizers, etc. could be considered as environment friendly option in bio-intensive management of agricultural pests and diseases. In the recent past, plant biodiversity has been extensively investigated for pest control properties and some of the potential botanicals based on them are highlighted below.

BOTANICAL BIOPESTICIDES: CURRENT SCENARIO

Plant biodiversity has been extensively exploited for the development of botanical biopesticides during couple of decades now. Plants like *Azadirachta indica* (neem), *Melia azedarach* (darek), *Allium sativum* (garlic), *Calceolaria andina*, *Anethum sowa* (dill), *Curcuma longa* (turmeric), *Eupatorium adenophorum* (crofton weed), *Pongamia glabra* (karanj), *Madhuca indica*, *Tagetes,* and several essential oil bearing plants have been reported to exhibit significant pest control properties.

Neem Biopesticides

Neem, *Azadirachta indica* (Meliaceae) is the storehouse of a large number of bioactive natural products that have a potential use in pest management, medicines, cosmetic industry and other diverse applications (Schmutterer, 1990, 2002; Koul and Wahab, 2004; Walia and

Parmar, 2005). Farmers in India have been traditionally using neem cake to control soil pests- nematodes, fungi and insects. Neem leaves and neem oil have also been used to protect stored grains. It has been an age-old practice to put dried leaves between folds of clothes to ward off moths, and insects and other pests in stored rice, wheat and other grains. Although neem seeds and leaves have been used as traditional insecticides for centuries, the scientific interest in neem as an insecticide dates back to the work of Pradhan *et al.* (1962) who reported that dilute aqueous kernel suspension (0.001%) prevented feeding of the desert and migratory locusts. It has now emerged as a single most important source of pesticides and allied products. A comprehensive account of neem research conducted at Indian Agricultural Research Institute has also been published (Walia and Parmar, 2005, Parmar *et al.*, 2008).

Neem biopesticides are compatible with other synthetic pesticides and bioagents and are harmless to non target and beneficial organisms like pollinators, honey bees, mammals and other vertebrates. They do not destroy natural predators and parasites of pests thus allowing such natural enemies to keep a check on the pest population. Unlike synthetic pesticides comprising of one or two active ingredients which are more likely to develop resistance in insect pests, neem biopesticides comprising of several active ingredients has lesser chance of developing resistance in insect pests. Neem products not only eradicate pests but also improve the fertility of the soil. In addition they also control bacterial and fungal diseases ensuring healthy vegetation. Being renewable, cost effective, biodegradable and safe to the environment, neem biopesticides can be promoted as effective tool of agricultural pest control in IPM.

Azadirachtin (AZA) from neem has been established as an active ingredient for pest management, though there are other limonoids also responsible for wide range of activities. Bioactive ingredients from neem are normally extracted in organic solvents from neem kernels and then appropriately formulated. The technology of making neem oil based emulsifiable concentrate (EC) containing 300 ppm azadirachtin contains neem oil, a commercial formulation solvent, surfactant and/or a stabilizer. Several other EC formulations containing AZA content, varying from 0.15 to 5.0%, have also been developed. An oil in water neem oil micro-emulsion, micro-emulsion forming concentrates and EW are relatively new introduction in neem formulation which are characterized by a stabilized emulsion employing water and/or minimum quantity of an organic solvent in the formulation. A large number of natural and synthetic additives have been evaluated to stabilize the otherwise photolabile azadirachtin molecule under sunlight, in solution and under accelerated storage conditions (Kumar *et al.*, 2003). Azadirachtin has also been encapsulated/entrapped to provide greater shelf life and slow release. Products based on monolithic dispersion of azadirachtin in the matrices of different polymers or embodied in capsules, providing considerable protection against its photo- and thermal degradation have been developed. A capsule matrix provides considerable protection to azadirachtin. Neem based seed coats/dressing materials using azadirachtin have also been developed to protect seeds from pest infestation during storage. A study conducted on the effect of azadirachtin and aqueous neem seed kernel concentrate on germination and seedling growth of soybean indicated that neem products are inhibitory to seedling growth (Chowdhury *et al.*, 2002). It is well known now that AZA in neem does occur as various isomers (AZA-A to AZA-K) (Fig. 10.1). A simple method has been devised for isolation, purification and characterization of AZA A, B and H which exhibit moderate nematicidal and anti-fungal activity (Sharma *et al.*, 2003). Since azadirachtin is unstable and degrades rapidly under sunlight, heat and moisture conditions, efforts have been made to stabilize the molecule by chemical modification of the labile moieties in the molecule. Azadirachtin has been hydrogenated to its more stable dihydro- and tetrahydroazadirachtin (Walia *et al.*, 2003, Sharma *et al.*, 2006). Reduced azadirachtin derivatives have since been included in the schedule of the Insecticides Act.

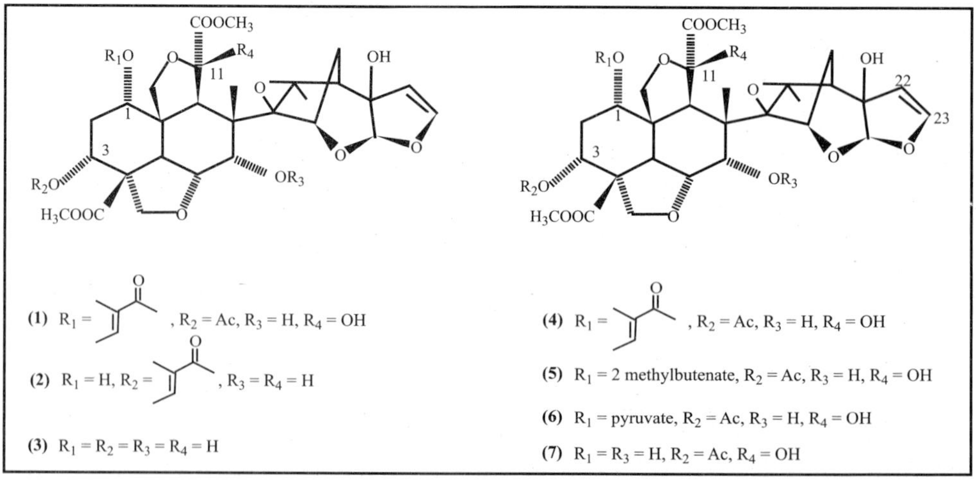

(1) R_1 = [tigloyl], R_2 = Ac, R_3 = H, R_4 = OH

(2) R_1 = H, R_2 = [tigloyl], R_3 = R_4 = H

(3) R_1 = R_2 = R_3 = R_4 = H

(4) R_1 = [2-methylbutanoyl], R_2 = Ac, R_3 = H, R_4 = OH

(5) R_1 = 2 methylbutenate, R_2 = Ac, R_3 = H, R_4 = OH

(6) R_1 = pyruvate, R_2 = Ac, R_3 = H, R_4 = OH

(7) R_1 = R_3 = H, R_2 = Ac, R_4 = OH

Fig. 10.1. Structure of azadirachtin and its various derivatives. Azadirachtin (1) and Dihydroazadirachtin (4) are most effective anti-insect compounds.

It is established now that neem or its active ingredients are active against more than 400 insect species (Koul and Wahab, 2004). The insecticidal and insect antifeedant activity of neem oil is influenced by azadirachtin content in the neem seed pressed oil (Isman *et al.*, 1990). Other limonoids from neem are also effective against various insect species (Koul, 1992; Champagne, 1992). Studies conducted on the efficacy and mode-of-action of some neem based non-azadirachtin limonoids indicated that combination of such limonoids do play an important role in the multi-component defense system of plants against lepidopteran larvae (Koul *et al.*, 2004b,c). However, some studies have indicated that neem based products could have moderate phytotoxicity against crops and weeds (Chowdhury *et al.*, 2002; Xuan *et al.*, 2004).

Neem biopesticides have been extensively used in IPM modules for the management of insect pests of various crops. Studies indicate that in rice IPM, neem treatments did not allow the occurrence of brown planthopper, green leafhopper, yellow stem borer, and gall midge to exceed economic injury levels. On an average, rice yield was 20.5% more under neem based IPM strategy under farmers' practice module. IPM module developed by National Research Centre for Groundnut (NRCG) in India indicates that neem products could be considered as economically viable alternative to synthetic pesticides as an important component of IPM of groundnut and rice. The IPM module had resulted in 24–46% control of major insect pests (white grub, leaf minor, tobacco caterpillar, gram pod borer, jassids, thrips and aphids), and diseases from 24–48%. It also resulted in average increase in yield by 19% and higher monetary returns by 40% as against the farmers practice. Aqueous extract of neem leaf, seed and oil may can be used for the management of foliar diseases like leaf spots, rust, *Alternaria* diseases and seed borne pathogens. These extracts also control defoliators, leaf miners, and sucking insects (Koul and Walia, 2009). In one of the cotton IPM modules, neem oil mixed with detergent (2.5 lit/ha) has been found effective for the management of jassids, whiteflies and aphids. Similarly mixture of neem seed kernel extract (NSKE), neem oil and detergent powder, fish oil–rosin liquid and karanj oil can be used to ward off insect pests including mealy bugs. Need based use of neem biopesticides has also been recommended for the management of whitefly in cotton IPM during vegetative, flowering and fruiting stage. Use of Neem oil 0.5% + teepol 0.1% and NSKE (5%) have been effective against cotton bollworms and sucking pests. National Centre for Integrated Pest Management in New Delhi, India has recommended the use of neem preparations for the management of various insect pests on different crops. For example NSKE crude extracts (5%) have been recommended for the control of insect pests on cotton, pigeon pea, okra, capsicum. Similarly neem cake application as soil amendment has been found effective to ward of aphids, hoppers, whitefly and mites on groundnut and vegetables such as cauliflower, cabbage and capsicum (Singh *et al.*, 2004).

Essential Oils and Volatile Constituents

Essential oils are volatile oil(s) that impart strong aromatic odour or flavour or scent due to a complex mixture of volatile monoterpenes, sesquiterpines and/or higher terpenes. Essential oils are usually extracted by steam distillation, maceration, solvent extraction and adsorption on active carbon. The aromatic characteristics of essential oils provide the plant various functions that include attracting or repelling insects, protecting the plant from heat or cold and utilizing them as defense, or anti-stress agents. Several such essential oils are known for their pest control properties (Isman, 1999; Lawerence and Reynolds, 2001; Koul *et al.*, 2008). Essential oils obtained from lemon grass, *Cymbopogon winteriana*, *Eulcalyptus globulus*; rosemary, *Rosemarinus officinalis*, vetiver, *Vetiveria zizanoides*; clove, *Eugenia caryophyllus*; basil, *Ocimum basillicum* and thyme, *Thymus vulgaris* oils are known for their pest control properties. While peppermint, *Mentha piperita* repels ants, mice, flies, lice and moths; pennyroyal, *Mentha pulegium* wards off fleas, ants, lice, mosquitoes, ticks and moths. Spearmint, *Mentha spicata* is also effective in discouraging flies. Some other essential oil bearing plants namely geranium, *Pelargonium roseum*; lavender, *Lavandula angustifolia*; peppermint, *M. piperita*; red cedarwood, *Juniperus virginiana*, *Artemisia vulgaris* and *Melaleuca leucadendron* are also effective against various pests. Composition and insect repellent activity of the essential oil of marigold, *Tagetes erecta* flower have also been reported (Ray *et al.*, 2000).

Many essential oil constituents (Fig. 10.2) exhibit feeding deterrent, insect repellent and insecticidal properties. Several terpenoids isolated from tropical Rutales have been reported to exhibit antifeedant activity (Omar *et al.*, 2007), Among the most active are the phenylpropene eugenol from cloves, *Eugenia cryophyllus*; 1,8-cineole from *Eucalyptus globules*; citronellal from lemongrass, *Cymbopogon nardus*; pulegone from pennyroyal, *Mentha pulegium* and thymol and carvacrol from thyme (*Thymus vulgaris*). Some terpenoides like thymol and carvacrol were more effective for deterring oviposoition by *Aedes aegypti* mosquitoes than N,N-diethyl-m-toluamide (DEET). 1, 8-cineole, the major constituent of *Artemisia annua* is toxic, induces feeding deterrence and reduces progeny production of *Tribolium castanaeum* (Tripathi *et al.*, 2001). Essential oil constituents such as thymol, citronellal and α-terpineol have also been found effective as feeding deterrents against tobacco cutworm, *Spodoptera litura* (Hummelbrunner *et al.*, 2001). Sweet basil, *O. basillicum* leaves and seeds are rich in essential oils which are repellent, toxic or growth inhibitory to many insects (Grainge and Ahmed, 1988). Its principal constituent eugenol repels mosquitos. Linalool, another terpenoid found in *O. canum* is responsible for the toxic effect of the dried leaves to the bruchid, *Zabrotes subfasciatus* and other storage pests (Weaver *et al.*, 1991). The plant has also been evaluated for its possible use as stored grain protectant (Hassanali and Lwande, 1989). At specified doses, limonene and linalool are nerve toxins and show fumigant action The essential oil extracted from root of sweet flag, *Acorus calamus* and its major constituent β-asarone is also known for its insecticidal and antigonadal action (Koul *et al.*, 1990; Koul, 1995). Several other compounds like ocimene from *Tagetes minuta*, α-terpineol and piperitenone need further consideration and exploitation as sources of botanical pesticides.

Curcuma longa (Zingiberaceae) commonly known as turmeric, is well known for its colouring, flavouring and digestive properties. Turmerone and dehydroturmerone, the major constituents of turmeric oil are strong repellents to stored grain pests (Su *et al.*, 1982; Jillani and Su, 1983; Jilani and Saxena, 1990). Turmeric oil is also insect repellent and insect growth inhibitor for flour beetle, *Tribolium castaneum* (Jilani *et al.*, 1998). Antifeedant activity of leaf essential oil of *C. longa* against adult and larvae of grain borer, *Rhyzopertha domestica*, rice weevil, *Sitophilus oryzae* and red flour beetle, *Tribolium castaneum* has been attributed to the presence of carvone, and dihydrocarvone (Tripathi *et al.*, 2003). Products isolated/ derived from *C. longa* (turmeric) and *Zingiber officinale* (ginger) are insect growth regulators and insect antifeedant (Agarwal *et al.*, 2000; Agarwal and Walia, 2003). α-phellandrene (70%) from turmeric leaf oil inhibited growth of *Spilosoma obliqua* and caused larval mortality (Agarwal *et al.*, 1999). The leaf oil also showed ovicidal and nymphicidal action against *Dysdercus koenigii* and exhibited moderate knockdown effect against stored grain pests (Tripathi *et al.*, 2002). Bioefficacy of azadirachtin against Bihar hair caterpillar *S. obliqua* could be enhanced using turmeric oil (Chowdhury and Walia, 2001)

Curcumene and ginger oil obtained from *Zingiber officinale* rhizomes also inhibit growth of plant pathogens like *R. solani* and several other phytopathogenic fungi (Chowdhury and Walia, 2001).

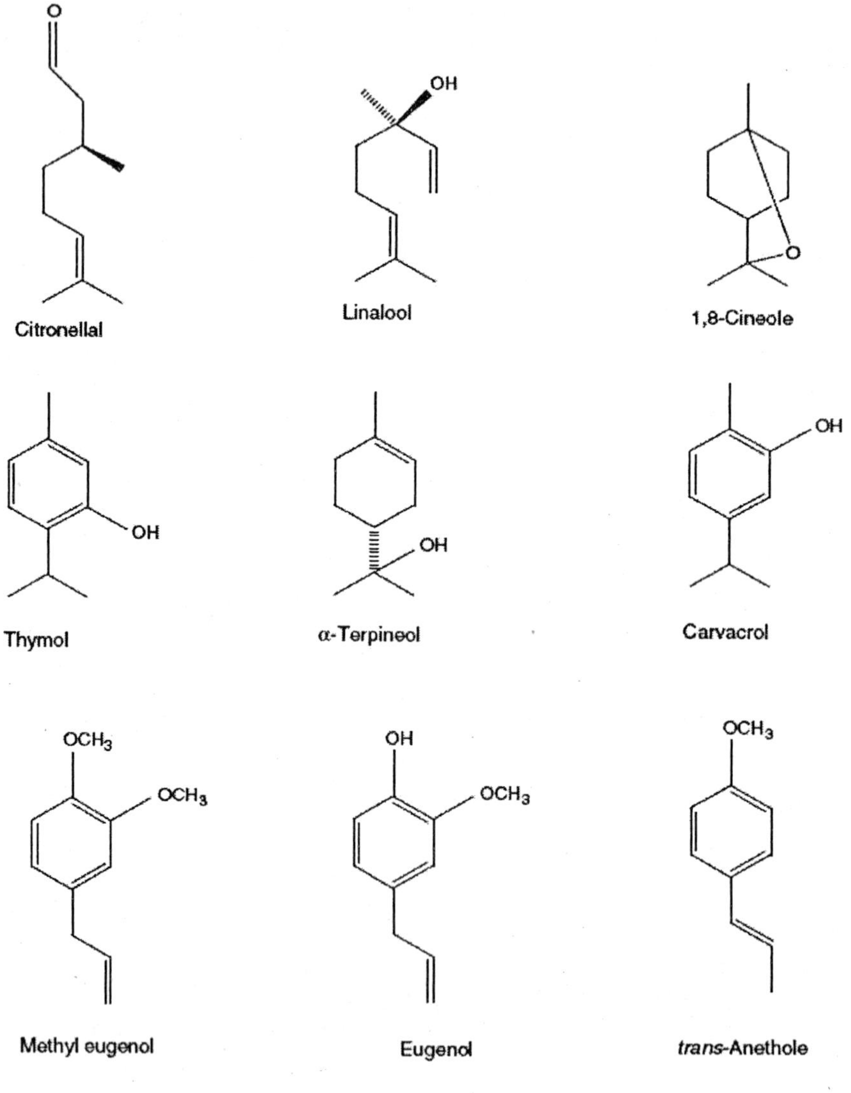
O
OH
Citronellal
Linalool
1,8-Cineole
OH
OH
OH
Thymol
α-Terpineol
Carvacrol
OCH3
OCH3
OH
OCH3
OCH3
Methyl eugenol
Eugenol
trans-Anethole

Fig. 10.2. Insecticidal compounds from essential oils.

Plant volatiles particularly derived from neem, garlic, etc. are also effective against a broad spectrum of pests presumably through a combination of contact, fumigant and repellent action. Because of their modest insect toxicity, minimal mammalian toxicity, and widespread availability, such products are considered as potential ingredients in botanical insecticides, and as potential alternatives to conventional insecticides. The alliaceous odor and pest control properties of neem seed and leaf volatiles have been attributed to the volatile organosulfur compounds. Several sulfurous constituents like di-n-propyl and propyl propenyl di-, tri-, and tetrasulphides among others have been reported from crushed neem seeds. These volatiles adversely affect oviposition, egg hatchability and viability, post-embryonic development and adult emergence of storage and other important agricultural pests (Koul, 2004b).

The composition of the neem leaf and seed volatiles obtained after steam distillation was determined by capillary gas chromatography mass spectroscopy (GCMS) as well as electrospray ionization mass spectroscopy (ESI-MS) in direct infusion mode (Khatavkar and Walia, 2005). Some of the major constituents identified by ESI-MS include tetracosane, tricosane, triacontane, triacosanol, hexacosanol, heptacosanol and dotriacontanoic acid along with trace quantities of dipropyl trisulphide and propyl propenyl tetrasulphides. Other constituents identified by GCMS include cyclic/acyclic alkyl alkenyl sulfides, trithiacyclodecanone, trithiocyclohexane, tetrathiacyclodecane and longifolene. These volatiles have been found effective as insect fumigants against the stored grain pests, *Callosobruchus maculatus* and *Tribolium castaneum*. It is believed that the cocktail of organosulphur and other constituents present in neem leaf and seed volatiles is responsible for causing adult mortality of the stored grain pests.

Bioactive organosulphur compounds have also been reported from garlic. Such bioactives have antibacterial, antifungal, nematicidal, amoebicidal, insecticidal, larvicidal, anti-termitic and insect repellent properties. It is the storehouse of a large number of bioactive molecules which include allin, allicin, garlicin, ajoine and allyl sulfides. From the aqueous ethanol extract, bis-2-propenyl trisulfide, bis-2- propenyl tetrasulfide, bis-2-propenyl pentasulfide, bis-2-propenyl thiosulfonate, 3-allyl sulfanyl-allyl ester, *trans*-sulfurous acid allyl ester and thiosulfinates have been reported (Qing *et al.*, 2002). Among the various constituents, diallyl trisulfide is the most toxic, followed by diallyl disulfide (Park and Sangchul, 2005). During storage, chemical composition of garlic oil tends to alter. While content of dimethyl thiophene, dipropyl trisulphide, methyl propyl trisulphide, and diallyl disulphide increased others like methyl thiophene, allyl thiophene and dimethyl thiophene disappeared after four months of storage. The commercial garlic product based on dilute garlic juice

(Mosquito BarrierO® at 1% concentration is not only repellent to target agricultural pests but also repel adult mosquitoes at repellency rate of 93–95%. Garlic preparations also keep away nematodes from infecting roots in the soil rhizosphere. Intercropping with garlic plants for insect control is a long established practice in tropical and sub-tropical regions.

Other Botanicals

During last decade lot of emphasis has been on the use of darekh, *Melia azedarach* (Meliaceae) allelochemicals in pest management as it contains several active limonoids. Several insect antifeedants have been reported from this plant which includes meliacarpinins, azedarachins, ammorastatins, amoorastatone, trichilins and nimbolidins. Among the meliacarpinins, 1-cinnamoyl-3-acetyl-11-methoxymeliacarpinin with antifeedant index (AI_{50}) of 50 ppm was the most active. Of the four azedarachins, azedarachin-A was twice as active (200 ppm) as other azedarachins against *Spodoptera eridania* larvae. At least thirty six limonoids including toosendanin have been isolated from stem and root bark of *Melia toosendan* (now reported as synonymous variety of *M. azedarach*) as antifeedant constituents (Nakatani *et al.*, 1999). Based on toosendanin a product has been developed and registered in China for use against fruit and vegetable pests. Cytotoxicity guided chromatographic fractionation has led to the isolation of a number of toxic tetranortriterpenoids namely meliantriol, 12-hydroxy amoorastatone, 12-hydroxy amoorastatin and 12-acetoxy amoorastatin from dried stem bark. Since these meliatoxins exhibit mammalian toxicity, it needs to be ensured that any potential product based on *M. azedarach* must be devoid of such harmful toxicants (Oelrichs *et al.*, 1983).

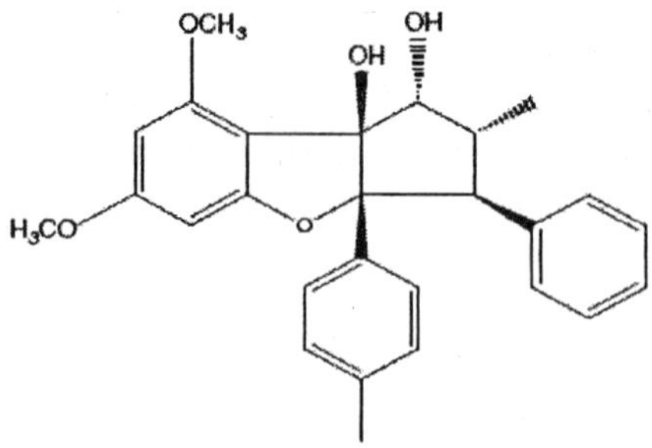

Fig. 10.3. Structure of rocaglamide, an effective insect toxin and growth inhibitor.

Ryania, Aglaia, and *Annona* species are well known for their insect control properties. Ryanodine and dehydroryanodine, the principal insecticidal components from *Ryania* species are relatively non-toxic to mammals and are capable of controlling important larval lepidopteran pests (Ruest *et al.*, 1985, Jefferies *et al.*, 1992a,b; Gonzalez Coloma *et al.*, 1993,1996). Among the botanicals isolated from *Aglaia odorata, A. elliptica*, and *A. duppereana* (Meliaceae), rocaglamide (Fig. 10.3) is the most effective (EC_{50} = 0.8 ppm). It is slightly more potent than azadirachtin (EC_{50} = 1.0 ppm) (Janprasert *et al.*, 1993). As growth inhibitors, rocaglamide and methyl rocaglate are almost similar in their activity (EC_{50} = 0.9 ppm) and quite comparable to azadirachtin (0.26 ppm) (Isman *et al.*, 1990). Rocaglamide analogues isolated from *A. elaeagnoidea* (syn. *A. roxburghiana*) also showed outstanding insecticidal activity against the gram pod borer *Helicoverpa armigera* (Hübner) (Koul *et al.*, 2004a). A new group of plant derived naphthoquinone isolated from a Chilean plant *Calceolaria andina* exhibited outstanding insecticidal activity pesticides. Several potential synthetic derivatives based on the lead naphthquinone have since been synthesized (Khambay *et al.*, 1997, 999).

Annona squamosa (custard apple) and other Annona species are well known for their pesticidal activity (Alkofahi *et al.*, 1989; Rupprecht *et al.*, 1990). Annonins (acetogenins) and related compounds namely squamocin, asimicin, and annonacins occur widely in twigs and branches, unripe fruits and seeds of several *Annona species*. Most of the acetogenins performed better than the conventional insecticides. Another abundantly available plant Karanj, *Pongamia glabra*, syn. *P. pinnata* is an important tree rich in fixed oils and other bioactive secondary metabolites. Karanj seed oil is rich in karanjin, a furano-flavonoid and a host of other polyphenolics like pongamol, pongapin, glabrin, karanj ketone, karanjone, and pongaglarone (Parmar *et al.*, 1976; Kumar *et al.*, 2006). Karanjin has been found effective against mites, scales, chewing and sucking insect pests (Biswas and Biswas, 2006). It is a potent deterrent to many different genera of insects and mites and is effective against whitefly, thrips, leafminers, caterpillers, aphids, jassids, beetles, mealybugs, etc. targeting a wide range of crops. Insect antifeedant, growth reduction, and miticidal activity of karanj oil have been attributed to the presence of high concentration of karanjin and pongamol in the oil (Singh and Kaidhar, 2006). Some of the products are produced by companies like AgriLife and Som Pharma in India.

Madhuca indica (syn. *M. lattifolia, M. longfolia*), commonly known as mahua, is an Indian topical tree found largely in the central and north Indian plains and forests. Mahua saponins are known for their fish toxicity, pesticidal activity and detergent action. *M. indica* extracts have been found effective against insect pests, fungi and nematodes. The extracts caused a significant mortality of *Spodoptera litura, Achaea janata* and other insect pests (Peta and Pathipati, 2008). Extracts of *M. indica* also inhibit feeding and development of *S. litura* larvae. (Loganathan *et al.*, 2005). The insect control properties of mahua extracts have been attributed to triterpenic saponins (Fig. 4) which have been characterized as 3-O-[β-D-gluco-β-D-glucopyranosyl]–16-α-hydroxyprotobassicacid-28-O-[arabino-gluco-xylo pyranosyl]-arabinopyranoside and 3-O-β-D-gluco-gluco-glucopyranosyl-16-α-hydroxylprotobassic acid-28-O-[arabinopyranosyl-xylopyranosyl-arabinopyranosyl]-apiofuranoside. Recent studies have shown that the antifeedant activity of *Madhuca* saponins increased with decrease in number of monosaccharide units in the molecule. Prosapogenins (Fig. 10.4) obtained after hydrolysis of *Madhuca* saponins were found to be more active than the saponins against *S. litura*. Plant derived saponins are known for their diverse use applications. Triterpenic saponins isolated from *Diploknema butyracea* and *Sapindus mukorossi* exhibit feeding deterrent and insect growth inhibitory activity (Saha *et al.*, 2010a,b). Structure-bioactivity relationships studies revealed that partially hydrolyzed prosapogenins were more active than the parent saponins (Saha *et al.*, 2010c). Synergistic and potentiation effects were observed when *Madhuca* and *Sapindus* saponins were mixed with azadirachtin (Saha *et al.*, 2010d). The insecticidal activity of the commercial products based on *Madhuca indica* and *Pongamia glabra* (RD-8 Repelin) were comparable with the commercial insecticide dimethoate (Dhawan *et al.*, 1992).

Piper species (Piperaceae) contain more than 200 secondary metabolites (Boll *et al.*, 1994; Scott *et al.*, 2008) of which the major piperamides and related isobutylamides provide pungent taste and biological activity. The most active pipercide is reportedly 100-fold more active than piperine (Miyakado *et al.*, 1979, 1989; Dev and Koul, 1997). Some of the isobutylamides from *Piper nigrum* fruits are mosquito larvicides (Park *et al.*, 2002). *P. nigrum* formulations also have been found effective against stored product insects such as *Sitophilus oryzae* and *Rhyzopertha dominica*. Three isobutyl amides isolated from *P. nigrum*, pipercide, pellitorine, and piperine were toxic at 0.15, 2, and 20 μg/male *Callosobruchus chinensis*, respectively. Isobutylamides like piperine, piperettine, tricostacine, peepuloidin, piplartin and trichonine have also been found toxic to fruit flies, weevils, cockroaches and many other insect species (Scott *et al.*, 2008). A. Structure-activity relationships indicate that the N-isobutylamine moiety plays a crucial role in the larvicidal activity.

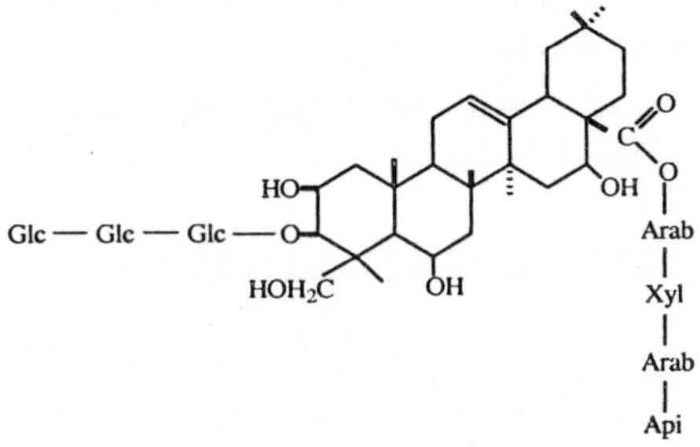

M. indica saponin

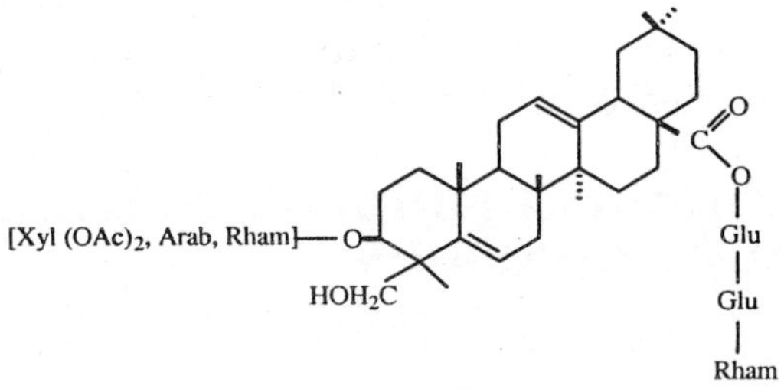

S. mukorossi saponin

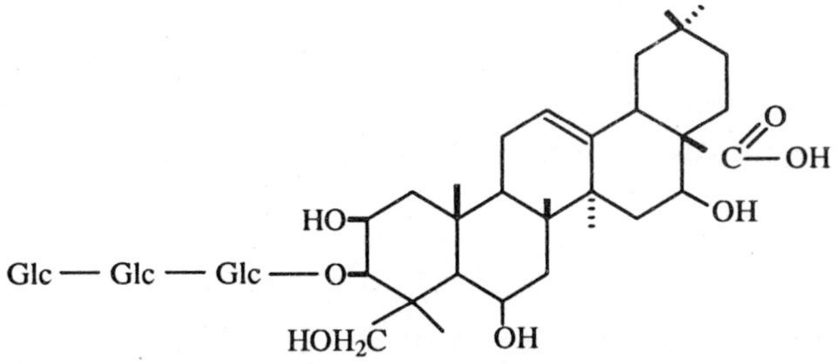

M. indica prosapogenin

Fig. 10.4. Structure of active saponins and prosapogenins.

Phytochemical investigations on different *Tagetes* species have revealed that flavonoids and terpene constituents in the plant have both pharmacological and insecticidal properties (Perich *et al.*, 1995; Tereschuk *et al.*, 1997). The oil of marigold, *T. minuta* reportedly contained β-ocimene, dihydrotagetone, tagetone, tagetenone, and ocimenone as the major constituents (Lawrence, 1992, 2000; Baserand and Malyer, 1996). *T. minuta*, African marigold, *T. erecta* and French marigold, *T. patula* extractives are effective for controlling pests. The exudates from the roots of growing plants are toxic to soil borne insects and nematodes. For instance, the Mexican bean weevils, *Zabrotes subfasciatus* is controlled at LC_{50} = 138–803 μg/cm^2 (Weaver *et al.*, 1994). A-terthienyl (αT), is one of the most active phototoxins found in abundance in the floral, foliar and root extracts of *T. minuta*. It has been demonstrated that irradiation of αT with UV-light, generates a reactive singlet oxygen species responsible for enhanced nematicidal activity. Under photosensitizing conditions αT is more toxic (LC_{50} = 20 ppb) to *A. aegypti* larvae than malathion (LC_{50} = 62 ppb); and less toxic than chlorpyriphos (LC_{50} = 1 ppb) and temephos (LC_{50} = 3 ppb). An acetylenic thiophene isolated from the *Tagetes* roots namely 5-(3-buten-1ynyl)-2,2-bithienyl not only exhibited nematicidal activity but also showed insecticidal activity against insect pests and mosquitoes. Similarly another acetylenic compound 2-(non-1-en-3,5,7-trinyl) furan exhibited excellent mosquitocidal activity againsy *A. aegypti* exhibiting LC_{50} of 0.079 ppm under UV-light.

CHEMICAL MODIFICATION AND STRUCTURE-ACTIVITY RELATIONSHIPS

What is obvious from the plant based biopesticidal studies today is that active ingredients identified are generally occurring in very low quantitites, stability is low, in many cases the activity is moderate in comparison to the conventional pesticides, and even procurement of enough raw material is a combursome process. Therefore, efforts have been made to develop develop stable and more effective compounds based on the chemical modifications of isolated active compounds. Once a bioactive lead molecule has been identified in an extract, it is subjected to structure modification and optimization studies to develop compound(s) with enhanced activity, stability and safety. Several lead phytochemicals have been subjected to chemical modification and structure-bioactivity relationship (SAR) studies to develop cost effective products. For example, recent SAR studies on *Calceolaria andina* provided a lead molecule naphthoquinone, which has been modified to give new products with outstanding activity against mites and whitefly (Khambay *et al.*, 1997,1999). Similalrly, derivatives of neem (*A. indica*) allelochemicals or compounds from dill (*Anethum sowa*), karanja (*Pongamia glabra*), ritha (*Sapindus muko-rossi*), mahua (*Madhuca indica*), turmeric (*Curcuma longa*), etc have provided some new molecules of good potential. Essential oils and some of their major constituents such as carvone, citral, citronellal, pulegone and fenchone have also been the subject of SAR studies to develop new semi-synthetic derivatives with significant insecticidal, insecticide synergistic, antifungal and/or nematicidal properties.

Azadirachtin and its Derivatives

Azadirachtin, the major bioactive constituent in neem seed kernel has limited shelf life due to its ready degradability under sunlight, heat, moisture and during storage. These constraints were overcome by its chemical modification to more stable reduced azadirachtins comprising of dihydroazadirachtin and tetrahydroazadirachtin. Thus hydrogenation of the labile moieties not only render the molecule stable but also make it equally or more effective. The efficacy of tetrahydroazadirachtin concentrates in inhibiting the feeding and growth of *Helicoverpa armigera* (Hübner) larvae has been compared with that of azadirachtin concentrates. They were in general more active and deterred larvae from feeding at all concentrations. Tetrahydroazadirachtin-A (90%) and azadirachtin-A (90%) with respective IC_{50} values of 280 and 390 mg/l were effective as insect growth regulators, while tetrahydroazadirachtin-A (90%) displayed higher antifeedant activity (AI_{50} 14 mg/l) against the test insect (Sharma *et al.*, 2006).

Dillapiole and its Derivatives

Insecticides-natural or synthetic are usually detoxified by enzymes inside the target pests. As a result of their rapid detoxification, insecticides are not capable of providing the desired effect. Insecticide synergists are generally added to such compounds to inhibit detoxification enzymes in the insects to enhance insecticidal action of the product, Besides increasing bioactivity, synergists also help to counter resistance development in insects, increase effective commercial lives and reduce application rate and dose. High cost insecticides such as pyrethrins are often mixed with the commercial insecticide synergists such as piperonyl butoxide (PBO) to enhance their efficacy. In view of the suspected acute and chronic toxicity of PBO, interest in natural synergists has substantially increased. Dillapiole (5-allyl-6,7-dimethoxy-1,3-benzodioxole), the chief constituent of Indian dill seed oil isolated from *Anethum sowa*, its more stable reduced derivative dihydrodillapiole (5-n-propyl 6,7-dimethoxy-1,3-benzodioxole), are known for their superior insecticide synergist activity towards natural pyrethrins.

A large number of dillapiole based synthetic and semisynthetic compounds have been developed that include furapiole, nonanoyl dihydrodillapiole, nonenyl dihydrodillapiole (Mukerjee *et al.*, 1982; Walia *et al.*, 2004) and piperonal oxime N-O-butyl and pentyl ethers (Walia *et al.*, 1985). These compounds exhibit superior synergistic action. The latter with synergistic factor of more than six was found to be three times more active than PBO. The structure-activity relationships studies in a given series had revealed that long chain, as present in PBO was not a pre-requisite for synergistic activity. Some dillapiole derivatives also show ovicidal action against the yellow stem borer *Scirpophaga incertulas* eggs (Vidyasagar and Walia, 1987). The semi synthetic derivatives of dillapiole namely heptanoyl dihydrodillapiole and dillapiole dichlorophenyl cyclopropane inhibited egg hatching and were effective even against older egg masses.

Karanjin and its Derivatives

Karanjin the lead furanoflavonoid constituent of *Pongamia glabra* (Malik *et al.*, 1977, Meera *et al.*, 2003) has been structurally altered to compounds with enhanced activity (Shailja, 2007). Chemical modification of karanjin had yielded karanj ketone, karanj ketone oxime, karanjic acid and a series of karanjic acid and karanj ketone oxime esters. The antifungal activity was significantly increased after conversion of karanjin to karanj ketone oxime ester derivatives. The most active karanj ketone oxime N-O-decanoyl ester exhibited ED_{50} in the range of 129.4–176.08 ppm against *Metarhizium phaseolina, Fusarium oxysporum* and *Sclerotium rolfsii*. Antifungal activity was surprisingly enhanced following conversion of karanjin to karanjic acid ester of octanoic, nonoic and decanoic acid (ED_{50} 41.0–42.0 ppm) against three phytopathogenic fungi.

Essential Oil Constituent Derived Molecules

Essential oils and their constituents are known for their low persistence and moderate pest control properties. Pest control properties of some of the potential monoterpenic constitunts make them excellent lead compounds for the develoment of safe and effective pest control agents. In the past essential oil constituents like camphor, geraniol, longifolene, thymol, dillapiole, etc. have been chemically transformed to increase their activity.(Phadnis *et al.*, 1987, 1988; Sawalikar *et al.*, 1995; Pawar *et al.*, 1995). Some monoterpenes and their derivatives have been subjected to QSAR evaluation (Grodnitzky and Coats, 2002). Efforts have also been made to improve/enhance pest control potential of some of the oxygenated monoterpenic constituents by chemical modification of the lead molecule. Very recently insect antifeeedant activity of fenchone, the major constituent of *Foeniculum vulgare* (Fennel) and *Xanthium brasilicum* essential oil was significantly enhanced following its conversion to oxime esters. Among the various semi-synthetic derivatives, fenchone oxime N-O-isovalerate exhibited AI_{50} of 16 ppm against *Spodoptera litura* followed by other equally effective nonanoate (AI_{50} = 79 ppm), octanoate (AI_{50} = 131 ppm) and pivaloate ester (AI_{50} = 141 ppm). Chemical modification of carvone similarly resulted in oxime esters with improved activity. Carvone oxime N-O-propanoyl ester, and N-O-cinnamoyl ester (EC_{50}

60–62 ppm) were found most effective against the phytopathogenic fungus *S. rolfsii*. Another monoterpenic constituents citronellal and its reduced derivative dihydrocitronellal were converted to citronellal oxime N-O-pentanoyl ester (EC_{50} 77.29 ppm) and carvone oxime N-O-isobutanoyl ester (EC_{50} 107.08 ppm) which exhibited significant nematicidal activity against the root knot nematode *M. incognita* (Choudhary, 2006; Paul, 2008). In view of the natural origin of the raw materials, such oxime esters hold promise and need to be further evaluated for their efficacy under field conditions

Naturally Occuring Alkanone Derivatives

Naturally occurring methyl ketones such as 2-tridecanone in wild tomato (*Lycopersicon hirsutum* f. *glabratum*) reportedly provide resistance to the plant against arthropod pests like tomato fruit worm *Heliothis zea* (Williams *et al.*, 1982), tomato pin worm, *Tuta absoluta*, (Gilardon *et al.*, 2001) and *Chilo sonorensis* (Kauffman and Kennedy, 1989). Such defensive plant secondary metabolites are located in the tips of the glandular trichomes of the plant leaves (Dimock *et al.*, 1982; Antonious, 2001). These have also been found effective against the two spotted spider mite *Tetranychus urticae* (Chatzivasileiadis and Sabelis, 1997), green peach aphid *Myzus persicae* (Antonious *et al.*, 2003), tobacco budworm *Helicoverpa virescens* (Antonious and Snyder, 2006) and cowpea weevil *Callosobruchus maculatus* (Braga *et al.*, 2007). 2-tridecanone the lead molecule occurring on the leaf surface has been chemically modified to more effective 2-tridecanone oxime and its various oxime esters. The semi-synthetic 2-tridecanone oxime N-O-nonanoate with LC_{50} of 73.16 µg/ml was identified as the most active followed by octanoyl ester (LC_{50} of 84.22 µg/ml). The activity of the most active compound though significant is less than the commercial nematicide triazophos (LC_{50} 22.35 µg/ml). However, in view of the natural origin and environmentally benign nature of the 2-tridecanone, the resultant oxime esters appear to be of considerable importance.

ISSUES, CONSTRAINTS AND STRATEGIES

Although many compounds have been isolated, characterized and evaluated for their efficacy against pests yet substantial headway is to be made for their large scale production and commercialization. Due to multitude of the active ingredients, standardization of anti-pest botanical preparations has really been one of the biggest constraints.

Some other potential issues hindering the growth of botanical pest control products are listed below:

- Indequate availability and supply of quality raw materials.
- Varying level of bioefficacy of identical preparation.
- Slow action and rapid breakdown in outdoor situations lead to repeated applications and increased cost of pest control.
- Lack of indigenous know-how for production of botanical concentrates, technical ingredients and formulations.
- Non-availability of standard and stable quality technical materials, concentrates.
- Poor shelf life and unpredictable performance.
- Cumbersome registration requirements.
- Varying recommendations of different producer and poor enforcement of quality control.

The advantages of botanical pesticides over chemical insecticides in terms of reduced threats to non-target species and limited potential for environmental degradation are compelling reasons for increasing efforts to increase their utility in IPM programmes. Following strategies may be adopted to fully explore their potential in pest control.

Exploration of Plant Biodiversity for Pesticidal Traits

Plant biodiversity remains a vast reservoir of new pesticidal molecules. However, deriving new biopesticidal principles from plants remains a complex and time consuming task as it needs interdisciplinary skills and expertise on isolation, purification, characterization, and their screening for biological effects. While plant extracts may afford additive/synergistic action and strong biopesticidal activities, their purification and structure determination is essential for standardization and bioefficacy improvement. In view of the wide variation and multifarious uses of plant derivatives, it is desirable to identify pesticidal traits. In the grim scenario of mounting hazards and cost of synthetic chemical pesticides, natural chemistry of plants shows a ray of hope for environment and human friendly and sustainable pest management.

Integration of Botanicals with Nonchemical Strategies

Nonchemical pest solutions are effective in certain situations but may fail in other situations. When they do fail, other alternatives need to be considered and included to provide safe pest control strategies. While chemical or nonchemical approaches are comparatively cheaper, pests often adapt to single system approaches and eventually become unmanageable and require a combination of methods to achieve satisfactory control. Most pest management programs in the agricultural farms, parks and/or in domestic areas, aboard the ships, planes, trains, trucks, etc., rely on a combination of nonchemical and chemical control methods. Incorporation of botanical pest management alternatives may not completely eliminate all the pests and diseases and provide permanent solutions, but they surely provide safer option to conventional highly toxic pesticides. They significantly lessen the need to use synthetic pesticides.

Overcoming the Problem of Inadequate Efficacy and Stability

Botanical pesticides and their formulations suffer from certain drawbacks in term of their inadequate persistence and inactivation by ultra-violet, sun light, heat, pH, moisture, etc. Because of these undesirable traits, repeated applications are necessitated outdoor. High cost, poor quality and varying level of bioefficacy in identical preparations are some other limiting factors that had hindered the growth and acceptance of botanical pesticides. Thermal and sunlight mediated inactivation has emerged as the major environmental factors affecting stability and efficacy. Suitable stabilizers, UV-screens and antioxidants, therefore, need to be identified for incorporation in the formulations to increase stability. Similarly novel adjuvants like insecticide synergists, potentiators or other botanical ingredient needs to be developed to enhance efficacy. In view of the fact that the supply of botanicals and other biopesticides will always be in limited quantity, appropriate extraction technologies novel formulations will have to be developed to improve both residual and shelf life.

Environment and Human Safety Concerns

In spite of numerous advantages, the limited public acceptance about the food containing pesticide residues is attributed to the fear of possible toxic residues and their possible adverse effects on human health. The phytochemical biopesticides in general appear exceptionally safe to non-target organisms. However, their use must take into account possible adverse effects on beneficial organisms like parasitoids, predators and pollinators. Before botanical pesticides are put to use, it is essential that their safety to the environment is fully established. The possible negative effects on non-target organisms should be weighed against alternative control strategies. Like chemical pesticides their continuous exposure to the target pest is likely to develop resistance in the target species

CONCLUSIONS

The search for naturally occurring crop protectants represents a challenge to pest-management specialists. From environment sustenance point of view, a potential bioactive ingredient rich plant should be perennial, should have a wide distribution and present in large quantity and should have the potential to be grown on large scale by standard agricultural practices. The plant parts of interest should be renewable and removable/ harvestable like leaves, flowers fruits, etc. In case of the tree, use of bark, roots, stem, etc should be avoided. Effective technologies need to be developed for the production of cost effective and more efficacious biopesticidal ingredients and formulations that have enhanced shelf life. The natural pesticides should be of high economic value and should be effective at low rates. Industry, regulatory agencies, research scientists, policy makers, and other stakeholders must accelerate their efforts to promote safe and judicious use of natural pest control chemicals in agriculture, veterinary and public health. Industry needs to establish necessary in-house R&D infrastructure for development and promotion of botanical biopesticides.

REFERENCES

Agarwal, M. and Walia, S. (2004) Pest control potential of phytochemicals derived from *Curcuma longa* and *Zingiber officinale*. *In* P. Dureja, D.B. Saxena, J. Kumar, S.B. Singh, M. Gopal and R.S. Tanwar (eds.), *Pesticide: Environment and Food Security*, Pesticide Science, New Delhi, pp. 110–119.

Agarwal, M., Walia, S. and Dhingra, S. (1999) Pest control properties of turmeric leaf oil against *Spilosoma oblique*, *Dysdercus koenigii* and *Tribolium castaneum*. *Proceedings 2nd All India People's Technology Congress,* Calcutta, pp. 1–7.

Agarwal, M., Walia, S. and Dhingra, S. (2001) Insect growth inhibition, antifeedant and antifungal activity of compounds isolated/derived from *Zingiber officinale* rhizomes. *Pest Manage. Sci.*, **37**, 289–300.

Alkofahi, A., Rupprecht, J.K., Anderson, J.E., McLaughlin, J.L., Mikolajczak, K.L. and Scott, B.A. (1989) Search for new pesticides from higher plants, *In* J.T. Arnason, B.J.R. Philogene and P. Morand (eds.), *Insecticides of Plant Origin*, American Chemical Society, Washington DC, pp. 25–43.

Antonious, G.F. (2001) Production and quantification of methyl ketones in wild tomato accessions. *J. Environ. Sci. Health* B, **36,** 835–888.

Antonious, G.F. and Snyder, J.C. (2006) Natural products: repellency and toxicity of wild tomato leaf extracts to the two-spotted spider mite, *Tetranychus urticae* Koch. *J. Environ. Sci. Health B.*, **41**, 43–55.

Antonious, G.F., Dahlman, D.L. and Hawkins, L.M. (2003) Insecticidal and acaricidal performance of methyl ketones in wild tomato leaves. *Bull. Environ. Contaminat. Toxicol.*, **71**, 400–407.

Arnason, J.T., Philogene, B.J.R. and Morand, P. (eds.) (1989) *Insecticides of Plant Origin.* ACS Symposium Series 387, American Chemical Society, Washington D.C.

Baserand, K.H.C. and Malyer, H. (1996) Essential oil of *Tagetes minuta* from Turkey. *J. Essen. Oil Res.*, **8**, 337–338.

Biswas, N.P. and Biswas, A.K. (2006) Use of non-edible oils as grain protectant against rice weevil (*Sitophilus oryzae* L.) and their subsequent effect on germination. *Adv. Plant Sci.*, **19**, 653–656.

Boll, P.M., Parmar, V.S., Tyagi, O.D., Prasad, A., Wengel, J. and Olsen, C.E. (1994) Some recent isolation studies from potential insecticidal *Piper* species. *Pure Appl. Chem.*, **66**, 2339–2342.

Braga, Y.F., Grangeiro, T.B., Freire, E.A., Lopes, H.L., Bezerra, J.N., Andrade-Neto, M. and Lima, M.A. (2007) Insecticidal activity of 2-tridecanone against the cowpea weevil *Callosobruchus maculatus* (Coleoptera: Bruchidae). *Ann. Acad. Bras. Cienc.*, **79**, 35–39.

Casida, J.E. and Quistad, G.B. (1994) *Pyrethrum Flowers, Producton, Chemistry, Toxicology and Uses.* Oxford Yniversity Press, Oxford.

Champagne, D.E., Koul, O., Isman, M.B., Towers, G.H.N. and Scudder, G.G. E. (1992) Biological activity of limonoids from the Rutales. *Phytochemistry*, **31**, 377–394.

Chatzivasileiadis, E.A. and Sabelis M.W. (1997) Toxicity of methyl ketones from tomato trichomes to *Tetranychus urticae* Koch. *Exp. Appl. Acarol.*, **21**, 473–484.

Choudhary, G. (2006) Synthesis and bioactivity of monoterpene oxime esters against phytopathogenic fungi and nematodes. Ph.D. Thesis, Post Graduate School, Indian Agricultural Research Institute, New Delhi.

Chowdhury, H., Walia, S. and Saxena, V.S. (2000) Isolation, characterization and insect growth inhibitory activity of major turmeric constituents and their derivatives against *Schistocerca gregaria* (F.) and *Dysdercus koenigi* (W). *Pest Manage. Sci.*, **56**, 1086–1092.

Chowdhury, H., Walia, S. and Dhingra, S. (2001) Bioefficacy of azadirachtin, turmeric oil and their mixture against Bihar hairy caterpillar (*Spilosoma [illegible] ehavio* Walk.). *Pestic. Res. J.*, **3**, 165–173

Chowdhury, H., Walia, S. and Kumar, P. (2002) Effect of azadirachtin and aqueous neem seed kernel concentrate on germination and seedling growth of soybean (*Glycine max* L. Merril) and Maize (*Zea mays* L.). *Pestic. Res. J.*, **14**, 120–124.

Copping, L.G. and Duke, S.O. (2007) Natural products that have been used commercially as crop protection agents. *Pest Manage. Sci.*, **63**, 524–554.

Dev, S. and Koul, O. (1997) *Insecticides of Natural Origin*. Harwood Academic Publishers, Amsterdam, The Netherlands

Dhawan, A.K., Simat, G.S., Tauro, P. and Narwal, S.S. (1992) Management of sucking pests of cotton with natural plant products. *Abstract Proc. Allel. Agroecosys* (Febryary 12–14), CCS Haryana Agricultural University, Hisar, pp. 154–155.

Dimock, M.B., Kennedy, G.G. and Williams, W.G. (1982) Toxicity studies of analogs of 2-tridecanone, a naturally occurring toxicant from a wild tomato. *J. Chem. Ecol.*, **8**, 837–842.

Gilardon, E., Pocovi, M., Hernandez, C., Collavino, G. and Olsen, A. (2001) Role of 2-tridecanone and type VI glandular trichome on tomato resistance to *Tuta absoluta*. *Pesquisa Agro. Brasil.*, **36**, 929-933.

Gonzalez-Coloma, A., Cabrera, R., Soccorro Monzon, A.R. and Fraga, B.M. (1993). *Persea indica* as a natural source of the insecticidal ryanodol. *Phytochemistry*, **34**, 397–400.

Gonzalez-Coloma, A., Terrero, D., Perales, A., Escoubas, P. and Fraga, B.N. (1996). Insect antifeedant ryanodine diterpenes from *Persea indica*. *J. Agric. Food Chem.*, **44**, 296–300.

Grainge, M. and Ahmed, S. (1988) *Handbook of Plants with Pest-Control Properties*. John Wiley & Sons, New York.

Grewal, P. and Georgis, R. (1999). Entomopathogenic nematodes, In F.R. Hall and J.J. Men (eds.), *Biopesticides: Use and Delivery*, Humana Press, Totowa, New Jersey, pp. 271–299.

Grodnitzky, J.A. and Coats, J.R. (2002) QSAR evaluation of monoterpenoids' insecticidal activity. *J. Agric. Food Chem.*, **50**, 4576-4580.

Hummelbrunner, A.L. and Isman, M.B. (2001) Acute, sublethal, antifeedant and synergistic effects of monoterpenoid essential oil compounds on the tobacco cut worm (Lep: Noctuidae). *J. Agric. Food Chem.*, **49**, 715–720.

Isman, M.B. (1999). Pesticides based on plant essential oils. *Pestic. Outlook*, **10**, 68–72.

Isman, M.B., Koul, O., Luczynski, A. and Kaminski, J. (1990) Insecticidal and antifeedant bioactivities of neem oils and their relationship to azadirachtin content. J. *Agric. Food Chem.*, **38**, 1406–1411.

Janprasert, J., Satasok, C., Sukumalanand, P., Champagne, D.E., Isman, M.B., Winyachitra, P. *et al.* (1993) Rocaglamide, a natural benzofuran insecticide from *Aglaia adorata*. *Phytochemistry*, **32**, 67–69.

Jefferies, P.R., Tora, R.F., Brannigan, E., Pessah, I. and Casida, J.E. (1992a). Ryania insecticide: Analysis and biological activity of 10 natural ryanoids. *J. Agric. Food Chem.*, **40**, 142–146.

Jefferies, P.R., Lam, W., Tora, R.F. and Casida, J.E. (1992b). Ryania insecticide: Structural assignment of four natural 8ax-hydroxy-10-eparyanoids. *J. Agric. Food Chem.*, **40**, 509–512.

Jilani, G. and Su, H.C.F. (1983) Laboratary studies on several plant materials as insect repellants for protection of stored grains. *J. Econ. Entomol.*, **76**, 154–157.

Jilani, G., Saxena, R.C. and Rueda, B.P. (1998) Repellant and growth inhibiting effects of turmeric oil, sweetflag oil, and 'Margosan-O' on red flour beetle (Coleoptera: Tenebrionidae). *J. Econ. Entomol.*, **81**, 1226–1230.

Kauffman,W.C. and Kennedy, G.G. (1989) Toxicity of allelochemicals from wild insect-resistant tomato *Lycopersicon hirsutum* f. *glabratum* to *Campoletis sonorensis*, a parasitoid of *Heliothis zea*. *J. Chem. Ecol.*, **15**, 2051–2060.

Khambay, B.P.S., Batty, D., Beddy, D.G., Denholm, I. and Cahill, M.R. (1997) A new group of plant derived naphthoquinone pesticides. *Pestic. Sci.*, **50**, 291–296.

Khambay, B.P.S., Batty, D., Beddy, D.G., Cahill, M., Denholm, I., Mead-Mriggs, M., Vinall, S., Niemeyer, H.M. and Simmonds, M.S.J. (1999) Isolation, characterization and biological activity of naphthaquinones from *Calceolaria andina* L. *J. Agric. Food Chem.*, **47**, 770–775.

Khatavkar, V., Walia, S., Srivastava, C., Kumar, J. and Parmar, B.S. (2005) Isolation characterization and insect fumigant action of neem leaf volatiles against two storage pests *Callosobruchus maculatus* and *Tribolium castaneum*. *Conference of biopesticides: Emerging trends* (Nov. 11-13, 2005), Society of Biopesticides Science, Jalandhar.

Koul, O. (1992) Neem allelochemicals and insect control. In Rizvi, S. J. H. and Rizvi, V. (eds.), *Allelopathy : Basic and Applied Aspects*, Chapman & Hall, London, pp. 389–413.

Koul, O. (1995) *Acorus* allelochemicals : Chemistry and bioefficacy against insects. In B.S. Parmar and S. Walia (eds.), *Pesticides, Crop Protection and Environment*, Oxford & IBH Publishing Co. Pvt. Ltd., New Delhi, pp. 99–113.

Koul, O. (2004a) *Insect Antifeedants.* CRC Press LLC, Boca Raton, Florida.

Koul, O. (2004b) Bioactivity of volatile di-n-propyl disulfide from seeds of neem, *Azadirachta indica* (Meliaceae) to two species of stored grain pests, *Sitophilus oryzae* (L.) and *Tribolium castaneum* (Herbst.). *J. Econ. Entomol.*, **97**, 1142–1147.

Koul, O. and Dhaliwal, G.S. (2001) *Phytochemical Biopesticides.* Harwood Academic Publishers Gmbh, Amsterdam, The Netherlands.

Koul, O. and Wahab, S. (eds.) (2004) *Neem : Today and in the New Millennium.* Kluwer Academic Publishers, Dordrecht, The Netherlands.

Koul, O. and Walia, S. (2009) Comparing impacts of plant extracts and pure allelochemicals and implications for pest control. *CAB Reviews: Perspectives in Agriculture, Veterinary Science, Nutrition and Natural Resources*, **4**, 049, 1–30.

Koul, O., Smirle, M.J. and Isman, M.B. (1990) Asarones from *Acorus calamus* L. oil : their effect on feeding behavior and dietary utilization in *Peridroma saucia*. *J. Chem. Ecol.*, **16**, 1911–1920

Koul, O., Kaur, H., Goomber, S. and Wahab, S (2004a) Bioefficacy and mode of action of rocaglamide from *Aglaia elaeagnoidea* (syn. *A. roxburghiana*) against gram pod borer, *Helicoverpa armigera* (Hübner) *J. Appl. Entomol.*, **128**, 177–181.

Koul, O., Multani, J. S., Goomber S., Daniewski, W.M. and Berlozecki, S. (2004b) Activity of some non-azadirachtin limonoids from *Azadirachta indica* against lepidopteran larvae. *Aust. J. Entomol.*, **43**, 189–195.

Koul, O., Singh G., Singh, R., Singh, J., Daniewski, W.M. and Berlozecki, S. (2004c) Bioefficacy and mode-of-action of some limonoids of salannin group from *Azadirachta indica* A. Juss and their role in a multicomponent system against lepidopteran larvae. *J. Biosci.*, 29, 409–416.

Koul, O., Walia, S. and Dhaliwal, G.S. (2008) Essential oils as green pesticides: Potential and constraints. *Biopestic. Int.*, **4**, 63–84.

Kumar, J., Rupak Raja., Walia, S., and Parmar, B.S. (2003) Effect of natural and synthetic additives on the stability of azadirachtin under accelerated storage. *Pestic. Res. J.*, **15**, 10–13.

Kumar, V., Chandrashekar, K. and Sidhu, O.P. (2006) Efficacy of karanjin and different extracts of *Pongamia pinnata* against selected insect pests. *J. Entomol. Res.*, **30**, 130–108.

Lawrence, B.M. (1992) Labiatae oils: mother nature's chemical factor. In B.M. Lawrence (ed.), *Essential Oils 1988-1991*, Allured Publ., Stream, IL, pp. 188–206.

Lawrence, B.M. (2000) Progress in essential oils. Turmeric oil. *Perf. Flav.*, **25**, 32–38.

Lawerence, B.M. and Reynolds, R.J. (2001) Progress in essential oils. *Perf. Flav.*, **26**, 44–52.

Loganathan, J., Walia, S. and Dhingra, S. (2005) Efficacy of various extracts of *Madhuca indica* on feeding and development of *Spodoptera litura* F. *Pestic. Res. J.*, **17**, 10–15.

Malik, S.B., Sharma, P. and Sheshadri, T.R. (1977) Furanoflavonoids from the leaves of *Pongamia glabra*. *Indian J. Chem.*, **15B**, 536–538.

Meera, B., Kumar, S. and Kalidhar, S.B. (2003) A review of the chemistry and biological activity of Pongamia pinnata. *J. Med. Aroma. Plant Sci.*, **25**, 441–446.

Miyakado, M., Nakayama, I., Yoshioka, H. and Nakatani, N. (1979) The piperaceae amides I: Structure of pipericide, a new insecticidal amide from *Piper nigrum* L. *Agric. Biol. Chem.,* **43**, 1609–1611.

Miyakado, M., Nakayama, I. and Ohno N. (1989) Insecticidal unsaturated isobutylamide from natural products to agrochemical leads. *In* J.T. Arnson, B.J.R. Philogene and P. Morand (eds.), *Insecticides of Plant Origin*, ACS Symposium Series 387, American Chemical Society, Washington DC, pp. 173–187.

Mukherjee, S.K., Walia, S. and Saxena, V.S. (1982) New pyrethrum synergists from dihydrodillapiole and furapiole. *Agril. Biol. Chem.*, **46**, 1277–1283.

Nakatani, N., Shimokoro, M., Zhou, J.B., Okamura, H., Iwagawa, T., Tadera, K., Nakayama, N. and Naoki, H. (1999) Limonoids from *Melia toosendan*. *Phytochemistry*, **52**, 709–714.

Oelrichs, P.B., Hill, M.W., Vallely, P.J., MacLeod, J.K. and Molinski, T.F. (1983) Toxic tetranortriterpenoids of the fruit of *Melia azedarach*. *Phytochemistry*, **22**, 531–534.

Omar, S., Marcotte, M., Fields, P., Sanchez, P.E., Poveda, L., Mata, R., *et al.* (2007) Antifeedant activities of terpenoids isolated from tropical Rutales. *J. Stored Prod. Res.*, **43**, 92–96.

Park, K. and Shin Sang-Chul (2005) Fumigant activity of plant essential oils and components from garlic (*Allium sativum*) and clove bud (*Eugenia caryophyllata*) oils against the Japanese termite (*Reticulitermes speratus* Kolbe). *J. Agric Food Chem.*, **53**, 4388–4392.

Park, K., Sang-Gil, L., Sang-Chul, S., Ji-Doo, P. and Young-Joon, A (2002) Larvicidal activity of isobutylamides identified in *Piper nigrum* fruits against three mosquito species. *J. Agric. Food Chem.*, **50**, 1866–1870.

Parmar, B.S. and Walia, S. (2001) Prospects and problems of phytochemical pesticides. *In* O. Koul and G.S. Dhaliwal (eds.), *Phytochemical Biopesticides*, Harwood Academic Publishers Gmbh, The Netherlands, pp. 133–210.

Parmar, B.S. and Walia, S. (2006) Biopesticides and Mycorrhiza. In: M.S. Swaminathan and K.L. Chadha (eds.), *Agriculture and Environment*, Malhotra Publishing House, New Delhi, pp. 249-278.

Parmar, B.S., Sahrawat, K.L. and Mukherjee, S.K. (1976) *Pongamia glabra*, constituents and uses. *J. Sci. Ind. Res.*, **35**, 608–611.

Parmar, B.S., Walia, S., Anupama, and Kumar, J. (2008) *Neem Pesticides in India, An Update of the Recent Developments*. Technical Bulletin. SPS Publication No. 15, Society of Pesticide Science, India.

Paul, R. (2008) Synthesis and pesticidal activity of acyloxyimino derivetives from naturally occurring fenchone and methyl ketones, Ph.D Thesis, Post Graduate School, Indian Agricultural Research Institute, New Delhi.

Pawar, P.V., Tungikar, V.B. and Sharma, R.N. (1995) Bioactivity of oxime ethers against mosquito, *Aedes aegypti* and termite, *Odontotermes brunnes*. *Pestic. Res. J.*, **7**, 128–131.

Perich, M., Wells, C., Bertsch, W. and Tredway, K. (1995) Isolation of the insecticidal components of *Tagetes minuta* (Compositae) against mosquito larvae and adults. *J. Am. Mosq. Cont. Assoc.*, **11**, 307–310.

Peta, D. and Pathipati, U.R. (2008) Biological potency of certain plant extracts in management of two lepidopteran pests of *Ricinus communis* L. *J. Biopestic.*, **1**, 170–176.

Phadnis, A.P., Patwardhan, S.A., Gund, P. and Sharma, R.N. (1987) Biologicalactivity of some new geraniol based ethers on insect pests and vectors. *Pestic. Sci.*, **21**, 93–103.

Phadnis, A.P., Nanda, B., Patwardhan, S.A., Pawar, P.V. and Sharma, R.N. (1988) Products active on mosquitoes Part III, Synthesis of biologically active dimethyl octane diol ethers. *Indian J. Chem.*, **27B**, 867–870.

Pradhan, S., Jotwani, M.G. and Rail B.K. (1962) The neem seed deterrent to locusts. *Indian Fmg.*, 12, 711.

Qing Yang, Qiuhui Hu, Osamu Yamato, Keun-Woo Lee, Yoshimitsu Maede and Teruhiko Yoshihara (2003) Organosulfur compounds from garlic (*Allium sativum*) oxidizing canine erythrocytes. *Z. Naturforsch.*, **58C**, 408–412.

Ray, D.P., Walia, S., Dureja, P and Singh, R.P. (2000) Composition and repellent activity of the essential oil of marigold (*Tagetes erecta*) flower. *Indian Perf.*, **44**, 267–270.

Rosell, G., Quero, C., Coll, J. and Guerrero, A. (2008) Biorational insecticides in pest management. *J. Pestic. Sci.*, **33**, 103–121.

Ruest, L., Taylor, D.R. and Deslongchamps, P. (1985) Investigation of the constituents of *Ryania speciosa*. *Can. J. Chem.*, **63**, 2840–2843.

Rupprecht, J.K., Hui, Y.H. and Mclaughlin, J.L. (1990) Annonaceous acetogenins, a review. *J. Nat. Prod.*, **53**, 237–278.

Saha, S., Walia, S., Kumar, J., Dhingra, S. and Parmar, B.S. (2010a) Screening for feeding deterrent and insect growth regulatory activity of triterpenic saponins from Diploknema butyracea and Sapindus mukorossi. *J. Agric. Food Chem.*, 58, 4334–440.

Saha, S., Walia, S., Kumar, J., Dhingra, S. and Parmar, BS. (2010b) Triterpenic saponins as regulator of plant growth. *J. Appl. Bot. Food Quality*, **83**, 189–195.

Saha, S., Walia, S., Kumar, J. and.Parmar, B.S. (2010c) Structure-biological activity relationships in triterpenic saponins: the relative activity of protobassic acid and its derivatives against plant pathogenic fungi. *Pest Manage Sci.*, **66**, 825–831.

Saha, S., Walia, S. and Prasad, D. (2010d) Synergistic/potentiation interaction between nematostatic constituents from *Azadirachta indica* and phytochemicals from *Madhuca indica* and *Sapindus mukorossi*. *Arch. Phytopathol. Plant Prot.*, **43**, 357–367.

Schmutterer, H. (1990) Properties and potential of natural pesticides from the neem tree. *Ann. Rev. Entomol.*, **35**, 271–292.

Schmutterer, H. (2002) *Neem.* Neem Foundation, Mumbai, India.

Scott, I.M., Jensen, H.R., Philogene, B.J.R. and Arnason, J.T. (2008) A review of *Piper* spp. (Piperaceae) phytochemistry, insecticidal activity and mode of action. *Phytochem. Rev.*, **7**, 65–75.

Shailja, V. (2007) Chemical investigation and modification of compound(s) isolated from *Pongamia glabra* as potential nitrification inhibitors. Ph.D. Thesis, Jamia Millia Islamia, New Delhi.

Sharma, V., Walia, S., Kumar, J., Nair, M.G. and Parmar, B.S. (2003) An efficient method for the purification and characterization of nematicidal azadirachtins A, B, and H, using MPLC and ESIMS. *J. Agric. Food Chem.* **51**, 3966–3972.

Sharma, V., Walia, S., Kumar, J., Goswami, B.K. and Parmar, B.S. (2004) Nematicidal and antifungal activity of azadirachtin-A concentrates, and azadirachtins B and H. In: P. Dureja, D.B. Saxena, J. Kumar, S.B. Singh, M. Gopal and R.S. Tanwar (eds.), *Pesticide: Environment and Food Security*, Society of Pesticide Sciences, New Delhi, pp. 121–125.

Sharma, V., Walia, S., Dhingra, S., Kumar, J. and Parmar, BS. (2006) Azadirachtin-A and tetrahydroazadirachtin-A concentrates: Preparation, LC-MS characterisation and insect antifeedant/IGR activity against *Helicoverpa armigera* (Hubner). *Pest Manage. Sci.*, **62**, 965–975.

Singh, A., Sardana, H.R. and Sabir, N. (eds.) (2004) *Validated IPM Technologies for Selected Crops*. National Centre for Integrated Pest Management, New Delhi.

Singh, R. and Kalidhar, S.B. (2006) Evaluation of Karanj (*Pongamia pinnata* Vent) compounds against lepidopteran pests. *Biopestic. Int.*, **2**, 153–159.

Swalkar, D.D., Sinha, B., Hebbalkar G.D. and Sharma, R.N. (1995) Product active on mosquitoes Part VIII: Synthesis and biological activity of longifolene drivatives. *Indian J. Chem.*, **34B**, 832–835.

Su, H.C.F., Robert, H. and Gilani, G. (1982) Isolation, purification and characterization of insect repellents from *Curcuma longa* L. *J. Agric. Food Chem.*, **30**, 290–292.

Tereschuck, M., Riera, M., Castro, G. and Abdala, L. (1997) Antimicrobial activity of flavonoids from leaves of *Tagetes minuta*. *J. Ethnopharmacol.*, **56**, 227–232.

Tripathi, A.K., Prajapati, V. and Kumar, S. (2003) Bioactivity of l-carvone, d-carvone and dihydrocarvone towards three stored product beetles. *J. Econ. Entomol.*, **96**, 1594–1601.

Tripathi, A.K., Prajanpati, V., Aggarwal, K.K. and Kumar, S. (2001) Toxicity, feeding deterrence, and effect of activity of 1, 8-cineole from *Artemisia annua* on progeny production of *Tribolium castanaeum*. *J. Econ. Entomol.*, **94**, 979–983.

Tripathi, A.K., Prajapathi, V., Verma, N., Bhal, J.R., Bansal, R.P., Khanuja, S.P.S and Kumar, S. (2002) Bioactivities of the leaf essential oils of *Curcuma longa* (Var.Ch.66) on three species of stored product beetles (Coleoptera). *J. Econ. Entomol.*, **95**, 183–189.

Ujvary, I. (1999) Nicotine and other insecticidal alkaloids. In: I. Yamamoto and J.E. Casida (eds.), *Nicotinoid Insecticide and the Nicotinic Acetylcholine Receptor*, Springer-Verlag, Tokyo, pp 229–269.

Vidyasagar, P.S.P.V. and Walia, S. (1987) Effect of certain dillapiole derivatives on rice stem borer *Scirpophaga incertulas* (Walker) eggs. *Entomon*, **12**, 345–348.

Walia, S (2005) Allelochemicals as biopesticides. In: *Souvenir, Biopesticides: Emerging Trends Conference* (Nov 11-13, 2005) Institute of Himalayan Bioresource Technology, Palampur, Society of Biopesticide Sciences, Jalandhar, pp. 19-32.

Walia, S. and Parmar, B.S. (2005) *Neem-A Success Story: Advances in Neem Research*. Indian Agricultural Research Institute, New Delhi.

Walia, S. and Parmar, B.S. (2006) Recent developments in botanical pesticides in India with special reference to neem. *Agribiotech: Proceedings Ist International Conference on Biotechnology for Sustainable Agriculture & Biotechnology*, Hyderabad, India, pp. 23–32.

Walia S and Koul, O. (2008) Exploring plant biodiversity for botanical insecticides. In: O. Koul, G.S. Dhaliwal and V.K. Koul (eds.), *Sustainable Crop Protection: Biopesticide Strategies*, Kalyani Publishers, New Delhi, pp. 191-206.

Walia, S., Saxena, V.S. and Mukherjee, S.K. (1984) Cyclopropyl MDP compounds as new pyrethrum synergists. *Agri. Biol. Chem.*, **48**, 2675–2679.

Walia, S., Saxena, V.S. and Mukherjee, S.K. (1985) Synthesis and synergistic activity of oxime ethers containing a benzo-1,3-dioxide group. *J. Agric. Food Chem.*, **33**, 308–310.

Walia, S., Kumar, J. and Parmar, B.S. (2001) Ecologically sound botanical pesticides: Development and use. *In* O. Koul, G.S. Dhaliwal, S.S. Marwaha and J.K. Arora (eds.), *Biopesticides and Pest Management*, Vol. 1, Campus Books International, New Delhi, pp. 56–63.

Walia, S., Saha, S. and Parmar, B.S. (2004) Liquid chromatographic method for the analysis of two plant based insecticide synergists dillapiole and dihydrodillapiole. *J. Chromatogr. A*, **1047**, 229–233.

Walia, S., Sharma, V., Kumar, J. and Parmar, B.S. (2003) Improvements in/or relating to the preparation of reduced azadirachtin(s). (Indian Patent number 226204 (2003).

Weaver, D.K., Wells, C.D., Dunkel, F.V., Bertsch, W., Sing, S.E. and Sriharan, S. (1994) Insecticidal activity of floral, foliar, and root extracts of *Tagetes minuta* (Asterales: Asteraceae) against adult Mexican bean weevils (Coleoptera: Bruchidae). *J. Econ. Entomol.*, **87**, 1718–1725.

Vohra, S. (2007) Chemical investigation and modification of compound(s) isolated from *Pongamia glabra* as potential nitrification Inhibitors. Ph D. Thesis. Jamia Millia Islamia, New Delhi.

Xuan, T.D., Eiji, T., Hiroyuki, T., Mitsuo, M., Khanh, T.D. and Chung, I. (2004) Evaluation of phytotoxicity of neem (*Azadirachtin indica* A. Juss) to crops and weeds. *Crop Prot.,* **23**, 335–345.

11

Pesticidal Plants: A Viable Alternative Insect Pest Management Approach for Resource-Poor Farming in Africa

PHILIP C. STEVENSON[1,2*], STEPHEN P. NYIRENDA[3], BRIGHTON MVUMI[4], PHOSISO SOLA[5], JOHN F. KAMANULA[6], GUDETA SILESHI[7] AND STEVEN R. BELMAIN[1]

[1]Natural Resources Institute, University of Greenwich
Chatham Maritime, Kent, UK
[2]Royal Botanic Gardens, Richmond, Surrey, UK
[3]Department of Agricultural Research Services, Mzuzu, Malawi
[4]Faculty of Agriculture, University of Zimbabwe, Mount Pleasant
Harare, Zimbabwe
[5]Southern Alliance for Indigenous Resources, Belvedere, Harare, Zimbabwe
[6]Department of Chemistry, Mzuzu University, Luwinga, Mzuzu 2, Malawi
[7]World Agroforestry Centre (ICRAF),
Chitedze Agricultural Research Station, Lilongwe, Malawi
*E-mail: p.c.stevenson@gre.ac.uk

INTRODUCTION

The principal need for most African farmers is food security, so the production of sufficient food is paramount. Additional yields, however, provide a potential route out of poverty by generating income. While productivity is affected by numerous biotic and abiotic constraints, arthropods (including insects and mites) are perhaps the most important due to their widespread and severe damage and because even the poorest farmers can have some direct control over them. Synthetic pesticides are usually effective but may have limited distribution in rural areas, are frequently adulterated and inappropriately applied and are often ineffective because the pest insect has become resistant to the limited number of insecticides available in remote areas.

When they are available, insecticides are applied with little personal protection which puts farm workers at risk. Furthermore, there is no process for ensuring the safety of food for consumers or for assessing chronic outcomes from exposure despite guidance on correct use from manufacturers. In environmental terms, the impact on wildlife, natural enemies and pollinators is a serious problem, and finally, the cost of pesticides can be prohibitive for poor farmers and can cost a substantial percentage of the overall product value. Recent surveys highlight these problems but also indicate that they are well understood by farmers (Kamanula *et al.*, 2011; Nyirenda *et al.*, 2011) and have led to many avoiding commercial products altogether. Thus an alternative management strategy for controlling insect pests of crops and livestock is essential since severe losses in most field crops and storage are otherwise guaranteed and the impact on poor farmers can be catastrophic.

Pesticidal plants are an effective alternative and their promotion, particularly with optimised application protocols and effective extension services, would have enormous impact on the ability of farmers to manage insect pests. Here we refer to pesticidal plants as crude materials that have undergone limited processing that can be carried out by farmers as compared to botanical insecticides that are generally more sophisticated and commercialised products and thus may come a price that farmers can't or don't wish to afford.

Farmers are familiar with the concept of plant materials as pesticides and recognise that they are environmentally benign, less toxic and cost-effective compared to synthetic pesticides. They also overcome the problems of adulteration particularly if produced or harvested by the farmers themselves. Perhaps the most important factor is their cost to farmers is calculated in time rather than cash or credit.

Generations of farmers have used plants in this way, making the technology familiar, trusted and acceptable, but their priority in agricultural policy is low (Belmain and Stevenson, 2001) and consequently, despite the overall interest in plant materials as pesticides, few farmers actually use them (Kamanula *et al.*, 2011). This may also be due to knowledge gaps or because there is inadequate policy to drive commercialisation as is the case with commercial synthetic pesticides; yet the hurdles to register products appear to be the same for plants as they are for synthetics.

Nevertheless, some plant species are effective at controlling some pests and so help secure crop production, storage and livestock production. Optimising their full potential, however, especially for poor farmers, is constrained by inadequate evaluation and development of materials, which is required to increase the reliable options available to farmers. This paper aims to identify the hurdles to the successful deployment of plants as pesticides in Africa and ways to overcome them.

Farmer surveys carried out in Ghana have highlighted that many farmers do not use commercial synthetics (Belmain and Stevenson, 2001) and, instead, use plant-based products. More recently, surveys in Malawi and Zambia in 2007/2008 (Kamanula *et al.*, 2011; Nyirenda *et al.*, 2011) reported that farmers were knowledgeable about plant materials as environmentally benign, safer and cost effective alternatives to synthetic pesticides. Surveyed farmers recognize pesticidal plants as reliable and if collected or produced themselves, that their cost can be calculated in terms of time rather than in cash. However, despite a wide knowledge of which plants are effective and how to use them, surprisingly few farmers in these countries were doing so. The study concluded that there was a need for targeted research to optimize their use rather than simply evaluate yet another plant against yet another species of insect with little concern for how the material might ultimately be used.

The target beneficiaries of the information in this paper include researchers at universities, governmental research stations, NGOs, policy makers, community-based organisations and small and medium enterprises such as farmer cooperatives who may be able to generate revenue from the technologies developed through pesticidal plant research. The collective knowledge of the stakeholders about the use and optimisation of pesticidal plants is important for the establishment of multi-disciplinary and multi-country research partnerships who are needed in Africa to develop, optimise and promote pesticidal plant use widely. An analysis of published African-based science on pesticidal plants indicates that much research remains largely ill-informed of the critical knowledge gaps and is repeated work or delivers outputs that are unlikely to have any impact on agricultural development and poverty alleviation. It does not address the problems that limit the development, uptake and promotion of pesticidal plants. Furthermore, >80% of published work in this research area in Africa reports laboratory bioassays that simply evaluate pesticidal plants against insects with little consideration of the information needed to optimise use, facilitate up-scaling, enhance field efficacy or address health and safety issues (vertebrate toxicity), propagation, cultivation or conservation. This is particularly well illustrated by the example of biological activity reported in non-polar extracts of a plant against a pest species that farmers will attempt to control in the field using crude water extracts that will unlikely contain any of the active components reported in the scientific study.

PESTICIDAL PLANTS AS ALTERNATIVES TO SYNTHETIC PRODUCTS IN AFRICA

The use of plants and their crude extracts for the protection of crops and stored products from insect pests has been a part of African agriculture for generations (Thacker, 2002). Globally, plants were the major technology for insect pest control before the advent of synthetic pesticides (Isman, 2008). They have been reviewed countless times in the literature and this paper does not attempt to repeat this. For example, Prakash and Rao (1997) have published a comprehensive review of pesticidal plants in which they describe the biological activities and practical applications for 150 species with some agricultural potential. Koul and Dhaliwal (2000) and Koul (2005) describe in detail numerous examples of phytochemical biopesticides and their role in integrated pest management, while discussing the problems and prospects of biopesticides for commercial exploitation. One sobering thought, however, is that despite tens of thousands of research articles in the literature there are only a handful of successful commercial botanical products used for insect control including pyrethrum, rotenoids, neem, and essential oils. Plants as commercial products have arguably fallen short of their purported potential (Isman, 2006). However, in Africa their use still has a major importance for farmers despite its decline elsewhere in the world and it is perhaps for the rural poor that plants have the greatest value in agriculture (Isman, 2008). Stoll (2005) provides a comprehensive list – specifically of African species – and presents suggested practical applications for a wide variety of different tropical pests and crops that are low tech requiring the minimum facilities for their preparation. Similarly in this paper, we consider pesticidal plants that comprise crude plant materials including leaves, stem bark, roots, fruits and seeds that are effective but need only rudimentary preparation and are thus more suitable for the resource poor small holder farmers across Africa. This processing includes drying, crushing and mixing with stored products (Belmain and Stevenson, 2001) or producing crude extracts in water for application with a sprayer or applied using a brush (Stoll, 2005). The less processing required the more universally appropriate the plant material.

A survey in Malawi and Zambia showed that all farmers were aware of at least one plant that could be used as a pesticide but less than half in Malawi and less than 20% in Zambia actually used them (Kamanula *et al.*, 2011; Nyirenda *et al.*, 2011). It is not immediately clear why this is but relatively low efficacy or high variability in efficacy could explain this anomaly. For many farmers some efficacy even if only moderate is better than none, provided a reduction in damage can be perceived that is quantifiable and of economic benefit to them. There may also be scope to optimize efficacy by understanding the mechanisms of the effect and developing processes for their application that exploit this knowledge as discussed below.

Pesticidal plants are low cost compared to synthetic products and some have multiple uses which can help promote their use. For example, *Tephrosia vogelii* is a well studied pest control species (Ogendo *et al.*, 2005; Boeke *et al.*, 2004a,b; Koona *et al.*, 2005) and is particularly popular among farmers in south eastern Africa where 80% of farmers report its use (Kamanula *et al.*, 2011; Nyirenda *et al.*, 2011). *Tephrosia vogelii* is also cultivated for soil improvement (Mafongoya and Kuntashula, 2005), and this means that farmers can cultivate their own pesticidal material rather than spend time harvesting from the wild. This additionally reduces pressure on declining natural environments. Indeed, efforts to optimize the propagation of pesticidal plants that cannot be easily cultivated are required. Furthermore, propagation needs to be assisted through understanding the chemical mechanisms involved in their use for the selection of elite material (Sarasan *et al.*, 2011). Home grown plants may provide an opportunity for entrepreneurial farmers to acquire additional income from their preparation and sale of excess material to other farmers. Commercialization of this process would ultimately drive an effective promotion strategy, but this is likely to be complicated by the regulatory bodies that approve of use of pesticides. Materials that are collected from natural woodlands can cost extra time which can make such plant materials less attractive, so the scope to develop the most basic commercial processing may have potential to increase the extent of their use.

SAFETY OF PESTICIDAL PLANTS

While there are clear economic benefits for the use of pesticidal plants, the greatest benefit from their use may be in terms of human health. Most acute human poisonings from pesticides are recorded in developing countries, particularly Africa where in some regions they are a major cause of mortality (Ecobichon, 2001) – this is very rare event in western Europe or North America. Despite dramatic improvements in the safety of agrochemicals with dramatically reduced health and environmental impacts, the overall perception of the public remains largely negative with strong memories of damaging products from the past, such as DDT. Plant products, however, are relatively safe but this cannot be assumed. Some of the most toxic substances known to man e.g., aconitine and ricin are produced by plants. While *Aconitum* spp. and *Ricinus communis* have no place in pest control some popular and commonly used plant materials in Africa such as *Tephrosia vogelii* have well-known environmental impacts, particularly against fish (Neuwinger, 2004; Agbon *et al.*, 2004). Many highly toxic plants are particularly well-known for their activity and their use, while impossible to curb, should be promoted with caution and good advice. Some plant compounds that are important in pest control such as nicotine and rotenone have acute mammalian toxicity but in reality, the human health risks associated with the application of these compounds are largely mitigated owing to the concentrations of the active substances being typically very low in crude preparations (Isman, 2008).

In an ideal world, the production of extracts should be standardized to ensure safety and moderate efficacy, but in reality this is an unrealistic expectation in many of the poorer locations of Africa. Regulations are in place for at least some of the most common materials elsewhere in the world such as rotenone in the USA and these standards should be adopted. What is clear is that plant compounds, e.g., rotenone, azadirachtin and pyrethrum, break down, particularly in sunlight, into environmentally benign products, leaving no dangerous residues behind as often occurs with synthetics. Pesticidal plants certainly have the popular vote. Organic growers and regulators accept pesticidal plants as organic production, an increasingly important market share in agricultural produce. Certain materials should be promoted with caution, however, since some plant compounds previously considered acceptable and indeed popular in organic farming are now listed as hazardous under new European directives and thus may not be acceptable contaminants on food entering Europe, including rotenone. This could have consequences for African farmers whose produce may contribute to European exports.

From a resource poor farmers perspective, pesticidal plants are appealing because they cannot be adulterated (at least if they are collected by the farmer), and are cost effective. Although they do require time to collect and prepare on top of the input required to apply them. However, their efficacy can vary across seasons and locations and the application procedure is not always as efficient or effective as it could be. This highlights the importance of understanding the chemistry of their activity to optimize their application. Nowhere is this demonstrated more emphatically than with Neem, although other species are now benefitting from increased knowledge about their activity.

KEY RESEARCH PRIORITIES FOR OPTIMIZING THE USE OF PESTICIDAL PLANTS IN AFRICA

Understanding the Chemistry of Activity

The sustainable use of pesticidal plants could be improved substantially through a deeper understanding of the chemistry that governs their activities. This information could be used to enhance application methods, improve harvesting strategies and even identifying novel plant alternatives for threatened species (Koul and Dhaliwal, 2000). For example, an abundant plant with similar chemistry to a scarce but popular and over harvested species could be promoted as an environmentally benign alternative. Furthermore, plants in which active components are water soluble, such as saponins, might be more efficiently applied as water extracts than powdered plant material, as has been proposed for the securidacaside saponins present in *Securidaca longepedunculata* (Stevenson *et al.*, 2010). The chemistry of some plants varies according to the season, the plant age and location so chemical analysis is essential to determine the best time to harvest or for identifying elite material for propagation (Sarasan *et al.*, 2011). Of course, the need to use fresh plant material over dry might determine harvesting priorities but chemical knowledge of the active components would certainly optimise harvesting times. The ability of many African laboratories to carry out this analytical work is limited to a handful of institutes. Cooperation through research networks is essential to move this area of work forward, and expertise and equipment in natural product chemistry needs to be expanded.

Chemical analysis using tandem linked detection techniques such as mass spectrometry can maximise information acquisition of compounds in each plant and can identify many components quickly, particularly seasonal variations and genotypic differences between specimens of a single species to better understand the potential variation in efficacy. These techniques can also be used to authenticate specimens before promoting widely. While these facilities are beyond the reach of many research institutes, many materials could be authenticated by less costly techniques such as thin layer chromatography.

Safety of Key Pesticidal Plants

It is invariably assumed that natural is safe, and this received wisdom applies particularly to pesticidal plants. This is, however, not the case with some illustrative plant species having notoriously high toxicity such as *Taxus* spp., *Aconitum* spp. and perhaps most notorious of all *Ricinus communis* (Hernandez *et al.*, 2010; Bonnici *et al.*, 2010; Kuca and Pohanka, 2010). Several well-known pesticidal plants are also known to have toxic properties that could present a significant health risk to users when extracting, pounding or concentrating active ingredients, including rotenoid containing species of *Derris*, *Tephrosia* and *Lonchocarpus*. Yet little work has been published internationally on the vertebrate toxicity of African pesticidal plants. One study on West African plants used to protect stored grain from insect pests has shown that some of those most commonly found and used by farmers can adversely affect growth and development (when fed to rodents), with potentially long term effects (Belmain *et al.*, 2001). Ideally plants that have other users such as in foods can indicate relative safety and this approach is being used in North America to focus attention on essential oils from herbs (Isman, 2006). Despite this, even plant species that are traditionally consumed – such as *Lippia javanica* which is drunk as a green tea in Malawi – showed acute oral toxicity in mice when consumed at very high concentrations (Madzimure *et al.*, 2011).

In reality, plant materials are normally used for pest control at concentrations that are too low for them to pose any significant acute toxicity to users (Isman, 2008) even when this is intended (Chesneau *et al.*, 2009). Perhaps the greatest risks exist when pesticidal plants are used for post-harvest treatment particularly for the storage of produce that will be used for food. Current post-harvest pest management practices in Africa rely on pirimiphos-methyl, an anticholinesterase targeting organophosphate, and in cases where pest complexes include the larger grain borer, *Prostephanus truncatus*, pirimiphos-methyl is combined with permethrin (Sekyembe *et al.*, 1993). Syngenta who market Actellic EC50 (pirimphos-methyl) warn that this product is moderately toxic to mammals if consumed directly and is extremely toxic to aquatic invertebrates and highly toxic to algae and fish (http://www.syngenta-crop.co.uk/pdfs/products/ Actellic50EC_uk_environmental_information.pdf). Thus, fears over the toxicity of plant species like *Tephrosia vogelii* that are very popular alternatives to pesticides in southeastern Africa (Kamanula *et al.*, 2011; Nyirenda *et al.*, 2011), but have known toxicity to fish, are perhaps excessively cautious. Such plants present an effective no cost product for very poor farmers that cannot be adulterated if produced by the farmers themselves and is less toxic and hazardous. Unfortunately, appropriate cell based procedures for mammalian toxicity testing for plant materials are not currently available and would be very useful to understand the potential risks associated with pesticidal plant materials.

Farmer Participatory Rural Appraisals

Much research published in international journals on African pesticidal plants has been carried out in laboratories or on agricultural research stations. While relevant, these trials fail to assess the use of plant materials under farm conditions which may differ from the uniform environment of a research station. Consequently, they do not assess many of the other factors, besides efficacy, which are important to farmers when they choose to use plant materials for insect pest control (e.g. labour, availability, ease of use, marketability of treated produce). For example, Stevenson *et al.* (2010) showed that saponins were active components in the root bark of *Securidaca longepedunculata* against stored product pests and postulated that the unsustainable use of the root bark could be made more sustainable by making a water extract of the bark and using this to treat grain prior to storage. The thinking behind this is that current practice involves pounding bark to a powder and mixing this with the grain, a technique that is inefficient for several reasons and which could be made more efficient since the application of an extract would optimise the use of the active compounds. Anecdotal reports from current field trials, however, indicate that this application method is unpopular with farmers since the process of making the extract and then dipping the grain and drying before storage is too labour intensive despite the fact that it is as effective as using the whole root but only requires half as much to material treat the same amount of commodity (P. Stevenson, unpublished).

The small and irregular sized plots that are common among resource poor farmers in Africa may require specialised knowledge about how best to apply materials and some farmers may apply using a watering can rather than a specialised device like a knapsack sprayer. Farm trials will also engage farmers with a promotion strategy and help to better understand how applications can be optimised based on the available technology. For example, most farmers make cold water extracts of pesticidal plant materials yet the components may be non-polar and, therefore, inefficiently extracted by this process. Field trials with farmers will enable the evaluation of appropriate novel approaches such as the use of hot water or soaps that might enhance extraction and further optimise the application by acting as a surfactant.

Ultimately, the goal of pesticidal plant research in Africa should be simple, safe and environmentally conscientious protocols that can be easily understood and distributed widely by extension services and NGOs.

Sustainable Harvesting and Local Production of Pesticidal Plants

The use of pesticidal plants collected from the wild is only sustainable if small numbers of people use the plant or if the plant is abundant and ubiquitous or propagated easily. However, demand for some plants is outstripping supply (Shackleton *et al.*, 2005; Harnischfeger, 2000), particularly as overgrazing or bush fires also reduce supply not withstanding the fact that overgrazing can actually be a benefit when it comes to tree recruitment (Chidumayo, 1997). There is little knowledge about growing and propagating some pesticidal plant species making attempts to cultivate them difficult while a lack of phytochemical knowledge makes it difficult to know which the optimal methods for collecting plants are (Sarasan *et al.*, 2011).

Propagation. The conditions for germination vary considerably and plant seeds often need to be stimulated. Recent work has shown that some species can be propagated from seeds which could be important to establish large scale cultivation of some species particularly tree species such as *Securidaca longepedunculata* and *Bobgunnia madagascariensis* (Zulu *et al.*, 2011; Thokozani *et al.*, 2011) Timing of seed collections, seed drying, storage, and germination are also required.

Sustainable harvesting and processing. In some cases, roots, bark, seeds or entire plants of pesticidal species are used, and the collection of these parts is not sustainable (Belmain and Stevenson, 2001; Stoll, 2005). For example, where roots are harvested, often killing the plant, active compounds may also occur in leaves which would be more sustainable. Modified and limited root harvesting methods could also reduce the impact on species abundance. The compounds in pesticidal plants can vary according to geographic location and season and, since their efficacy as pest control agents depends on these compounds, harvesting times can be crucial in optimising their efficacy. Once harvested, plant materials are usually dried, ground or extracted to give a product which is used immediately, stored or sold. Exposure to sunlight and oxidation can change chemistry and thus affect efficacy, so trials are required that evaluate differences in the way in which farmers process pesticidal plant species.

Maximising availability of plant pesticides to small-scale farmers. If techniques to propagate pesticidal plants can be established, then this may open up the possibility of marketing pesticidal plants as a cash crop for small-scale farmers. This could provide very important additional income for poor farmers (Sarasan *et al.*, 2011) while developing a way to upscale and promote a technology. Up-scaling technologies are limited by funding programmes, particularly when such activities are paid through short-term extension initiatives with little follow-up. However, sustainable promotion can be more assured if the practice can be handed over to farmers with the incentive of income generation. This is an essential step towards the formalisation of pesticidal plant use for agricultural pest management in Africa, particularly if government and non-governmental organisations continue to promote the use of wild plants. Demand for pesticidal plants will continue to grow, which can only realistically be met through their cultivation and marketing. One problem that must be overcome, however, is the leniency with which governments are willing to allow plant based products to be registered as approved pest control products.

Currently in most African countries any pesticidal product whether a synthetic mass produced chemical or a plant material that is packaged for the specific purpose of controlling insects are required to satisfy the same level of interrogation regarding their efficacy, toxicity and environmental hazard. The costs of producing most of the information are prohibitive and will continue to be an insurmountable hurdle to their wide-scale use and production until the registration procedure is modified for natural products. Interestingly the laws governing the sale of herbal remedies in South Africa, products that are intentionally sold to consume internally, are very relaxed owing the perceived importance of this trade as a traditional practise. In the same country, however, if a plant product is sold as a pest control product it is quite simply considered as a pesticide and needs to be registered as such. Consequently, while the production and sale of pesticidal plants does occur, it all happens under the radar of the formal pesticide sector. Market surveys are required to establish which plants are already sold in markets, prices, demand, and the marketing chains involved. Perceptions of the target users is also required and could be obtained through surveying local and regional markets using questionnaires and structured discussion with individuals and groups of market traders, purchasers, suppliers and producers.

PESTICIDAL PLANTS IN MIOMBO WOODLANDS OF AFRICA: CURRENT RESEARCH AND DEVELOPMENTS

Numerous species, many with only local use and knowledge, are found across Africa. But through our recent work combining surveys (Kamanula *et al.*, 2011; Nyirenda *et al.*, 2011), databases (e.g., http://epic.kew.org) and scouring the literature, we identified key pesticidal plant species found in Miombo woodland. Miombo woodland is dominated by trees in the genus *Brachystegia* along with *Colophospermum mopane* (Mopane) and *Baikiaea plurijuga* which is misleadingly referred to as Rhodesian teak. Miombo derives from the Bemba pleural for *Brachystegia longifolia*. The below list is just a snapshot of plants of interest in the Miombo woodlands and is by no means exhaustive. However, it provides an opportunity to update the research on a select few species and raises questions and issues that are generally relevant to all pesticidal plant species. Interestingly, several of the species we encountered as of value in pest control are virtually unknown in the literature for the reported applications and present interesting new research and development opportunities. These include, *Aloe ferox* which is reportedly burned and the dust used to treat stored grain in Zimbabwe and Zambia and *Euphorbia tirucalli* which has been shown to have activity against larvae of two species of mosquitoes (Rahuman *et al.*, 2008; Yadav *et al.*, 2002) but has potent skin irritant properties associated with the latex (Kinghorn, 1979) which may in practice dissuade use.

Another species commonly reported is the poison apple *Solanum panduriforme* (now included in *S. incanum* although sufficiently different to warrant distinction) but apart from some work on its antimicrobial activity (e.g. Pesewu *et al.*, 2008) there is no evidence to support insect activity. Some exotic species have widespread uses such as *Cymbopogon nardus* and *C. citratus* and Neem, but the latter highlights the importance of good extension and training in use. Most farmers who report using Neem use the leaves which are notoriously low in their bioactive components. And in some parts of southern Africa, the climate is not conducive to flowering in *Azadirachta indica*, thus the trees do not produce seeds where the greatest quantities and diversity of pesticidal and deterrent compounds can be found. Other species such as *Lantana camara* that are popular in other parts of Africa do not appear to be used in southern Miombo regions where they are considered highly invasive and receive high priority for local eradication programmes.

Securidaca longepedunculata

Securidaca longepedunculata (Polygalaceae) is a small tree often referred to as the African Violet tree and is found throughout sub-Saharan Africa. Several studies have corroborated the effect of the plant. For example, Belmain *et al.* (2001) showed that *S. longepedunculata* was the most toxic to *Rhyzopertha dominica*, *Callosobruchus maculatus*, *Sitophilus zeamais*, and *Prostephanus truncatus* when compared to *Cassia sophera*, *Chamaecrista nigricans*, *Mitragyna inermis*, *Ocimum americanum* and *Synedrella nodiflora* which are other species reported in the Northern Region of Ghana to be used traditionally for pest control in stores. When evaluating 33 West African species for toxicity to *C. maculatus*, Boeke *et al.* (2004a,b) showed *S. longepedunculata* along with *Nicotinia tabacum* and *T. vogelii* to reduce F1 progeny, indicating a level of toxicity to the beetles or oviposition deterrence. In the same study, *Clausena anisata*, *Dracaena arborea*, *T. vogelii*, *Momordica charantia* and *Blumea aurita* were shown in a linear olfactometer to be repellent to the beetle. Surprisingly, *Chamaecrista nigricans*, *Azadirachta indica* and *Hyptis suaveolens* were attractive to the beetles. Root bark was also shown to reduce the FI progeny of *R. dominica*, *C. maculatus*, *S. zeamais*, and *P. truncatus* (Jayasekera *et al.*, 2003).

Methyl salicylate is the principal volatile component in the root of *S. longepedunculata* (Jayasekera *et al.*, 2002) and it is this compound that causes at least some of the toxic effects (Jayasekera *et al.*, 2005). However, toxicity is also attributable to the saponins (Fig. 11.1) that are found in abundance in the root extract (Stevenson *et al.*, 2009). These compounds might then explain the toxicity of *S. longpedunculata* non-polar extracts to the mosquito *Ochlerotatus triseriatus* and to a lesser extent the whitefly, *Bemisia tabaci* reported by Georges *et al.* (2008). This activity was also shown to be high against the mosquito in other species in the same study including, *Datura inoxia*, *C. nigricans*, *Jatropha curcas* (skin and seeds) *Strophanthus hispidus*, and *Sapium grahamii*. If the active components against whitelfy and mosquitoes in *S. longepedunculata* are the saponins then they would be conducive to use as a spray by making a water extract since saponins are water soluble.

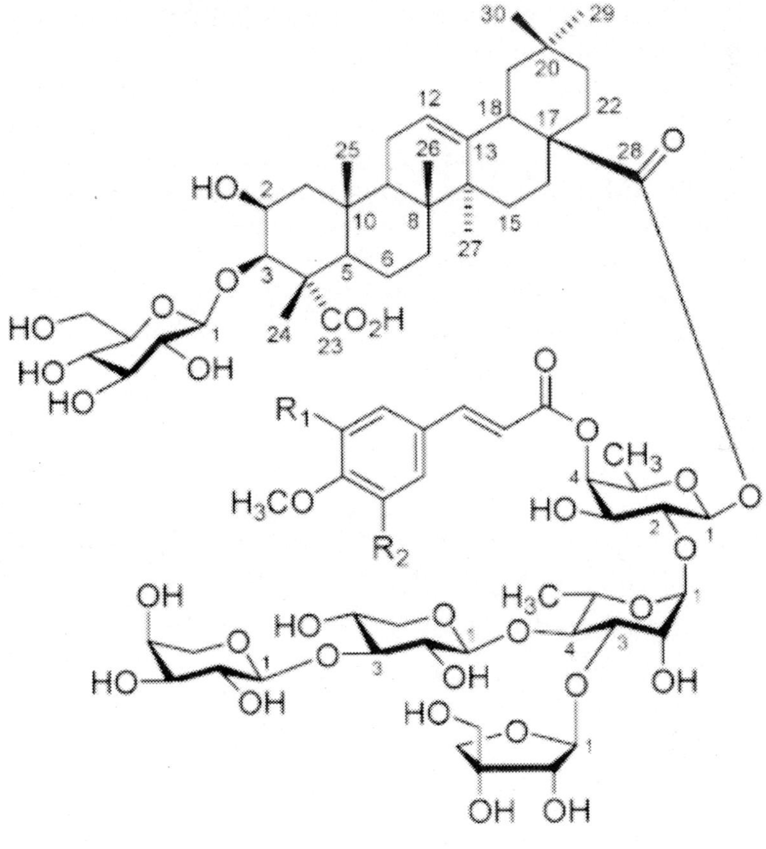

1 $R_1 = R_2 = H$; Securidacaside A

2 $R_1 = R_2 = OCH_3$; Securidacaside B

Fig. 11.1. Saponins isolated from *Securidaca longepedunculata*.

Bobgunnia madagascariensis

Bobgunnia madagascariensis (syn. *Swartzia madagascariensis*; Kirkrbide and Wiersema, 1997) (Leguminosae) is reportedly used for protection of stored products from beetles in Zambia and other parts of southern Africa. However, there is surprisingly little scientific evidence to substantiate activity for this reported application but considerably more reports of medicinal and molluscicidal activity (Kone *et al.*, 2004; Borel and Hostetmann, 1984; Marston *et al.*, 1993). There are some reports of activity against mosquitoes and whitefly. For example, ethyl acetate extract of *B. madagascariensis* caused 80% mortality in one report with other plants (Georges *et al.*, 2008) whereas other plants in the same bioassays exhibited only 30-50% mortality. In the report, antifeedant assays against *Helicoverpa zea* and *Heliothis virescens*, showed that the methanol extracts of *B. madagascariensis*, *C nigricans*, and *S. hispidus* were effective against *H. zea* causing a reduced weight gain of test insects. Earlier, the extracts of *B. madagascariensis* were shown to be active against mosquito larvae (Minjas *et al.*, 1986) and repellent to termites (Crombie *et al.*, 1971). The active components are likely to be saponins (Fig. 11.2) which occur in the pods and bark.

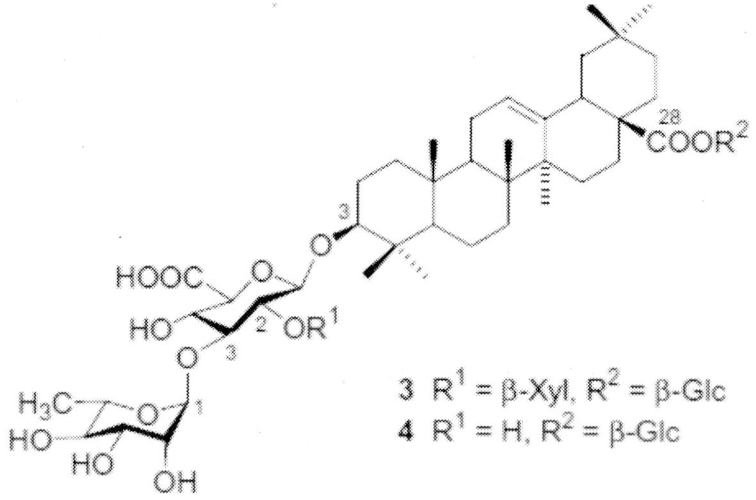

28
COOR^2
3
HOOC
HO
2
OR^1
3
H_3C
HO
HO
OH
1
3 R^1 = β-Xyl, R^2 = β-Glc
4 R^1 = H, R^2 = β-Glc

Fig. 11.2. Saponins from the pods of *Bobgunnia madagascariensis*.

Stevenson *et al.*, 2010; Marston *et al.*, 1993 since the only other components found in the pods are highly glycosylated flavonoids which are not biologically active to insects (Stevenson *et al.*, 2010). The presence of these saponins, however, does vary between locations, thus the selection of appropriate progeny for propagation or simply as a source of bioactive material is critical and should be based on chemical analysis (Sarasan *et al.*, 2011).

Tephrosia vogelii

T. vogelii (Leguminosae) is possibly the best known and widely used pesticidal plant in Africa, particularly in southern and eastern Africa (Nyirenda *et al.*, 2011; Kamanula *et al.*, 2011) but is widespread across the continent. The species is cultivated widely for soil improvement (Mafongoyo *et al.*, 2005) as well as for its pesticidal use and fish poisoning properties (Neuwinger, 2004) although the latter application is now prohibited in virtually all regions of the continent. It is clearly introduced in many locations so the extent of its natural coverage is unclear. A related species, *T. candida* is widespread in parts of Asia where it is also used as a mulch and cover crop (Fagerstrom *et al.*, 2001) and this is being increasingly used in Africa since it is considered to be more effective at soil improvement than *T. vogelii* (Mafongoyo and Kuntashula, 2005). This immediately presents a possible problem in that farmers, who know about the insecticidal properties of *T. vogelii* leaves now might assume the same of *T. candida*, which while not well studied, is not well-known for its use as a plant pesticide. While some chromene rotenoids are reported from the roots (Andrei *et al.*, 1997) and the stems and leaves (Kole *et al.*, 1992); our recent studies of *T. candida sensu stricto* herbarium vouchers indicate rotenoids to be absent from this species indicating poor scope for use as a pesticide (Stevenson, 2010) and any biological activity towards insects that they express is not associated with rotenoids (Lapointe *et al.*, 2003). Leaves of *T. vogelii, however,* contain at least four rotenoids (Fig. 11.3) with deguelin and tephrosin the major isoflavonoid components and rotenone and dehydrorotenone as minor components (Lambert *et al.*, 1993). The highest concentration of the active compounds is found in the leaf which makes it ideal for use since the foliage is the most abundant and sustainably harvested plant part. Besides this, it is relatively easily cultivated so is the species for which its use is least likely to have any environmental impact.

There is a need to determine the biological activity of the two main rotenoids tephrosin and deguelin to fully understand the potential activity of the species. Since much of the flag waving about the value of *Tephrosia* rests on its content of active rotenoids, and that rotenone is the only one that has been evaluated in laboratories much work still needs to be done in evaluating the other components to determine optimised applications. While much 'grey' literature cites rotenoids in the leaves of *Tephrosia* to be insecticidal, surprisingly there is no published work to corroborate this. Indeed the bioactivity of *T. vogelii* against bruchids and weevils was even reportedly not associated with rotenoids (Koona and Dorn, 2005) although this was largely based on the expectation that rotenoids would not occur in methanol extracts. Our recent research can confirm that rotenoids are found in methanol extracts and that rotenoids are the active principles, at least against bruchids.

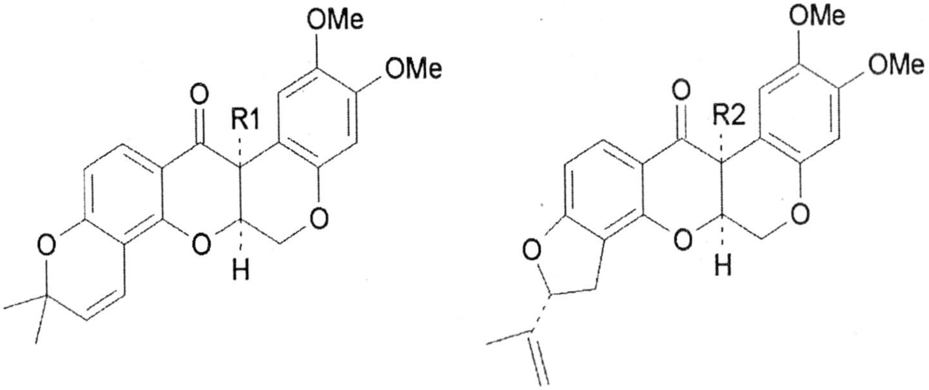

Deguelin R1 = H
Tephrosin R1 = OH

Rotenone R2 = H
Rotenonlone R2 = OH

Fig. 11.3. Rotenoids from *Tephrosia vogelii*

The active components are generally reported to be in the leaves and roots and the reported effects are antifeedant, insecticidal, acaricidal, ovicidal, icthyotoxic and as a contact and stomach poison to insects (Gaskins *et al.*, 1972). Unpublished work indicates that the addition of liquid soaps can greatly optimise the efficiency of extraction (Stevenson unpublished) of rotenoids. For example, 5% teepol extracts 8 times more rotenoids from dried leaves than water alone, which is almost as efficient as using 100% methanol. If concentrated extracts are made using 5% liquid soap, these can then be diluted prior to spraying so that soap concentrations are more appropriate for field applications e.g., < 0.1%. The incorporation of soaps early on in the process may help to ensure that surfactants are ultimately used when sprayed to help disperse the active chemicals on the plants – an inclusion that is all too often overlooked by farmers using pesticidal plant extracts or commercial products.

Neorautanenia mitis

N. mitis (Leguminosae) is related chemically to *Tephrosia* spp. and other rotenoid producing plants. It produces a very large dense tuber that is high in flavonoids including rotenoids, isoflavones and pterocarpans. The size of the tuber can weigh tens of kilos and is relatively easy to propagate from tuber cuttings, thus like Tephrosia is conducive to cultivation by farmers. The principal rotenoid component of the root tuber in East and southern African specimens is rotenone (Stevenson, 2010) although earlier studies have identified a variety of pterocarpans with potential antifungal activity (Sakurai *et al.*, 2006) and earlier still reporting neotenone as the principal isoflavonoid component (Vanpuyvelde *et al.*, 1987). Indeed extracts of *N. mitis* root are biologically active against larvae of *Ancpheles gambiae* and *Culex quinquefaciatus* mosquitoes, as well as against the adults of *An. gambiae* with the isoflavonoids neotenone and neorautanone and the pterocarpans neoduline, nepseudin and 4-methoxyneoduline reported to be responsible for these effects with activities comparable to deltamethrin and cypermethrin (Joseph *et al.*, 2004). With reports of this efficacy, it is surprising that more research on this species with respect to its possible role in pest control has not been conducted. Chimbe and Galley (1996) reported activity of the petroleum extract of the plant material against *Sitophilus oryzae* and *Prostephanus truncatus* but it was less effective than *Dicoma sessiliflora*.

Tithonia diversifolia

Tithonia diversifolia (Asteraceae) (Mexican marigold or Tree marigold) is a noxious exotic weed of southern and eastern Africa growing from South Africa to Uganda and from sea level to at least 1500m. It occurs in abundance in many places and is now listed in South Africa as an invasive plant that is illegal to grow and subject to national eradication programmes. On the brighter side, there may be scope to control its weed status by using it as a pesticidal plant. Indeed it is reported by farmers to be one of the pesticidal plant options available and used by them in southeastern Africa (Nyirenda *et al.*, 2011). Adedire and Akinneye (2004) showed that the leaf powder and the leaf extract reduced oviposition, adult emergence and increased mortality of *C. maculatus* although this effect was greatest for the leaf extracts. One hundred percent mortality was reported for adults after 48 h at all concentrations tested indicating a potent effect of compounds in the leaves. This effect could be associated with the sesquiterpene lactones that are reported to exude from the leaf hairs, particularly on the abaxial leaf face and are considered to be responsible for the deterrent effect of the leaves against bordered patch larvae, *Chlosyne lacinia*. *T. diversifolia* along with *Montanoa hibiscifolia* was also known to be a deterrent to *Bemisia tabaci* (Bagnarello *et al.*, 2009). Additionally there is evidence that *T. diversifolia* is deterrent to the leaf cutting ants, *Atta cephalotes*, although the greatest impact on colonies is associated with the effects of the plant on the fungus colony which they feed the leaves to (Rodriguez *et al.*, 2008; Valderrama-Eslava *et al.*, 2009).

Lippia javanica

The genus *Lippia* (Lamiaceae) has received a lot of attention in the last few years with respect to its potential uses in medicine and as an insect repellent owing to its broad spectrum activity and wide distribution in Africa and South America. It is also perhaps motivated by its use as a medicinal tea against the symptoms of fever, flu and cold (Viljoen *et al.*, 2005). This use suggests a low acute toxicity to mammals and favours promotion. Examples of its biological activity against various arthropods occur in the literature. Madzimure *et al.*, (2011) recently reported its biological activity against cattle ticks while mosquito repellency has also been reported in Zimbabwe (Lukwa *et al.*, 2009) and Kenya (Omolo *et al.*, 2004). In the case of mosquito repellency, the active components were reported to be essential oils including perillyl alcohol, cis-verbenol, *cis*-carveol, geraniol, citronellal, perillaldehyde and caryophyllene oxide whereas the compounds responsible for tick activity are not known but were present in a water extract thus unlikely to be essential oils. The work of Madzimure *et al.* (2011), however, reported acute toxicity of *Lippia javanica* when fed at high concentrations to mice so any promotion of this material needs to be done with caution.

Several related species are also known to be effective. For example, *Lippia alba* has been reported to be a potent deterrent to *Tribolium castaneum* with benzyl benzoate, β-myrcene, and carvone reported to be responsible for the effects (Caballero-Gallardo *et al.*, 2011). *Cymbopogon citrates*, *Lippia sidoides*, *Ocimum americanum* and *O. gratissimum* were reported along to have potent deterrent effects against *Aedes egypti* (Cavalcanti *et al.*, 2004) with similar effects reported for two other species of *Lippia*; *L. integrifolia* and *L. junelliana* (Gleiser *et al.*, 2011). Another species, *L. multiflora*, which is reportedly important in West Africa (Belmain and Stevenson, 2001) has recently been shown to be toxic against *C. maculatus* at around

5ul/l, although in these trials, *O. americanum* was ten times more effective and more persistent (Ilboudo *et al.*, 2010).

If there is one thing that is consistent about *L. javanica*; its chemistry is highly inconsistent. Numerous chemical studies report a variety of chemical compositions, and this identifies a critical area of work required before the material can be promoted widely. Even within Malawi, our recent work indicates that samples from 3 sites are chemically diverse. If there is a role for *L. javanica* in pest control, it needs to be based on a progeny of wild material that is validated by biological studies and chemical analysis. Since most of the compounds that are reported are likely to occur in foods its registration as a pest control material may be straight forward.

Vernonia amygdalina

V. amygdalina (Asteraceae) is well known for its healthy properties as a food supplement and perhaps best known as a choice supplement of chimpanzees who use it to deparasitise themselves (Yeap *et al.*, 2010). However, the species is reported to be a pesticidal plant by some farmers and appears to have some evidence supporting this potential use. Adeniyi *et al.* (2010) reported that organic extracts of *V. amygdalina* were more toxic to the bean weevil *Acanthoscelides obtectus* than *Sida acuta*, *Ocimmum gratissimum* and *Telfaria occidentalis*, while the combined essential oil of *V. amygdalina* was toxic against *S. oryzae* (Asawalam *et al.*, 2008).

CONCLUSIONS

Historically, pesticidal plant use in Africa has been an important part of traditional pest management practices by farmers. We believe that pesticidal plants should remain an important part of the future pest management options available for the many small-scale farmers in Africa who continue to face major impacts on their livelihoods caused by arthropod pests and who have been ill served by commercial synthetics for the reasons of cost, safety and availability as discussed earlier in this article. To ensure the future for pesticidal plants, many bottlenecks need to be addressed by the scientific community, policy makers and institutions involved in knowledge dissemination. Scientists need to provide better information that explains how pesticidal plants work, which arthropod species are affected, how the bioactive chemicals may vary according to season, locality or variety and how best plants should be harvested and processed to conserve and deliver bioactivity. Such information will increase the reliability of pest control and the expectations that farmers can make when they choose to use a pesticidal plant. Furthermore, scientists need to engage with policy makers to tackle issues such as conservation of wild habitats and the implications of "professionalising" pesticidal plant use as has occurred in many countries concerning Traditional Healers and medicinal plant use.

If African governments wish to encourage their farmers to use pesticidal plants, they need to recognise that useful plants need to be encouraged through conservation and propagation, and ultimately allow farmers to generate income by selling pesticidal plants (as individuals or through collectives) to their neighbours or urban farmers with limited access to pesticidal plants. In our opinion, it makes no sense to permit Africans to legally buy, sell and drink concoctions of medicinal plants, whereas to buy, sell and use the same plant species to control agricultural pests is prevented through regulatory burdens designed for synthetic pesticides. A strong future for indigenous knowledge and use of pesticidal plants in Africa will need policy changes to be made to current regulatory frameworks. Scientists and policy makers must work together to establish basic safety information data that are sensible and relatively easy to regulate, particularly for pesticidal plant species that are already being widely used by African farmers. Improving our knowledge on variability of efficacy, conservation and regulation remain the big challenges for the future of pesticidal plants in Africa. We remain confident that such a strategy will increase farmer usage of pesticidal plants and ultimately help pull farmers out of poverty, increase family incomes and health.

REFERENCES

Adedire, C.O. and Akinneye, J.O. (2004) Biological activity of tree marigold, *Tithonia diversifolia*, on cowpea seed bruchid, *Callosobruchus maculatus* (Coleoptera: Bruchidae). *Ann. Appl. Biol.*, **144**, 185–189.

Adeniyi, S.A., Orjiekwe, C.L., Ehiagbonare, J.E. and Arimah, B.D. (2010) Preliminary phytochemical analysis and insecticidal activity of ethanolic extracts of four tropical plants (*Vernonia amygdalina, Sida acuta, Ocimum gratissimum* and *Telfaria occidentalis*) against beans weevil (*Acanthscelides obtectus*). *Int. J. Phys. Sci.*, **5**, 753–762.

Agbon, A., Ofojekwu, C. and Ezenwaka, I. (2004) Acute toxicity of water extract of *Tephrosia vogelii* Hook to species relevant in aquaculture ponds: rotifers, Cyclops, mosquito larvae and fish. *J. Appl. Ichthyol.*, **20**, 521–524.

Andrei, C.C., Viera, P.C., Fernandes, J.B., daSilva, M. and Fo, E.R. (1997) Dimethylchromene rotenoids from *Tephrosia candida*. *Phytochemistry*, **46**, 1081–1085.

Asawalam, E.F., Emosairue, S.O. and Hassanali, A. (2008) Contribution of different constituents to the toxicity of the essential oil constituents of *Vernonia amygdalina* (Compositae) and *Xylopia aetiopica* (Annonaceae) on maize weevil, *Sitophilus zeamais* Motschulsky (Coleoptera: Curculionidae). *African J. Biotechnol.*, **7**, 2957–2962.

Bagnarello, G., Hilje, L., Bagnarello, V., Cartin, V. and Calvo, M. (2009) Phagodeterrent activity of the plants *Tithonia diversifolia* and *Montanoa hibiscifolia* (Asteraceae) on adults of the pest insect *Bemisia tabaci* (Homoptera: Aleyrodidae). *Revista De Biol. Trop.*, **57**, 1201–1215.

Belmain, S.R., Neal, G.E., Ray, D.E. and Golob, P. (2001) Insecticidal and vertebrate toxicity associated with ethnobotanicals used as post-harvest protectants in Ghana. *Food Chem. Toxicol.*, **39**, 287–291.

Boeke, S.J., Sinzogan, A.A.C., de Almeida, R.P., de Boer, P.W.M., Jeong, G., Kossou, D.K. and van Loon, J.J.A. (2003) Side-effects of cowpea treatment with botanical insecticides on two parasitoids of *Callosobruchus maculatus*. *Entomol. Exp. Appl.*, **108**, 43–51.

Boeke, S.J., Barnaud, C., Van Loon, J.J.A., Kossou, D.K., Van Huis, A. and Dicke, M. (2004a) Efficacy of plant extracts against the cowpea beetle, *Callosobruchus maculatus*. *Int. J. Pest Manage.*, **50**, 251–258.

Boeke, S.J., Baumgart, I.R., van Loon, J.J.A., van Huis, A., Dicke, M., Kossou, D.K. (2004b) Toxicity and repellence of African plants traditionally used for the protection of stored cowpea against *Callosobruchus maculatus*. *J, Stored Prod. Res.*, **40**, 423–438.

Bonnici, K., Stanworth, D., Simmonds, M.S.J., Mukherjee, E. and Ferner, R.E. (2010) Flowers of evil. *Lancet*, **376**, 1616–1616.

Borel, C. and Hostettmann, K. (1987) Molluscicidal saponins from *Swartzia madagascariensis* desvaux. *Helv. Chim. Acta*, **70**, 570–576.

Breytenbach, J.C. (1986) Isoflavone glycosides from *Neorautanenia amboensis*. *J. Nat. Prod.*, **49**, 1003–1009.

Caballero-Gallardo, K., Olivero-Verbel, J. and Stashenko, E.E. (2011) Repellent activity of essential oils and some of their individual constituents against *Tribolium castaneum* Herbst. *J. Agric. Food Chem.*, **59**, 1690-1696.

Cavalcanti, E.S.B., de Morais, S.M., Lima, M.A.A. and Santana, E.W.P. (2004) Larvicidal activity of essential oils from Brazilian plants against *Aedes aegypti* L. *Mem. Do Inst. Oswaldo Cruz*, **99**, 541–544.

Chesneau, P., Knibiehly, M., Tichadou, L., Calvez, M., Joubert, M., Hayek-Lanthois, M. and De Haro, L. (2009) Suicide attempt by ingestion of rotenone-containing plant extracts: one case report in French Guiana. *Clinical Toxicol.*, **47**, 830–833.

Chidumayo, E.N. (1997) *Miombo Ecology and Management: An Introduction.* Stockholm Environment Institute, London.

Chimbe, C.M. and Galley, D.J. (1996) Evaluation of material from plants of medicinal importance in Malawi as protectants of stored grain against insects. *Crop Prot.*. **15**, 289–294.

Crombie, L., Ham, P.J. and Whiting, D.A. (1971) Sapogenin of termite-repellent fruit of *Swartzia madagascariensis. Chem. Ind.*, 176–180..

Fagerstrom, M.H.H., van Noordwijk, M., Phien, T. and Vinh, N.C.(2001) Innovations within upland rice-based systems in northern Vietnam with *Tephrosia candida* as fallow species, hedgerow, or mulch: net returns and farmers' response. *Agric. Ecosyst. Environ.*, **86**, 23–37.

Gadzirayi, C.T., Mutandwa, E., Mwale, M. and Chindundu, T. (2009) Utilization of *Tephrosia vogelii* in controlling ticks in dairy cows by small-scale commercial farmers in Zimbabwe. *Afr. J. Biotechnol.*, **8**, 4134–4136.

Gaskins, M.H., White, G.A., Martin, F.W., Delfel, N.E., Rupel E. G. and Barnes, D.K. (1972) *Tephrosia vogelii: A Source of Rotenoids for Insecticidal and Piscidal Use.* Technical Bulletin No. 1445, Agricultural Research Service, United States Department of Agriculture.

Georges, K., Jayaprakasam, B., Dalavoy, S.S. and Nair, M.G. (2008) Pest-managing activities of plant extracts and anthraquinones from *Cassia nigricans* from Burkina Faso. *Bioresource Technol.*, **99**, 2037–2045.

Gleiser, R.M. and Zygadlo, J.A. (2007) Insecticidal properties of essential oils from *Lippia turbinata* and *Lippia polystachya* (Verbenaceae) against *Culex quinquefasciatus* (Diptera: Culicidae). *Parasitol. Res.*, **101**, 1349–1354.

Harnischfeger, G. (2000) Proposed guidelines for commercial collection of medicinal plant material. *J. Herbs, Spices Med. Plants*, **7**, 43–50

Hernandez, J.L.H., Teran, F.Q. and Macias, J.G. (2010) *Taxus* poisoning. *Med. Clinica*, **135**, 575–576.

Ilboudo, Z., Dabire, L.C.B., Nebie, R.C.H., Dicko, I.O., Dugravot, S., Cortesero, A.M. and Sanon, A. (2010) Biological activity and persistence of four essential oils towards the main pest of stored cowpeas, *Callosobruchus maculatus* (F.) (Coleoptera: Bruchidae). *J. Stored Prod. Res.,* **46**, 124–128.

Isman, M.B. (2006) Botanical insecticides, deterrents, and repellents in modern agriculture and an increasingly regulated world. *Annu. Rev. Entomol.*, **51**, 45–66.

Isman, M.B. (2008) Botanical insecticides: for richer, for poorer. *Pest Manage. Sci.*, **64**, 8–11.

Jayasekera, T., Stevenson, P.C., Belmain, S.R. and Hall, D.R. (2002) Methylsalicylate isomers in the roots of *Securidaca longepedunculata*, *J. Mass Spectr.*, **37**, 577–580.

Jayasekera, T.K. Belmain, S.R., Stevenson P.C.and Hall D.R. (2003) *Securidaca longepedunculata* as a control for stored product insect pests. In: P.F. Credland, D.M. Armitage, C.H. Bell., P.M. Cogan and E. Highley (eds.), *Advances on Stored Product Protection*, CAB International, Wallingford, UK, pp. 596–599.

Jayasekera, T., Stevenson, P.C., Hall, D.R. and Belmain, S.R., (2005) Effect of volatile constituents from *Securidaca longepedunculata* on stored grain insect pests. *J. Chem. Ecol.*, **31**, 303–313.

Joseph, C.C., Ndoile, M.M., Malima, R.C. and Nkunya, M.H.H. (2004) Larvicidal and mosquitocidal extracts, a coumarin, isoflavonoids and pterocarpans from *Neorautanenia mitis*. *Trans. Royal Soc. Trop. Med. Hyg.*, **98**, 451–455.

Kamanula, J., Sileshi, G.W., Belmain, S.R., Sola, P., Mvumi, B.M., Nyirenda, G.K.C., Nyirenda, S.P. and Stevenson, P.C. (2011) Farmers' insect pest management practices and pesticidal plant use in the protection of stored maize and beans in Southern Africa. *Int. J. Pest Manage.*, **57**, 41–49.

Kinghorn, A.D. (1978) Isolation of a novel irritant from *Euphorbia tirucalli*. *J. Nat. Prod.*, **41**, 648–649.

Kite, G.C., Lawrence, T.J. and Dauncey, E.A. (2000) Detecting *Taxus* poisoning in horses using liquid chromatography/mass spectrometry. *Vet. Human Toxicol.*, **42**, 151–154.

Kole, R.K., Satpathi, C., Chowdhury, A., Ghosh, M.R. and Adityachaudhury, N. (1992) Isolation of amorpholone, a potent rotenoid insecticide from *Tephrosia candida*. *J. Agric. Food Chem.*, **40**, 1208–1210.

Kone, W.M., Atindehou, K.K., Terreaux, C., Hostettmann, K., Traore, D. and Dosso, M. (2004) Traditional medicine in North Cote-d'Ivoire: screening of 50 medicinal plants for antibacterial activity. *J. Ethnopharmacol.*, **93**, 43–49.

Koona, P. and Dorn, S. (2005) Extracts from *Tephrosia vogelii* for the protection of stored legume seeds against damage by three bruchid species. *Ann. Appl. Biol.*, **147**, 43–48.

Koona, P., Malaa, D. and Koona, O.E.S. (2007) Hexane extracts from *Tephrosia vogelii* Hook. f. protect stored maize against the weevil *Sitophilus zeamais* Motschulsky (Coleoptera : Curculionidae). *Entomol. Sci.*, **10**, 107–111.

Koul, O. (2005) *Insect Antifeedants*. CRC Press, Boca Raton, Florida.

Koul, O. and Dhaliwal, G.S. (eds.) (2000) *Phytochemical Biopesticides* (Advances in Biopesticide Research). CRC Press, Boca Raton, Florida.

Kumar, A., Prasad, M., Mishra, D., Srivastav, S.K. and Srivastav, A.K. (2010) Toxicity of aqueous extract of *Euphorbia tirucalli* latex on catfish, *Heteropneustes fossilis*. *Ecotoxicol. Environ. Safety*, **73**, 1671–1673.

Kuntashula, E.and Mafongoya, P.L. (2005) Farmer participatory evaluation of agroforestry trees in eastern Zambia. *Agric. Syst.*, **84**, 39–53.

Lambert, N., Trouslot, M.F., Nefcampa, C. and Chrestin, H. (1993) Production of rotenoids by heterotrophic and photomixotrophic cell-cultures of *Tephrosia vogelii*. *Phytochemistry*, **34**, 1515–1520.

Lapointe, S.L., McKenzie, C.L. and Hunter, W.B. (2003) Toxicity and repellency of *Tephrosia candida* to larval and adult diaprepes root weevil (Coleoptera: Curculionidae). *J. Econ. Entomol.*, **96**, 811–816.

Lukwa, N., Molgaard, P., Furu, P. and Bogh, C. (2009) *Lippia javanica* (Burm F) Spreng: its general constituents and bioactivity on mosquitoes. *Trop. Biomed.*, **26**, 85–91.

Madzimure, J., Nyahangare, E.T., Hamudikuwanda, H., Hove, T., Stevenson, P.C., Belmain, S.R. and Brighton, M. (2011) Acaricidal efficacy against cattle ticks and acute oral toxicity of *Lippia javanica* (Burm f.). *Trop. Anim. Health Prod.*, **43**, 481–489.

Mafongoya, P.L. and Kuntashula, E. (2005) Participatory evaluation of *Tephrosia* species and provenances for soil fertility improvement and other uses using farmer criteria in eastern Zambia. *Exp. Agric.*, **41**, 69–80.

Marston, A., Maillard, M. and Hostettmann, K. (1993) Search for antifungal, molluscicidal and larvicidal compounds from African medicinal-plants. *J. Ethnopharmacol.*, **38**, 215–223.

Neuwinger, H.D. (2004) Plants used for poison fishing in tropical Africa. *Toxicon*, **44**, 417–430.

Minjas, J.N. and Sarda, R.K. (1986) Laboratory observations on the toxicity of *Swartzia madagascariensis* (leguminosae) extract to mosquito larvae. *Trans. Royal Soc. Trop. Med. Hyg.*, **80**, 460–461.

Nyirenda, S.P.N., Sileshi, G., Belmain, S.R., Kamanula, J.F., Mvumi, B., Sola, P., Nyirenda, G.K.C. and Stevenson, P.C. (2011) Farmers' ethno-ecological knowledge of vegetable pests and their management using pesticidal plants in Northern Malawi and Eastern Zambia. *Afr. J. Agric. Res.*, **6**, 1525–1537.

Ogendo, J.O., Belmain, S.R., Deng, A.L. and Walker, D.J. (2003) Comparison of toxic and repellent effects of *Lantana camara* L. with *Tephrosia vogelii* Hook and a synthetic pesticide against *Sitophilus zeamais* motschulsky (Coleoptera: Curculionidae) in stored maize grain. *Insect Sci. Applic.*, **23**, 127–135.

Omolo, M.O., Okinyo, D., Ndiege, I.O., Lwande, W. and Hassanali, A. (2004) Repellency of essential oils of some Kenyan plants against *Anopheles gambiae*. *Phytochemistry*, **65**, 2797–2802.

Pesewu, G.A., Cutler, R.R. and Humber, D.P. (2008) Antibacterial activity of plants used in traditional medicines of Ghana with particular reference to MRSA. *J. Ethnopharmacol.*, **116**, 102–111.

Prakash, A and Rao, J. (1997) *Botanical Pesticides in Agriculture.* CRC Press, Boca Raton, Florida.

Rahuman, A. A., Gopalakrishnan, G., Venkatesan, P. and Geetha, K. (2008) Larvicidal activity of some Euphorbiaceae plant extracts against *Aedes aegypti* and *Culex quinquefasciatus* (Diptera : Culicidae). *Parasitol. Res.*, **102**, 867–873.

Rodriguez, J., Calle, Z. and Montoya-Lerma, J. (2008) Herbivory of *Atta cephalotes* (Hymenoptera : Myrmicinae) on three plant substrates. *Revista Colomb. Entomol.*, **34**, 156–162.

Sakurai, Y., Sakurai, N., Taniguchi, M., Nakanishi, Y., Bastow, K.F., Wang, X.H., Cragg, G.M. and Lee, K.H. (2006) Rautandiols A and B, pterocarpans and cytotoxic constituents from *Neorautanenia mitis*. *J. Nat. Prod.*, **69**, 397–399.

Sarasan, V., Kite G.C., Sileshi, G.W. and Stevenson P.C. (2011) The application of phytochemistry and in vitro tools to the sustainable utilisation of medicinal and pesticidal plants for income generation and poverty alleviation. *Plant Cell Reports* (in press)

Silva, A.C.P., de Faria, D.E.P., Borges, N.B.D., de Souza, I.A., Peters, V.M. and Guerra, M.D. (2007) Toxicological screening of *Euphorbia tirucalli* L.: Developmental toxicity studies in rats. *J. Ethnopharmacol.*, **110**, 154–159.

Sekyembe, S., Maurer, G., Gatimu, J., Nkanya, J. and Injairu, S. (1993) Training manual for the control of the greater (larger) grain borer (*Prostephanus truncatus* Horn)., FAO, Rome. http://www.fao.org/ wairdocs/ X5421E/x5421e00.htm#Contents.

Shackleton, C.M., Guthrie, G. and Main, R. (2005) Estimating the potential role of commercial over-harvesting in resource viability: A case study of five useful tree species in South Africa. *Land Degrad. Develop.*, **16**, 273–286.

Sirrine, D., Shennan, C., Snapp, S., Kanyama-Phiri, G., Kamanga, B. and Sirrine, J.R. (2010) Improving recommendations resulting from on-farm research: agroforestry, risk, profitability and vulnerability in southern Malawi. *Int. J. Agric. Sustain.*, **8**, 290–304.

Stevenson, P.C., Nyirenda, S.P. and Veitch, N.C. (2010) Highly glycosylated flavonoid glycosides from *Bobgunnia madagascariensis*. *Tetrahedron Lett.*, **51**, 4727–4730.

Stevenson, P.C., Jayasekera,T.K., Belmain, S.R. and Veitch, N.C. (2009) Bisdesmosidic saponins from *Securidaca longepedunculata* (Polygalaceae) with deterrent and toxic properties to Coleapteran storage pests. *J. Agric. Food Chem.*, **57**, 8860–8867.

Stoll, G. (2005) *Natural Crop Protection in the Tropics.* Margraf Publishers, West Germany.

Thokozani, B.L.K., Zulu, D, Sileshi, G.W., Teklehaimanot, Z., Gondwe, D.S.B., Sarsan, V. and Stevenson, P.C. (2011) Seed germination and in vitro regeneration of the pan African medicinal and pesticidal plant *Bobgunnia madagascariensis*. *Afr. J. Biotech.* (in press)

Valderrama-Eslava, E.I., Montoya-Lerma, J. and Giraldo, C. (2009) Enforced herbivory on *Canavalia ensiformis* and *Tithonia diversifolia* and its effects on leaf-cutting ants, *Atta cephalotes*. *J. Appl. Entomol.*, **133**, 689–694.

Vanpuyvelde, L., Geysen, D., Ayobangira, F.X., Hakizamungu, E., Nshimiyimana, A. and Kalisa, A. (1985) Screening of medicinal plants of Rwanda for acaricidal activity. *J. Ethnopharmacol.*, **13**, 209–215.

Vanpuyvelde, L., Dekimpe, N., Mudaheranwa, J.P., Gasiga, A., Schamp, N., Declercq, J.P. and Vanmeerssche, M. (1987) Isolation and structural elucidation of potentially insecticidal and acaricidal isoflavone-type compounds from *Neorautanenia mitis*. *J. Nat. Prod.*, **50**, 349–356.

Viljoen, A.M., Subramoney, S., van Vuuren, S.F., Baser, K.H.C. and Demirci, B. (2005) The composition, geographical variation and antimicrobial activity of *Lippia javanica* (Verbenaceae) leaf essential oils. *J. Ethnopharmacol.*, **96**, 271–277.

Vongtau, H.O., Amos, S., Binda, L., Kapu, S.D., Gamaniel, K.S., Kunle, O.F. and Wambebe, C. (2000) Pharmacological effects of the aqueous extract of *Neorautanenia mitis* in rodents. *J. Ethnopharmacol.*, **72**, 207–214.

Vongtau, H.O., Abbah, J., Chindo, B.A., Mosugu, O., Salawu, A.O., Kwanashie, H.O. and Gamaniel, K.S. (2005) Central inhibitory effects of the methanol extract of *Neorautanenia mitis* root in rats and mice. *Pharm. Biol.*, **43**, 113–120.

Waka, M., Hopkins, R.J. and Curtis, C. (2004) Ethnobotanical survey and testing of plants traditionally used against hematophagous insects in Eritrea. *J. Ethnopharmacol.*, **95**, 95–101.

Yadav, R., Srivastava, V.K., Chandra, R. and Singh, A. (2002) Larvicidal activity of latex and stem bark of *Euphorbia tirucalli* plant on the mosquito *Culex quinquefasciatus*. *J. Commun. Dis.*, **34**, 264–269.

Yeap, S.K., Ho, W.Y., Beh, B.K., Liang, W.S., Ky, H., Yousr, A.H.N. and Alitheen, N.B. (2010) *Vernonia amygdalina*, an ethnoveterinary and ethnomedical used green vegetable with multiple bio-activities. *J. Med. Plants Res.*, **4**, 2787–2812.

Zulu, D, Thokozani, B.L.K., Sileshi, G.W., Teklehaimanot, Z., Gondwe, D.S.B., Sarsan, V. and Stevenson, P.C. (2011) Towards successful propagation of the Africa medicinal and pesticidal plant *Securidaca longipendunculata*. *Afr. J. Biotechnol.* (in press).

12

Biologically-Based Stored-Product Insect Pest Management

THOMAS W. PHILLIPS*

Department of Entomology, 123 Waters Hall, Kansas State University
Manhattan, Kansas USA 66502
*E-mail: twp1@ksu.edu

INTRODUCTION

Stored-product insects are represented by specialized guilds or associations of certain mostly phytophagous or mycophagous insects adapted to utilizing human-stored foods and exploiting the storage habitats. Many of these species are fully domesticated pests and do not persist in natural populations outside of human storages (Lindsley, 1944). "Stored-products" as a food source are represented by any dried animal or plant agricultural products that can be safely stored for weeks to years without refrigeration. Thus stored-products for the context of this paper include all dried cereal grains, legumes, milled and value-added grain products, dried meats, dried dairy products and other dried animal products containing keratin such as furs, dried skins and feathers.

Stored-product insects can cause postharvest losses, estimated from up to 9% in developed countries to 20% or more in developing countries (Pimentel, 1991). Storage insects and mites also contribute to contamination of food products through the presence of live arthropods, insect products such as chemical excretions or silk, dead insects and insect body fragments, potential human allergens from insect and mite proteins, general infestation of buildings and other storage structures, and accumulation of chemical insecticide residues in food or human exposure to dangerous chemicals as a result of pest control efforts against them. The broad objectives behind current research on stored product integrated pest management, IPM, are those that have lead the field since the beginning. Recent, more specific objectives are driven by government regulations and influenced by public and commercial needs. The traditional objectives are to store grain and food in a wholesome way with minimum impact from insects or from chemical insecticides that may be used in pest control. More recently the worldwide phase-out and ban of the fumigant insecticide methyl bromide, an extremely effective compound for killing post harvest insects, under the international agreement of the Montreal Protocol has spurned research into various alternatives to replace it. In addition to regulatory pressures for low-risk control of stored-product insects, consumers and governments around the world have set standards for "organic" food, which should be derived from raw products that are free of human-made chemicals, among other requirements. Thus research on chemical-free or biologically-based methods to control stored-product insects is encouraged and supported. Interest in alternatives to conventional insecticides for controlling stored-product insects is driven also by evolution of insecticide resistance in pest populations and because of increasing consumer demand for products that are free of insects and insecticide residues. This article give a broad overview of biologically based methods hat have been reviewed recently (Phillips and Throne, 2010), as well as some recent and previously unpublished work of my colleagues and me, and focuses on a few key methods that are most effective and/or are most likely to be commercially adopted and expanded in the near future.

INTEGRATED PEST MANAGEMENT IN STORED PRODUCTS

Integrated pest management, IPM, in storage systems has the same general components found in IPM with other agricultural systems: prevention, monitoring and control actions when needed. Infestation prevention derives mostly from the practice of sanitation and is the first line of defense for grain stored at farms or elevators and for food-processing and warehouse facilities. Some of the most promising biologically based management tools for farm-stored grain are temperature management and use of natural enemies. New tools for computer-assisted decision-making and insect sampling at grain elevators appear most promising (Hagstrum *et al.*, 2010). Processing facilities and warehouses usually rely on trap capture data for decision-making, which needs further research to optimize this process (Arthur and Phillips, 2003). IPM is a decision-making process that relies on accumulating high quality monitoring data on the commodities and the pest populations coupled with decision matrices based on thresholds for taking action. IPM for stored cereal grains now has a sophisticated, though simply applied, program for collecting important grain environment and insect population data for use in predictive tools and decision-making (Flinn *et al.*, 2007) In stored-products the tolerance for insect numbers and product quality loss decrease as value is added to the product.

PHYSICAL PREVENTION AND CONTROLS

Sanitation

Cereal grains and pulses can be stored safely for long periods if the storage structures are clean, free from residual commodity and insect infestation, and if the new commodity is kept cool and at a safe moisture content. Mills, other food processing facilities and warehouses for value-added products also can be kept clean, dry and at reasonable temperatures, if possible, to prevent or suppress populations of pest insects. Sanitation is overall one of the best preventive measures fore storage pests. Exclusions of pest insects from storage and processing areas is also key. Thus, well-sealed grain silo with no entry-ways for insects and food plants with insect screening on window and other exclusion features will prevent infestation better than structures with numerous opening for insect entry.

Packaging

Finished food products that are free of insects when packaged can be further protected by use of insect-proof packaging (Mullen and Mowry, 2006). Stored-product insects found in packaged products gained entry either trough direct chewing through the packaging material or by finding their way inside the package via existing openings, poor seals or opening resulting from damage of the package after production. Thus insect-resistant packaging must be very well sealed, its seals must tolerate routine handling and damage without opening, and the materials used for package construction must be impenetrable by chewing insects. The value and/or size of certain food packages, such as large bags of pet food, bird seed or animal feed products may not be conducive to insect-proof packages, therefore, these must be carefully monitored and not stored too long in one place. Higher value products such as breakfast cereals, cake mixes, pastas or baked products should have food in relatively thick-gauged polypropylene that is thorough heat sealed, and in many cases have another package of cardboard that well sealed at all openings and will protect the inner package from being breached if dropped or damaged.

Air Temperature

Insects are "cold-blooded" and as such can develop and survive within a range of acceptable temperature and will suffer sickness and death at warmer and colder extreme temperatures. Air temperature in the storage environment can be manipulated to cause direct death by extreme exposure or to limit pest population growth by managed temperature (Fields, 1992). Bulk cereal grains and pulses stored in temperate climates can be cooled with winter temperatures to stop all insect activity (less than 5°C), or can be maintained at cooler-than-ambient temperatures with night-time air by using controlled aeration to slow insect development (Arthur *et al.*, 2008). The most effective configurations have cables with thermocouples spaced vertically through the grain mass couples with fan controllers that turn aeration fans on when outside temperatures fall below 5°C less than that of the grain mass, and bring in cool air directly to the grain mass too cool it. Freezing temperatures will kill all insect life stages within hours to minutes and they can be applied to isolated containers or packages of high value foods that can tolerate freezing. Extreme high temperatures of 42° to 50°C will kill insects in minutes to hours and, although not safe for bulk commodities, can be applied to empty storage facilities or processing facilities to kill insects surviving on food debris and/or within structural refugia. Building-wide heat treatments have now become alternatives of choice for food processors wanting to avoid chemical fumigants (Tilley *et al.*, 2007)

Controlled Atmosphere

Controlled or modified atmospheres are viable bio-rational alternatives to traditional chemical control methods. Chemical fumigants such as phosphine gas, hydrogen phosphide, remain the industry standard for definitive control of pest outbreaks with no residues or residual activity. However, many pest populations are evolving resistance to phosphine and the organic food market requires non-chemical alternatives. Biologically based alternatives to chemical fumigants include the use of controlled or modified atmospheres applied to bulk stored products. Controlled atmospheres change the composition of atmospheric gases by addition or deletion of key gases to create an insecticidal atmosphere. Thus, addition of carbon dioxide to levels of 60-80% can be insecticidal if held for several days at an appropriate temperature. Additionally, displacement of air by the addition of an inert gas like nitrogen, or by removing air via vacuum (Fig. 12.1), to achieve oxygen levels of 3% or less will kill all life stages of key insect pests in a matter of hours to days. Hermetic or gas-tight storage of infested commodities results in a modified atmosphere in which atmospheric gas concentrations vary dynamically via biological processes over time to become insecticidal. Thus, aerobic insects and microbes on commodities in hermetic storage will consume oxygen and increase carbon dioxide resulting in an insecticidal atmosphere in a matter of days to weeks, and can be stored as such very safely for months.

Fig. 12.1. Stack of cocoa beans being treated under low pressure applied with a vacuum pump (to right in photo) to achieve an insecticidal low oxygen atmosphere inside a flexible polyethylene “cocoon”.

Desiccation

Desiccation from the application of natural or synthetic silicon dioxide-based drying agents represent a biorational alternative for controlling stored product insect pests. Diatomaceous earth, DE, is mined from naturally occurring deposits of the fossilized silicon dioxide skeletons of aquatic diatom algae, and is a highly effective desiccant for use as an insecticide. The apparent mode of action for DE is to absorb the hydrocarbon infusion from the insect cuticle facilitating desiccation of the insect and death in an appropriately dry environment (Kuronic, 1998). DE is completely non-toxic to humans and other animals, and is commonly used as a food-safe additive. Admixture of DE to entire grain masses can be very effective for protection against insect infestation, but is not recommended as DE lowers the bulk density of the grain, putting airspaces between kernels lying flat, and thus severely lowers the value from equivalent volumes of untreated grain. Treatment of empty grain silos containing grain debris and insects can be very effective (Fig. 12.2). Similarly, DE has been promoted for surface treatments in food plants and warehouses where the dustiness can be tolerated and traditional chemical insecticides are to be avoided.

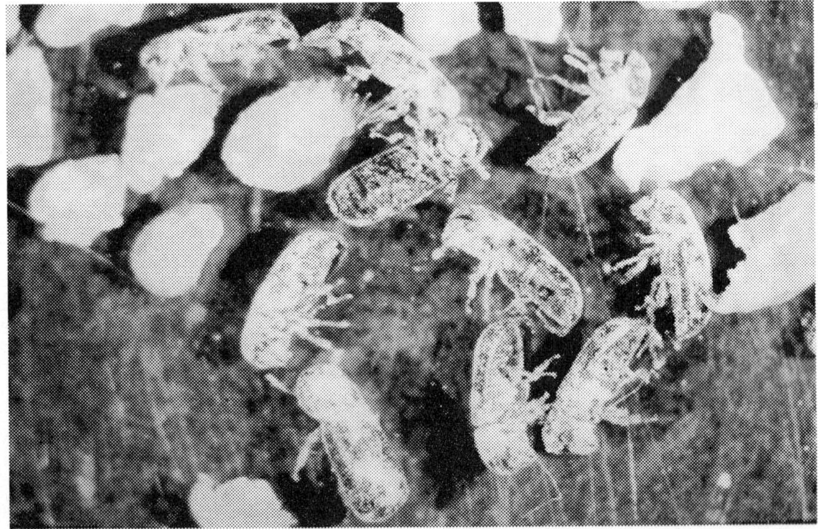

Fig. 12.2. Adults of *Rhyzopertha dominica*, the lesser grain borer, killed from desiccation following treatment of an empty grain bin containing grain debris and insects with diatomaceous earth.

BIOLOGICALLY BASED INSECTICIDES

Biologically based insecticides have received considerable research and some recent products show promising efficacy. A number of entomopathogens such as the bacterium *Bacillus thuringiensis* and the fungus *Beauveria bassiana* have been tested for control of stored-product insects, but none are in common use because of lack of sufficient efficacy (reveiwed in Moore *et al.*, 2000). Many tests have been conducted to synergize pathogens with other control technologies, particularly those that might be expected to increase efficacy of pathogens, such as the addition of diatomaceous earth (DE) to abrade the cuticle or absorb cuticular hydrocarbons, both of which might possibly make the insect more susceptible to the pathogen. Spinosad is an insecticide that is derived from metabolites from the fermentation of the actinomycete bacterium, *Saccharopolyspora spinosa* (Thompson and Sparks, 2000). There is much interest in the use of Spinosad on stored grain because other residual insecticides have limited efficacy against the major pest of stored wheat and rice, *Rhyzopertha dominica*, either because of simple lack of efficacy or because of development of resistance. Spinosad is very effective for season-long control of *R. dominica* in stored wheat and rice, and it may be an organically-compliant compound for this use.

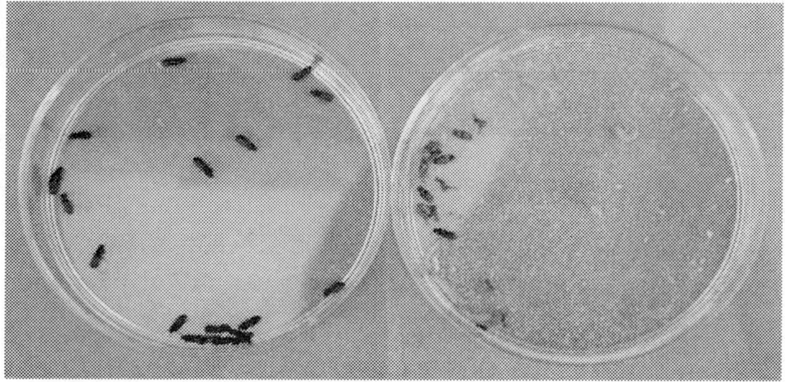

Fig. 12.3. *Tribolium castaneum*, red flour beetles, in an experiment with the insect growth regulator hydroprene. Insects in both dishes were exposed to treated surfaces as last instar larvae. Those on the left were exposed to a surface treated with water as an experimental control and all developed into normal adults. Beetles on the right were exposed to a surface treated with the label-rate of hydroprene and 95% failed to develop normally after four weeks (see Arthur *et al.*, 2009)

Insect growth regulators (IGRs) are a general grouping of synthetic compounds that structurally, but more so biologically, mimic the activity of insect endocrine regulators such that normal growth and development is disrupted and insects die (Oberlander and Silhacek, 2000). The historically classic IGR is methoprene, which is chemically similar to insect juvenile hormone and has the resulting effect of prolonging larval development, altering sexual maturation in adults and ultimately disrupting normal development leading to insect death. The chemically similar and slightly more volatile hydroprene has found adequate adoption in stored-product pest management due to its very low vertebrate toxicity and high level of insecticidal efficacy (e.g. Fig. 12.3). Several other IGRs have been developed that, although lacking chemical similarity with insect hormones, they have "juvenilizing" or other developmental effects on insects, and they have been widely registered for use on or around food due to their essential non-toxic status toward vertebrates (e.g. Arthur *et al.*, 2009).

Fig. 12.4. An adult female of the parasitoid wasp *Anisopteromalus calandrae* probing into a cereal grain kernel that contains one or more internal-feeding host larvae.

ARTHROPOD NATURAL ENEMIES

Arthropod predators and parasitoids of stored-product pests have been known for decades and their use in systematic IPM for stored products has been discussed (Brower *et al.*, 1996). The stored product ecosystems is like any other in nature, such that there are the primary producers, in this case the cereal grains and/or the food products made from grains or pulses, the primary concumers and the secondary consumers, many of which act as biological control agents against the pests. Predatory beetles in the Cleridae and pirarte bugs in the Athocoridae are well known and have significant impacts an many life stages of pest insects. Parasitoid wasps in the families Braconidae, Icheumondiae and Pteromalidae (e.g. Fig. 12.4) and Trichogramatidae utilize eggs, larvae and pupae of serious moth and beetle pests of stored products. Inundative biological control, in which large numbers of parasitoids or predators are released into a storage system at one time, much like the use of a chemical fumigant, has been tested with apparent efficacy. Unfortunately, the cost of such as inundative biological control for large commercial settings such as bulk grain storages or large commercial food processing or product storage facilities, is simply not practical or cost-effective (Schöeller and Flinn, 2000). A few key species or parasitoid wasps, and occasionally predatory bugs, are commercially reared and provided for sale as biological control agents, this practice is spotty and known at this time from Europe and sometime North America.

SEMIOCHEMICALS

Pheromones and other behavioral chemicals have been identified for over forty species of stored product insects and commercial formulations for use in monitoring tools are available for nearly 20 species (Phillips *et al.*, 2000). Food-based attractants have been identified that can act alone to attract pests or provide additive or synergistic increases to pheromone response. By far the predominant use of pheromones in IPM of storage pests is for lures in traps to detect and monitor pest populations. Trapping data can help pest managers make decision about sanitation, prevention and direct control, and can help them target specific locations in a facility these practices should be implemented. The synthetic sex pheromone of the Indian meal moth, *Plodia interpunctella*, was recently registered by the US Environmental Projection Agency for use in mating disruption against *P. interpunctella* and related species of storage moths (Phillips, 2006; Figs. 12.5 and 12.6). Additional experimental studies on the use of synthetic sex pheromones for suppression of storage moths using the "attract-and-kill" method have been conducted (Campos, 2008; Campos and Phillips, 2010) and show promise for good effectiveness by mass-killing of males without the need for multiple and high-maintenance sticky traps. Recent and ongoing (unpublished) studies indicate the mating disruption of the cigarette beetle, *Lasioderma serricorne* is possible using its synthetic sex pheromone. Studies on the response of the lesser grain borer, *Rhyzopertha dominica*, to its aggregation pheromone suggest that this behavior can be manipulated with other semiochemicals, such as known pheromones of pest bark beetles (Coleoptera: Scolytidae) (Edde *et al.*, 2011). Thus semiochemical-based pest management can go beyond the effective use of traps in monitoring and control decision-making to actual population suppression with these behavioral compounds as alternatives to traditional insecticides.

Fig. 12.5. Pheromone-based suppression of *Plodia interpunctella*, the Indian meal moth (IMM), with two different methods. On the left is a plastic sphere containing a synthetic sex pheromone lure and coated with a contact insecticide for use in the attract-and-kill method (see Campos, 2008; Campos and Phillips, 2010). On the right, hanging from equipment in a food plant is a high release dispenser for the synthetic sex pheromone of IMM that is releasing approximately 2 mg per day for purposes of mating disruption and multiple dispensers are deployed throughout the treated area (Phillips, 2006)

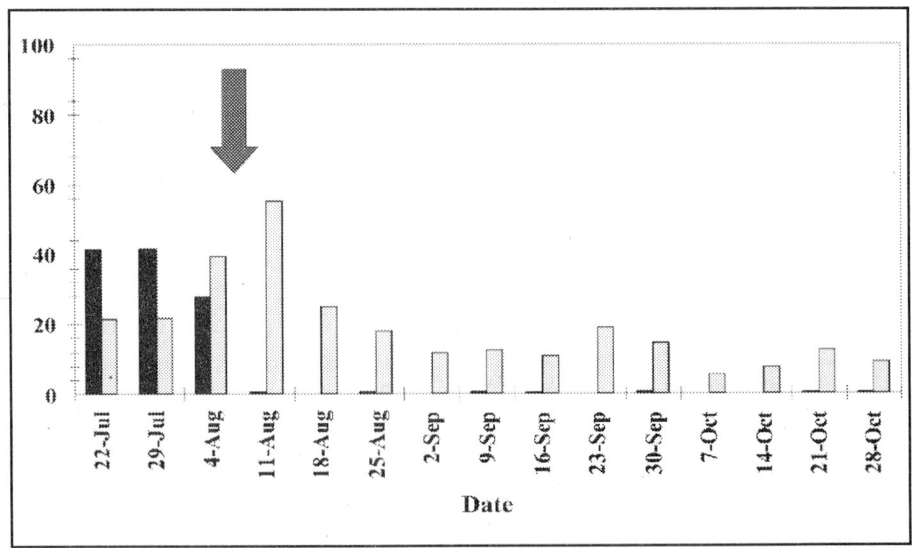

Fig. 12.6. Average captures of *Plodia interpunctella*, Indian meal moth (IMM), males (y-axis) in pheromone-baited monitoring traps at two sites, one of which (black bars) received a mating disruption (MD) treatment after traps were checked on 4 August (arrow) in the treated site vs average trap capture in an untreated control facility (light bars) Additional monitoring of egg-laying at these and several other paired sites in the same year indicated that moth reproduction was substantially suppressed at MD-treated sites (Phillips, 2006)

CONCLUSIONS

The broad objectives behind current research on stored product integrated pest management, IPM, are those that have lead the field since the beginning; recent, more specific objectives are driven by government regulations and influenced by public and commercial needs. Research on chemical-free or biologically-based methods to control stored-product insects is encouraged and supported. Prevention in the practice of sanitation is the first line of defense for grain stored at farms or commercial bulk storage facilities and for food processing and warehouse sites. Some of the most promising biologically based management tools are temperature management, biological insecticides and semiochemical-based approaches. New tools for computer-assisted decision-making and insect sampling at grain elevators appear most promising. Processing facilities and warehouses usually rely on trap capture data for decision-making, which needs further research to optimize this process. There is much interest in use of Spinosad, an insecticide that is derived from metabolites from the fermentation of the actinomycete bacterium, *Saccharopolyspora spinosa*, because other residual insecticides have limited efficacy against key major pests. Biologically-based alternatives to chemical fumigants include the use of controlled or modified atmospheres applied to bulk stored products. Insect growth regulators such as methoprene or pyriproxyfen are effective bio-rational control agents as they are active and persistent at low concentrations to disrupt insect development and have virtually no harm to vertebrates. Pheromones are known for over forty species of stored product insects and are routinely used in monitoring traps. Pest suppression using pheromone -based mating disruption and attract-and-kill methods show promise as effective alternatives to traditional insecticides.

Therefore, overall challenges are driven by increasing government regulations on the use of traditional chemical insecticides, particularly fumigants, increased genetically based resistance to chemical insecticides in pest populations, and consumer demand for high quality grains and food products with minimal insect damage and little or no chemical residue. Biologically based methods for IPM and pest suppression will continue to gain favor in the research community and especially among government regulators and commercial food and pest control organizations.

ACKNOWLEDGEMENTS

I am very grateful for collaboration and discussions with Edmond Bonjour, Peter Edde, Paul Flinn, David Hagstrum, Rizana Mahroof, George Mbata, James Throne and Phil Villers. Financial support was provided by various programs of the U.S. Department of Agriculture and the Kansas Agricultural Experiment Station. I greatly appreciate the invitation of Dr. Opender Koul to attend the conference (BET 2010), the work of Dr. Koul and his colleagues in organizing and executing an excellent event, and especially to Dr. Koul for his patience and encouragement during the preparation of this document.

REFERENCES

Arthur, F.H., Liu, S., Zhao, B. and Phillips, T.W. (2009) Residual efficacy of pyriproxyfen and hydroprene applied to wood, metal and concrete for control of stored-product insects. *Pest Manage. Sci.*, **65**, 791–797.

Arthur, F.H. and Phillips, T.W. (2003) Stored-product insect pest management and control. In: Y.H. Yui, B.L. Bruinsma, J.R. Gorhap, W. Nip, P.S. Tong, and P. Ventresca (eds.), *Food Plant Sanitation*, Marcel Dekker, New York, pp. 341–348.

Arthur, F.H., Yang, Y., Wilson, L.T. and Siebenmorgen, T.J. (2008) Feasibility of automatic aeration for insect pest management for rice stored in East Texas. *Appl. Eng. Agric.*, **24**, 345–350.

Brower, J.H., Smith, L., Vail, P.V. and Flinn, P.W. (1996) Biological control, In: B. Subramanyam and D.W. Hagstrum (eds.), *Integrated Management of Insects in Stored Products,* Marcel Dekker, New York, pp. 295–222.

Campos, M. (2008) Attract-and-kill methods for *Plodia interpunctella* (Lepidoptera: Pyralidae) and comparisons with other pheromone-based methods. Ph.D. thesis. Oklahoma State University. Stillwater, OK.

Campos, M. and Phillips, T.W. (2010) Contact toxicity of insecticides for attract-and-kill applications against adult *Plodia interpunctella* (Hübner) (Lepidoptera: Pyralidae). *Pest Manage. Sci.*, **66**, 752–761.

Edde, P.A., Toews, M.D. and Phillips, T.W. (2011) Effects of various semiochemicals on the responses of *Rhyzopertha dominica* (F.) to pheromone traps in the field. *Ann. Entomol. Soc. Am.*, **104,** 1297-1302.

Fields, P.G. (1992) The control of stored-product insects and mites with extreme temperatures. *J. Stored Prod. Res.* **28**, 89–118.

Flinn, P.W., Hagstrum, D.W., Reed, C.R. and Phillips, T.W. (2007) Stored grain advisor pro: Decision support system for insect management in commercial grain elevators. *J Stored Prod. Res.*, **43**, 375–383.

Hagstrum, D.W., Flinn, P.W., Reed, C.R. and Phillips, T.W. (2010) Ecology and IPM of insects at grain elevators and flat storages. *Biopestic. Int.*, **6**, 1–20.

Korunic Z. (1998) Diatomaceous earths, a group of natural insecticides. *J. Stored Prod. Res.*, **34**, 87–97.

Linsley, E.G. (1944) Natural sources, habitats, and reservoirs of insects associated with stored food products. *Hilgardia*. **16**, 187–224.

Moore, D., Lord, J.C. and Smith, S.M. (2000). Pathogens. In: Bh. Subramanyam and D.W. Hagstrum (eds.), *Alternatives to Pesticides in Stored-Product IPM,* Kluwer Academic Publishers, Boston, pp. 193–228.

Mullen, M.A. and Mowery, S.V. (2006) Insect-resistant packaging. In: J. Heaps (ed.), *Insect Management for Food Storage and Processing*, 2nd Ed., American Association of Cereal Chemistry, St. Paul, pp. 35–38.

Oberlander, H. and Silhacek, D. (2000) Insect growth regulators. In: Bh. Subramanyam and D.W. Hagstrum (eds.), *Alternatives to Pesticides in Stored-Product IPM,* Kluwer Academic Publishers, Boston, pp. 147–164,

Pimentel, D. (1991) World resources and food losses to pests. In: J.R. Gorham (ed.), *Ecology and Management of Food Industry Pests*, Assoc. Offic. Anal. Chemists, Arlington, pp. 5–11.

Phillips, T.W. (2006) The science and technology of postharvest insect control: Challenges, accomplishments and future directions. In: J. Heaps (ed.), *Insect Management for Food Storage and Processing*, 2nd Ed., American Association of Cereal Chemistry, St. Paul, pp. 211–222.

Phillips, T.W. and Throne, J.E. (2010) Biorational approaches to managing stored-product insects. *Ann. Rev. Entomol.*, **55**, 375–397.

Phillips, T.W., Cogan, P.M. and Fadamiro, H.Y. (2000) Pheromones. In: Bh. Subramanyam and D.W. Hagstrum,(eds.), *Alternatives to Pesticides in Stored-Product IPM*, Kluwer Academic Publishers. Boston, pp. 273–302.

Schöeller, M. and Flinn, P.W. (2000) Parasites and predators. In: Bh. Subramanyam and D.W. Hagstrum (eds.), *Alternatives to Pesticides in Stored-Product IPM,* Kluwer Academic Publishers, Boston, pp. 229–272.

Tilley, D.R., Langemeier, M.R., Casada, M.E. and Arthur, F.H. (2007) Cost and risk analysis of heat and chemical treatments. *J. Econ. Entomol.*, **100**, 604–612.

Thompson, G.D., Dutton, R. and Sparks, T.C. (2000) Spinosad-a case study: An example from a natural products discovery programme. *Pest Manage. Sci.*, **56**, 696–702.

13

From the Lab to the Field: Entomopathogens, their Success, and Limitations for Widespread Use

SURENDRA K. DARA

University of California Cooperative Extension, San Luis Obispo
California- 93401, USA
E-mail: skdara@ucdavis.edu

INTRODUCTION

Advances in basic scientific research always helped the applied science. The more we understand about the interactions of insects and their pathogens, the better the chances of developing good biopesticides. Each year, hundreds of scientific presentations are made at professional conferences around the world about the biochemical, biological, molecular, ecological, and other aspects of the complex interactions of the entomopathogens, their hosts, and the environment. With a lot of these research findings published in scientific journals, there is a wealth of knowledge about entomopathogens. When basic research is extended into applied research in the laboratory, greenhouse or field trials, many a time there are very promising results. But only a minor part of the applied research becomes practical or commercially realistic. Whether it is public concern, regulatory restrictions, environmental safety, commercial challenges or practical limitations, a lot of promising laboratory research does not translate into field application. This praper briefly reviews the global trends in biopesticides with specific emphasis on entomopathogens, some success stories in the laboratory research, their limitations for commercial application, and some challenges that biopesticide industry generally faces.

Biopesticide usage is widely distributed around the world. The table below shows worldwide availability of biopesticides based on some major entomopathogenic bacteria, fungi, viruses and nematodes is shown in Table 13.1.

Table 13.1. Availability of biopesticides at global level

Category	Organism	Number of products	Number of countries
Bacteria	*Bacillus thuringiensis* subsp. *aizawai*	36	45
	Bt subsp. *israelensis*	50	79*
	Bt subsp. *kurstaki*	220	113*
	Bt subsp. *tenebrionis*	7	20
	Bt subsp. *thuringiensis*	10	6
Fungi	*Beauveria bassiana*	94	44
	Metarhizium anisopliae	73	63*
	Paecilomyces fumosoroseus	18	20
	Lecanicillium lecanii	43	24
Nematodes	*Heterorhabditis bacteriophora*	19	25
	H. megidis	9	21
	Steinernema carpocapsae	37	25
	S. feltiae	39	23
Viruses	*Cydia pomonella* GV	13	30
	Helicoverpa spp. NPV	18	9
	Spodoptera spp. NPV	16	14

*Includes up to 34 Sub-Saharan African nations.
Source: CPL Business Consultants (2010)

Despite the widespread usage, the value of global pesticide market is anywhere from 26-40 billion US dollars annually, according to different sources. Microbial biopesticides account for a small fraction of this market with 1-1.5% share. Increasing restrictions on some chemical pesticides, resistance to certain chemicals, and consumer preference for non-chemical alternatives helped the growth of biopesticide industry in recent years. However, microbial control is vastly underutilized around the world. Here are a few limitations for fully exploiting the potential of microbial organisms.

COMMERCIAL UTILIZATION

A lot of promising laboratory findings does not make a strong business case when it is considered for commercial utilization. Any new technology or product has to make economic sense to be considered for commercial production. It has to be better than what is currently available in terms of the performance and financial returns. The following examples provide details of some of the scientific advancements with bacterial, fungal and viral entomopathogens.

Bacteria

Cadherin-like proteins, aminopeptidases, alkaline phosphatases are among those that act as toxin-binding receptors on the insect midgut epithelium of lepidopteran and dipteran larvae and facilitate pore formation in the midgut membrane (Vadlamudi *et al.*, 1993; Sangadala *et al.*, 1994; Jurat-Fuentes and Adang, 2004; Abdullah *et al.*, 2006; Fernandez, *et al.*, 2006; Hua *et al.*, 2008). Cry toxins of *B. thuringiensis* (*Bt*) bind to cadherin in order to form a prepore oligomer, which upon binding to aminopeptidase. an alkaline phosphatase is inserted into lipid rafts in the membrane forming pores (Zhuang *et al.*, 2002). Based on some research findings, Adang and co-workers patented (Patent No. US 7,396,813 B2, 2008) the technology of feeding a fragment of cadherin-like protein from *Manduca sexta* with *Bt* Cry and other insecticidal proteins to enhance the toxicity (Dorsch *et al.*, 2002; Hua *et al.*, 2004). When insect consumes the toxic proteins and receptor proteins, the latter are readily available to increase the binding of the toxin to the midgut membrane. It is expected to have improved speed of kill especially for older larvae. Cadherin-like protein fragments from one insect species can synergize the toxicity of insecticidal proteins in other species (Abdullah *et al.*, 2009). As a result, multiple species of insects can be targeted with toxins that otherwise have a narrow spectrum of specificity. Applying this protein to transgenic *Bt* plants or combining it with *Bt* products are potential ways of practical use.

Adang and others founded InsectiGen to market BtBooster™ (BtB), a technology that boosts the efficiency of *Bt* products. When science meets with economics new challenges can surface. Use of BtB increases the efficiency of *Bt* products and potentially reduce the volume of *Bt* product usage. One would expect that a higher cost of BtB would compensate for the reduced volume of *Bt* sale. However, with only a couple or so of *Bt* producers in the US and generally lesser efficacy of biopesticide products in the field compared to chemical insecticides, a strong and viable business strategy needs to be developed. Although very promising, BtB technology took several years of trials and business planning. InsectiGen is currently partnering with Pioneer/DuPont to commercialize an array of their products that are effective against lepidopteran, coleopteran and dipteran pests (C. Baile, Co-founder of InsectiGen, personal communication).

Fungi

Pathogenicity of *Metarhizium anisopliae* was improved when it was genetically engineered to express an insect-specific scorpion toxin in insect hemolymph (Wang and St. Leger, 2007). When neurotoxin AaIT of the scorpion *Androctonus australis* was expressed in the treated insects, potency of the fungus increased by 22 times in *M. sexta* larvae and nine times in *Aedes aegypti* adults along with a significant reduction in their survival time. Such a technique is helpful in improving the efficacy of entomopathogenic fungi which is often questioned in the fields. Since environmental safety of the scorpion neurotoxin is a concern, authors suggested application methods that target pest species and further genetic engineering of the fungus to arrest spore production from the cadavers. Additional studies with this engineered strain against coffee berry borer, *Hypothenemus hampei,* showed increased mortality, especially at lower concentrations, and reduced survival times (Pava-Ripoll *et al.*, 2008). It was also found that sporulation from cadavers killed by the modified strain was significantly lower compared to the wild strain. This, on the one hand, could reduce the spread of the inoculum and infection. On the other hand, reduced spore production means limited environmental risk that may be associated with a genetically modified organism.

Over-expression of a toxic protease in *M. anisopliae*, which increased mortality of the host insect was achieved more than a decade ago (St. Leger *et al.*, 1996). Similar results were observed in recent studies *Beauveria bassiana* was genetically engineered with insecticidal genes from *M. anisopliae* and a scorpion (Gongora, 2004; Lu *et al.*, 2008). Commercial interest for products based on this technology is yet to be seen.

Sweet whey based entomopathogenic fungal formulations were reported to increase spore production after field application (Costa and Grassano, 2007). This is a considerable development towards increased fungal inoculums and possibly improved efficacy post-application.

However, there is no evidence this technology is being considered for commercial utilization.

In many other studies, new and naturally occurring isolates or commercial formulations of entomopathogenic fungi demonstrated pest management potential against the Diaprepres root weevil, *Diaprepes abbreviatus* (Quintela and McCoy, 1997), the red imported fire ant, *Solenopsis invicta* (Bextine and Thorvilson, 2002), the tobacco spider mite, *Tetranychus evansi* (Wekesa *et al.*, 2005), the sugarbeet root maggot, *Tetanops myopaeformis* (Campbell *et al.*, 2006), the glassy-winged sharpshooter, *Homalodisca coagulata* (Dara *et al.*, 2007), the emerald ash borer, *Agrilus planipennis* (Castrillo *et al.*, 2009), and the pea leafminer, *Liriomyza huidobrensis* (Migiro *et al.*, 2010). However, cost of application, practical issues related to the pest or its habitat, and difficulties in developing some of the isolates into commercial products limited the widespread application of these findings.

Nematodes

Entomopathogenic nematodes are effective against many pests, but they are sensitive to high temperature inactivation. Thermotolerance of *Heterorhabditis bacteriophora* Hp88 was improved by transferring a *Caenorhabditis elegans* heat-shock protein *hsp70 A* (Hashmi *et al.*, 1998). However, due to regulatory and other difficulties, this strain was never commercialized.

Viruses

Genetic engineering of baculoviruses for their improved performance is not a new concept. Meada (1989) introduced a synthetic diuretic gene from *M. sexta* into *Bombyx mori* NPV, which hastened the mortality of silkworm larvae. In later studies around the world, baculoviruses have been genetically altered either by deleting certain genes or adding toxic genes from other organisms. Removal of the viral ecdysteroid UDP-glucosyltransferase gene that hinders molting or pupation of the infected host and/or expressing scorpion neurotoxin (AaIT) gene has improved the insecticidal activity of viruses in some lepidopterans (O'Rielly and Miller, 1991; Stewart *et al.*, 1991; Chen *et al.*, 2000). Fusing *Bt* Cry1Ac toxin protein with viral polyhedrin increased speed of kill and pathogenicity by *Autographa californica* nucleopolyhedrovirus (Chang *et al.*, 2003). Other examples include expression of genes that encode juvenile hormone esterase (Hammock *et al.*, 1990), eclosion hormone (Eldridge *et al.*, 1991), fly, human and rat proteases (Harrison and Bonning, 2001), insect-specific toxins (Tomalski and Miller, 1992), and prothoracicotropic hormone (O'Reilly *et al.*, 1995) among a long list of such studies that increased the speed of kill up to 60% compared to the wild types. On the other hand, introducing a gene responsible for DNA repair to protect from UV inactivation has also resulted in increased virulence of baculovirus (Petrik *et al.*, 2003)

In several field studies, genetically altered baculoviruses produced results comparable to the use of chemical pesticides (Kamita *et al.*, 2005). Additionally several studies demonstrated the safety of these viruses to natural enemies like the social predatory wasp, *Polistes metricus* (Mcnitt *et al.*, 1995), carabid beetle *Pterostichus madidus* (Possee *et al.*, 1993), the green lacewing, *Chrysopa carnae*, minute pirate bug, *Orius insidiosus* or honeybees, *Apis mellifera* (Heinz *et al.*, 1995), The inability of these modified viruses to replicate and persist in the environment like their wild type counter parts (Cory *et al.*, 1994, Hoover *et al.*, 1995) is an advantage from the environmental safety perspective. While this can be a disadvantage for *in vivo* production of viruses for commercial use, some techniques have been developed to overcome this hurdle. Although baculoviruses have promising future for commercialization compared to other modified entomopathogens, several practical limitations including high cost of production, sensitivity to environmental factors, public acceptance, and regulatory challenges discouraged the industry to pursue commercialization of modified baculoviruses.

REGISTRATION REQUIREMENTS

Depending on the country, registration of a biopesticide can be a lengthy and expensive procedure. In the US, for example, it can take anywhere from 1-3 million dollars for registering a biopesticide product. Efficacy trials to add pests or crops to the label are also very expensive.

Registration of any microorganism for pest control involves detailed data on its mode of action, efficacy, formulation, environmental risk assessment, safety to the natural enemies like predators and parasitoids, beneficial insects like honeybees, and other organisms like birds, terrestrial and aquatic animals, and human risk assessment. If it is a genetically modified organism, stringent requirements include additional information on the superiority of the modified organism over the wild strain, potential of the genetic material transferring to other organisms, and possible transmission vectors. All these regulatory requirements increase the research and develop-ment expenditure and become a major risk factor.

FIELD EFFICACY

Compared to chemical pesticides, microbial pesticides are expensive, slow acting and field efficacy is inconsistent. Farmers generally prefer faster and inexpensive pest management options. Biopesticides are slow to act and depending on the entomopathogen, it can take 3-5 days for the target organism to die. Another limitation is a wide variation in the mortality achieved based on the crop, pest, and environment.

Successful laboratory findings could not be repeated in many field studies. For example, entomopathogenic fungi were highly pathogenic to many species of ants in the laboratory assays, but in the wild, infected members are isolated from rest of the colony before fungal sporulation and reduced infection rates (Siebeneicher *et al.*, 1992).

Synergism between sublethal doses of insecticides and entomopathogenic fungi has been known in some studies (Quintela and McCoy, 1997; Dara and Hountondji, 2001), but there is no evidence of its widespread use in practical agriculture.

Environmental persistence is also a major factor in the efficacy of biopesticides. Exposure to UV light inactivates many entomopathogens and development of formulation technology that protects the inoculum and increases the exposure to the target pest in the field is need. This can be a significant progress in improving the efficiency of biopesticides.

SPECIAL PROCEDURES

Biopesticides require special procedures for formulation, packaging, storage and application. Their efficacy is also influenced by environmental conditions. Since the active ingredient is a live organism, biopesticides need special equipment, materials and protocols for their manufacturing and formulation. Vacuum sealed packing enhances the shelf life of certain products, but it is not ideal for all formulations. In some cases, biopesticides need to be stored at cooler temperatures so that the active ingredient remains viable. Application of multiple tank-mixed pesticides is a common practice to save on application costs. Compatibility of biopesticides with other products can be an issue in some cases. For example, entomopathogenic fungi may not be compatible with certain fungicides. Once applied, persistence of the microorganism in the pest environment and infection rate largely depends on environmental factors. High humidity and cooler temperatures are important for efficient control with entomopathogenic fungi. Exposure to the UV light can also inactivate the inoculum and reduce the efficacy. All these factors add complexity to the utilization of biopesticides.

MALPRACTICES

Certain practices render the biopesticides ineffective. In some markets around the world where regulations are not as stringent as in the developed countries, selling expired products in new packaging, adulteration with inert or ineffective material, storing in unfavorable conditions, or even mixing chemical pesticides with biopesticides damage the reputation of biopesticides.

These limitations, however, should not overshadow the potential of entomopathogens and their successful use in managing pests. *Beauveria bassiana* for managing pine caterpillars (*Dendrolimus* spp.), major forest pests in China (Li, 2007), *M. anisopliae* var. *acridum* for controlling locusts and grasshoppers in Africa (Langewald and Kooyman, 2007), *B.thuringiensis* subsp. *israelensis* for controlling mosquito larvae in the United States (USEPA, 2007), widespread use of NeabNPV against balsam fir sawfly, *Neodiprion abietis* (Harris) in Canadian forest systems (Lucarotti *et al.,* 2007), use of CpGV for managing larvae of the codling moth, *Cydia pomonella* (L.) (Lacey and Huber, 2007), and worldwide use of entomopathogenic nematodes for controlling several agricultural, horticultural and veterinary pests (Grewal *et al.*, 2005; Georgis *et al.*, 2006) and other widely used entomopathogenic bacteria, baculovirus, nematode and fungus-based products against various pests are excellent examples of biopesticides playing a major role for pest management.

CONCLUSIONS

The share of biopesticides - based on entomopathogenic fungi, bacteria and viruses - is very small in the global pesticide market. Modern scientific techniques facilitated innovative research in understanding and manipulating these pathogens for improved pest management. Researchers were able to genetically modify some pathogens that resulted in increased pathogenicity and speed of kill with minimal environmental risk. Despite these advancements, a majority of promising laboratory research has not seen the light of field application. Production issues, environmental concerns, inconsistent field efficacy, public perception and regulatory challenges are among many limiting factors for practical application of such research advancements.

However, biopesticide industry continues to grow and expected to double its market share in the next 5–10 years. This is mainly due to growing awareness and need for environmentally sustainable pest management practices. A strong collaboration between the scientific community and industry both to promote research and commercialization of the scientific advancements is very important to realize the benefits of biopesticide potential. Continuing outreach efforts to educate farmers about these developments and improved and pest management options also plays a vital role. Poor quality or adulteration of the biopesticide products in some markets taints the reputation of this industry, in general. Strong investment in developing and marketing efficacious products is a certain way to increase the marketability of biopesticides.

REFERENCES

Abdullah, M.A., Valaitis, A.P. and Dean, D.H. (2006) Identification of a *Bacillus thuringiensis* Cry11Ba toxin-binding aminopeptidase from the mosquito *Anopheles quadrimaculatus*. *BMC Biochem.*, **7**, 16.

Abdullah, M.A., Moussa, S., Taylor, M.D. and Adang, M.J. (2009) *Manduca sexta* (Lepidoptera : Sphingidae) cadherin fragments function as synergists for Cry1A and Cry1C *Bacillus thuringiensis* toxins against noctuid moths *Helicoverpa zea*, *Agrotis ipsilon* and *Spodoptera exigua*. *Pest Manage. Sci.*, **65**, 1097–1103.

Bextine, B.R. and Thorvilson, H.G. (2002) Field applications of bait-formulated *Beauveria bassiana* alginate pellets for biological control of the red imported fire ant (Hymenoptera: Formicidae). *Environ. Entomol.*, **31**, 746–752.

Campbell, L.G., Boetel, M., Jonason, N.B., Jaronski, S. and Smith, L. (2006) Grower-adoptable formulations of the entomopathogenic fungus *Metarhizium anisopliae* (Ascomycota: Hypocreales) for sugarbeet root maggot (Diptera: Ulidiidae) Management. *Environ. Entomol.*, **35**, 986–991.

Castrillo, L.A., Bauer, L.S., Liu, H., Griggs, M.H. and Vandenberg, J.D. (2009) Natural infections of *Beauveria bassiana* in *Agrilus planipennis* populations in Michigan. In: K.A. McManus and K.W. Gottschalk (eds.), *Proceedings of 20th U.S. Department of Agriculture Interagency Research Forum on Invasive Species*, USDA, Michigan.

Chang, J.H., Choi, J.Y., Jin, B.R., Roh, J.Y., Olszewski, J.A., Seo, S.J., O'Reilly, D.R. and Je, Y.H. (2003) An improved baculovirus insecticide producing occlusion bodies that contain *Bacillus thuringiensis* insect toxin. *J. Invertebr. Pathol.*, **84**, 30–37.

Chen, X., Sun, X., Hu, Z., Li, M. O'Reilly, D.R., Zuidema, D. and Vlak, J.M. (2000) Genetic engineering of *Helicoverpa armigera* single-nucleocapsid nucleopolyhedrovirus as an improved pesticide. *J. Invertebr. Pathol.*, **76**, 140–146.

CPL Business Consultants (2010) *Worldwide Biopesticides Market Summary*. CPL Business Consultatns, Wallingford, U.K.

Cory, J.S., Hirst, M.L., Williams, T., Hails, R.S., Coulson, D., Green, B.M., Carty, T.M., Possee, R.D., Cayley, P.J. and Bishop, D.H.L. (1994) Field trial of a genetically improved baculovirus insecticide. *Nature*, **370**, 138–140.

Costa, S.D. and Grassano, S. (2007) Whey-based fungal micro-factories for in situ production of biological control fungi. *Annual Meeting of the Society for Invertebrate Pathology*, Warwick, UK, p. 71.

Dara, S.K. and Hountondji, F.C.C. (2001) Effects of formulated imidacloprid on two mite pathogens, *Neozygites floridana* (Zygomycotina: Zygomycetes) and *Hirsutella thompsonii* (Deuteromycotina: Hyphomycetes). *Insect Sci. Applic.*, **21**, 133–138.

Dara, S.K., McGuire, M.R. and Kaya, H.K. (2007) Isolation and evaluation of *Beauveria bassiana* (Deuteromycotina : Hyphomycetes) for the suppression of the glassy-winged sharpshooter, *Homalodisca coagulate* (Homoptera: Cicadellidae). *J. Entomol. Sci.*, **42**, 56–65.

Dorsch, J.A., Candas, M., Griko, N.B., Maaty, W.A.A., Midboe, E.G., Vadlamudi, R.K. and Bulla L.A. (2002) Cry1A toxins of *Bacillus thuringiensis* bind specifically to a region adjacent to the membrane-proximal extracellular domain of BT-R_1 in *Manduca sexta*: involvement of a cadherin in the entomopathogenicity of *Bacillus thuringiensis. Insect Biochem. Mol. Biol.*, **32**,1025–1036.

Eldridge, R., Horodyski, F.M., Morton, D.B., O'Reilly, D.R., Truman, J.W., Riddiford, L.M. and Miller, L.K. (1991) Expression of an eclosion hormone gene in insect cells using baculovirus vectors. *Insect Biochem.*, **21**, 341–351.

Fernandez, L.E., Aimanova, K.G., Gill, S.S., Bravo, A. and Soberon, M. (2006) A GPI-anchored alkaline phosphatase is a functional midgut receptor of Cry11Aa toxin in *Aedes aegypti* larvae. *Biochem. J.*, **394**, 77–84.

Georgis, R., Koppenhöfer, A.M., Lacey, L.A., Bélair, G., Duncan, L.W., Grewal, P.S., Samish, M., Tan, L., Torr, P. and van Tol, W.W.H.M. (2006) Successes and failures in the use of parasitic nematodes for pest control. *Biol. Contr.*, **38**, 103–123.

Gongora, C.E. (2004) Transformacion de *Beauveria bassiana* cepa Bb9112 con les genes de la proteina verde fluorescente y la protease pr1A de *M. anisopliae. Rev Colomiana Entomol.*, **30**, 1–5.

Grewal, P.S., Ehlers, R.U. and Shapiro-Ilan, D.I. (2005) *Nematodes as Biological Control Agents*. CAB International, Wallingford, UK.

Hammock, B.D., Bonning, B.C., Possee, R.D., Hanzlik, T.N. and Meada, S. (1990) Expression and effects of the juvenile hormone esterase in a baculovirus vector. *Nature,* **344**, 458–461.

Harrison, R.L. and Bonning, B.C. (2001) Use of proteases to improve the insecticidal activity of baculoviruses. *Biol. Contr.*, **20**, 199–209.

Hashmi, S., Hashmi, G., Glazer, I. and Gaugler, R. (1998) Thermal response of *Heterorhabditis bacteriophora* transformed with the *Caenorhabditis elegans hsp 70* encoding gene. *J. Exp. Zool.*, **281**, 164–170.

Heinz, K.M., McCutchen, B.F., Herrmann, R., Parrella, M.P. and Hammock, B.D. (1995) Direct effects of recombinant nuclear polyhedrosis viruses on selected nontarget organisms. *J. Econ. Entomol.*, **88**, 259–264.

Hoover, K., Schultz, C.M., Lane, S.S., Bonning, B.C., Duffey, S.S., McCutchen, B.F. and Hammock, B.D. (1995) Reduction in damage to cotton plants by a recombinant baculovirus that knocks moribund larvae of *Heliothis virescens* off the plant. *Biol. Contr.*, **5**, 419–426.

Hua, G., Jurat-Fuentes, J.L. and Adang, M.J. (2004) Fluorescent-based assays establish *Manduca sexta Bt*-R_{1a} cadherin as a receptor for multiple *Bacillus thuringiensis* toxins in *Drosophila* S2 cells. *Insect Biochem. Mol. Biol.*, **34**, 193–202.

Jaronski, S.T., Grace, J.A. and Schlothauer, R. (2005). *Metarhizium anisopliae* for biocontrol of sugarbeet root maggot: constrains and challenges. *Proceedings of 33rd Biennial Meeting, American Society of Sugar Beet Technologists*, pp 185-187.

Jurat-Fuentes, J.L. and Adang, M.J. (2004) Characterization of a Cry1Ac-receptor alkaline phosphatase in susceptible and resistant *Heliothis Virescens* larvae. *Eur. J. Biochem.*, **271**, 3127–3135.

Kamita, S.G., Kang, K.D., Inceoglu, A.B. and Hammock, B.D. (2005) Genetically modified baculoviruses for pest insect control. In: L.I. Gilbert, K. Iatrou and S.S. Gill (eds.), *Comprehensive Molecular Insect Science*, Vol. 6, Elsevier, Oxford, pp. 271–322.

Lacey, L.A. and Huber, J. (2007) Microbial control of lepidopteran pests of apple orchards. In: L.A. Lacey and H.K. Kaya (eds.), *Field Manual of Techniques in Invertebrate Pathology*, Springer, Dordrecht, the Netherlands, pp. 527–546.

Langewald, J. and C. Kooyman (2007) Green Muscle™, a fungal biopesticide for control of grasshoppers and locusts in Africa. In: C. Vincent, M.S. Goettel and G. Lazarovits (eds.), *Biological Control: A Global Perspective*, CABI, Oxfordshire, pp. 311–318.

Li, Z.Z. (2007) *Beauveria bassiana* for pine caterpillar management in the People's Republic of China. In: C. Vincent, M.S. Goettel and G.

Lazarovits (eds.), *Biological Control: A Global Perspective*, CABI, Oxfordshire, pp. 300–310.

Lu, D., Pava-Ripoll, M., Liz, Z., and Wang, C. (2008) Insecticidal evaluation of *Beauveria bassiana* engineered to express a scorpion neurotoxin and a cuticle degrading protease. *Appl. Microbiol. Biotechnol.*, **81**, 515–522.

Lucarotti, C.J., Moreau, G., and Kettela, E.G. (2007) Albietiv™ a viral biopesticide for control of the balsam fir sawfly. In: C. Vincent, M.S. Goettel and G. Lazarovits (eds.), *Biological Control: A Global Perspective*, CABI, Oxfordshire, pp. 353–361.

Mcnitt, L., Espelie, K.E. and Miller, L.K. (1995) Assessing the safety of toxin-producing baculovirus biopesticides to a nontarget predator, the social wasp *Polistes metricus* Say. *Biol. Contr.*, **2**, 267–278.

Meada, S. (1989) Increased insecticidal effect by a recombinant baculovirus carrying a synthetic diuretic hormone gene. *Biochem. Biophys. Res. Commun.*, **165**, 1177–1183.

Migiro, L.N., Maniania, N.K., Chabi-Olaye, A. and Vandenber, J. (2010) Pathogenicity of entomopathogenic fungi *Metarhizium anisopliae* and *Beauveria bassiana* (Hypocreales: Clavicipitaceae) isolates to the adult pea leafminer (Diptera: Agromyzidae) and prospects of an autoinoculation device for infection in the field. *Environ. Entomol.*, **39**, 468–475.

O'Reilly, D.R. and Miller, L.K. (1991) Improvement of a baculovirus pesticide by deletion of the *egt* gene. *Bio/Technology*, **9,** 1086–1089.

O'Reilly, D.R., Kelly, T.J., Masler, E.P., Thyagaraja, B.S., Robson, R.M., Shaw, T.C. and Miller, L.K. (1995) Overexpression of *Bombyx mori* prothoracicotropic hormone using baculovirus vectors. *Insect Biochem. Mol. Biol.*, **25**, 475–485.

Pava-Ripoll, M., Posada, F.J., Momen, B., Wang, C. St. Leger, R.J. (2008) Increased pathogenicity against coffee berry borer, *Hypothenemus hampei* (Coleoptera: Curculionidae) by *Metarhizium anisopliae* expressingthe scorpion toxin (AaIT) gene. *J. Invertebr. Pathol.*, **99**, 220–226.

Petrick, D.T., Iseli, A., Montelone, B.A., Van Etten, J.L. and Clem, R.J. (2003) Improving baculovirus resistance to UV inactivation: Increased virulence resulting from expression of a DNA repair enzyme. *J. Invertebr. Pathol.*, **82**, 50–56.

Possee, R.D., Hirst, M., Jones, J.D., Bishop, D.H.L. and Cayley, P.J. (1993) Field tests of genetically engineered baculoviruses. *Brit. Crop Prot. Council*, **55**, 23–36.

Quintela, E.D. and McCoy, C.W. (1997) Pathogenicity enhancement of *Metarhizium anisopliae* and *Beauveria bassiana* to first instars of *Diaprepes abbreviatus* (Coleoptera: Curculionidae) with sublethal doses of imidacloprid. *Environ. Entomol.*, **26**, 1173–1182.

Sangadala, S., Walters, F., English, L.H. and Adang M.J. (1994) A mixture of *Manduca sexta* aminopeptidase and alkaline phosphatase enhances

Bacillus thuringiensis insecticidal Cry1A(c) toxin binding and $^{86}Rb^+$-K^+leakage *in vitro*. *J. Biol. Chem.*, **269,** 10088–10092.

Siebeneicher, S.R., BradleighVinson, S. and Kenerley C.M. (1992) Infection of the red imported fire ant by *Beauveria bassiana* through various routes of exposure. *J. Invertebr. Pathol.*, **59**, 280–285.

St. Leger R.J., Joshi. L., Bidochka, M.J. and Roberts, D.W. (1996) Construction of an improved mycoinsecticide over-expressing a toxic protease. *Proc. Nat. Acad. Sci. USA*, **93**, 6349–6354.

Vadlamudi, R.K., Ji, T.H. and Bulla, L.A.Jr. (1993) A specific binding protein from *Manduca sexta* for the insecticidal toxin of *Bacillus thuringiensis* Berliner. *J. Biol. Chem.*, **268**, 12334–12340.

Stewart, L.M.D., Hirst, M., Ferber, M.L., Merrywheather, A.T.,Caylay, P.J. and Possee, R.D. (1991) Construction of an improved baculovirus insecticide containing an insect-specific toxin gene. *Nature*, **352,** 85–88.

Tomalski, M.D. and Miller, L.D. (1992) Expression of a paralytic neurotoxin gene to improve insect baculoviruses as biopesticides. *Bio/Technology*, **10**, 545–549.

U.S. Environmental Protection Agency (2007) *Larvicides for Mosquito Control*. EPA Publication 735-F-02-007.

Wang, C. and St. Leger, R.J. (2007) A scorpion neurotoxin increases the potency of fungal insecticide. *Nat. Biotechnol.*, **25**, 1455–1456.

Wekesa, V.W., Maniania, N.K., Knapp, M. and Boga, H.I. (2005) Pathogenicity of *beauveria bassiana* and *Metarhizium anisopliae* to the tobacco spider mite *Tetranychus evansi*. *Exp. Appl. Acarol.*, **36**, 41–50.

Zhuang, M., Oltean, D.I., Gomez, I., Pullikuth, A.K., Soberon, M., Bravo, A. and Gill, S.S. (2002) *Heliothis* Virescens and *Manduca sexta* lipid rafts are involved in Cry1A toxin binding to the midgut epithelium and subsequent pore formation. *J. Biol. Chem.*, **277**, 13863–13872.

14

Registration Related Issues in Effective use of Biopesticides in Pest Management

O.M. Bambawale and Someshwar Bhagat

National Centre for Integrated Pest Management, L.B.S. Building, IARI Campus, Pusa, New Delhi-110 012, India
E-mail: om.bambawale@ncipm.org.in

INTRODUCTION

Biopesticides use naturally occurring organisms, such as fungi, bacteria, viruses and nematodes to control plant diseases and arthropod pests. A compound that kills organisms by virtue of specific biological effects rather than as a broader chemical poison which differ from biocontrol agents in being passive agents, whereas biocontrol agents actively seek the pest. The rationale behind replacing conventional pesticides with biopesticides is that the latter are more likely to be selective and biodegradable. Integrated management strategies are designed to tackle some of the key pests, which are not readily amenable to chemical pesticide way of management and microbes offer a large array of alternatives/supplements and hope to our efforts in IPM. Biopesticides as pest management agents based on living micro-organisms or natural products have proven potential for pest management and they are being used across the world. However, they are regulated by systems designed originally for chemical pesticides that have created market entry barriers by imposing burdensome costs on the biopesticide industry. There are also significant technical barriers to making biopesticides more effective (Chandler *et al.*, 2011). There is a hope that IPM as part of agricultural policy could lead to innovations in the way that biopesticides are regulated. There are also new opportunities for developing biopesticides in IPM by combining ecological science with post-genomics technologies. The new biopesticide products that will result from this research will bring with them new regulatory and economic challenges that must be addressed through joint working between social and natural scientists, policy makers and industry (Chandler *et al.*, 2011).

BIOPESTICIDES VIS-À-VIS INSECTICIDE ACT 1968

A number of fungi, viruses, bacteria, predators, parasitoids, etc., are naturally occurring and organisms, which are able to manage different categories of pests such as insects, disease pathogens, nematodes, weeds, etc. The Central Insecticides Board (CIB) in its meeting of January 15, 1994 deliberated the inclusion of such biocontrol agents in the schedule to the Insecticides Act and the Board constituted a sub-committee to look into various technological aspects of these biocontrol agents etc., before bringing them under the purview of the Act. Subsequently, the sub-committee meetings held in June, 1994 and October, 1995 and their recommendations were further deliberated by the Board (March, 1995; December, 1995). The board accepted the recommendations of the subcommittee and agreed to include the microbial products like entomogenous fungi (e.g., *Beauveria bassiana, Verticillium lecanii, Metarrhizium anisopliae, Nomuraea rileyi*); antagonistic fungi and bacteria (*Trichoderma* spp., *Pseudomonas fluorescens, Bacillus subtilis* etc.); mycoherbicides, insect repellents, nuclear polyhedrosis virus and granulosis virus, the pheromones and biochemicals in the Schedule of the Act. The subcommittee did not recommend to include parasitoids and predators in the Schedule to the Insecticides Act. Further, the CIB in its 28th meeting held on January 18, 1999 decided that inclusion of insect repellents, mycoherbicides, and pheromones should be put up to CIB for approval as individual chemicals/substances with their complete chemical/ biological identity and not on the basis of their mode of action.

As per the Gazette Notifications No.GSR 200 (E) dated 29/03/78, GSR 839 (E) dated 14/11/85, GSR 13 (E) dated 05/01/90, GSR 621 (E) dated 27/10/97, G.S.R.224 (E) dated 26/03/99, G.S.R. 868 (E) dated 15/11/2000, G.S.R. 69 (E) dated 5/02/2001, G.S.R. 378 (E) dated June 24/06/2004, G.S.R. 198 (E) dated 31/03/2005, G.S.R. 51(E) dated 6/02/2006, G.S.R. 683(E) dated 31/10/2006 and G.S.R. 650(E) dated 11/09/2008 of the Central Insecticide Board, Department of Agriculture and Cooperation, Ministry of Agriculture, Govt. of India, the following biopesticides have been included under the Insecticide Act 1968.

Botanicals

(a) Herbal extract (mixture of diallyl disulphide, allyl propyl disulphide and allyl isothiocyanate), (b) Neem products (neem leaves, neem oil, neem seed kernel, neem cake), (c) Karanjin (isolated from *Pongamia glabra*), (d) Extract of *Cymbopogon* species, (e) Oxymatrine (plant derived pesticide - *Sophora flavescens*), (f) *Tripterygium wilfordii* Hook GTW - Plant extract, (g) Bitterbarkomycin (plant extract of *Apocynum venetum*), (h) Squamocin (seed extract of plant *Annona squamosa* Linn) and (i) Eucalyptus leaf extract.

Antagonistic Fungi and Bacteria

(a) *Bacillus subtilis* Cohen, (b) *Bacillus* spp., (c) *Streptomyces griseoviridis,* (d) *Streptomyces lydicus,* (e) *Gliocladium* spp., (f) *Pseudomonas fluorescens* Migul, *(g) Trichoderma* spp., *(h) Burkholderia cepacia,* (i) *Agrobacterium radiobacter* strain 84, *(j) Agrobacterium tumefaciens,* (k) *Erwinia amylovora* (Hairpin protein), *(l) Alcaligenes* spp., (m) *Ampelomyces quisqualis, (n) Candida oleophila, (o) Coniothyrium minitans,* (p) *Pythium oligandrum, (q) Phlebia gigantean,* (r) *Penicillium islanidicum,* (s) *Aspergillus niger*-strain AN27, (t) VAM-Vesicular arbuscular mycorrhiza, (u) *Piriformospora indica,* (v) *Photorhabdus luminescens* akhurustii Strain K-I, and (w) *Serratia marcescens* GPS 5.

Entomopathogenic Fungi

(a) Beauveria bassiana (Bals.) Vuill., *(b) Metarrhizium anisopliae* (Metschnikoff) Sorokin, *(c) Nomuraea rileyi* (Farlow) Samson, *(d) Verticillium lecanii* (Zimm.), *(e) Verticillium chlamydosporium* and *(f) Hirsutella* species.

Baculoviruses

(a) Granulosis Virus (GVs) and (b) Nuclear Polyhedrosis Virus (NPVs) Apart from these, sex pheromones are also under the Insecticide Act, 1968, which are being used for pest monitoring purposes in India. Since these are used in minute quantities, registration is not required. However, if pheromones are to be used in relatively large quantities for mating disruption, registration is mandatory. The naturally occurring parasitoids and predators are not under the Insecticide Act, 1968 and do not require registration. The mass production technologies for these biological control agents and their hosts have been developed and standardized and being used for the control of insect pests.

The revised guidelines/data requirement for provisional registration and regular registration of biopesticides are made under *Section 9(3B)* and *Section 9(3)* of the Insecticide Act, 1968, respectively. Further details may be seen at http://cibrc.gov.in/rcpage.htm.

LIST OF REGISTERED BIOPESTICIDES IN INDIA

Botanicals

Azadirachtin (Neem products) and Pyrethrins (pyrethrum)

Bacillus thuringiensis/Bacillus subtilis

Bt. var. *israelensis* 164, serotype H-14 WP formulation, *Bt.* var. *kurstaki* strain A-97, serotype H3a, 35 WP, *Bt.* var. *israelensis* serotype H-14, strain VCRC B-17 slow release granules, *Bt.* var. israelensis serotype H-14, strain VCRC B-17 WP formulation, *Bt.* var. israelensis (WS), *Bt.* var. *israelensis* serotype H-14 12(AS).

Trichoderma spp.

Trichoderma viride WP, *Trichoderma viride* 1% WP, *Trichoderma viride* 0.50% WS, *Trichoderma harzianum* 0.50% WS, *Trichoderma1*.15% WP.

Pseudomonas spp.

Pseudomonas fluorescens 1.25% WP.

Entomogenous fungi

Verticillium lecanii, Beauveria bassiana (WP).

Baculoviruses

NPV of *Helicoverpa armigera* (AS), NPV of *Helicoverpa armigera* 0.43% (AS) and NPV of *Spodoptera litura* 0.50% (AS).

The desirous applicant can apply on prescribed Form I along with prescribed fee of Rs. 5,000/- to the Registration Committee for registration above mentioned biopesticides under section 9(3 B) as per guidelines for data generation. The certificate for registration is issued for a period of two years under section 9(3B). Permanent certificate of registration is issued under section 9(3) as per guidelines. Permanent certificate of registration is also issued under section 9(4) on the same conditions on which biopesticides were registered originally under Section 9(3) in case of neem based biopesticides. In case of biopesticides based on *Bacillus* and other living organisms, the certificate of registration is issued only under Section 9(3), i.e. certificate of registration under section 9(4) is not issued due to very specific strains of biopesticides.

The registration guidelines were revised in 2009 (http://cibrc.gov.in/rcpage.htm, 58 pages) and are in force from January 2010. The guidelines/data requirements for registration of Baculoviruses – Neculear Polyhedrosis Viruses (NPV) and Granulosis Viruses (GV), antagonistic fungi, entomopathogenic fungi, antagonistic bacteria and entomotoxic bacteria along with Indian Standards have been elaborated in this publication. For *Helicoverpa armigera* and *Spodoptera litura*, NPVs 1×10^9 Polyhedral Occlusion Bodies (POBs)/ ml or gm (minimum) and for *Chilo infuscatellus, Plutella xylostella* and *Achaea janata* GVs a standard of 5×10^9 capsules/ml or g (minimum) has been prescribed. Antagonistic fungi include only *Trichoderma* spp. and have a minimum indian standard of 2×10^6 CFUs per ml or gram of the product. Entomopathogenic fungi in the registration list include *Beauveria bassiana, Metarrhizium anisopliae* and *Nomuraea rileyi* against the rearing insects *H. armigera, S. litura* and *Plutella xylostella* with a minimum indian standard of 1×10^8 CFUs/ml or gm. Antagonistic bacteria include *Pseudomonas fluoresc-ens* and *Bacillus subtilis* with a minimum indian standard of 1×10^8 CFUs/ ml or gm. The last category of entomotoxic bacteria includes *Bacillus thuringiensis* and needs to have a minimum of 2 per cent δ-endotoxin content in the product.

THE GUIDELINES

The guidelines for any pesticide provide national policy-makers with critical elements to develop and strengthen national policy for the management of public health pesticides, such as vector control pesticides, household pesticides, and pesticides used by pest control operators. Issues and driving forces that may instigate national policy development are discussed and guidance is provided on the process of policy formulation, implementation and evaluation. The guidelines focus on public policy development, formulated by a government and implemented under the overall responsibility of a government. The guidelines do not provide specific advice on technical aspects relating to use and management of public health pesticides, for which separate guidance documents exist. References to such technical guidelines are provided. While it is recognized that the large majority of products used for public health purposes are insecticides, biopesticides are being used and accordingly the guidlenies are required for eco-friendly pesticides as well. In recent past many such guidelines were framed for biopesticides in India too and now we have some latest revised versions available.

REVISED GUIDELINES FOR VIRUSES

Revised Guidelines/data requirements for baculoviruses. nclear polyhedrosis virus (NPV) and granulosis virus (GV) under Section 9(3) of the Insecticide Act, 1968 have been issued.

Biological Characteristics and Chemistry

Systematic name (genus and species, strain if any), common name, if any, source and origin as per Annexure – I, specification of the product as per Annexure –I, and composition of the product are required. The parameters required are viral unit (POB/capsule count/ ml or g of the product), content (content % of the biocontrol organism in the formulation and nature of biomass, carrier/filler (%), wetting/ dispersing agent, stabilizers/emulsifiers, containments/ impurities and moisture content.

Appropriate test. Procedure and criteria used for identification by DNA test (Restriction enzyme analysis test).

Method of analysis. (as detailed under Appendix-I)

Viral unit: NPVs 1×10^9 POB/ml or g, minimum; GVs: 5×10^9 capsules/ml or g, minimum

Biological assays for determining the LC_{50} /LD_{50} of the formulation

Bioassay for NPV by the diet surface contamination method as detailed in Appendix-II

Bioassay for GV against *Chilo infuscatellus* as detailed in Appendix-III

Bioassay for GV against *Plutella xylostella* as detailed in Appendix-IV

Bioassay for GV against *Achaea janta* as detailed in Appendix-V

(Appendix attachments can be seen at http://cibrc.gov.in/ rcpage.htm)

Contaminants. Pathogenic contaminants (*Salmonella, Shigella, Vibrio*, etc.) should be absent, other microbial contaminants (1×10^4 maximum) and no chemicals and botanical pesticide contaminants in the formulated products are acceptable.

Shelf-life claim. Not less than 6 months; a sample for verification (100 ml or g) is required.

Bioefficacy

Field test. Results of two trials in two agroclimatic conditions conducted by ICAR/CSIR/SAU/ICMR institutes are required. Either the university authorities or head of the institute concerned should certify data.

Laboratory studies. Results of two trials in two agroclimatic conditions conducted by ICAR/CSIR/SAU/ICMR institutes. Either the university authorities or head of the institute concerned should certify data.

Toxicity

For primary culture. (i) Single dose oral (rat and mouse), (ii) Single dose pulmonary, (iii) Single dose intravenous, (iv) Cell culture and (v) Human safety records.

For formulation. (i) Data for primary culture with technical specification, (ii) Single dose oral (rat and mouse), (iii) Single dose pulmonary, (iv) Primary skin irritation, (v) Primary eye irritation and (vi) Human safety records.

For formulated product to be directly manufactured. Mammalian toxicity testing of formulations will include (i) single dose oral (rat and mouse) for toxicity/infectivity/pathogenicity; (ii) single dose pulmonary toxicity/infectivity/pathogenicity (intra-tracheal prefe-rred); (iii) single dose intravenous toxicity /infectivity/p athogenicity; (iv) human safety records (effect of lack of effects); (v) primary skin and eye irritation and (vi) cell culture.

Environmental safety testing: core information requirem-ents (for formulation only). This is required for (i) non-target vertebrates, i.e. mammals, birds (two species) and fresh water fish and (ii) non-target invertebrates, i.e terrestrial invertebrates and soil invertebrates.

Processing, Packaging and Labelling

Processing, packaging and labelling is applicable to formulations and several setps are very important in the process development:

Manufacturing process/process of formulation. This includes raw material, plant and machinery, unit process operation/ unit process, and output (finished product and generation of waste)

Packaging. Packaging depends up on classification of formulation, i.e. solid, liquid or other type of products; unit pack size (in metric system); specification details of primary, secondary and transportation of packages and lastly the compatibility of primary pack with the product (glass bottles are not recommended).

Labels and leaflets. As per Insecticides rules, 1971 and as per existing norms the common name, composition, antidote/storage statements, etc are necessary to be indicated on the labels and leaflets.

Notes

1. Applicants are required to submit an undertaking that strain is indigenous, naturally occurring, not exotic in origin, and not genetically modified as per Annexure 1.1.
2. Additional three months data for up to one year shelf-life claim and additional six months data for more than one year claim at two different agro climatic locations at ambient temperature along with meteorological data should be submitted.
3. Considering the fact that many small entrepreneurs are engaged in the business of cultivation of NPV/GV, the following mentioned simplification has been suggested.

3.1 If same mother culture is used for making formulation by different entrepreneurs then the information submitted once on mother culture will be sufficient. All entrepreneurs need not to generate data.

3.2 If same mother culture, same method and same adjutants, stabilizers etc. are used for making formulation, then data once submitted will be sufficient for subsequent registrants.

4. The packaging material should also be ensured free from contamination from handling, storage and transportation.

REVISED GUIDELINES FOR FUNGI

The revised guidelines/data requirements for antagonistic and entomogenous fingi are listed in Table 14.1.

Table 14.1. Revised guidelines/data requirements for antagonistic and entomogenous fungi under section 9(3B) and 9(3) of the Insecticide Act, 1968

Items	Antagonistic fungi	Entomopathogenic fungi
Standard of formulations	Colony Forming Unit (CFU) count on selective medium should be minimum of 2x 10^6 per ml or gm for *Trichoderma* spp. Pathogenic contaminants such as gram negative bacteria *Salmonella, Shigella, Vibrio* and such other microbials should not be present. Other microbial contaminants should not exceed 1 x 10^4 count per ml or per g of formulation. Chemical / botanical pesticide contaminants should not be present.	Colony Forming Unit (CFU) count on selective medium should be minimum of 1x 10^8 per ml or gm for entomopathogenic fungi, Pathogenic contaminants such as gram negative bacteria *Salmonella, Shigella, Vibrio* and such other microbials should not be present. Other microbial contaminants should not exceed 1 x 10^4 count per ml or per g of formulation. Chemical/botanical pesticide contaminants should not be present and
	Stability of CFU counts at 30°C and 65% RH	Stability of CFU counts at 30°C and 65% RH
Biological characteristics and chemistry	Systematic name (genus and species), strain name, common name, if any, source of origin as Annexure-1.1, and habitat and morphological description	Systematic name (genus, species and strain), common name, if any, source of origin as per Annexure-1.1, and natural occurrence of the organism and morphological description,
	Composition of the product CFU/g of the product, percent content of the biocontrol organism in the formulation & nature of biomass, percentage of carrier/filler, wetting/ dispersing agent, stabilizers/ emulsifiers, contaminants/ impurities, etc. and moisture content	**Composition of the product** CFU/g of the product, percent content of the biocontrol organism in the formulation & nature of biomass, percentage of carrier/filler, wetting/ dispersing agent, stabilizers/ emulsifiers, contaminants/ impurities, etc. and moisture content
	Qualitative analysis: CFU on selective medium Specification of the product as per Annexure-I, Manufacturing process including type of fermentation and biological end products: The microbial cultures are multiplied by liquid/ solid	**Qualitative analysis: CFU on selective medium** Specification of the product as per Annexure-I, Manufacturing process including type of fermentation and biological end products: The microbial cultures are multiplied by liquid/solid fermentation and information

	fermentation and information pertaining to use of entire mycelial mats with spores separated must be provided in terms of biomass.	pertaining to use of entire mycelial mats with spores separated must be provided in terms of biomass.
	Method of analysis CFU counts by serial dilution and examination under regular compound research microscope with bright field optics. Antagonistic mycolytic capability on target organism by bioassay on plants (laboratory test). Bioassay procedure based on diseased severity and root colonization as detailed in Appendix-I **(http://cibrc.gov.in/ rcpage. htm)**	**Method of analysis** Pathogencity test on insect as per Appendix-I Bioassay procedure for *Plutella xylostella* as detailed in Appendix-II (http://cibrc.gov.in/rcpage.htm)
	Shelf life claims : Not less than 6 months Data on storage stability as per shelf life claims as detailed in Note-2 A sample for verification (100 g)	**Shelf life claims**: (Not less than 6 months): Data on storage stability as per shelf life claims as detailed in Note-2 A sample for verification (100g)
Bioefficacy	**Field studies** Data from SAU's/ICAR Institute certified by Director Research of SAU or Head of the ICAR Institute	**Field tests** Data on bioeffectiveness from SAU's/ICAR Institute certified by Director Research or Head of the Institute Data on non-target organism: One season/one year on effect on product against natural parasites/ predators
	Laboratory studies The product should be tested at a laboratory under ICAR/ SAU/ CSIR/ICMR.	**Laboratory tests** The product should be tested at a laboratory under ICAR/ SAU/ CSIR/ICMR.
Toxicity	**For mother culture** Single dose oral (rat and mouse) Single dose pulmonary Single dose dermal Single dose intraperitoneal Human safety records.	**For mother culture** Single dose oral (rat and mouse) Single dose pulmonary Single dose dermal Single dose intraperitoneal Human safety records.
	For formulation Data on mother culture as in above	**For formulation** Data on mother culture as in above

	Single dose oral (rat & mouse) Single dose pulmonary Primary skin irritation Primary eye irritation Human safety records	Single dose oral (rat & mouse) Single dose pulmonary Primary skin irritation Primary eye irritation Human safety records
	For formulated product to be directly manufactured (Mammalian toxicity testing of formulations) Single dose oral (rat & mouse) Toxicity/Infectivity/ Pathogenicity Single dose pulmonary Toxicity/Infectivity/ Pathogenicity (Intratracheal preferred) Single dose dermal infectivity Single dose intraperitoneal (Infectivity) Primary skin irritation Primary eye irritation Human safety records (effect/lack of effects)	**For formulated product to be directly manufactured** (Mammalian toxicity testing of formulations) Single dose oral (rat & mouse) Toxicity/Infectivity/Pathogenicity Single dose pulmonary Toxicity/Infectivity/Pathogenicity (Intratracheal preferred) Single dose dermal infectivity Single dose intraperitoneal (Infectivity) Primary skin irritation Primary eye irritation Human safety records (effect/lack of effects)
	Environmental safety testing **Core Information requirements** (For formulation only) **Non-target vertebrates** Mammals Birds(two species)[b] Fresh water fish[c] **Non-target invertebrates** Soil invertebrates	**Environmental safety testing** **Core Information requirements** (For formulation only) **Non-target vertebrates** Mammals Birds(two species)[b] Fresh water fish[c] **Non-target invertebrates** Terrestrial invertiberates
Packaging and labelling	**Manufacturing process/ process of formulation** Raw material Plant and machinery Unit process operation/unit process Out-put (finished product and generation of waste)	**Manufacturing process/process of formulation** Raw material Plant and machinery Unit process operation/unit process Out-put (finished product and generation of waste)
	Packaging	**Packaging**

	Classification-solid, liquid or other types of product. Unit pack size–in metric system Specification – details of primary, secondary and transport pack Compatibility of primary pack with the product (glass bottles are not recommended). **Labels and leaflets** As per Insecticides Rules, 1971/ as per existing norms indicating the common name, composition, antidote, storage, statements, etc	Classification-solid, liquid or other types of product. Unit pack size – in metric system Specification – Details of primary, secondary and transport pack Compatibility of primary pack with the product (glass bottles are not recommended). **Labels and leaflets** As per Insecticides Rules, 1971/as per existing norms indicating the common name, composition, antidote/storage, statements, etc
Notes	Applicants are required to submit an undertaking that strain is indigenous, naturally occurring, not exotic in origin, and not genetically modified as per Annexure 1.1. Additional two months data for six months shelf-life claim, three months additional data for one year shelf-life claim at two different agro climatic locations at ambient temperature along with meteorological data should be submitted. Considering the fact that many small entrepreneurs are engaged in the business of cultivation of antagonistic/entomogenic fungi the following simplification has been considered. If same microbial strain is used for making formulation by different entrepreneurs then the information submitted once on the said strain will be sufficient. All entrepreneurs need not to generate relevant data If same microbial strain, same method and same adjutants, stabilizers etc. are used for making the given formulation, data once submitted for validitating these claims will be sufficient for subsequent registrants, as substantiated by the relevant supportive documents. The packaging material should also be ensured to be free from contamination from handling, storage and transportation and is as per prescribed standards, as the case may be. **a**=Information on infection and pathogencity in mammals will be available from mammalian safety testing. **b**=Information on infection and pathogenicity, suggested test: single-dose, oral test. Suggested test species: pigeon and chicken. **c**=Information on infection and pathogencity. Suggested test species: *Tilapia mossambica* or other appropriate spp. **d**=Information on mortality effects. It is recommended that information be obtained for honey bee and *Bombyx mori* (silk worm). **e**=Information on mortality effects. It is recommended that test species include an earthworm (*Lumbricus terrestris*) or other appropriate macro invertebrates of ecological significance.	

REVISED GUIDELINES FOR BACTERIA

The revised guidelines/data requirements for anatagonistic bacteria and entomotoxic bacteria are listed in Table 14.2.

Table 14.2. Revised guidelines/data requirements for antagonistic bacteria and entomotoxic bacteria under section 9(3B) and 9(3) of the Insecticide Act, 1968

Items	Antagonistic bacteria	Entomogenous bacteria
Standard of formulations	Colony Forming Unit (CFU) count on selective medium should be minimum of 1x 10^8 per ml or gm for antagonistic bacteria.	Delta endotoxin content (minimum 2.0%) – estimation as per Appendix-V Adjuvants Beta Exotoxin content – negative through housefly bioassay test as per Appendix-IV (http://cibrc.gov.in/rcpage.htm)
	Pathogenic contaminants such as gram negative bacteria *Salmonella, Shigella, Vibrio* and such other microbials should not be present. Other microbial contaminants should not exceed 1 x 10^4 count per ml or per g of formulation.	Human pathogens (gram negative bacateria *Salmonella, shigella* & *vibrio* etc) - absent Other microorganisms (not more than 10^4 / g)
	Chemical / botanical pesticide contaminants should not be present	Chemical/botanical pesticide contamination – absent
	Stability of CFU counts at 30°C and 65% RH	Stability of CFU counts at 30°C and 65% RH
Biological characteristics and chemistry	**Composition of the product** Percent content of the Biocontrol organism in the formulation & nature of biomass. CFU/g or ml of the product, Percentage of other components: carrier / filler, wetting / dispersing agent, stabilizers / emulsifiers, contaminants / impurities etc. , moisture content, and specification of the product as per Annexure-I,	**Composition of the product** Delta endotoxin content – through housefly bioassay test. Beta Exotoxin content – to be ruled out through bioefficacy test (housefly bioassay method), Adjuvants, Human pathogens (culture method), Other microorganisms (not more than 10^4 / g), chemical and botanical pesticide contaminants and natural occurrence of the organism.

Systematic name (genus and species), strain name, common name, if any, source of origin as Annexure-1.1,	Common name of **Entomotoxic bacteria**, systematic name: (genus, species, serotype and strain)*, Cry toxin classification (delta endotoxin), physical specification, form and appearance, moisture content, pH, particle size, suspensibility, and miscibility,
Natural occurrence of the organism and morphological description	Natural occurrence of the organism- its relationship of the organisms, history (exotic or indigenous strain) and the isolate should not be genetic ally modified organism (GMO).
Manufacturing process including type of fermentation and biological end products: The microbial cultures are multiplied by liquid/ solid fermentation. Information pertaining to use of entire mats with spores separated must be provided in terms of biomass.	**Manufacturing process:** A technique for the separation and purification of the crystals (Appendix III) is to be used by the manufacturer and the antisera to be raised using solublized toxin. Toxin content (3.5%) to be standardized in the formulation using this antisera (ELISA /Dot blot assay). Separation and purification of crystals required if antisera is to be developed for the strains delta endotoxin (http://cibrc.gov.in/rcpage.htm)
Test method Dual culture for antagonistic bacteria to attain at least 35% reduction in target organism **Bioassay**: based on disease severity and root colonization as detailed in **Appendix-I**	**Test method** Test procedure and criteria used for identification – morphology, biochemistry, serology/ immunology, morphological description and particle size. **Potency** of product by bioassay method (LC_{50} on target larvae) and potency against a reference using artificial diet or leaf disc method or in the water for mosquito larvae as detailed in Appendix-II). Housefly Bioassay method for Beta-exotoxin content (Appendix-IV)

		Determination of toxin content by ELISA / Dot Blot Assay Method (Appendix-V) (http://cibrc.gov.in/rcpage.htm)
	Contaminants:	**Contaminants**:
	Pathogenic contaminants such as *Salmonella, Shigella, Vibrio* and such other microbials	Pathogenic contaminants such as *Salmonella, Shigella, Vibrio* and such other microbials
	Other microbial contaminants	Other microbial contaminants
	Chemical and botanical pesticide contaminants	Chemical and botanical pesticide contaminants
	Shelf life claims (Not less than 6 months):	Shelf life claim(not less than 6 months)
	Data on storage stability as per shelf life claims as detailed in Note-2	Shelf life data in support of shelf life claim as detailed in Note-2
	A sample for verification (100 g)	A sample for test (100 g)
Bioefficacy	**Field tests**	**Field tests**
	Data from SAU's/ICAR Institute certified by Director Research of SAU or Head of the ICAR Institute	Data on bioeffectiveness and phytotoxicity generated at ICAR, SAUs, CSIR / ICMR institutes. The data should be certified either by the Director or Head of the Institute.
	Data on non-target organism: One season/one year on effect on product against natural parasites/ predators	Data on non-target organisms: one season / one year data on the effect of the product on natural predators / parasites
	Laboratory tests	**Laboratory tests**:
	The product should be tested at a laboratory under ICAR/ SAU/ CSIR/ICMR.	LC_{50} values for each insect species under laboratory conditions should be generated at least at two institutes of ICAR, SAUs, CSIR and ICMR
Toxicity	**For mother culture**	**Single exposure studies**
	Single dose oral (rat and mouse)	Oral toxicity / pathogenicity
	Single dose pulmonary	Dermal toxicity / pathogenicity
	Single dose dermal	Inhalation toxicity / pathogenicity
	Single dose intra-peritoneal	Primary skin irritation
	Human safety records.	Mucous membrane irritation
		Allergy / sensitization / immuno suppression

	For formulation Data on mother culture as in above Single dose oral (rat & mouse) Single dose pulmonary Primary skin irritation Primary eye irritation Human safety records **Environmental safety testing: Core Information requirements** (For formulation only) **Non-target Vertebrates** Mammals[a] Birds (two species)[b] Fresh water fish[c] **Non-target invertebrates** Terrestrial invertibrates[d] Soil invertebrates [e]	**For formulation** Data on mother culture as in above Single dose oral (rat & mouse) Single dose pulmonary Primary skin irritation Primary eye irritation Human safety records **Eco-toxicity** Toxicity to birds[a] Toxicity to fish[b] Toxicity to honeybees Toxicity to silkworm
Packaging and labelling	**Manufacturing process/ process of formulation** Raw material Plant and machinery Unit process operation/unit process Out-put (finished product and generation of waste)	Packaging requirements as per ISI or as per the approval of RC New packaging system approved by FAO / ASPM / other global standards (relevant)
	Manner of Packaging Classification-solid, liquid or other types of product. Unit pack size–in metric system Specification – details of primary, secondary and transport pack Compatibility of primary pack with the product (glass bottles are not recommended).	**Manner of Packing** Specification of primary packing, Secondary packing and transport packing Detailed information for completely filled transport packing containing quantity of primary and secondary packing
	Labels and leaflets As per Insecticides Rules, 1971/as per existing norms indicating the common name, composition, antidote, storage, statements etc.	**Manner of labelling** Specification of primary packing, secondary packing and transport packing
		Container content compatibility If BIS methodology is not available, then as per protocols being approved by RC and

	validation of analytical method by CIL **Labels and leaflets** 7 copies of L/Ls upto 250ml 7 copies of L/Ls upto 500ml
Notes for Antagonistci/ Entomotoxic Bacteria	1. Applicants are required to submit an undertaking that strain is indigenous, naturally occurring, not exotic in origin, and not genetically modified as per Annexure 1.1. 2. Additional two months data for six months shelf-life claim, three months additional data for one year shelf-life claim at two different agro climatic locations at ambient temperature along with meteorological data should be submitted. 3. Considering the fact that many small entrepreneurs are engaged in the business of cultivation of antagonistic bacteria/ entomotoxic fbacteria the following simplification has been considered. 3.1. If same microbial strain is used for making formulation by different entrepreneurs then the information submitted once on the said strain will be sufficient. All entrepreneurs need not to generate relevant data 3.2. If same microbial strain, same method and same adjutants, stabilizers etc. are used for making the given formulation, data once submitted for validitating these claims will be sufficient for subsequent registrants, as substantiated by the relevant supportive documents. 4. The packaging material should also be ensured to be free from contamination from handling, storage and transportation and is as per prescribed standards, as the case may be.
Notes for Entomotoxic Bacteria	5. The percentage of ingredient relative to total material is required to be stated and may vary from 2-7 per cent, the balance being inert ingredients. In addition, the labels will have to contain a measurement of toxin protein as percent protein, referring to the Lepidopteran-active toxin(s) present in the crystal. 6. Bt products should be labeled with biopotency and (or) toxin content. 7. The presently used Bt var. *kurstaki* standard is HD-1-S-1980 and its potency was calculated at 16,000 IUS per milligram of powder (Beegle et al. 1986. Standardization of HD-1-S-1980: US Standard for Lepidopterous-active *Bacillus thuringiensis*. Bulletin Ent. Soc. America 32: 44-45.). This standard strain is now available with PDBC, Bangalore and DOR, Hyderabad. 8. Defined potency and toxin concentration – Bioassay would require the use of an insect species. Normally manufacturers could select *Trichoplusia ni* / *Helicoverpa armigera* for Lepidopteran specific Bt formulations. *Spodoptera* Units (SPU), *Leptinotarsa* Units (LTUs) or International Toxin Units (ITUs) are to be used for denoting a specific insect.

9. No test for beta exotoxin is required for *Bacillus sphaericus*, because this species is not known to produce exotoxins.
10. The biopotency of products based on *B. thuringiensis* subsp. *israelensis* (*Bti*) is compared against a reference strain IPS82, 1884 using early fourth-instar larvae of *Aedes aegypti* (strain Bora Bora). The toxicity of IPS82 has an arbitrarily assigned toxicity of 15,000 ITU/mg powder.
11.The bioefficacy of products based on *B. sphaericus* (*Bsh*) is determined against a reference standard SPH88, strain 2362 using early fourth-instar larvae of *Culex pipiens pipiens* (strain Montpellier). The toxicity of SPH88 has an arbitrarily assigned toxicity of 1,700 ITU/mg of the powder (Guidelines for laboratory and field testing of mosquito larvicides, WHO 2005 pp 45).
12. The use of alternative bacterial reference powders and / or strains must be approached cautiously. Such alternatives must be the subject of careful cross-calibration against the reference powders and should be conducted by recognized laboratories. The alternative powders/strains and the cross-calibration data which support them, should be made available to anyone who wishes to use, or check, the test with the alternative powders/strains.
13. Water content should not exceed 5%, to preclude premature degradation of the product.

Antagonistic Bacteria

a = Information on infection and pathogenicity in mammals will be available from mammalian safety testing.

b =Information on infection and pathogenicity: suggested test: single-dose, oral test. suggested test species: pigeon and chicken.

c = Information on infection and pathogenicity: suggested test species: *Tilapia mossambica* or other appropriate spp.

d = Information on mortality effects. It is recommended that information be obtained for honey bee and silk worm (*Bombyx mori*).

e= Information on mortality effects. It is recommended that test species include an earthworm (*Lumbricus terrestris*) or other appropriate macro nvertebrates of ecological significance

Entomotoxic Bacteria

If H-Serotype is not known, it is mandatory to provide the details of Cry toxin to confirm that it is *Bacillus thuringiensis.*

Two seasons/years data on bioeffectiveness from minimum two agro-climatic conditions

Two seasons/years data on bioeffectiveness from minimum three agro climatic conditions

a = Information on infection and pathogenicity: suggested test: single-dose, oral test. suggested test species: pigeon and chicken.

b= Information on infection and pathogenicity: suggested test species: *Tilapia mossambica* or other appropriate species.

COMMERCIALIZATION OF BIOPESTICIDES

Since early nineties, a good number of entrepreneurs have entered into producing microbials and now the number stands at around 400. Since 2001, the microbials have come under the CIB pesticide registration act and there are now a little more than 100 registered products of *Trichoderma*, around 50 of *Pseudomonas fluorescens*, a few of NPVs and other insect pathogens.

It is well known that only 10% of these production units have good facility with regard to fermentation equipment, trained staff including qualified microbiologists, sufficient space for each operation, rolling capital and enough buffering capacity. They are providing high standard quality products, though there is a good scope for further improvement. The rest of the 90% of the biopesticide enterprises are mostly 'one-man-shows' which look mostly at short term gains. These lack proper infrastructure and often end up with contamination of their nucleus culture obtained from public institutes.

These two types of set ups have some major problems. In general, they lack Research and Development facility. A strain or culture is obtained from some public sector institute / foreign laboratory and the production starts even without preliminary evaluation against the target pest (pathogens, insects, nematodes). The label claim says that it can manage several different insects / diseases, etc. There exists a high level of specificity with respect to *Beauveria bassiana*, *P. fluorescens*, *Trichoderma* spp., etc. It has been found that even a pathogen like *Fusarium ciceri* of chickpea is antagonistic to several isolates of *Trichoderma*. It is, therefore, required to develop tailor made strains of microbes for specific pest problems. CIB registration is a cumbersome process, which is costly for each product registered, including the generation of toxicological data, shelf life, bio-efficacy etc. The need is to relax registration requirements for different isolates of the same microbe so that suitable label claims can be generated.

Another problem facing this industry is marketing. Several issues are involved, but primarily the industry does not know where to focus their efforts, in which crop and in which geographical location. They also lack the aggressiveness of the chemical pesticide industry, because they are not confident about the performance of their product. In subsistence farming of pigeon pea/chickpea, even though the industry is product-confident, the resource poor farmers can not afford the cost.

There are serious concerns on minimum CFU counts for *Trichoderma* spp. and other biopesticides. CFU count of *Trichoderma* and field efficacy in India shows that a minimum of 1 x 10^8 cfu/ml or g of the product is necessary (ICAR, 2002). The existing standard for *Trichoderma* is 2 x 10^6 CFUs/g product. This standard appears to be low. It may be understood that in a normal and commonly used semi-solid fermentation procedure for *Trichoderma* mass production, the production levels of CFU are in the range of 2 × 10^9–1 × 10^{11} (In molasses medium the population of *T. harzianum* and *T. virens* attained 3.76 × 10^{10} and 7 × 10^{10} CFU / ml, respectively (IISR, 2002). To get a count of 2 × 10^6, only a miniscule quantity of the biomass, say about 0.5–1 g/kg of the commercial product is required, which comes to only 0.05 to 0.1% active ingredient/kg of the commercial product. The perusal of count of some of the standard formulations of *Trichoderma* registered worldwide show that these are at least 10 times higher than the Indian Standards.

The transfer of antagonists from public research institutions to private firms is like a relay race with the ultimate goal of achieving adequate control of a disease. Those at the beginning of a relay must have a baton that meets certain size and weight specifications, or the team might be disqualified at the end. A number of characteristics are desirable for the biocontrol agent to become commercial reality. Researchers need to demonstrate through preliminary screening tests that the antagonist has potential uses at appropriate CFU counts.

All India Coordinated Project on Biological Control, Bangalore is one of the major institutions dealing with the field screening aspects of biopesticides. These reports indicate that very high levels of CFUs have been used not only for *Trichoderma* field evaluations but also for other microbials such as *Beauveria bassiana*, *Metarrhizium anisopliae*, *Nomuraea rileyi*, *Verticillium lacanii*, SlNPV, HaNPV, etc. If the preliminary screenings require very high doses of 10 or even 100 kg/ha (as per the existing standards) to achieve field efficacy, it cannot be expected that these biopesticides would work at existing standards and rates of applications. It may be necessary to appropriately revise the existing commercial standards.

A few more points to address commercial aspects of the biopesticides are as follows: (i) A good infrastructure for quality check by public institutes and supply chain for nucleus cultures and sale needs to be established. (ii) Stable ecosystems like that those of orchards/plantation crops and subsistence farming such as pulses and chickpea need to be supported on war-footing through developing the missions on project mode. (iii) Secrecy of production protocols is common to both the large and small scale/cottage industries. But they need to make their procedures amenable to the public sector scientists' scrutiny so that any serious lapses can be rectified or simple improvements suggested in the protocols can substantially improve the output. (iv) Contract farming would offer right opportunity in implementing the IPM of several crops/fruits where biopesticides would be the most crucial and efficient components.

CONCLUSIONS

Strong synergies between Ministry of Agriculture,Govt. of India, CIPMCs, State Agriculture Departments, SAUs and ICAR Institutes, Private Biopesticide Industries, KVKs, NGOs etc. need to be developed to attain the overall goal of promotion of use of biopesticides in the country. Certain quality issues like the original and true to the type of biocontrol strains, efficacy, minimum spore concentration, physical parameters of formulated products and free from contaminants and chemical pesticides are of paramount importance for the registration of biopesticdes. The quality of biopesticide product should not be compromised at any cost during production and storage if required, for ensuring the supply of authenticated products to the farmers. At the same time the registration process of biopesticides should be simplified and relaxed as compared to chemical pesticides to encourage more number of applicants come forward for registration as well as more number and varieties of biopesticide products in the market.

REFERENCES

Chandler, D., Bailey, A.S., Tatchell, G.M., Davidson, G., Greaves, J. and Grant, W.P. (2011) The development, regulation and use of biopesticides for integrated pest management. *Phil. Trans. R. Soc. B.*, **366**, 1933-1942.

ICAR (2002) *Final Report of ICAR Network Project- National Network Project on Phytophthora Diseases of Horticultural Crops* (1997–2002). Indian Council of Agriculturtal Research, New Delhi.

IISR (2002) *Phytonet Project Final Report.* Indian Institute of Spices Research, Calicut.

15

Acarines as Biocontrol Agents in Pest Management

RACHNA GULATI

Directorate of Research, CCS Haryana Arigcultural University
Hisar-125 004, India
E-mail: rgulati@hau.ernet.in

INTRODUCTION

Biological control provides an environmentally safe, cost-effective and energy efficient means of pest control, either alone or as a component of integrated pest management. The predatory mites in the family Phytoseiidae are examples of biological control agents that have been recognized only recently as effective components of agricultural systems (Gerson *et al.*, 2003). The benefits of using predatory mites are many, including reducing or eliminating the need for miticides, no re-entry intervals for workers, nor chemical exposure. They do not cause phytoxicity and can eliminate mite pests before any plant damage occurs. They can disperse on their own, including under the leaves and into tight spaces. They can increase their populations over time and overwinter in nurseries, giving control not just for the season but possibly seasons to come.

Studies demonstrate the potential of acarines as biocontrol agents in orchards, ornamental plants and greenhouse crops. Use of acarines as biocontrol agents dates back to 1868 when Shimer (1868) recorded *Hemisarcoptes malus* feeding on apple bark louse, *Lepidosaphes ulmi*, which was later on, transferred from eastern to western Canada in 1917 to control the same pest. At that time it was rated as a successful biological control project. Riley (1874) found *Tyroglyphus phylloxerae* predating on major root infesting pest of grapes, *Daktulosphaira vitifollii* in France. *Pyemotes* was recognized in the 1880s as a natural enemy of many insect pests. A further Australian discovery of *Bdellodes lapidaria* as predator of collembolan, *Sminthurus viridis*, a major pest of pastures (Womersley, 1933) stimulated interest in acarine biocontrol agents. It was later introduced to South Africa to exert control on *Bourletiella arvalis*, another collembolan pest. Predaceous mites such as *Euseius* species have been recognized as highly important predators in regulating phytophagous mites and other pests (Mcmurtry *et al.*, 1992). Predaceous mites are often released in field to control phytophagous pests, for instance, their releases have been made on crops and *Euseius tularensis* was released to control citrus thrips (Grafton-Cardwell and Ouyang, 1995). Some insect predators like *Orius laevigatus* have been found to be less effective to control pests on sweet pepper as compared with the phytoseiid natural enemies such as *Iphiseius* (*Amblyseius*) *degenerans* (Rubin *et al.*, 1996). Two phytosei-idae genera, *Neoseiulus* and *Euseius* were found as predators associated with *T. urticae* in sweet pepper (Gallardo *et al.*, 2005). Some recent studies demonstrate the effects of 4 mite predators as biocontrol agents of *T. urticae* on sweet pepper grown in protected environment. In addition, the distribution of the predators and their *T. urticae* prey was also monitored on crop. Specific objectives were to examine and evaluate thc predator's effectiveness in suppressing *T. urticae* population infesting sweet pepper and to observe the allocations of the introduced predators with respect to plant canopy (Sarwar *et al.*, 2011).

ACARINES AS PREDATORS IN AGRO ECOSYSTEM

Acarine biocontrol agents usually are voracious by nature and possess well-developed searching and dispersal mechanisms. Many are fast runners and can be transported by wind, others locate their prey by kairomones and some utilize their hosts for dispersal. Another important criterion, which increases their chances for inclusion in successful biocontrol programme, is their greater reproductive potential than that of their pests (Gerson and Smiley, 1990). Species of *Blattisocius, Macrocheles, Acarophenax* and *Hemisarcoptes* reproduce more rapidly than their prey. The significance of sheer size of acarine predator as compared to prey is also considered a promising character in pest control.

Most of these mites show distinct preferences for prey of one group; like, *Bdellodes* spp. mainly attacks collembolans, *Blattisocius* sp. prefers stored product moths, *Hemisarcoptes* sp. are restricted to scale insects and *Macrocheles* sp. feed on nematodes or housefly eggs (Chhillar *et al.*, 2007). However, most of these can survive in the field on alternate diets in absence of preferred diet, which according to Van Emden (1989) is an important characteristic to make biological control self-perpetuating. Many phytoseiid mites can be considered as important biological control agents for phytophagous mites (McMurtry, 1982). Generalist predators can feed and develop on various kinds of food, viz. phytophagous mites (such as eriophyids, tarsonemids, tydeids), eggs or immature stages of insects (including scale crawlers, thrips, aleyrodids), pollen, nectar and (other) plant exudates, honeydew, and fungi (McMurtry and Croft, 1997). *Euseius victoriensis* was, for example, successfully used in Australia to control the mite pests *P. oleivora, P. latus, Tegolophus australis* (Eriophyidae), *Brevipalpus* spp. (Tenuipalpidae) and *Tetranychus urticae* (Tetranyc-hidae) in citrus orchards (Beaulieu and Weeks, 2007).

Many acarine agents are quite sensitive to prey density and do best when pest populations are strongly clumped whereas others showed enhanced feeding activity with increased pest population. Predatory mites may lay their eggs in protective locations (eg. phytoseiids) or where the hatched larvae (eg. *Pergamasus* sp.) will have easy access to the prey. This behavior increased their survival and less energy is spent by hatched larvae in searching food.

Influence of Host Plant Characteristics on Predatory Mite Survival

The performance of natural enemies is also influenced by host plant characteristics (Lester *et al.,* 2000). Leaf surface texture and vestiture affect the performance of acarine predators associated with plants. Leaf domatia, minute cavities frequently located at the junctions of primary and secondary veins on the underside of woody dicot leaves often harbour various predaceous mites (Karban *et al.*, 1995; Pemberton and Turner, 1989). Phytoseiids show higher efficacy on hairy leaves, leaves with more pronounced veins and fruits with rough spurs which provide them overwintering shelter and better protection from their enemies. Anystids, on the other hand, provide better control on puberscent or glabrous leaves. In addition, many plants provide floral as well as extra floral nectars, pollen and even nutrients from other plant feeders (i.e. homopteran honeydew), natural foods known to be of importance for many generalist predators (McMurtry and Rodriguez, 1987). The quality and quantity of such natural foods differs from plant to plant with oblivious effects on predators. It was demonstrated that a stigmaeid ingesting palm date pollen will produce more than twice as many progeny as one feeding on castor bean pollen.

Plant density and plant architecture influence the distribution of spider mites on a plant species and the ease with which the predators can find patches of prey. For example, *P. persimilis* is very efficient on cucumbers that have large leaves and vines that intermingle, but less so on peppers with smaller leaves that don't touch. *P. persimilis* is also less effective on cut rose varieties with fewer leaves because the mites can't move around as easily on these plants. *N. californicus* is a better choice for control of spider mites on roses, if introduced early. Arranging plants so that their leaves are touching may improve biological control on some plant species.

Crop Reservoirs of Predatory Mites

The importance of vegetal diversity in agroecosystems was emphasized by various workers for promoting the occurrence of natural enemies and enhancing biological control (Landis *et al.*, 2000). Studies conducted in vineyards and orchards showed that high densities of many species of phytoseiid mites are observed in surrounding vegetation (Tuovinen and Rokx, 1991; Duso *et al.*, 2004; Toyoshima *et al.*, 2008). Ground cover could also be a reservoir for natural enemies by providing refugia and alternative food, as well as places to breed (Gravena *et al.*, 1993).

Naturally occurring plant species such as *Achyranthes aspera, Amaranthus dubius, Eleutheranthera ruderalis* and *Neonotonia wightii* act as a reservoir of phytoseiid mites sustaining abundant and diverse populations all year round. Higher numbers of phytoseiid mites were observed in citrus orchards with cover crops than without the cover crops (Muma, 1961). Later, Gravena *et al.* (1993) showed that the presence of *Ageratum conyzoides* (Asteraceae), a common wild plant and *Eupatorium pauciflorum* (Asteraceae) in citrus orchards in Brazil led to an increase of phytoseiid mite densities, compared to a bare ground management. As a consequence of increased phytoseiid mite densities, damage caused by *P. oleivora* decreased. *A. conyzoides* was shown to release airborne chemicals that maintain the *Amblyseius newsami* population for a long time (Kong *et al.*, 2005).

Herbicide application or mowing seems to reduce phytoseiid densities. Reducing mowing frequency could be a good alternative to let phytoseiid mites develop in the ground cover. The characterization of dispersal from cover crop to trees is also a key point to determine how cover crop management could directly affect tree biological control (Tixier *et al.*, 2006).

Suitability of Environmental Conditions to Predatory Acarines

Predatory mites have effectively controlled spider mites on chrysanthemum, rose, and other ornamental crops under experimental conditions. However, the need to prevent cosmetic damage on floral or foliage crops may make biological control of mites difficult, especially when pesticides are used to suppress other pests and/or diseases also kill predatory mites. Species selection and release rates vary considerably depending on the plant species and the environmental conditions such as temperature and humidity which influence the growth rate of both predator and prey. *P. persimilis* is an excellent predator of spider mites on low growing plants in humid greenhouses with moderate temperatures. There are a few crops on which *P. persimilis* cannot be used. For example, *P. persimilis* slips off the stems and leaves of carnations. It does not do well on tomato because the mites become trapped on glandular hairs on the leaf petioles and stems, and are also affected by toxic compounds in the tomato leaf. *P. macropilis* performs better than *P. persimilis* on ornamental plants, such as dieffenbachia, dracena, parlor palm and schefflera, under warm, humid conditions. *Mesoseiulus longipes* is frequently used to control spider mites in hot, dry greenhouses on taller plants because it tolerates lower humidity than does *P. persimilis*. *N. californicus* does well on most potted plants in greenhouses with moderate temperatures and average humidity. *G. occidentalis* and *N. californicus* may be better suited for use on semi-permanent greenhouse crops such as rose or gardenia than on short-term vegetable crops. A combination of predators released at regular intervals works best in greenhouses with a variety of plants and growing conditions.

Various studies have shown the effectiveness of introducing multiple predatory mite species versus single predatory mite species to control tetranychid mites (Rhodes *et al.*, 2006; Fitzgerald *et al.*, 2007). Phytoseiid mites have been shown to be effective in controlling tetranychids on many cultivated crops (Helle and Sabelis, 1985; van Lenteren *et al.*, 1992). Single predatory mite species are able to suppress spider mites rapidly if they have a high numerical response

(Janssen *et al.*, 2006). However, some predatory mite species tend to disappear from the crop after reducing pest mites and thus provide only short-term control (Walzer and Schausberger, 1999). Therefore, combined release of different predatory mite species has been investigated for long-term spider-mite control, but the results vary from negative effects and no effects to positive effects, and the underlying causal mechanisms leading to these different outcomes are not clear in most cases (Cakmak *et al.*, 2009). However, it has been shown that some acarines have a capacity to persist in the crop for a longer time. Because their lower densities arose in the end phase of the experiments with no clear advantage for a mite to manifest with respect to other-mite control. In case lower densities arise earlier in the crop growing season or locally in the space occupied by a crop, the predatory mite might add to overall control by its ability to persist longer at lower densities (Cakmak *et al.*, 2009). Perhaps this is the reason why Rhodes *et al.* (2006) found better performance of *Neoseiulus californicus* with and without *Phytoseiulus persimilis* when considered over two crop seasons.

The western predatory mite, *Galendromus occidentalis*, is smaller than *P. persimilis* and develops best at cooler temperatures, but it tolerates a wide range of relative humidity (40–80%). It has the capability of regulating spider mite populations at lower densities and for longer periods of time than *P. persimilis,* although it will not control spider mite populations as quickly as *P. persimilis* can. It can also survive long periods without prey. Several different strains are commercially available, including non-diapausing strains that allow control of spider mites through the winter, when days are short, and strains resistant to a number of organophosphate insecticides. The commercially available *Mesoseiulus longipes* is similar to *P. persimilis* in activity, but can tolerate warmer temperatures-up to 70–90°F and relative humidity as low as 40% (http://www.entomology.wisc.edu/mbcn/kyf505.html).

Use of Predatory Mites for Insect Control

Mites and insects co-exist in a variety of habitats which provides an opportunity for predation between these groups. Acarine predators generally restrict their feeding to the eggs or the immature stages of insects. Several families including Acaridae, Anystidae, Ascidae, Bdellidae, Camerobiidae, Cheyletidae, Cunaxidae, Erythraeidae, Eupalopsellidae, Laelapidae, Limnesiidae, Macrochelidae, Parasitidae,

Phytoseiidae, Pionidae, Trombidiidae, and Tydeidae are associated with insect control as predators (Table 15.1).

Table 15.1. Acarine families associated with insects

Acarine family	Insect pests	Pest order
Acaridae	Corn rootworm, grape phylloxera	Coleoptera, Homoptera
Acarophenaxidae	Stored product insects	Coleoptera
Anystidae	Citrus thrips, grape leafhopper, pine needle aphid, soil insects	Thysanoptera, Homoptera, Collembola
Arrenuridae	Mosquitoes	Diptera
Ascidae	Stored product insects, mushroom fly, mosquitoes	Hemiptera, Coleoptera, Lepidoptera, Diptera
Bdellidae	Lucerne flea	Collembola
Camerobiidae	Armoured scale insects	Homoptera
Cheylitidae	Stored insects	Coleoptera
Cunaxidae	Book lice, scales	Psocoptera, Homoptera
Erythraeidae	Bugs, pear psylla, apple tortricid	Hemiptera, Lepidoptera
Eupalopsellidae	Scale insects	Homoptera
Hemisarcoptidae	Armoured scale insects	Homoptera
Hydryphantidae	Aquatic arthropods	Diptera
Laelapidae	Corn rootworm	Coleoptera
Macrochelidae	Flies	Diptera
Parasitidae	Garden symphylan, flies	Symphyla, Diptera
Phytoseiidae	Thrips, white flies, scale crawlers	Thysanoptera, Homoptera
Pionidae	Mosquitoes	Diptera
Podapolipidae	Locust, grasshoppers, beetles	Orthoptera, Coleoptera
Pterygosomatidae	Blood sucking bugs	Hemiptera
Pyemotidae	Pink boll worm, stored insects, thrips, fire ants	Coleoptera, Lepidoptera, Hymenoptera
Trombidiidae	Aphids	Homoptera
Uropodidae	Mosquitoes	Diptera

Reports on aphids, beetles, collembolans, grasshopper, housefly, leafhoppers, mosquito, scales, stored insect pests, symphylans, and thrips control through acarines are available in literature. Among these, *Amblyseius (Neoseiulus) cucumeris* for thrips, *Blattisocius tarsalis* for stored product moths, *Macrocheles muscaedomesticae* for house fly, *Adactylidium* sp. for thrips, *Bdellodes lapidaria* for collembolans are exploited successfully in different parts of the world. Mites of family Anystidae are large, reddish, soft bodied and fast runners whose tarsus is longer than its claw. *Anystis agilis* is an important predator of grape leafhopper, *Erythroneura elegantula* and woolly pine needle aphid, *Schizolachnus piniradiata*. The mite feeds on all instars and morphs of aphids on alfalfa and prefers citrus thrips, *Scirtothrips citri* to other pests. But long generation time (two generations/ year) and cannibalistic habits of this mite are a major hindrance in its mass rearing.

Blattisocius tarsalis (Ascidae), an important predator of stored grain pest, was recorded from 18 insect hosts but it prefers eggs and first instar larvae of flour moth (*Ephestia* sp.) and *Tribolium* spp. The other important hosts are *Plodia interpunctella, Rhyzopertha dominica, Ips grandicollis,* and *Sitotroga* sp. High potential rate of increase as compared to moth, highly mobile feeding stages, phoresy on host, high speed (52-60 cm distance in 5 min.; 3 m distance in 30 min.) make it an effective biocontrol agent against moths in many parts of the world. Weekly releases of mite population are recommended in stored grains. The mite destroyed 3-5 eggs per day whereas gravid female consumed 5-6 eggs per day. Another family associated with stored grain pests is Cheyletidae. *Acaropsis docta* male and female feed on stored product insects like *T. granarium, S. granarium, R. dominica, T. castaneum, Lasioderma sericornae* and consume an average of 10 and 20 eggs in 10 days. *A. sollers* is reported to feed on *Trogoderma* eggs and consumes 1-2 eggs/day and 37-40 eggs in its life. *Cheletogenes ornatus*, an orangish coloured mite, is a predator of first instar of armoured scale insects. It consumes 90 crawlers during adult life and produces about a dozen progeny (Gerson *et al.*, 1990).

Mites of family Bdellidae are red, brown or greenish with snout like mouth parts, and long palpi. These mites actively hunt for their prey, sometimes tether them with silken lines. *Bdellodes lapidaria* feeds on collembolan, *Sminthurus viridis, Bourletiella arvalis* (pests of pastures in Australia and South Africa). Presence of 20 mites/m^2 early in winter prevented pest outbreaks in Australia and South Africa. Cunaxid mites are yellow, red or brown in colour, fast runners. Their palps are ornamented with strong spines. They hunt and fasten their prey with silken threads excreted through their mouth parts. One *Cunaxa capreolus* mite consumed 230 booklice during its development. *Cunaxoides parvus* is considered an important predator of oystershell scale, *Lepidosaphes ulmi* on apples in North America. Hydrophantids are normally found in ponds, springs and seepage areas and associated with mosquitoes (Mullen, 1977), blackflies, biting midges (Smith and Oliver, 1986) and horse flies (Smith, 1983). *Panisopsis* sp. is known to curtail the flight activity and host seeking of *Aedes ventrovittis*. Some larvae attach to young emerging mosquitoes thus parasitizing both the sexes whereas others attack only gravid females (Mullen, 1977).

Soil borne *Androlaelaps* and *Stratiolaelaps* spp. (Laelapidae) usually accounted for 20% natural control of corn rootworms (*Diabrotica longicornis, D. virigifera*). In field experiments, manure application increased mite population three fold, which led to 63%t pest control. Another soil borne predator, *Hypoaspis* sp. is currently being exploited against fungus gnats, springtails, thrips pupae and plant parasitic nematodes. This tiny (0.5 mm) light-brown mite inhabits the top 1/2" layer of soil and consumes 5-20 prey or eggs per day. It was also recorded to feed on thrips pupae, collembolans and beetles and brought about 70% reduction in *Oryctes* (beetle) egg hatch. Commercially available cultures are available for which details are provided in Table 15.2.

Mites belonging to family Macrochelidae are fast moving, free living predators common in habitats, which are rich in decaying organic material including manure. *Macrocheles muscaedomesticae* feeds on 20 house flies immature per day. It gave effective control in cattle manure by reducing the house fly population by 90% (Rodriguez *et al.*, 1970). *M. glaber* predates on Australian bush fly (*Musca vetustissima*) and 100% control was obtained at 50 mites/1000 ml dung pad with 300–400 fly eggs. *M. peregrinus* (generation in 3 days), which was introduced from Australia to South Africa to control bush fly and buffalo fly, provided good control (Wallace and Holm, 1983) and colonized in 180,000 km^2 area around the release sites. *Poecilochirus monospinosus* (Parasitidae) also feeds on house fly immature in poultry manure (Wise *et al.*, 1988) and provided an effective control in poultry manures against house fly. *Parasitus hituherosus,* a member of Parasitidae is considered a possible biocontrol agent of *Heteropeza pygmaea* in mushroom compost (Al-Amidi and Downes, 1990).

Table 15.2. Commercially available predatory mite cultures for pest control

Mite species	Host pest	Host plant	Development	Environmental conditions	Release rate	Storage period
Phytoseiulus persimilis	Tetranychidae mites	Greenhouse crops, low growing field crops (straw berries), ornamentals.	egg to adult: 7–8 days at 20°C	Best suited to moderately warm humid climates (20–30°C.) at 70% RH. Ideal in protected cropping. Not appro-priate in hot, dry environments.	1/sq ft., 20–40,000/acre, 100,000/ha. Recommended pest/predator ratio at time of release 10/1	Highly perishable should be used immediately upon delivery. If storage is absolutely necessary, refrigerate at 10°C. Not to exceed 5 days to mini-mize to mortality
Mesoseiulus longipes	spider mites	Greenhouse crops	One week, depending on temperature.	best in warm to hot greenhouses. Tolerant of lower humilities (40% RH at 70°F).	one per plant or one per square foot at the first sign of spider mites. Later release will require much higher numbers to be effective.	
Neoseiulus californicus	Spider mites, Broad mite, Cyclamen mite	Strawberries, Corn, Grapes, Roses, Vegetables, Ornamentals	One to two weeks depending on temperature.	best in warm humid conditions, but will also tolerate low humidity (40–80% RH at 50°-105°F	release rates are 1/sq ft., 20–40,000/ acre, 100,000/ha.	
Neoseiulus fallacis	Tetranychus urticae, *T. pacificus*, *Panonychus ulmi*	Fruit trees, strawberries, corn, hops, mint, other field crops	life cycle 7–9 days.	Does best in moderate to cooler humid conditions, but will also tolerate warmer temperatures	Release at least one per plant or one per square foot in green houses at the first sign of spidermites. 1:10 ratio optimum	

Mite species	Host pest	Host plant	Development	Environmental conditions	Release rate	Storage period
Galendromus (=Typhlodromus, =Metaseiulus) occidentalis	All stages of Spider mites	Fruit trees, grapes, corn, cotton, ornamentals, apple, plum, peach, cherry strawberries	Egg to Adult: 7 to 14 days depending on temperature	Does best in warm weather (80–110°F). Tolerates low humidity of inland valleys. Does not do well in cool coastal areas	2000 to 5000/ acre at the **first** sign of spider mites. In greenhouses, 2/ft^2 or 1 predator per 5 spider mites at the **first** sign of spider mites	
Galendromus helveolus	Persea mite, Six-spotted mite	Avocado, Citrus, and others.	It completes a generation in one to two weeks depending on the average temperature	well adapted to the Persea mite *Oligonychus perseae.*	releasing 5000/ acre. Approximately 100 predators should be placed on each tree in the grove	
P. macropilis	Two-spotted spider mites, yellow spider mites, apple and pear rust mites, Prunus rust mites, blister mites, and European red mites	Green house crops	Total development time is 6–12 days, depend on the temp. there may be 8–10 generations per year	can thrive in hot, dry climates.		
Galendromus (=Typhlodromus) pyri	European red mite, two-spotted spider mite, apple rust mite	orchard crops	They live about 20 days, laying approximately 20–30 eggs	thrives in cool, humid climates		
Zetzellia mali	Apple rust mite, European red mite, two-spotted spider mite	Apples orchards	It completes about four life cycles per year,			*Z. mali* is not available commercially
Hypoaspis sp.	fungus gnats, springtails, thrips pupae and plant parasitic nematodes	Soil application	7–11 days	shipped as adults, packed in a medium of bran, vermiculite and peat for easy dispersal	10,000-25,000/per acre	Cultures can be kept for 2-3 weeks at 45–55°F

Mite species	Host pest	Host plant	Development	Environmental conditions	Release rate	Storage period
Macrocheles robustulus	thrips pupae, nematodes	Soil application			250 mites per square metre	

Phytoseiidae holds the best-known predators among the Acari and may easily be mass reared and shipped. These fast runner mites are cultured on acarid mites, spider mites or on alternate diets (pollen). IPM technology is developed for cucumber and pepper using *Amblyseius (Neoseiulus) cucumeris* culture (Jacobson *et al.*, 2001). Diapause resistant strain of phytoseiid is used as one sachet/plant at the time of planting so that phytoseiid mites multiply in time and was there when thrips attack appeared. Complete suppression of thrips population is achieved in Brazil and Egypt and culture remained active for nine weeks. Details are provided in Table 15.2 for ready reference.

ACARINES FOR MITE CONTROL

The predatory mites belong to the families Phytoseiidae, Ascidae, Anystidae, Bdellidae, Cheyletidae, Cunaxidae, Erythraeidae, Raphignathidae, Stigmaeidae, Tydeidae, etc. and are voracious feeders of phytophagous mites. Among these, Phytoseiidae has received maximum attention globally due to high fecundity, abundant availability, good searching ability, dispersal rate, adaptability to different ecological niches and high degree of prey specificity. Among the 173 species that are so far known from India, 18 species have been found actively feeding on phytophagous mites. Details on commercially available formulations are provided in Table 15.2. Stigmaeidae are probably next to phytoseiidae as far as predatory efficiency is concerned. *Agistemus* spp. and *Eryngiopus coimbatorensis* are found to be of high promise as biocontrol agents for controlling slow moving mites and for destruction of mite eggs. Table 15.3 depicts the important predatory mites associated with mite pests of agricultural importance.

SHIPMENT, HANDLING AND RELEASE PROCEDURES

Predatory mites are commercially available. They are shipped in plastic vials as nymphs, adults and eggs (approximately 1,000 active predators plus predator eggs) in a dry carrier material such as bran or vermiculite, often with spider mite eggs as prey to prevent cannibalism. The carrier-mite mixture can be sprinkled directly onto the foliage of infested plants or in small paper lunch bag around a cluster of several leaves at the end of a branch, staple the bag to the leaves and the mites will disperse on their own. Predatory mites are most effective when introduced while spider mite populations are low.

Table 15.3. Important predatory mites associated with mite pests of agricultural importance

Predator	Prey species and stage consumed	Developmental period	Feeding rate
Amblyseius vannai	Larva, nymphs	85–107 h for ♂ 84–113 h for ♀	14-19 eggs of *Raoiella indica* by ♂ 15-38 eggs of *R. indica* by ♀
A. alstoniae	Nymphs, eggs of *Eutetranychus orientalis*. All stages of *Tetranychus macfarlanei, Brevipalpus phoenicis*		10.16 protonymph of *E. orientalis* by ♀ 46.26 adults of *T. macfarlanei* by ♀ 1:10 & 1:20 ratio effective for *B. phoenicis* control
A. concordis	Castor pollen preferred over tetranychids, tenuipalpids, eriophyids	120 h at 20°C, 63% RH	
A. finlandicus	Nymphs, eggs of *E. orientalis, B. phoenicis*	Varied with temperature	When only *B. phoenicis* eggs wee fed, the progency was female only
A. ovalis	Tetranychids	127 h on castor pollen 12.3 d	10.14 larvae/female/day 3.86 adult/female/day
A. rhododendronis	*Tetranychus ludeni*		Effective in checking *T. ludeni* population within 4 weeks of release
A. longispinosus	Tetranychids	99.11 h for ♀ 95.30 h for ♂	Efficient predator 250 predators per 15 plants
A. tetranychivorus	*T. ludeni* stages eggs and nymphs of *R. indica*	140±7.5 h for ♀ 136.2±6 h for ♂	26.74 preys/female 1:6 ration gave full control in 10 d in okra
Phytoseiulus persimilis	Tetranychids	5.03 d at 25°C 3.84 d at 29°C	Release technique available in French bean
Agistemus industani	*T. ludeni* eggs		15 prey eggs/day
Cheletogenes ornatus	*Larvcarus. transitans*		5-60 mites in 12 h

In greenhouses with a history of spider mite problems, the first releases should be made one week after plant emergence. Most failures of biological control occur when the predator is released too late. Spider mite populations should be monitored by observing foliage of susceptible plants at least weekly. Additional predator releases may be necessary every 2–4 weeks to achieve good control within 6 weeks. Release rates are provided in Table 15.2 for ready reference.

Predator mites can be released uniformly throughout the greenhouse, or concentrated in infested patches. Uniform distribution of predators throughout the greenhouse is the most common method of introduction. It provides predictable levels of control. Introduction in patches of mite damage will often result in better control than uniform distribution. Misting the plants with water before application of the predators may help the carrier material (and predatory mites) adhere to the foliage. However, avoid watering plants immediately after application of the predators to allow them time to find appropriate niches in the foliage. Spraying the crop one week before or after releasing predators should be avoided. It is advisable not to release predators at temperatures below 45°F or above 85°F and during dry windy conditions.

RELATIVE EFFECTIVENESS

Predatory mites can seek out prey in places that may be inaccessible to chemical sprays. These beneficial mites are used against spider mites and several species of thrips by greenhouse flower, ornamental, and vegetable producers in Canada, Europe, United States and other countries growers. Predatory mites are particularly successful in greenhouses partly because of the high degree of control that the grower has over the environment. However, several predatory species also have been used successfully to suppress spider mites in United States apple, citrus, and avocado orchards and also in field-grown strawberries and raspberries.

Some species of predatory mites, particularly *Phytoseiulus*, are among the few biological control agents that will eat all their prey and starve if they cannot locate new colonies of spider mites. Scouting is, therefore, important to ensure proper timing and placement when releasing the predators. It may be necessary to reintroduce the predatory mites if the pest population reoccurs. The ability of predatory mites to seek out prey can be hindered on plants with hairy stems or leaves.

Neoseiulus fallacis being a voracious consumer of mites and because of rapid population increase in relation to its prey, can overtake an expanding pest population. It develops into the adult stage in about one third the time required by *Galendromus pyri.* However, when *N. fallacis* has reduced the prey population, it will leave the tree in search of more tetranychid mites whereas *G. pyri* thrives on alternate foods. Unlike *N. fallacis, G. pyri* does not increase in population as quickly as an expanding pest mite population. However, *G. pyri* survives the winter in greater numbers than *N. fallacis.* In addition, it is able to thrive and will remain in the tree at low prey density, surviving on alternate foods whereas *N. fallacis* will leave. Thus, once the prey population is low, *G. pyri* will prevent its increase. A mixed population of *G. pyri* and *N. fallacis* is desirable in orchards. In orchards, *G. pyri* also works well with *G. occidentalis.* In early and late season when it is cool, *G. pyri* is more efficient, whereas, *G. occidentalis* is more efficient during the hot, dry summers. It has also been established that *N. fallacis* but not *G. occidentalis* are attracted to volatiles from *T. urticae*-damaged hop plants when paired with undamaged plants (Gardiner *et al.*, 2005).

Z. mali, another predatory mite is not likely to control pest mite infestations alone. It prefers to feed on the apple rust mite and does not eat many tetranychid mites. In addition, it is slow to increase in population and may prey on other predator mite eggs. However, *Z. mali* can be beneficial over the long run in these ways: it can survive on several alternate foods when its preferred prey is not present and thus can live for long periods without prey; because of its feeding versatility, it can eventually reach population densities capable of controlling pest mites; it can complement biological control by other predator mites by feeding on stationary forms of pest mites such as eggs or immature stages (*N. fallacis* and *G. pyri* prefer the mobile forms); it can serve as another form of prey for *N. fallacis* and *G. pyri.*

PESTICIDE SUSCEPTIBILITY

Pest mite problems are most common where pesticides are heavily used because predatory mite populations are killed along with target species It may take up to three years to establish a population of predators high enough to control pest mites. Strains of some beneficial mites that are tolerant of organophosphorous and carbamate insecticides, sulfur, and pyrethroids have been selected for mass production. Some of these strains are commercially available. Spot spraying of localized outbreaks of thrips and spider mites with short-residual chemicals that are compatible with the predatory mites or, on a small scale, with insecticidal soap or horticultural oil may help to keep infestations under control while the predatory mite population establishes. A strain of *G. pyri* that is resistant to pyrethroids has been patented in New Zealand. *G. occidentalis* is highly susceptible to permethrin, methomyl, and fenvalerate. *Z. mali* has shown some resistance to organophosphate insecticides, it is susceptible to endosulfan. This suggests the variability of the performance of mites under different insecticide regimes (Hardman *et al.*, 2000).

ACARINES FOR NEMATODE CONTROL

Soil acarines are instrumental in keeping the populations of plant parasitic nematodes under check in some of the undisturbed ecosystems. Nematodes sometimes provide a high nutrient rich diet to nematophagous acarines, which develop faster and reproduce more under such conditions. In *Macrocheles muscaedomesticae*, 23.11 ± 5.58 progeny/female/day is produced in presence of both, nematodes and house fly eggs (Rodriguez *et al.*, 1962). *Gamasellodes vermivorax* completes its life cycle on rhabditid nematode just in 10 days at 24°C and population increases four fold after 16 days depending upon nematode densities in microcosms (Walter, 1987). In greenhouse cultures of *Tylenchulus semipenetrans* and root-knot nematode at Florida, USA, the densities of nematophagous mites often exceeded 1000 mites per liter (Walter *et al.*, 1993).

Acarines belonging to all the four sub-orders, viz. Astigmata, Cryptostigmata, Mesostigmata and Prostigmata have shown varying degrees of nematophagous activity (Gulati *et al.*, 2005). Among astigmatid mites, *Tyrophagus putrescentiae, Caloglyphus* sp. and *Rhizoglyphus echinopus* completely devoured the small plant parasitic nematodes, whilst larger ones were cut into pieces and sucked. *Pergalumna* sp. (oribatid mite) voraciously fed upon free living nematodes, *Chiloplacus* sp. and *Rhabditis* sp. *Lasioseius* spp., *Hypoaspis* spp. and *Macrocheles* sp. (mesostigmatid mites) have been found to significantly reduce the plant parasitic nematodes population in greenhouse trials.

Tyrophagus putrescentiae, Caloglyphus sp. and *Rhizoglyphus echinopus* was recorded as feeding on several plant parasitic nematodes (Bilgrami, 1994; Bilgrami and Tahseen, 1992; Walia and Mathur, 1995). Small nematodes were completely devoured, whilst larger ones were cut into pieces and sucked (Sturhan and Hampel, 1977; Walter *et al.*, 1986; Sell, 1988; Walia and Mathur, 1995). *T.*

putrescentiae, T. similis and *T. zachvatkini* could be cultured on a nematode diet in addition to their regular diet. There are few reports which suggest preference for nematode diet over other diets (Walter and Ikonen, 1989). *T. putrescentiae* also multiplied on root-knot nematode, *Meloidogyne javanica* and seedgall nematode, *Anguina tritici,* successfully (Ranjana, 2002). The population of the nematode decreased in the presence of the mite. The size of the prey (Ranjana *et al.*, 2003), prey and predator density (Ranjana *et al.*, 2005) also influenced the predatory potential of acarid mites. Consumption of small sized nematodes (*M. incognita*, 0.3 mm) by mites was maximum (50%) followed by medium sized nematodes (*A. tritici,* 0.7 mm) (59%) and large sized nematodes (*Longidorus pisi*, 2-4 mm) (9%) in 24 h. Prey and predator density exhibited a positive response on predatory potential of nematophagous mites.

Feeding behaviour of cryptostigmatid mites indicated that the mite neither searched nor preferred nematodes. There are limited reports of predation by oribatid mites on nematodes as they are heavily sclerotised and slow moving in nature. The time required for a tritonymph or adult pergalumnid to grasp and completely devour an average sized nematode (± 500 μm) varied considerably from 15 s to 25 min. As many as three nematodes could be consumed successively. Oribatids as nematophagous mites are also documented by Rockett (1980) who observed individual mites of *Pergalumna* spp. feeding on plant parasitic nematodes.

The most of the nematode predacious acarines belong to Mesostigmata and are quite abundant and frequent geographically in the soil ecosystem. *Lasioseius* species reduced the seedgall nematode population by more than 82% in *in vitro* trials with in 72 h of its inoculation (Ranjana, 2002). It was successfully multiplied on *M. incognita,* fungivorous nematode, *Aphelenchoides composticola* and *A. tritici* (Ranjana, 2002). *L. penicilliger, L. scapulatus* have also been reported to feed upon nematodes, *Tylenchorhynchus dubius, Paratylenchus* sp., *Pratylenchus* sp., *Rotylenchus* sp., *Cephalobus* sp., *Heterodera glycines, Meloidogyne* spp., *A. avenae, Acrobeloides* sp., *Caenorhabditis* sp., *Seinura* spp., etc. (Sharma, 1971; Van de Bund, 1972; Muraoka and Ishibashi, 1976; Imbriani and Mankau, 1983; Inserra and Davis, 1983; Ishibashi *et al.*, 1987). *L. thermophilus* is found associated with mushroom compost and feeds on the eggs of *Ditylenchus myceliophagus.* The mite population increased from 10 and 50 to 603 and 1611, respectively in the presence of nematodes whereas nematode population declined to 18,550 (50 mites inoculated) from 504,000 in control (absence of mites). Its possible use as a biocontrol agent was also suggested by Cayrol (1970) against *D. myceliophagus* in mushroom beds.

Hypoaspis miles are tiny (0.5 mm) light-brown mites that naturally inhabit the top 1/2" layer of soil where fungus gnats, as well as springtails and thrips pupae dwell. The female *Hypoaspis* mites lay their eggs in the soil, which hatch in 1-2 days, and the nymphs and adults feed on the soil-dwelling pests. Each *Hypoaspis* mite will consume 5-20 prey or eggs per day. Under certain field ecosystems the mesostigmatic mite, *H. aculeifer* could limit the populations of *M. chitwoodi, M. hapla* and *Heterodera schachtii* (Inserra and Davis, 1983). Addition of organic manures, leaf litter, intercropping in field crops favours the development and multiplication of the mites which, in turn, helps in regulating the population of plant parasitic nematodes (Karg, 1983).

In greenhouse trials, the citrus nematode, *Tylenchulus semipenetrans* could be restricted by *Macrocheles muscaedomesticae* successfully. The mesostigmatic mites have been judged as better predators in soil ecosystem than astigmatids which are rare in soil (Walia and Mathur. 1996). These also hold potential to be used for the nematode pests of mushroom. The experiments with *Hypoaspis calcuttaensis* against *A. composticola* (Walia and Mathur, 1994) and *L. thermophilus* against *D. myceliophagus* are encouraging (Cayrol. 1970). *H. calcuttaensis*, *L. athiasae* and *Protogamasellus discorus* have been found to reduce soil population and root galls caused by root-knot nematodes. The mite population increased eight times after 45 days in the presence of nematodes (Walia and Mathur, 1996; Amin *et al.*, 1999). Sharma (1971) and Mankau (1980) while reviewing the control of nematode pests by natural enemies, believed that mites, which have greater searching ability than other predators, should be further investigated, despite some drawbacks, such as non-specificity, being limited to the upper soil layer, which is only a portion of the vertical range of some pest nematodes, restriction by size to soils with pore spaces which allow for their passage and initiation of attack only by fortuitously coming upon the nematodes. Few encouraging reports suggest that under certain conditions predatory mites could protect field crops from nematode damage.

ACARINES AS PARASITES IN AGRICULTURAL SYSTEM

In parasitic mites, structural and biological adaptations are evident. In some ectoparasitic mites, only larval stage is parasitic (water mites and earth mites) but in others, all postembryonic stages are parasites on host (*Pyemotes* sp., *Podapolipus* sp.). The life cycle shows reduction in number of active developmental stages, e.g. in Erythraeidae and Trombidiidae only one active nymphal instar is produced while in *Pyemotes* sp., the female gives birth to mature adults. In parasitic mites, the chelicerae are needle like or hooked shape. In latter case, a feeding tube is produced by mite, which enters the host skin. These very small acarines (*Pyemotes* sp.) can subdue their prey by injecting them with quick action paralyzing toxin. Some parasitic mites (*Dicrochelus* sp.) use host tissue for oviposition. Mites of family Acarophenaxidae, Arrenuridae, larvae of Erythraeidae, Hemisarcoptidae, Hydryphantidae, Podapolipidae, Pterygosomatidae, Pyemotidae, and larvae of Trombidiidae are associated with insect and mite control.

Reports on beetles, mosquitoes, blackflies, biting midges, horse flies, blood sucking bugs, booklice, cockroaches and locusts control through acarines are available in literature. Among these, *Acarophenax manhunkai* for lesser meal worm, *Pyemotes herfsi* for *Tribolium confusum* larvae, *P. tritici* for *Plodia interpunctella*, *Cadra cautella* (Lepidoptera), *Oryzaephilus mercati*, *Lasioderma serricorne* (Coleoptera) and *Hemisarcoptes* spp. for armoured scale insects are exploited successfully in different parts of the world.

Acarophenax lacunatus (Acarophenaxidae) is an important parasite of *Rhizopertha dominica* (Matioli, 1997), *Cryptolestes ferrugineus* (rusty grain beetle) and *A. manhunkai* on *Alphitobius diaperinus* (lesser meal worm) (Sternkraus and Cross, 1993). *A. manhunkai* has short life cycle, high fertility and no evidence of harmful effects on humans and higher animals. The mite exhibits female physogastry and phoresy. Phoretic mite leaves the host body, parasitizes its eggs and feeds on egg contents. A female produces 17 off springs. Twenty-four hours after emergence, female undergo physogastry. Numerous mites develop within the female body, reach adulthood and mate within themselves. After release from mother's body females track host whereas both male progeny and mother die in this process. Four individuals/ 500ml jar containing 50 adults of pests, results in complete suppression of pest population after 45 days (Faroni *et al.*, 2001). Reduction in loss of stored wheat was estimated as 15–25%.

Larvae of *Arrenurus* sp. (Arrenuridae) are the most common parasites on mosquitoes. The larvae locate host pupae and parasitize the adult mosquitoes as they emerge. Female of the species produces 55–88 eggs/year (Stechmann, 1978). *A. madaraszi* caused high mortality in preoviposition *Aedes* and *Culex* females (Reisen and Mullen, 1978). The larvae of Erythraeid mites (*Leptus* and *Bochartia* sp.) are parasitic on other arthropods like mango pest, *Drosicha mangiferae*. They killed the pest by sucking the fluid. On one host, 1–60 mites were recorded by Tandon and Lal (1976). Hemisarcoptidae is the only family presently exploited for armoured scale insects control in many parts of the world. These are whitish, soft-bodied mites having wide range of climatic tolerances. *Hemisarcoptes coccophagus, H. cooremani* and *H. malus* are the major mortality factor of the oystersheel scale *Lepidosaphes ulmi* on apple, date palm scale *Parlatoria blanchardi* and citrus scale insects. Their successful introduction in New Zealand (1987-88) led to 80% reduction in scale insect population (1991) in orchards.

Panisopsis sp. (Hydryphantidae) is known to curtail the flight activity and host seeking of *Aedes ventrovittis*. Some larvae attach to young emerging mosquitoes thus parasitizing both the sexes whereas others attack only gravid females (Mullen, 1977). *Locustacarus* sp. (Podapolipidae) parasitizes the tracheae of acridids and their feeding is found to be detrimental to host health. Another species, *Podapolipoides grassi* is a common, most wide spread ectoparasite of locusts. It infests third instar nymphs and subsequent stages and normally occurs on thorax or anterior abdominal segment and at the base of wings. The parasite spends its entire life cycle on the locust, dispersing during host mating. Their feeding leads to decrease in host vigour (Gauchat, 1972).

Pterygosomatidae family is associated with blood sucking bugs and cockroaches. These red, small to medium sized mites are parasitic in nature. *Pimeliaphilus plumifer* parasitizes blood-sucking bugs of the family Triatomidae. When present in large numbers (25 mites/insect), they kill their host within one hour. These are not very host specific and can survive on any triatomine bugs within their range (Anderson, 1968). Death of the host is due to injected toxin. Furthermore, egg capsules produced by parasitized cockroaches fail to hatch; hence host population is reduced considerably. Mites also serve as vectors for pathogenic virus and bacteria which enhances the possible use of this group for the biological control of blood sucking bugs.

Pyemotidae is an important acarine family exploited for insect control. *Pyemotes tritici,* the straw itch mite attacks all stages of *Plodia interpunctella, Cadra cautella* (Lepidoptera), *Oryzaephilus mercati, Lasioderma serricorne* (Coleoptera) and red fire ant, *Solenopsis invicta* (Bruce and LeCato, 1979, 1980). It is easy to rear (synthetic diet is available; Bruce, 1989), population can be synchronized with pest population, high reproductive potential (100-200 progeny/ female), short life cycle (7 days), female makes up 98% population and seeks out host immediately upon emergence and shows phoresy, physogastry behaviour. It injects toxin, paralyzes the host before killing. Till mid 80s not much progress was made regarding its exploitation as biocontrol agent because it causes dermatitis in humans. Moser *et al.* (1987) observed two groups of *Pyemotes* spp., one was parasitic on Scolytid beetle and venom less while the other group was polyphagous and venomous. The latter group was found to be very lethal to insects but show reduced dermatological risks to humans.

P. herfsi parasitizes *Tribolium confusum* larvae and can locate the beetle 7 cm beneath of bran. The progeny of a single mite kills all larvae of a single beetle within ten days. It also parasitizes over wintering caterpillars of pink bollworm, *Pectinophora gossypiella* (85% mortality). Mite induced mortality reached up to 85% in the first generation larvae of pink bollworm, an important pest of cotton. It was found to parasitize the soft scale *Saissetia oleae*, a citrus pest, in Argentina (Mallea *et al.*, 1983). *Adactylidium* sp. is another mite of this family which parasitizes the eggs of thrip, *Gynaikothrips ficorum*, a pest of ornamentals in Egypt and Brazil. It is quite abundant in nature and is associated with all stages of thrips, which is a contributory factor for thrips control.

Tombidiid mites are large, usually red velvet whose larvae are parasites of insects whereas their nymphs and adults are predatory in nature. Mites preferred the ventral sides of small and medium sized hosts whereas lateral sides of large sized hosts were preferred. *Allothrombium pulvinum* larvae are ectoparasites on aphids whereas nymphs and adults predate upon aphids and lepidopteran eggs. Two or more larvae per aphid kill the host in 1–3 days (Zhang and Xin, 1989; Dong *et al.*, 1996). *Dinothrombium pandora* larvae parasitize grasshoppers, however, its nymphs and adults feed on termites. One female lays approximately 80,000 eggs during her lifetime (Newell, 1979). *Eutrombidium locustarum* larvae parasitize first instar to adult stage of host at the rate of 35 mites/host. More than 75% parasitization is observed in grasshopper population. Mite feeding interferes with normal adult flight by causing grasshopper wings to become battered and even broken. Each female lays about 4700 eggs and each generation requires 60 days to complete in the laboratory.

CONCLUSIONS

Most of the work related with acarine predators and parasites is still at exploratory stage but beginning has been made in this direction with few species being commercially mass reared to control thrips, stored insects and plant parasitic nematodes. Seasonal inoculative release of acarine biocontrol agents with proper release strategies and techniques provide an effective control of insects, mites and nematodes in greenhouse crops and orchards. Once releases are complete, it is critical to monitor pest and acarine predator/parasite population weekly. Further research on development of chemical and abiotic resistant strains, feasibility of 'banker' plants or cover plants in fallow season is required to maintain the stock. Awareness on the side effects of pesticides will definitely play an important role in popularization of biocontrol methods.

REFERENCES

Al-Amidi, A.H.K. and Downes, M.J. (1990) *Parasitus hituherosus* (Acari: Parasitidae), a possible agent for biological control of *Heteropeza pygmaea* (Diptera: Cecidomyiidae) in mushroom compost. *Exp. Appl. Acarol.*, **8**, 13–25.

Amin, A.W., Mowafe, M.H. and Fatma, S.A. (1999) Effect of predaceous mesostigmatid mites in the control of *Meloidogyne javanica,* root -knot nematode on kidney bean. *Pak. J. Nematol.*, **17**, 91–96.

Anderson, R.C. (1968) Ecological observations on three species of *Pimeliaphilus* parasites of Triatominae in the United States. (Acarina : Pterygosomidae) (Hemiptera: Reduviidae). *J. Med. Entomol.*, **5**, 459–464.

Beaulieu, F. and Weeks, A.R. (2007) Free-living mesostigmatic mites in Australia: their roles in biological control and bioindication. *Aust. J. Exp. Agric.*, **47**, 460–478.

Bilgrami, A.L. 1994. Predatory behavior of a nematode feeding mite *Tyrophagus putrescentiae* (Sarcoptiformes : Acaridae): *Funda. Appl. Nematol.* **17**: 293-297.

Bilgrami, A.L. and Tahseen, Q. (1992) A nematode feeding mite *Tyrophagus putrescentiae* (Sarcoptiformes: Acaridae). *Fundam. Appl. Nematol.*, **15**, 477–478.

Bruce, W.A. (1989) Artificial diet for the parasitic mite *Pyemotes tritici* (Acari. Pyemotidae). *Exp. Appl. Acarol.*, **6**, 11–18.

Bruce, W.A. and Le Cato, J.L. (1979) *Pyemotes tritici*: Potential biological control agent of stored product insects. In: J.G. Rodriguez (ed.), *Recent Advances in Acarology*, Academic Press, New York, pp. 213–220.

Bruce, W.A. and LeCato, J.L. (1980) *Pyemotes tritici*: A potential new agent for biological control of the red imported fire ant, *Solellopsis invicta* (Acari: pyemotidae). *Int. J. Acarol.*, **6**, 271–274.

Cakmak, I., Janssen, A., Sabelis, M.W. and Baspinar, H. (2009) Biological control of an acarine pest by single and multiple natural enemies. *Biol. Contr.*, **50**, 60–65.

Cayrol, J.C. (1970) Action des autres composants de la biocenose du champignon de couche sur le nematode mycophage, *Ditylenchus myceliophagus* J.B. Goodey, 1958, et etude de son anabiose: forme de survie en conditions defavourables, *Rev. Ecol. Biol. Sol.*, **7**, 409–440.

Chhillar, B.S., Gulati, R. and Bhatnagar, P. (2007). *Agricultural Acarology.* Daya Publishing House, New Delhi.

Dong, Y.C., Ran, R.B. and Xiang, J.Y. (1996) Biology of *Allothrombium ovatum* (Acari: Trombidiidae) and its controlling effect on *Aphis gossypii* (Homoptera: Aphididae). *Syst. Appl. Acarol.*, **1**, 35–40.

Duso, C., Fontana, P. and Malagnini, V. (2004) Diversity and abundance of phytoseiids mites (Acari: Phytoseiidae) in vineyards and the surrounding vegetation in northeastern Italy. *Acarologia,* **44**, 31–47.

Faroni, L.R.D.A., Guedes, R.N.C. and Matioli, A.L. (2001) Effect of temperature on the development and population growth of *Acarophenax lacunatus* (Cross and Krantz) (Prostigmata : Acarophenaeidae) on *Rhyzopertha dominica* (F.) (Coleoptera : Bostrichidae). *Biocontr. Sci. Technol.*, **11**, 5–12.

Fitzgerald, J., Pepper, N., Easterbrook, M., Pope, T. and Solomon, M. (2007) Interactions among phytophagous mites, and introduced and naturally occurring predatory mites, on strawberry in the UK. *Exp. Appl. Acarol.*, **43**, 33–47.

Gallardo, A., Vasquez, C., Morales, J. and Gallardo, J. (2005) Biology and natural enemies of *Tetranychus urticae* in sweet pepper. *Manejo Integrado de Plagas y Agroecologia*, **74**, 34–40.

Gardiner, M.M., Eigenbrode, S.D., Cervantes, D.E. and Barbour, J.D. (2005) Response of *Neoseiulus fallacis* Garmen and *Galendromus occidentalis* Nesbitt (Acari: Phytoseiidae) to *Tetranychus urticae* Koch (Acari: Tetranychidae)-Damaged Hop *Humulus lupulus* (L.) (Urticales: Cannabaceae). *Agric. Forest Emntomol.*, **7**, 245–251.

Gauchat, C.A. (1972) A note on *Podapolipoides grassi* Berlese (Acarina: Podapolipidae), a parasite of *Chortoicetes terminifera* Walker, the Australian plague locust. *J. Aust. Entomol. Soc.*, **11**, 259–265.

Gerson, U. and Smiley, R.L. (1990) *Acarine Biocontrol Agents: An Illustrated Key and Manual*. Chapman & Hall, London.

Gerson, U., O'Connor, B.M. and Houck, M.A. (1990) Acari. In: D. Rosen (ed.), *Armored Scale Insects, Their Biology, Natural Enemies and Control*, Vol. 4B, Elsevier Science, New York, pp. 77-97.

Gerson, U., Smiley, R.L. and Ochoa, R. (2003) *Mites (Acari) for Pest Control.* Blackwell Science, Oxford, UK.

Grafton-Cardwell, E.E. and Ouyang, Y. (1995) Augmentation of *Euseius tularensis* (Acari: Phytoseiidae) in citrus. *Environ. Entomol.*, **24**, 738–747.

Gravena, S., Coletti, A. and Yamamoto, P.T. (1993) Influence of green cover with *Ageratum conyzoides* and *Eupatorium pauciflorum* on predatory and phytophagous mites in citrus. *Bull. IOBC-SROP,* **16**, 104–114.

Gulati, R., Walia, K.K. and Rani, R. (2005) Acarines as biocontrol agents of plant parasitic nematodes. In: D. Prasad (ed.), *Crop Protection: Management Stratergies*, Eastern Book Corporation, New Delhi, pp. 526–534.

Hardman, J.M., Moreau, D.L., Snyder, M., Gaul, S.O. and Bent, E.D. (2000) Performance of a pyrethroid-resistant strain of the predator mite *Typhlodromus pyri* (Acari: Phytoseiidae) under different insecticide regimes. *J. Econ. Entomol.*, **93**, 590–604.

Helle, W. and Sabelis, M.W. (1985) *Spider Mites: Their Biology, Natural Enemies and Control*, Vol. 1B. Elsevier, Amsterdam, The Netherlands.

Imbriani, L. and Mankau, R. (1983) Studies on *Lasioseius scapulatus,* a mesostigmatid mite predaceous on nematodes. *J. Nematol.*, **15**, 523–528.

Inserra, R.N. and Davis, D.W. (1983) *Hypospis* nr. *aculenfer*: A mite predaceous on root knot and cyst nematodes. *J. Nematol.*, **15**, 324–325.

Ishibashi, N., Young, F., Nakashima, M., Abiru, C. and Horaguchi, N. (1987) Effect of application of DD-136 on silkworm, *Bombyx mori*, predatory insect, *Agriosphodrus dohrni,* parasitoid, *Trichomalus apanteloctenus,* soil mites and other non-target soil arthropods, with brief notes on feeding behavior and predatory pressure of soil mites, tardigrades and predatory nematodes on DD-136 nematodes. In: N. Ishibashi (ed.), *Recent Advances in Biological Control of Insect Pests by Entomogenous Nematodes in Japan,* Ministry of Education, Culture and Science, Sago, Japan, pp. 158–164.

Jacobson, R.J., Croft, P. and Fenlon, J. (2001) Suppressing establishment of *Frankliniella occidentalis* Pergande (Thysanoptera : Thripidae) in cucumber crops by prophylactic release of *Amblyseius cucumeris* Oudemans (Acarina : Phytoseiidae). *Biocontr. Sci. Technol.*, **11**, 27–34.

Janssen, A., Montserrat, M., Hille Ris Lambers, R., de Roos, A., Pallini, A. and Sabelis, M.W. (2006) Intraguild predation usually does not disrupt biological control. In: J. Brodeur and G. Boivin (eds.), *Trophic and Guild Interactions in Biological Control*, Springer, Dordrecht, pp. 21–44.

Karban, R., English-Locb, G., Walker, M.A. and Thaler, J. (1995) Abundance of phytoseiid mites on *Vitis* species: Effects of leaf hairs, domatia, prey abundance and plant phylogeny. *Exp. App. Acarol.*, **19**, 189–197.

Karg, W. (1983) Verbreitung and bedeutung von vau-bonilben der cohors gamasina als antagonisten von nematoden. *Pedobiologia*, **25**, 419–432.

Kong, C., Hu, F., Xu, X., Zhang, H. and Liang, W. (2005) Volatile allellochemicals in the *Ageratum conyzoides* intercropped citrus orchard and their effects on mites *Amblyseius newami* and *Panonychus citri*. *J. Chem. Ecol.*, **31**, 2193–2203.

Landis, D.A., Wratten, S.D. and Gurr, G.M. (2000) Habitat management to conserve natural enemies of arthropod pests in agriculture. *Annu. Rev Entomol.*, **45**, 175–201.

Lester, P.J., Thistlewood, H.M.A. and Harmsen, R. (2000) Some effects of pre-release host-plant on the biological control of *Panonychus ulmi* by the predatory mite *Amblyseius fallacies. Exp. Appl. Acarol.*, **24**, 19–33.

Mallea, A.R., Macola, G.S., Garcia Saez, J.G. and Lamate, S.J. (1983) Localizacion de *Pyemotes* sp. (Acarina: Pyemotidae) sobre *saissetra oleae* Bern. (Homoptera: Lecaniidae). *Intersectum*, **15**, 8–10.

Mankau, R. (1980) Biological control of nematode pests by natural enemies. *Annu. Rev. Phytopathol.*, **18**, 415–440.

Matioli, A.L. (1997) *Biologia de Acarophenax lacunatus (Cross and Krantz) (Prostigmata: Acarophennacidae) e Seu Potential de Parasitismo Sobre Rhizopertha dominica (F.) (Coleoptera: Bostrichidae).* Universidade Fedral de Vicosa, Vicosa, Brazil.

McMurtry, J.A. (1982) The use of phytoseiids for biological control: progress and future prospects: In: M.A. Hoy (ed.), *Recent Advances in Knowledge of the Phytoseiidae*, University of California, Division of Agricultural Sciences Publication No. 3284, Berkeley, pp. 23–48.

McMurtry, J.A. and Rodriguez, J.G. (1987) Nutritional ecology of phytoseiid mites. In: F. Slansky and J.G. Rodriguez (eds.), *Nutritional Ecology of Insects, Mites and Spiders*, Wiley, Chichester, pp. 609–644.

McMurtry, J.A. and Croft, B.A. (1997) Life-styles of phytoseiid mites and their roles in biological control. *Annu. Rev. Entomol.*, **42**, 291–321.

Mcmurtry, J.A., Morse, J.G. and Johnson, H.G. (1992) Studies of the impact of *Euseius* species (Acari: Phytoseiidae) on citrus mites using predator exclusion and predator release experiments. *Exp. Appl. Acarol.*, **15**, 233–248.

Moser, J.C., Smiley, R.L. and Otvos, I.S. (1987) A new *Pyemotes* (Acari : Pyemotidae) reared from the Douglas-fir cone moth. *Int. J. Acarol.*, **13**, 141–147.

Mullen, C.R. (1977) Acarine parasites of mosquitoes. IV. Taxonomy, life history and behavior of *Thyas barbigera* and *Thyasides sphagnorum* (Hydrachnellae: Thyasidae). *J. Med. Entomol.*, **13**, 475–485.

Muma, M.H. (1961) The influence of cover crop cultivation on populations of indigenous insects and mites in Florida citrus groves. *Fla. Entomol.*, **44**, 61–68.

Muraoka, M. and Ishibashi, N. (19760 Nematode feeding mites and their feeding behaviour. *Appl. Entomol. Zool.*, **11**, 1–7.

Newell, I.M. (1979) *Acarus tinctorius* Linnaeus 1767 (Trombidiidae). . In: J.G. Rodriguez (ed.), *Recent Advances in Acarology*, Vol. 2, Academic Press, New York, pp. 425–428.

Pemberton, R.W. and Turner, C.E. (1989) Occurrence of predatory and fungivorous mites in leaf domatia. *Ann. J. Bot.*, **76**, 105–112.

Ranjana, R. (2002) Studies on the nematophogy in acarine fauna and their feeding potential on seed gall nematode. M.Sc. Thesis, CCS Haryna Agricultural University, Hisar.

Ranjana, R., Gulati, R. and Walia, K.K. (2003) Influence of prey size, prey density and test area on predatory potential of two nematophagous mites. In: *5th Nat. Sym. Biocontrol Agents for Sustainable Management of Pests,* Abstracts, Pantnagar, Uttaranchal.

Ranjana, R., Gulati, R. and Walia, K.K. (2005) Feeding potential of *Tyrophagus putrescentiae* Schrank (Acarina: Acaridae) on seedgall nematode, *Anguina tritici* (Steinbuch) Chitwood. *J. Insect Sci.*, **19**, 32–35.

Reisen, W.K. and Mullen, G.R. (1978) Ecological observations on acarine associates (Acari) of Pakistan mosquitoes (Diptera: Culicidae). *Environ. Entomol.* **7**, 769–776.

Rhodes, E.M., Liburd, O.E., Kelts, C., Rondon, S.I. and Francis, R.R. (2006) Comparison of single and combination treatments of *Phytoseiulus persimilis*, *Neoseiulus californicus* and Acramite (bifenazate) for control of twospotted spider mites in strawberries. *Exp. Appl. Acarol.*, **39**, 213–225.

Riley, C.V. (1874) Description of two new subterranean mites. *Trans. Acad. Sci. St. Louis*, **25**, 215–216.

Rockett, C.L. (1980) Nematode predation by oribated mites (Acari: Oribatida). *Int. J. Acarol.*, **6**, 219–224.

Rodriguez, J.G., Wade, C.F. and Wells, C.N. (1962) Nematodes as a natural food for *Macrocheles muscaedomesticae* (Acarina: Macrochelidae) a predator of the housefly eggs. *Ann. Entomol. Soc. Ani.*, **55**, 507–511.

Rodriquez, J.G., Singh, P. and Taylor, B. (1970) Manure mites and their role in fly control. *J. Med. Entomol.*, **7**, 335–341.

Rubin, A., Ucko, O., Orr, N. and Offenbach, R. (1996) Efficacy of natural enemies of the western flower thrips, *Frankliniella occidentalis*, in pepper flower in Arava Valley, Israel. *IOBC/WPRS Bull.*, **19**, 139–142.

Sarwar, M., Kongming, W., Xuenong, X. and Endong, W. (2011) Evaluations of four mite predators (Acari: Phytosiidae) released for suppression of spider mite infesting protected crop of sweet pepper (*Capsicum annuum* L.). *Afr. J. Agric. Res.*, **6**, 3509–3514.

Sell, P. (1988) *Caloglyphus* sp. (Acarina : Acaridae), an effective nematophagous mite on root-knot nematodes (*Meloidogyne* spp.). *Nematologica*, **34**, 246–248.

Sharma, R.D. (1971) Studies on the plant parasitic nematode *Tylenchorhnchus dubius*. *Meded. Landbouww*. Wageningen. **71**, 1–154.

Shimer, H. (1868) Notes on the "apple bark louse" (*Lepidosaphes conchiformes*, Gmelin sp.) with a description of a supposed new *Acarus*. *Trans. Ann. Entomol. Soc.*, **1**, 361–374.

Smith, B.P. (1983) The potential of mites as biological control agents of mosquitoes. In: M.A. Hoy, C.L. Cunningham and L. Knutson (eds.), *Biological Control of Pests by Mites*, University of California Special Publication no. 3304, Berkeley, pp. 79–85.

Smith, I.M. and Oliver, D.R. (1986) Review of parasitic associations of larval water mites (Acari: Parasitengona: Hydrachnida) with insect hosts. *Can. Entomol.*, **118**, 407–472.

Stechmann, D.H. (1978) Eiablage, Parasitismus and post parasitische Entwicklung von *Arrenurus* – Arten (Hydrachnellae, Acari). *Z. Parasitenkd.*, **57**, 169–188.

Stemkraus, D.C. and Cross, E.A. (1993) Description and life history of *Acarophenax mahunkai* n. sr. (Acari, Tersonemina: Acarophenaciedae), an egg parasite of the leser meal warm (Coleoptera: Tenebrionidae). *Ann. Entomol. Soc. Am.*, **86**, 239–224.

Sturhan, D. and Hampel, G. (1977) (Plant parasitic nematodes as prey of the bulb mite, *Rhizoglyphus echinopus* (Acarina: Tyroglyphidae)]. *Anz. Schadlingskunde Pflanzen. Umweltschutz.*, **50**, 115–118.

Tandon, P.L. and Lal, B. (1976) New record of the predatory mites on mango mealy bug, *Drosicha mangiferae* Green (Margarodidae : Hemiptera). *Curr. Sci.*, **45**, 566–567.

Tixier, M.S., Kreiter, S., Cheval, B., Guichou, S., Auger, P. and Bonafos, R. (2006) Immigration of phytoseiid mites from surrounding uncultivated areas into a newly planted vineyard. *Exp. Appl. Acarol.*, **39**, 227–242.

Toyoshima, S., Ihara, F. and Amano, H. (2008) Diversity and abundance of phytoseiid mites in natural vegetation in the vicinity of apple orchards in Japan. *Appl. Entomol. Zool.*, **43**, 443–450.

Tuovinen, T. and Rokx, J.A.H. (1991) Phytoseiid mites (Acari: Phytoseiidae) on apple trees and in surrounding vegetation in southern Finland. Densities and species composition. *Exp. Appl. Acarol.*, **12**, 35–46.

Van de Bund, C.F. (1972) Some observations on predatory action of mites on nematodes. *Zeszyty Problemoux Postepow Nank Roliniczych*, **129**, 103–110.

Van Emden, H.F. (1989) *Pest Control*. Cambridge University Press, Cambridge.

Van Lenteren, J.C., Benuzzi, M., Nicoli, G. and Maini, S. (1992) Biological control in protected crops in Europe. In: J.C. van Lenteren, A.K. Minks, A.K. and O.M.B. de Ponti (eds.), *Biological Control and Integrated Crop*

Protection: Towards Environmentally Safer Agriculture, Pudoc, Wageningen, pp. 77–89.

Walia, K.K. and Mathur, S. (1994) Predatory potential of two nematophagous mites on fungivorous nematode, *Aphelenchoides composticola in vitro. Indian J. Nematol.*, **24**, 243–245.

Walia, K.K. and Mathur, S. (1995) Predatory behavior of two nematophagous mites, *Tyrophagus putrescentiae* and *Hypoaspis calcuttaensis*, on root knot nematode, *Meloidogyne javanica*. *Nematol. Medit.*, **23**, 255–261.

Walia, K.K. and Mathur, S. (1996) Potential of two nematophagous mites in suppressing root- knot nematode, *Meloidogyne javanica* on Okra. *Indian. J. Nematol.*, **126**, 172–175.

Wallace, M.M.H. and Holm, E. (1983) Establishment and dispersal of the introduced predatory mite, *Macrocheles peregrimus* Krantz, in Australia. *J. Aust. Entomol. Soc.*, **22**, 345–348.

Walter, D.E. (1987) Life history, trophic behaviour and description of *Gamasellodes vermivorax* n.sp. (Mesostigmata: Ascidae), a predator of nematodes and arthropods in semiarid grassland soils. *Can. J. Zool.*, **65**, 1689–1695.

Walter, D.E. and lkonen, E.K. (1989) Species, guilds and functional groups: Taxonomy and behaviour in nematophagous orthropods. *J. Nematol.*, **21**, 315–327.

Walter, D.E., Hudgens, R.A. and Freckman, D.W. (1986) Consumption of nematodes by fungivorous mites, *Tyrophagus* spp. (Acarina: Astigmata: Acaridae). *Oecologia*, **70**, 357–361.

Walter, D.E., Kaplan, D.T. and Davis, E.L. (1993) Colonization of greenhouse nematode cultures by nematophagous mites and fungi. *J. Nematol.* (Supplement) **25**, 789–794.

Walzer, A. and Schausberger, P. (1999). Cannibalism and interspecific predation in the phytoseiid mites *Phytoseiulus persimilis* and *Neoseiulus californicus*: Predation rates and effects on reproduction and juvenile development. *BioControl*, **43**, 457–468.

Womersley, H. (1933) A possible biological control of the clover spring tail or Lucerne flea (*Sminthurus viridis* L.) in Western Australia. *J. Aust. Counc. Sci. Ind. Res.*, **6**, 83–91.

Zhang, Z.Q. and Xin, J.L. (1989) Studies on the morphology and life history of *Allothrombium pulvinum* Ewing (Acariformes : Trombididae). *Acta Entomol. Sinica.*, **32**, 192–199.

16

Plant Growth Promoting Rhizomicroorganisms and Plant Health Management

SUSEELENDRA DESAI[1*] AND Y.R. SARMA[2]

[1]Central Research Institute for Dryland Agriculture, Santoshnagar
Hyderabad, AP- 500 059, India
[2] FAO Consultant and Former Director, Indian Institute of Spices Research
Aramam, M 10-5, KSHB Colony, Kozhikode, Kerala - 673 009, India
*E-mail: desai1959@yahoo.com

INTRODUCTION

Agriculture is the predominant occupation in India, employing more than 50% of the rural population. During 2009-10, agriculture and allied sectors contributed 14.6% to Gross Domestic Product of India out of which agriculture alone contributed 12.3%. To meet the growing demand for food, adequate production and even distribution of food has become a high priority global concern. With the present agricultural scenario and global competition, there is a need for enhancing the farm productivity through integrated plant health management. In India, during the year 2008-09, the total area under cultivation was 8.74 million ha with a production of about 8.83 million tonnes reflecting a productivity of 1011 kg/ha. In the last few years, contribution of agriculture to national economy is declining, primarily due to aberrant weather conditions, frequent occurrence of extreme weather events and increasing input costs.

The opportunities for horizontal agricultural growth through increase in area are relatively less as the total cropped area has remained static at around 140 million ha since 1970s. This increasing pressure on land resources has led to an imperative vertical growth by increasing crop productivity and intensity. The 6.5 billion human populations today consume about 25 million tonnes of protein nitrogen each year. By 2050, it is expected to reach 40-45 million tonnes. To meet such enormous nitrogen requirements, chemical fertilizers may not be able to cater to the needs and thereby becoming expensive. Excessive extraneous supplementation may also lead to severe degradation of soil health and environment and pollution of water bodies and other associated risks.

To meet these challenges, integrated plant health management packages should thus be appropriately configured to meet the specific demands of different agro-ecological conditions. An integrated plant health management package obviously includes appropriate components like organic and inorganic nutrients, biofertilizers, biopesticides and synthetic pesticides, in addition to agronomic practices. Plant growth promoting rhizomicroorganisms are one of the viable options that can supplement nutrients and pesticides and also help in sustaining agricultural production without much harm to environment.

PLANT GROWTH PROMOTING RHIZOMICROORGANISMS

Total bacteria in agricultural soils total about 109 colony forming units (cfu)/g. This population is made up of mostly Gram-positive eubacteria (108cfu/g), actinomycetes (108cfu/g), and cyanobacteria (106 cfu/g). Archeabacteria are not an important component because conditons are not extreme enough (halophiles) or the proper gaseous environment is not available (CH_4, H_2S, *etc.*). Bacteriophages, or bacterial viruses, make up a variable part of the biotic agents in soil (106^{-10} plaque forming units [pfu]/g). Eucaryotic soil microflora includes fungi (104^{-5}cfu/g), protozoa (104^{-6} viable cells/g), nematodes (104^{-5}worms/g), and arthropods, usually mites, 104^{-6}orgnisms/g. From the point of view of volume and surface area interacting with the soil, bacteria have greater contact with the soil environment than fungi.

The term plant growth-promoting rhizobacteria (PGPR) was first coined by Kloepper in 1976 to denote the microorganisms associated with plant roots and help in plant growth either directly or indirectly. The term PGPR specifically refers to bacteria whereas in the recent years many other microorganisms have also been shown to possess these trait and hence, the the term Plant Growth Rhizomicroorganisms has been thought to be appropriate. These PGPRs are naturally occurring soil bacteria that colonize roots, enhance plant growth, and activate plant defences against microbial pathogens.

PGPRs promote plant growth through influence of the plant hormonal balance by bacterial IAA or ethylene production; antagonisitic to pathogens through antibiotic production, competition for iron, and stimulation of the plant resistance/defence mechanisms; affect nutrient uptake by secretion of organic acids or protons to solubilize phosphorus, by atmospheric N_2 fixation, by structuration of the rhizospheric soil by producing exopolysaccharides. The direct promotion of plant growth by PGPR is through production of plant growth-promoting substances or facilitation of uptake of certain nutrients from the soil.

PGPRs as Nutrient Cyclers

The commercial history of biofertilizers began with the launch of 'Nitragin' by Nobbe and Hiltner, a laboratory culture of *Rhizobia* in 1895, followed by the discovery of *Azotobacter* and then the blue green algae and a host of other micro-organisms. *Azospirillum* and arbuscular mycorrhizal fungi (AMF) are fairly recent discoveries. In India the first commercial production started as early as 1956.

Bacterial inoculants that help in plant growth are of two types a) symbiotic, and b) free-living (Kloepper *et al.*, 1988; Frommel *et al.*, 1991). Beneficial free-living bacteria referred to as PGPR are found in the rhizosphere of the roots of many different plants (Kloepper *et al.*, 1989). Sajjad Mirza *et al.* (2006) demonstrated nitrogen fixing ability of strains of *Pseudomonas*. The effect of *Azotobacter* and *Azospirillum* is attributed not only to the amounts of fixed nitrogen but also to the production of plant growth regulators such as indole acetic acid, gibberellic acid, cytokinins and vitamins which show additional positive effects on the plants (Rodelas *et al.*, 1999). The association of diazotrophic rhizobacteria with grasses is well documented (Baldani *et al.*, 1997) and includes several bacterial genera and many important agricultural plants. Most cultivated tropical soils in India are reported to have relatively large populations (> 100 cfu/g dry soil) of rhizobia capable of nodulating the legumes and fixing nitrogen (Nambiar *et al.*, 1988; Venkateswarlu, 1992; Khurana and Dudeja, 1997). The tri-partite association composed of legume plant, rhizobia and *Pseudomonas* spp. has been reported to increase root and shoot weight, plant vigour, nitrogen fixation and grain yield in various legumes (Sindhu *et al.*, 1999). Sindhu *et al.* (2002) showed that dual inoculation of chickpea with *Pseudomonas* and *Mesorhizobium* enhanced plant growth. Dual inoculation increased yields in sorghum (Algawadi and Gaur, 1992), barley (Belimov *et al.*, 1995), black gram

(Tanwar *et al.*, 2002), soybean (Abdalla and Omer, 2001) and wheat (Galal, 2003). Sahin *et al.* (2004) documented increased sugar content in sugar beet by co-inoculation of N-fixing and PSB. Parmar and Dadarwal (1999) also observed that co-inoculation of *Pseudomonas* and *Bacillus* sp. with *Rhizobium* strains enhanced the nodule weight, root length, shoot biomass and total plant nitrogen in chickpea, when grown in sterilized jars or under pot culture conditions.

Phosphorus is one of the major nutrients that limit plant growth. Most of the soils around the world are P deficient (Batjes, 1997). The use of rock phosphate as a phosphate fertilizer and its solubilization by microbes (Kang *et al.*, 2002), through the production of organic acids (Maliha *et al.*, 2004), has become a valid alternative to chemical fertilizers. Several studies have shown that phosphate solubilizing microorganisms (PSMs) solubilize the fixed P in the soil resulting in higher crop yields (Gull *et al.*, 2004). The combined application of *Rhizobium* and PSM (Perveen *et al.*, 2002) or PSM and arbuscular mycorrhizal (AM) fungi (Zaidi *et al.*, 2003) showed enhanced plant growth as compared to their individual inoculation in 'P' deficient soils. Phosphate solubilizing bacteria are ubiquitous (Gyaneshwar *et al.*, 2002) and *Bacillus, Enterobacter, Erwinia* and *Pseudomonas* spp. are among the most potent strains. The importance of root growth and architecture for the efficient capture of P, well documented and in many cases, is a specific response of plants to P deficiency (Richardson *et al.*, 2009). Under controlled growth conditions, various studies have demonstrated enhanced growth and P nutrition of plants inoculated with PSM (Rodriguez and Fraga, 1999; Gyaneshwar *et al.*, 2002).

Zinc solubilizing potential of few bacterial genera has also been studied. Di Simine *et al.* have reported (1998) a zinc solubilizing strain of *Pseudomonas fluorescens* from forest soil. Hutchins *et al.* (1986) reported that *Thiobacillus thiooxidans, T. ferrooxidans* and facultative thermophilic iron oxidizers solubilized zinc from sulphide ore (sphalerite). Saravanan *et al.* (2007) reported a strain of *Gluconoacteobacter diazotrophicus* with zinc solubilization and its anti-nematode activity against *Meloidogyne incognita*.

PGPRs as Biocontrol Agents

Role of microbial biocontrol agents in pest management in crops has been well established long time ago. However, the science of biological control did not grow as it was expected due to many reasons. The science was virtually reinvented during late 70s due to growing concern about increasing environmental pollution and increasing input costs in agriculture. The boost also came from the modern science, which offered many, molecular and biotechnology tools that helped in better understanding the plant-microbe interactions. Thus, today there is an assured response from farmers for a promising biocontrol agent as the products entering the market are on a firm scientific footing.

The most promising biocontrol agents are species of *Trichoderma*, *Pseudomonas* and *Bacillus* that have been successfully formulated and marketed for the management of many diseases. Biocntrol agents exhibit several modes of biocontrol ability such as mycoparasitism, competition, production of hydrolases and antibiosis (Papavizas, 1985). Breakthrough research in the field of PGPR occurred in the mid 1970s with studies demonstrating the ability of *Pseudomonas* strains capable of controlling soil-borne pathogens and indirectly enhances plant growth and thereby increased the yield of potato and radish (Kloepper *et al.*, 1980; Howie and Echandi, 1983). The antibiotics produced by fluorescent pseudomonads controlled *Rhizoctonia solani* and *Pythium ultimum* causing damping off in cotton. Certain strains of *Pseudomonas* produce a multitude of inhibitory compounds such as pyrrolnitrin, Oomycin A, pyoleteorin, phenazine and diacetylphloroglucinol that contribute to the suppression of many pathogens. However, not all chemicals are effective against all pathogens, which indicate a recognition mechanism followed by an induction system. PGPR can prevent the proliferation of phytopathogens and thereby support plant growth. One of the mechanisms involved here is through their ability to produce siderophores for sequestering iron. The secreted siderophores bind to the Fe^{3+} that is available in the rhizosphere, and thereby effectively prevent growth of pathogens in that region. Certain PGPR degrade fusaric acid produced by *Fusarium* sp. causative agent of wilt and thus prevents the pathogenesis. *Pseudomonas stutzeri* produces extracellular chitinase and laminarinase, which lyses the mycelium of *Fusarium solani*.

The secondary metabolites produced by certain species and strains of *Bacillus* have antibacterial or antifungal activity against phytopathogenic microorganisms (Asaka and Shoda, 1996). The major antibiotics that play a vital role in the suppression of plant pathogens are grouped into non-volatile and volatile antibiotics. *Pseudomonas fluorescens* strain CHA0 suppressed *Pythium* root rot of cucumber by producing of DAPG and pyoluteorin (Fenton *et al.*, 1992; Maurhofer *et al.*, 1992). Seed bacterization of tomato and chilli with a talc based consortium comprising *P. fluorescens* and *P. chlororaphis* performed better in reducing the incidence of damping-off (Kavitha *et al.*, 2003). Recently, Upadhyay and Srivastava (2011) showed the importance of Phenazine-1-carboxylic acid as an important contributor in biocontrol by *P. fluorescens* strain Psd than pyrrolnitrin. Products for plant disease biocontrol containing *B. subtilis* and other *Bacillus* species have been used in the past as seed dresser in several crops (Schisler *et al.*, 2004). *Bacillus* strains have the advantage of forming endospores that confer them high stability as biofungicides or biofertilizers (Schisler *et al.*, 2004). Several mechanisms have been proposed to explain the inhibition of fungal pathogens by *Bacillus* spp., including production of antimicrobials, secretion of hydrolytic enzymes, competition for nutrients, or a combination of mechanisms (Compant *et al.*, 2005).

Pleban *et al.* (1997) have shown chitinolytic activity in a strain of *B. cereus* showing antibiosis against *Macrophomina phaseolina*, while Sela *et al.* (1998) have characterized a collagenolytic and proteolytic enzyme from *B. cereus* that has damaged the cuticles of *Meloidogyne javanica*. Inhibitory furanone, identified as 3-(2-hydroxypropyl)-4-(2-hexadienyl)-2(5H)-furanone, was isolated from the culture filtrate of *T. harzianum* that inhibited the growth of microorganisms (Ordentilich *et al.*, 1999). The production of antimicrobial components in soil is influenced by several environmental factors such as soil chemistry and microbial population. For example, presence of Zn^{2+} in soil increases the production of Phenazine and low temperature (12°C) influences production of 2,4-diacetyl phloroglucinol by *Pseudomonas* sp. It is, however, not possible to differentiate the strains based on production of these metabolites.

Kay and Stewart (1994) applied four BCAs as a soil amendment (sand:bran:fungal homogenate, 1:1:2) at the rate of 0.1% wheat bran/g dry soil in a soil-box method for the management of *Sclerotium cepivorum* which causes white rot of onions. Though all BCAs provided various levels of disease control equivalent to the fungicide (procymidone 0.5 g a.i./100 g seed) treatment, *Cheatomium globosum* and *Trichoderma* spp. were more effective. Successful biocontrol of *S. rolfsii* with cultures of *Trichoderma* spp. has also been reported (Sreenivasprasad *et al.*, 1993; Elad *et al.*, 1980). Dinakaran *et al.* (1995), Sankar and Jeyarajan (1996) used BCAs as seed dressers for the management of root rot of sesamum caused by *R. bataticola*. Endophytic *Bacillus* spp. *B. cereus, B. subtilis*, and *B. pumilus* protected cotton seedlings against *R. solani* in the greenhouse by 51, 46 and 56%, respectively (Pleban *et al.*, 1995). Similarly, 72, 79 and

26% reduction in seedling rot due to *S. rolfsii* in bean was observed when inoculated with *B. subtilis, B. cereus* and *B. pumilus*, respectively. *P. chitinolytica* was found to be effective in greenhouse, screen house, and micro plot tests against *M. javanica* on tomatoes (Spiegel *et al.*, 1991). Biocompetitive and non-toxigenic *Aspergillus parasiticus* (NRRL 13539) was introduced into the soil using sprinkler for successful inhibition of toxigenic *Aspergillus flavus* and thus reduce aflatoxin contamination in peanuts (Dorner *et al.*, 1992).

PGPRs as Resistance Inducers

Microbial products that are sold for biological control of soil-borne pathogens may also induce resistance to diseases of above ground parts. We suspect that induced resistance is involved because of cases in which the product was applied only to the soil and the effect was seen in the above ground parts of the plants. In a trial conducted on an organic farm in western NY during the summer of 2001, tomatoes drenched with a suspension of Plant Shield (*Trichoderma harzianum*) at transplanting showed the lowest levels of early blight at the end of the season compared with other treatments. In a greenhouse tomato trial, Mycostop (*Streptomyces griseoviridis*) applied as a soil drench provided control of gray mold comparable to foliar fungicide (Bravo) applications (Lamboy *et al.*, 1996).

Wilt was significantly reduced when carnation roots were bacterized with *Pseudomonas* sp. strain WCS417r one week prior to stem-inoculation with conidia of *Fusarium*. Heat-killed cells or lipopolysaccharides of *Pesudomonas* applied to roots were also protective suggesting that LPS could be an inducing signal. Similarly, LPS from bacterial cell surface of PGPRs was as effective as living bacteria to suppress radish wilt. Seed treatment with PGPRs resulted in a reduction of severity of cucumber anthracnose. Seed bacterization with PGPR reduced bacterial blight of bean. In all above cases, the beneficial bacterium and the pathogen were spatially separated to eliminate direct contact. *P. fluorescens* induced SAR-associated proteins while conferring systemic resistance against tobacco necrosis virus.

Mutants of CHA0 that does not produce the siderophore pyoveridin induced a lower level of resistance, suggesting a novel role of pyoveridin in disease suppression. More recently, salicycilic acid and pyocyanin produced by *P. aeruginosa* was shown to be inducing resistance against grey mold of bean. Similar role by pyochelin was not ruled out. Evidence was developed that indicated that infection sites are not blocked but subsequent colonization of susceptible radish cortex cells was blocked by pseudomonads. In this model, club root fungus was prevented from causing a high incidence or severe damage to a susceptible radish cultivar by a siderophore producing pseudomonad. This reduction mirrored exactly the events, which transpired in a resistant radish cultivar. Therefore, the blockage of colonization within the cortex may have been related to bacterial release of small molecular weight compounds that altered the resistance of the host an induced defense mechanism. *P. aeruginosa* strain 7NSK2 produces salicylic acid which induces systemic resistance against *Botrytis cinerea* on bean. Similarly, rhizomicroorganisms were reported to be activators of SAR in protecting *Arabidopsis thaliana* against *Fusarium oxysporium* f.sp. *raphani*. 2,3-butanediol and acetoin were released exclusively from two PGPR strains that triggered the greatest level of growth promotion and induced disease resistance. During the last few years, Impressive evidence has been shown that plant growth-promoting rhizobacteria (PGPR) can protect cucumber plants against fungal, bacterial, and viral diseases. PGPR-mediated ISR is an important mechanism of biological disease control, and it can be as effective as pathogen-induced systemic acquired resistance. Hence, it is important to characterize the physiological, biochemical, and molecular mechanisms involved to make optimal use of ISR in plant protection.

Signalling molecules activate specific sets of defence-related genes encoding pathogenesis-related proteins (Uknes *et al.*, 1992) with antimicrobial activity (Kombrink and Somssich, 1995). Rapid accumulation of phenylalanine ammonia lyase was involved in protection of AF1-treated pigeonpea seedlings against *Fusarium* wilt infection (Podile and Laxmi, 1998). ISR appears phenotypically similar to systemic acquired resistance (SAR). ISR is the phenomenon of enhanced resistance of plants that are challenged by same pathogen against which the plant is already showing some degree of resistance. The antibiotics (phenazine and phloroglucinol) produced by these bacteria played a significant role in both plant defense and ecological competence (Cook *et al.,* 1995). In an aseptic hydroponic system, *T. harzianum* (T-203) assisted in strengthening the epidermal and cortical cells of cucumber seedlings. Wall appositions contained large amounts of callose, while peroxidase and chitinase activities were triggered within 72 h after infection. Since all these activities are related to defence mechanisms, it was concluded that the fungus helps in the induction of such responses to protect from pathogenic microbes (Yedidia *et al.,* 1999).

PGPRs and Plant Growth Promotion Traits

The rhizosphere environment, at the interface between root and soil, is a major habitat for soil processes. A significant part of the carbon fixed by plants during photosynthesis is released into the soil during rhizodeposition. Root exudates, the soluble part of rhizodeposition, are used for the growth of microorganisms naturally occurring in the soil. In others words, rhizodeposition (quality and composition) has a great influence on the development of rhizospheric microflora, which contributes to soil and plant health. The rhizoplane is the two-dimensional interface between the root and the soil. The bacteriology of the rhizoplane is more difficult to study than that of the rhizosphere. The physical difficulty of separating the microbes associated with the mineral component of the rhizosphere from those on the rhizoplane discourages any but the most rigorous studies.

Numerous examples of plant growth stimulation by fluorescent *Pseudomonas* spp. have been reported (Kloepper *et al.*, 1980; Bashan, 1986; Suslow, 1980). Burr *et al.* (1978) reported significant increases in growth and yield of potatoes up to 367% in greenhouse experiments with specific fluorescent *Pseudomonas* strains. Van Peer and Schippers (1988) documented increases in root and shoot fresh weight for tomato, cucumber, lettuce and potato as a result of bacterization with *Pseudomonas* strains. Biological control with fluorescent pseudomonads offers an effective method of managing plant pathogens (Ramamoorthy *et al.*, 2001). The treatment of soybeans with *Bacillus cereus* has been shown to improve yield in field (Osburn *et al.*, 1995). Systemic resistance in sugar beet was elicited by a non-pathogenic, phyllosphere colonizing *Bacillus mycoides* (Bargabus *et al.*, 2002). Different strains *of B. subtilis, B. cereus, B. mycoides* and *B. thuringiensis* exhibit antifungal activity (Stabb *et al.*, 1994). A combination of *P. fluorescens* strains has been shown to increase rice yield and for the effective management of sheath blight (Nandakumar *et al.*, 2001). Similarly, a mixture of fluorescent *Pseudomonas* spp. was demonstrated for suppression of fusarium wilt of radish (Marjan de Boer *et al.*, 1999).

Increase in overall plant growth by soil application of PGPRs along with chitosan commercial formulations has been reported (Vasudevan *et al.*, 2002). *Bacillus pumilis* with chitosan and induction of resistance in tomato against *F. oxysporum* was demonstrated by Benhamou *et al.*, 1998. Niranjan Raj *et al.* (2003) showed that chitosan based mixed formulation of *B. subtilis* strain GB03+*B. pumilus* strain INR7 (LS256) and *B. subtilis* strain GB03+*B. pumilus* strain T4

(LS257) inoculation reduced downy mildew and increased plant growth promotion in pearl millet. Strains of the genus *Bacillus* are another most commonly reported PGPR (Compant *et al.*, 2005; Vessey, 2003). Strains which exhibited the potential to be PGPRs were identified as *Pseudomonas putida* biovar B, *P. fluorescens*, *Arthobacter citreus* and *Serratia liquefaciens* (Lifshitz *et al.*, 1986; Kloepper *et al.*, 1988). Salamone (2000) reported the growth-promoting effect of *P. fluorescens* strain G20-18 on wheat and radish plants by production of cytokinin phytohormones. As the effect of PGPR on plants was demonstrated, the concept of PGPR began to gain importance and a large number of bacterial strains have been isolated, screened (Cattelan *et al.*, 1999; Bertand *et al.*, 2001) and evaluated for plant growth promotion (Bent *et al.*, 2001; Salamone, 2000).

The PGPRs have different modes of action and they include production of growth promoting substances, release of nutrients that are in fixed state, production of siderophores, mycoparasitism and antibiosis. Research results indicate that the PGPRs can have one or more modes of action. PGPR may induce plant growth promotion by direct or indirect modes of action. Direct mechanisms include the production of stimulatory bacterial volatiles and phytohormones, lowering the ethylene level in plant, improvement of the plant nutrient status (liberation of phosphates and micronutrients from insoluble sources, nitrogen fixation etc.) and stimulation of disease resistance mechanisms (induced systemic resistance). Indirect effects originate when PGPR acts like biocontrol agents in disease reduction, stimulation of other beneficial symbioses and protects the plant by degrading xenobiotics in inhibitory contaminated soils (Jacobsen, 1997). Bashan and Holguin (1997) proposed the division of PGPR into two classes: biocontrol-PGPB (plant growth promoting bacteria) and PGPB. This classification may include beneficial bacteria that are not rhizosphere bacteria but it does not seem to have been widely accepted. Furthermore, in most studied cases, a single PGPR will often reveal multiple modes of action including biological control (Vessey, 2003).

A mutant strain of *Pseudomonas putida* with four-fold increase in IAA production lost its ability to induce root elongation in canola seedlings, though its growth rate and production of siderophores and 1-aminocyclopropone 1 carboxylate (ACC) deaminase remained unaltered (Xie *et al.*, 1996). Plant development of spring wheat (*Triticum aestivum* cv. Rubino) was affected by change in IAA production by *Azospirillum brasiliense* (Stijn Spaepen *et al.*, 2008). Bacteria like *Azospirillum* and *Pseudomonas* spp. produced cytokinins and gibberellins, in addition to IAA (Gaudin *et al.*, 1994). It has been suggested that gibberellins along with other microbial metabolites may affect plant growth and development (Arshad and Frankenberger, 1993). Ethylene, a gaseous phytohormone commonly induced by wounding in plants, causes root growth inhibition. ACC deaminase produced by PGPR cleaved ACC, the immediate precursor molecule of ethylene in plants, with a net result of increase in root growth (Glick *et al.*, 1998).

PGPRs for Abiotic Stress Management in Crops

Abiotic stresses are one of the major constraints in realizing the achievable yields of crops especially under rainfed conditions. Research has been carried out on the occurrence and functional diversity of AIMs in stressed environments (Bashan and Holguin, 1997). Tilak *et al.* (2005) have reported occurrence of *Rhizobium, Bradyrhizobium, Azotobacter, Azospirillum, Pseudomonas* and *Bacillus* from desert ecosystem, acid soils, saline and alkaline areas and highly eroded hill slopes in India. Venkateswarlu *et al.* (1984) studied distribution of *Azotobacter* and P solubilising organisms in Indian arid zone. In saline-alkaline soils of the Indo-gangetic plains, significant numbers of *Bradyrhizobium* could be recovered with much higher tolerance limit to these conditions (Bhardwaj, 1975). Nautiyal (2003) comprehensively reviewed aspects of isolation of *Bradyrhizobium, Azospirillum, Azotobacter* superior strains and PSB tolerant to drought, temperature, salinity, alkalinity and acidity and in most cases, such strains performed better than the normal ones.

Efforts have been made to protect the crops from abiotic stresses by inoculating with strains of microorganisms. Co-inoculation of *Rhizobium* and *Pseudomonas* can combat salt stress in maize by increasing the total uptake of K, P and Ca without affecting Na accumulation (Asghari and Mussarat, 2009). Sunita Rangarajan *et al.* (2003) have demonstrated biological control of rice diseases by *Pseudomonas* spp. under saline conditions. *Rhizobium, Azotobacter* and *Pseudomonas* produce copious exopolysaccharides (EPS) under natural environment but reports on the usefulness of EPS in desiccation tolerance are conflicting (Deaker *et al.*, 2004). Mayak *et al.* (2004a) reported that *Achromobacter piechaudii* having ACC deaminase activity significantly increased the fresh and dry weights of tomato seedlings grown in the presence of NaCl salt (up to 172 mM). Lixia *et al.* (2010) reported growth promotion of cotton under salt stress by a strain of *Pseudomonas putida*. Drought affects virtually all climatic regions of the world (Wilhite, 2000) and more than one-half of the earth is susceptible to drought every year (Hewitt, 1997). Mayak *et al.* (2004b) reported that ACC deaminase PGPR *Achromobacter piechaudii* ARV8 significantly increased the fresh and dry weights of both tomato and pepper seedlings exposed to transient water stress. Bensalim *et al.* (1998) revealed that *Burkholderia phytofirmans* strain PsJN helped potato plants in maintaining normal growth under heat stress. Recently, Barka *et al.* (2006) reported ACC deaminase activity *in vitro* and inoculation of grapevine (*Vitis vinifera* L.) explants with the same bacterium (*Burkholderia phytofirmans* strain PsJN) enhanced plant growth and physiological activity at both ambient (26°C) and low (4°C) temperatures. Ali *et al.* (2009) reported that *Pseudomonas* strain AKM-P6 enhanced tolerance of sorghum seedlings to high temperatures (50°C).

PGPRs as Biosensors

The adaptation of different strains of microbes in the rhizosphere could be used as an indicator for characterization of soil types. Such soil typing will help in classification of soils and develop suitable land use patterns. In order to adapt to the fluctuations in soil salinity/ osmolarity the bacteria of the genus *Azospirillum* accumulate compatible solutes such as glutamate, proline, glycine betaine, trehalose, etc. Proline seems to play a major role in osmoadaptation. With increase in osmotic stress the dominant osmolyte in *A. brasilense* shifts from glutamate to proline. Accumulation of proline in *A. brasilense* occurs by both uptake and synthesis. *A. brasilense* accumlated more intracellular glycine betaine with increasing osmolarity. Although a simlar protein is also present in *A. brasilense* Cd it does not help in osmoprotection. Except *A. halopraeferens*, all other species of *Azospirillium* lack the ability to convert choline into glycine betaine.

Research on the molecular mechanism of ISR was initiated on the role of PR-proteins, as the accumulation of these proteins was considered to be strictly correlated with induced disease resistance. However, radish roots treated with *P. fluorescens* did not accumulate PR-proteins although these plants showed enhanced resistance against wilt. Apart from the role in systemically induced resistance, the defense signal molecules like salicylic acid, jasmonic acid and ethylene have repeatedly been implicated in the regulation of primary resistance responses. Based on these results it was concluded that induced resistance is an enhancement of basal resistance. Both ISR and SAR could be combined to manage many diseases. Classic resistance depends on the availability of resistance genes, which often show limited durability. Induced disease resistance is an attractive alternative form of plant protection, as it is based on the activation of extant resistance mechanisms in the plant and is effective against a broad spectrum of pathogens. Therefore, a detailed knowledge about the molecular mechanisms underlying induced disease resistance would be instrumental in developing biologically-based, environm-entally-friendly and durable crop protection.

CONSERVATION OF BIODIVERSITY OF MICROBES

The changing crop production systems, globalization of agriculture, and changing climate have been significantly influencing the microbial biodiversity at global, national and regional levels. The epicentres of crop diversity offer vast microbial biodiversity as the crops and microbes have co-evolved over a long period. Similarly, the diverse production systems of the Indian subcontinent offer a scope to identify a variety of useful microorganisms. Climate change impacts have been shown to alter the morphology, physiology, pathogenicity, and biocontrol ability of microorganisms considerably. The adaptation of microorganism to a changing climate may prove to be one of the most important challenges under future climates (Garrett *et al.*, 2006), especially in terms of managing the productivity and sustainability of food crops. Climate changes will influence soil organisms both directly (warming) and indirectly (warming and elevated CO_2) via changes in quantity and quality of plant-mediated soil C inputs. Elevated atmospheric CO_2 commonly stimulates flow of organic C into the soil system, increases root production and exudation, but decreases litter quality. There is little evidence that atmospheric CO_2 enrichment will increase total soil organic matter content because greater C flow into soil stimulates the soil food web, often leading to equivalent increases in soil CO_2 efflux. Effects of warming on C allocation belowground, on the other hand, will depend largely on the temperature optima of different plant species. In the current IPR regime, the immediate requirement is to conserve these microbes, develop diagnostics tools and catalogue their useful traits. The Indian Council of Agricultural Research has given due importance to the agriculturally important microbes and established an independent National Research Centre for Agriculturally important microorganisms with a mandate to conserve the biodiversity, exploit their utility to its optimum and develop a national road map for the research in this direction.

CONCLUSIONS AND FUTURE PROSPECTS

With the growing concern for our immediate environment, there has been a renewed interest in exploring the microbiota for their beneficial traits. It has been proved beyond doubt that microorganisms are an essential component of plant growth and productivity. Their association with plant system continues to play a predominant role in deciding the establishment and yielding of a particular crop species in a given agro-ecosystem. Several examples have proved that microbes help in cycling of nutrients in soil environment and thereby promoting plant growth. They protect crops from pathogenic organisms and insect pests that cause severe yield losses. They produce various growth promoting substances that help in establishment of plants.

Excess usage of fertilizers and pesticides has lead to the deterioration of soil quality, loss of natural soil microbial biodiversity and pollution. For successful exploitation of microbes for sustainable agriculture, the research efforts should go into basic, applied and strategic aspects. The basic aspects should focus on the identification of potential microbes; screening, characterization and cataloguing the strains for their all desirable traits; enhancement of potential and understand the mechanism of action of these microbes to harness their optimum potential. The applied and strategic research aspects should include development of cheap and viable mass multiplication protocols; identify suitable carrier systems; develop methods for prolonged shelf-life; working out the economics to demonstrate the usefulness of the agents; to develop effective communication methods to disseminate the technology among the end-users to get feed back and refine technology. Despite many years of research on rhizosphere, our knowledge is fragmentary. This is mainly because of variations in methodology followed by various workers. For instance, the methods of counting soil populations do not give equal opportunity to all microbes to grow on certain media. Similarly, the variability among the microbes also makes it imperative to organize research activities to understand such and develop technologies after taking into account such variability. The recent advances in the taxonomy of bacteria using BIOLOG and FAME now offer a better opportunity for classification of isolates than the conventional methods.

As a long term objective, a promising approach to improve the efficiency of rhizomicroorganisms is combining several mechanisms of action in a single micro-organism by genetic modification after giving due consideration to the ecological safety of releasing large numbers of these genetically modified microorganisms. In addition to the plant growth promotion, the biocontrol ability of these microbes should be harnessed to manage weeds that are causing enormous loss in realizable yields and degrading the soil health.

Thus, future thrust should be on the following lines:

- To study the role of these microorganisms in production systems mode rather than in isolation and optimise their application systems
- Identify the candidate microbes as biosensors to assess soil health
- To immediately develop methods to conserve biodiversity of the microorganisms in the wake of IPR and WTO regime.
- To develop research networks and centres of excellence for the research on PGPRs
- Develop a national road map for research on PGPRs
- Since the success of the PGPR research is ultimately decided by the active participation of industry and other associated stakeholders, a transparent platform should be developed for participation of all stakeholders.

REFERENCES

Abdalla, M.H. and Omer, S.A. (2001) Survival of Rhizobia/Bradirhizobia and a rock phosphate solubilizing fungus *Aspergillus nigeron* various carriers from some agroindustrial wastes and their effect on nodulation and growth of fababean and soybean. *J. Plant Nutr.*, **24**, 261–272.

Alagawadi, A.R., and Gaur, A.C. (1992) Inoculation of *Azospirillum brasilense* and phosphate-solubilizing bacteria on yield of sorghum (*Sorghum bicolor* L. Moench). *Dryland Trop. Agric.*, **69**, 347–350.

Ali, S.Z., Sandhya, V., Grover, M., Kishore, N., Venkateswar Rao, L. and Venkateswarlu, B. (2009) *Pseudomonas* sp. strain AKM-P6 enhances tolerance of sorghum seedlings to elevated temperatures. *Biol. Fert. Soils*, **46**, 45–55.

Arshad, M., and Frankenberger, W.T. Jr. (1993) Microbial production of plant growth regulators. In: F.B. Metting, Jr. (ed.), *Soil Microbial Ecology. Applications in Agricultural and Environmental Management*, Marcel Dekker Inc., New York, pp. 307–343.

Asaka, O. and Shoda, M. (1996) Biocontrol of *Rhizoctonia solani* damping-off of tomato with *Bacillus subtilis* RB14. *Appl. Environ. Microbiol.*, **62**, 4081–4085.

Asghari, B., and Mussarat, F. (2009) Salt tolerance in *Zea mays* (L). following inoculation with *Rhizobium* and *Pseudomonas*. *Biol. Fert. Soils*, **45**, 405–413.

Baldani, J.I., Caruso, L., Baldani, V.L.D., Goi, S.R. and Dobereiner, J. (1997) Recent advances in BNF with non-legume plants. *Soil. Biol. Biochem.*, **29**, 911–922.

Bargabus, R.L., Zidack, N.K., Sherwood, J.E. and Jacobsen, B.J. (2002) Characterization of systemic resistance in sugar beet elicited by a nonpathogenic, phyllosphere-colonizing *Bacillus mycoides*, biological control agent. *Physiol. Mol. Plant Pathol.*, **61**, 289–298.

Barka, E.A., Nowak, J. and Clément, C. (2006) Enhancement of chilling resistance of inoculated grapevine plantlets with a plant growth promoting rhizobacterium, *Burkholderia phytofirmans* Strain PsJN. *Appl. Environ. Microbiol.*, **72**, 7246–7252.

Bashan, Y. (1986) Alginate beads as synthetic inoculant carriers for slow release of bacteria that affect plant growth. *Appl. Environ. Microbiol.*, **51**, 1089–1098.

Bashan, Y. and Holguin, G. (1997) Proposal for division of plant growth promoting rhizobacteria into two classes: biocontrol- PGPBs and PGPB. *Soil. Biol. Biochem.*, **30**, 1225–1228.

Batjes, N.H. (1997) A world data set of derived soil properties by FAO-UNESCO soil unit for global modelling. *Soil Use Mgmt.*, **13**, 9–16.

Belimov, A.A., Kojemakov, A.P. and Chuvarlijeva, C.V. (1995) Interaction between barley and mixed cultures of nitrogen fixing and phosphate-solubilizing bacteria. *Plant Soil*, **173**, 29–37.

Benhamou, N., Kloepper, J.W. and Tuzun, S. (1998) Induction of resistance against *Fusarium* wilt of tomato by combination of chitosan with an endophytic bacterial strain: ultrastructural and cytochemistry of the host response. *Planta*, **204**, 153–168.

Bensalim, S., Nowak, J. and Asiedu, S.K. (1998) A plant growth promoting rhizobacterium and temperature effects on performance of 18 clones of potato. *Am. J. Potato. Res.*, **75**, 145–152.

Bent, E., Tuzun, S., Chanway, C.P. and Enebak, S. (2001) Alterations in plant growth and in root hormone levels of lodgepole pines inoculated with rhizobacteria. *Can. J. Microbiol.*, **47**, 793–800.

Bertrand. H, Nalin, R., Bally, R. and Cleyet-Marel, J.C. (2001) Isolation and identification of the most efficient plant growth-promoting bacteria associated with canola (*Brassica napus*). *Biol. Fert. Soil.*, **33**, 152–156.

Bharadwaj, K.K.R. (1975) Survival and symbiotic characteristics of *Rhizobium* in saline-alkali soils. *Plant Soil*, **43**, 377–385.

Burr, T.J., Schroth, M.N. and Suslow, T. (1978) Increased potato yields by treatment of seed pieces with specific strains of *Pseudomonas fluorescens* and *Pseudomonas putida*. *Phytopathology*, **68**, 1377–1383.

Cattelan, A.J., Hartel, P.G. and Fuhrmann, J.J. (1999) Screening for plant growth-promoting rhizobacteria to promote early soybean growth. *Soil Sci. Soc. Am. J.*, **63**, 1670–1680.

Compant, S., Duffy, B., Nowak, J., Cle´ment, C. and Barka, E.A. (2005) Use of plant growth-promoting bacteria for biocontrol of plant diseases: principles, mechanisms of action, and future prospects. *Appl. Environ. Microbiol.*, **71**, 4951–4959.

Cook, R.J., Thomashow, L.S., Weller, D.M., Fujimoto, D., Mazzola, M. and Kim, D.S. (1995) Molecular mechanisms of defense by rhizobacteria against root disease. *Proc. Natl. Acad. Sci. USA*, **92**, 4197–4201.

Deaker, R., Roughley, R.J. and Kennedy, I.R. (2004) Legume seed inoculation technology – A review. *Soil Biol. Biochem.*, **36**, 1275–1288.

Di Simine, D.C., Sayer, J.S. and Gadd, G.M. (1998) Solubilization of zinc phosphate by a strain of *Pseudomonas fluorescens* isolated from a forest soil. *Biol. Fert. Soils*, **28**, 87–94.

Dinakaran, D., Sridhar, R., Jeyarajan, R. and Ramakrishnan, G. (1995) Management of sesamum root rot with biocontrol agents. *J. Oilseeds Res.*, **12**, 262–263.

Dorner, J.W., Cole, R.J. and Blankenship, P.D. (1992) Use of biocompetitive agent to control preharvest aflatoxin in drought stressed peanuts. *J. Food Prot.*, **55**, 888–892.

Elad, Y., Chet, I. and Katan, J. (1980) *Trichoderma harzianum*: A biocontrol agent effective against *Sclerotium rolfsii* and *Rhizoctonia solani*. *Phytopathology*, **70**, 119–121.

Fenton, A.M., Stephens, P.M., Crowley, J., Ocallaghan, M. and O'Gara, F. (1992) Exploitation of gene(s) involved in 2,4- diacetylphloroglucinol biosynthesis to confer a new biocontrol capability to a *Pseudomonas* strain. *Appl. Environ. Microbiol.*, **58**, 3873–3878.

Frommel, M.I., Nowak, J. and Lazarovits, G. (1991) Growth enhancement and development modifications of in vitro grown potato (*Solanum tuberosum* ssp. *tuberosum*) as affected by a nonfluorescent *Pseudomonas* sp. *Plant Physiol.*, **96**, 928–936.

Galal, Y.G.M. (2003) Assessment of nitrogen availability to wheat (*Triticum aestivum* L.) from inorganic and organic N sources as affected by *Azospirillum brasilense* and *Rhizobium leguminosarum* inoculation. *Egypt. J. Microbiol.*, **38**, 57–73.

Garrett, K., Dendy, S., Frank, E., Rouse, M. and Travers, S. (2006) Climate change effects on plant disease: genomes to ecosystems. *Annu. Rev. Phytopathol.*, **44**, 489–509.

Gaudin, V., Vrain, D. and Jouanin, L. (1994) Bacterial genes modifying hormonal balance in plant. *Plant Physiol. Biol.*, **32**, 11–29.

Glick, B.R., Penrose, D.M. and Li, J.P. (1998) A model for the lowering of plant ethylene concentrations by plant growth-promoting bacteria. *J. Theor. Biol.*, **190**, 63–68.

Gull, F.Y., Hafeez, I., Saleem, M. and Malik, K.A. (2004) Phosphorus uptake and growth promotion of chickpea by co-inoculation of mineral phosphate solubilizing bacteria and a mixed rhizobial culture. *Aust. J. Exp. Agric.*, **44**, 623–628.

Gyaneshwar, P., Naresh, K.G., Parekh, L.J and Poole, P.S. (2002) Role of soil microorganisms in improving P nutrition of plants. *Plant Soil*, **245**, 83–93.

Hewitt, K. (1997) *Regions at Risk: A Geographical Introduction to Disasters*. Addision Wesley Longman Limited, Harlow.

Howie, W. and Echandi, E. (1983) Rhizobacteria, influence of cultivar and soil type on plant growth and yield of potato. *Soil Biol. Biochem.*, **15**, 127–132.

Hutchins, S.R., Davidson, M.S., Brierey, J.A. and Brierley, C.L. (1986) Microorganisms in reclamation of metals. *Annu. Rev. Microbiol.*, **40**, 311–336.

Jacobsen, C.S. (1997) Plant protection and rhizosphere colonization of barley by seed inoculated herbicide degrading *Burkholderia* (*Pseudomonas*) *cepacia* DBO1(pRO101) in 2,4-D contaminated soil. *Plant Soil*, **189**, 139–144.

Kang, S.C., Ha, C.G., Lee, T.G. and Maheshwari, D.K. (2002) Solubilization of insoluble inorganic phosphates by a soil fungus *Fomitopsis* sp. PS 102. *Curr. Sci.*, **82**, 439–442.

Kavitha, K., Nakkeeran, S., Chandrasekar, G., Fernando, W.G.D., Mathiyazhagan, S., Renukadevi, P. and Krishnamoorthy, A.S. (2003) Role of antifungal antibiotics, siderophores and IAA production in biocontrol of *Pythium aphanidermatum* inciting damping off in tomato by *Pseudomonas chlororaphis* and *Bacillus subtilis*. In: Z.A. Siddiqui (ed.), *Proceedings of the 6th International Workshop on PGPR*, IISR, Calicut, pp. 493–497.

Kay, S.J. and Stewart, A. (1994) Evaluation of fungal antagonists for control of onion white rot in soil box trials. *Plant Pathol.*, **43**, 371–377.

Kloepper, J.W., Schroth, M.N. and Miller, T.D. (1980) Effects of rhizosphere colonization by plant growth-promoting rhizobacteria on potato yield and development. *Phytopathology*, **70**, 1078–1082.

Kloepper, J.W., Lifshitz, R. and Zablotowicz, R.M. (1989) Free living bacterial inocula for enhancing crop productivity. *Trends Biotechnol.*, **7**, 39–44.

Kloepper, J.W., Hume, D.J., Scher, F.M., Singleton, C., Tipping, B., Laliberté, M., Frauley, K., Kutchaw, T., Simonson, C., Lifshitz, R., Zaleska, I. and Lee, L. (1988) Plant growth-promoting rhizobacteria on canola (rapeseed). *Plant Dis.*, **72**, 42–45.

Kombrink, E. and Somssich, I.E. (1995) Defence responses of plants to pathogens. *Adv. Bot. Res.*, **21**, 1–34.

Lamboy, J.S., Cobb, A.C. and Dillard, H.R. (1996) Evaluation of two biocontrol agents for prevention of gray mold in pruning wounds of greenhouse tomatoes. Cornell Community Conference on Biological Control. http://www.nysaes.cornell.edu/ent/bcconf/abstracts2.html#lamboy.

Lifshitz, R., Kloepper, J.W., Scher, F.M., Tipping, E.M. and Laliberté, M. (1986) Nitrogen-fixing pseudomonads isolated from the roots of plants grown in the Canadian high arctic. *App. Environ. Microbiol.*, **51**, 251–255.

Lixia Yao, Zhansheng Wu, Yuanyuan Zheng, Imdad Kaleem, and Chun Li (2010) Growth promotion and protection against salt stress by *Pseudomonas putida* Rs-198 on cotton. *Eur. J. Soil Biol.*, **46**, 49–54.

Maliha, R., Samina, K., Najma, A., Sadia, A. and Farooq, L. (2004) Organic acids production and phosphate solubilization by phosphate solubilizing microorganisms under *in vitro* conditions. *Pak. J. Biol. Sci.*, **7**, 187–196.

Marjan de Boer, Sluis, I., van Loon Leendert, C. and Baker, P.A.H.M. (1999) Combining fluorescent *Pseudomonas* spp. strains to enhance suppression of fusarium wilt of radish. *Eur. J. Plant Pathol.*, **105**, 201–210.

Maurhofer, M., Keel, C., Schnider, U., Voisard, C., Haas, D. and Defago, G. (1992) Influence of enhanced antibiotic production in *Pseudomonas fluorescens* strain CHA0 on its disease suppressive capacity. *Phytopathology*, **82**, 190–195.

Mayak, S., Tirosh, T. and Glick, B.R. (2004a) Plant growth-promoting bacteria confer resistance in tomato plants to salt stress. *Plant Physiol. Biochem.*, **42**, 565–572.

Mayak, S., Tirosh, T. and Glick, B.R. (2004b) Plant growth-promoting bacteria that confer resistance to water stress in tomato and pepper. *Plant Sci.*, **166**, 525–530.

Nambiar, P.T.C., Rupela, O.P. and Kumar Rao, J.V.D.K. (1988) Nodulation and nitrogen fixation in groundnut (*Arachis hypogaea* L), chickpea (*Cicer arietinum*) and pigeonpea (*Cajanus cajaan* L Mill sp.). In: N.S. Subba Rao (ed.), *Biological Nitrogen Fixation-Recent Developments*, Oxford and IBH Publishing Co. Pvt. Ltd., New Delhi, pp. 53–70.

Nandakumar, R., Babu, S., Viswanathan, R., Raguchander, T. and Samiyappan, R. (2000) Induction of systemic resistance in rice against sheath blight disease by *Pseudomonas fluorescens*. *Soil Biol. Biochem.*, **33**, 603–612.

Nautiyal, C.S. (2003) Plant growth promoting rhizobacteria from stressed ecosystems. In: Z.A. Siddiqui (ed.), *Proceedings of 6th International Workshop on PGPR*, Indian Institute of Spices Research, Calicut, Kerala, pp. 47–49.

Niranjan Raj, S., Deepak, S.A., Basavaraju, P., Shetty, H.S., Reddy, M.S. and Kloepper, J.W. (2003) Comparative performance of formulations of plant growth promoting rhizobacteria in growth promotion and suppression of downy mildew in pearl millet. *Crop Prot.*, **22**, 579–588.

Ordentlich, A., Wiesman, Z., Gottlieb, H.E., Cojocaru, M. and Chet, I. (1992) Inhibitory furanone produced by the biocontrol agent *Trichoderma harzianum*. *Phytochemistry*, **31**, 485–486.

Osburn, R.M., Milner, J.L., Oplinger, E.S., Smith, R.S. and Handelsman, J. (1995) Effect of Bacillus cereus UW85 on the yield of soybean at two field sites in Wisconsin. *Plant Dis.*, **79**, 551–556.

Papavizas, G.C. (1985) *Trichoderma* and *Gliocladium* : Biology, ecology and potential for biocontrol. *Annu. Rev. Phytopathol.*, **23**, 23–54.

Parmar, N. and Dadarwal, K.R. (1999) Stimulation of nitrogen fixation and induction of flavonoid like compounds by rhizobacteria. *J. Appl. Microbiol.*, **86**, 36–44.

Perveen, S., Khan, M.S. and Zaidi, A. (2002) Effect of rhizospheric microorganisms on growth and yield of greengram (*Phaseolus radiatus*). *Indian J. Agric. Sci.*, **72**, 421–423.

Pleban, S., Ingel, F. and Chet, I. (1995) Control of *Rhizoctonia solani* and *Sclerotium rolfsii* in the greenhouse using endophytic *Bacillus* spp. *Eur. J. Plant Pathol.*, **101**, 665–672.

Pleban, S., Chernin, L. and Chet, I. (1997) Chitinolytic activity of an endophytic strain of *Bacillus cereus*. *Lett. Appl. Microbiol.*, **25**, 284–288.

Podile, A.R. and Laxmi, V.D.V. (1998) Seed bacterization with *Bacillus subtilis* increases phenyl alanine ammonia lyase and reduces the incidence of fusarial wilt of pigeon pea. *J. Phytopathol.*, **146**, 255–259.

Ramamoorthy, V., Viswanathan, R., Raguchander, T., Prakasam, V. and Samiyappan, R. (2001) Induction of systemic resistance by plant growth promoting rhizobacteria in crop plants against pests and diseases. *Crop Prot.*, **20**, 1–11.

Richardson, A.E., Simpson, R.J., George, T.S. and Hocking, P.J. (2009) Plant mechanisms to optimize access to soil phosphorus. *Crop Pasture Sci.*, **60**, 124–143.

Rodelas, M.B., González, J., Martínez, M.V., Pozo, C. and Salmeron, V. (1999) Influence of *Rhizobium/Azotobacter* and *Rhizobium/Azospirillum* combined inoculation on mineral composition of faba bean (*Vicia faba* L.). *Biol. Fert. Soil.*, **29**, 165–169.

Sajjad Mirza, M., Mehnaz, S., Normand, P., Prigent-Combaret, C., Moënne-Loccoz, Y., Bally, R. and Kauser, A.M. (2006) Molecular characterization and PCR detection of a nitrogen-fixing *Pseudomonas* strain promoting rice growth. *Biol. Fert. Soil.*, **43**, 163–170.

Salamone, I.E.G. (2000) Direct beneficial effects of cytokinin producing rhizobacteria on plant growth. Ph.D. Thesis, University of Saskatchewan, Saskatoon, SK, Canada

Sankar, P., and Jeyarajan, R. (1996) Seed treatment formulation of *Trichoderma* and *Gliocladium* for biological control of *Macrophomina phaseolina* in sesamum. *Indian Phytopathol.*, **49**, 148–151.

Saravanan, V.S, Kalaiarasan, P., Madhaiyan, M. and Thangaraju, M. (2007) Solubilization of insoluble zinc compounds by *Gluconacetobacter diazotrophicus* and the detrimental action of zinc ion (Zn^{2+}) and zinc chelates on root knot nematode *Meloidogyne incognita*. *Lett. Appl. Microbiol.*, **44**, 235–241.

Schisler, D.A, Slininger, P.J., Behle, R.W. and Jackson. M.A. (2004) Formulation of *Bacillus* spp. for biological control of plant diseases. *Phytopathology*, **94**, 1267–1271.

Sela, S., Schickler, H., Chet, I. and Spiegel, Y. (1998) Purification and characterization of a *Bacillus cereus* collagenolytic/proteolytic enzyme and its effect on *Meloidogyne javanica* cuticular proteins. *Eur. J. Plant Pathol.*, **104**, 59–67.

Selin, C., Habibian, R., Poritsanos, N., Athukorala, S.N.P., Fernando, D. and Teresa, R. (2010) Phenazines are not essential for *Pseudomonas chlororaphis* PA23 biocontrol of *Sclerotinia sclerotiorum* but do play a role in biofilm formation. *FEMS Microbiol. Ecol.*, **71**, 73–83.

Sindhu, S.S., Gupta, S.K. and Dadarwal, K.R. (1999) Antagonistic effect of *Pseudomonas* spp. on pathogenic fungi and enhancement of growth of green gram (*Vigna radiata*). *Biol. Fert. Soil.* **29**: 62–68.

Sindhu, S.S., Suneja, S., Goel, A.K., Parmar, N. and Dadarwal, K.R. (2002) Plant growth promotion effects of *Pseudomonas* sp. on co-inoculation with *Mesorhizobium* sp. Cicer strain under sterile and "wilt sick" soil conditions. *Appl. Soil Ecol.*, **19**, 57–64.

Spiegel, Y., Cohn, E., Galper, S., Sharon, E. and Chet, I. (1991) Evaluation of a newly isolated bacterium, *Pseudomonas chitinolytica* sp. nov. for controlling the root-knot nematode *Meloidogyne javanica*. *Biocontrol Sci. Technol.*, **1**, 115–125.

Sreenivasaprasad, S. and Manibhushanarao, K. (1993) Efficacy of *Gliocladium virens* and *Trichoderma longibrachiatum* as biocontrol agents of groundnut root and stem rot diseases. *Int. J. Pest Manage.*, **39**, 167–171.

Stabb E.V., Jacobson, L.M. and Handelsman, J.O. (1994) Zwittermicin A-producing strains of *Bacillus cereus* from diverse soils. *Appl. Environ. Microbiol.*, **60**, 4404–4412.

Stiju Spaepen, Dobbelaere, Sofie, Croonenborghs and Vanderleyden, J. (2008) Effects of *Azospirillum brasilense* indole-3-acetic acid production on inoculated wheat plants. *Plant Soil*, **312**, 15–23.

Rangarajan, S., Lilly, M., Vasudevan, S.P. and Nair, S. (2003) Biological suppression of rice diseases by *Pseudomonas* spp. under saline soil conditions. *Plant Soil*, **251**, 73–82.

Suslow, T.V. (1980) Growth and yield enhancement of sugarbeets by pelleting seed with specific *Pseudomonas* spp. (abstract). *Phytopathol. News*, **12**, 40.

Tanwar, S.P.S., Sharma, G.L. and Chahal, M.S. (2002) Effect of phosphorus and biefertilizers on the growth and productivity of black gram. *Anal. Agric. Res.*, **23**, 491–493.

Tilak, K.V.B.R., Ranganayaki, N., Pal, K.K., De, R., Saxena, A.K., Nautiyal, C.S., Mittal, S., Tripathi, A.K. and Johri, B.N. (2005) Diversity of plant growth and soil health supporting bacteria. *Curr. Sci.*, **89**, 136–150.

Uknes, A., Mauch-Mani, B., Moyer, M., Potter, S., Williams, S., Dincher, S., Chandler, D., Slusarenko, A., Ward, E. and Ryals, J. (1992) Acquired resistance in *Arabidopsis*. *Plant Cell*, **4**, 645–656.

Upadhyay, A. and Srivastava, S. (2011) Phenazine-1-carboxylic acid is a more important contributor to biocontrol *Fusarium oxysporum* than pyrrolnitrin in *Pseudomonas fluorescens* Strain Psd. *Microbiol. Res.*, **166**, 323–335.

Van Peer, R. and Schippers, B. (1988) Plant growth responses to bacterization with selected *Pseudomonas* spp. strains and rhizosphere microbial development in hydroponic cultures. *Can. J. Microbiol.*, **35**, 456–463.

Vasudevan, P., Reddy, M.S., Kavitha, S., Velusamy, R.S., Raj, A.P., Purushothaman, S.M., Brindha Priyadarsini, V., Bharath Kumar, S., Kloepper, J.W. and Gnanamanickam, S.S.. (2002) Role of biological preparations in enhancement of rice seedling growth and grain yield. *Curr. Sci.*, **83**, 1140–1144.

Venkateswarlu, B., Rao, A.V. and Raina, P. (1984) Evaluation of phosphorus solubilization by microorganisms isolated from Aridisols. *J. Ind. Soc. Soil Sci.*, **32**, 273–277.

Vessey, J.K. (2003) Plant growth promoting rhizobacteria as biofertilisers. *Plant Soil*, **255**, 571–586.

Wilhite, D.A. (2000) Drought as a natural hazard. In: D.A. Wilhite (ed.), *Drought: A Global Assessment,* Routledge, London. pp. 3–18.

Xie, H., Pasternak, J.J. and Glick, B.R. (1996) Isolation and characterization of mutants of the plant growth-promoting rhizobacterium *Pseudomonas putida* GR12-2 that over produce indole acetic acid. *Curr. Microbiol.*, **32**, 67–71.

Yedidia, I., Benhamou, N. and Chet, I. (1999) Induction of defense responses in cucumber plants (*Cucumis sativus* L.) by the biocontrol agent *Trichoderma harzianum*. *Appl. Environ. Microbiol.*, **65**, 1061–1070.

Zaidi, A., Khan, M.S. and Amil, M. (2003) Interactive effect of rhizotrophic microorganisms on yield and nutrient uptake of chickpea (*Cicer arietinum* L.). *Eur. J. Agron.*, **19**, 15–21.

17

Endophytic Fungal Entomopathogens in Insect Pest Management System

V. AMBETHGAR

All India Co-ordinated Research Project on Cashew, Tamil Nadu Agricultural University, Regional Research Station, Vridhachalam-606 001
Cuddalore District, Tamil Nadu, India
E-mail: drva_1965@yahoo.co.in

INTRODUCTION

Since the late 1940s, insect pest control has relied mostly on chemical insecticides, although pest management strategies have been shifting to use of transgenic plants in many industrialized nations. However, the replacement of chemicals with transgenic plants does not represent a fundamental change in approach (Vega *et al.,* 2009). In reality it is "like-for-like" replacement in which the tools are different but the "silver bullet" strategy is the same (Lewis *et al.,* 1996; Welsh *et al.,* 2002). Furthermore, traditional methods of chemically controlling crop pests can be expensive, ineffective, and have a negative impact on both environmental and human health. A true paradigm shift would be a change from a dependence on chemicals to a total system approach (Lewis *et al.,* 1996) or to ecological engineering (Gurr *et al.,* 2004). A basic component of both approaches is a better understanding of the various ecological components in an ecosystem, including biological control agents (Vega *et al.,* 2009). It is well known that biological control as part of integrated pest manage-ment has been suggested as the most sustainable long-term solution.

Additional interests in the biological control of crop pests led to the serendipitous discovery of another form of natural/induced crop protection mechanisms associated with the colonization of plant tissues by certain endophytic microorganisms. These have widespread existence in various tissues and organs of terrestrial and some aquatic plants (Azevedo, 1998; Crutchfied and Potter, 1994), whose infections are inconspicuous and the infected host tissue are at least transiently symptomless (Stone *et al.,* 2000). In otherwords, endophytic micro-organisms are those that inhabit interior of plants, especially leaves, branches and stems, showing no apparent harm to the host (Azevedo, 1998). Fungal endophytes live in intercellular space or inside cells of host plant causing no apparent damage (Saikkonen *et al.,* 1998). A guild of endophytic fungi has been isolated from leaves, twigs, barks and roots of several angiosperm and gymnosperm species (Butin, 1992; Stone *et al.,* 2000). Fungal endophytes receive nutrition and protection from the host plant while host plant may benefit from enhanced competitive abilities and increased resistance to various biotic and abiotic stresses in the same environmental conditions.

Exploitation of beneficial properties of endophytes is of great relevance at an applied level in order to increase production yields by mitigating biotic stress in agricultural crops. More specically, promising results for the control of insect pests have been obtained using fungal entomopathogens, which display a diverse array of adaptations in relation to plant fitness against a range of crop pests (Azevedo *et al.,* 2000). Furthermore, recent evidence shows that many fungi traditionally known as insect pathogens have been isolated as endophytes, including species of *Acremonium, Beauveria* (Pocasangre *et al.,* 2000), *Cladosporium, Clonostachys, Isaria, Metarhizium, Paecilomyces, Phomopsis* and *Verticillium* (Vega *et al.,* 2008), from diverse plant species, including many important agricultural commodities (Azevedo *et al.,* 2000). Fungal endophytes benefit plant by promoting plant growth (Azevedo *et al.,* 2000), improving resistance to multiple stress (Arnold and Lewis 2005), protection from pest and diseases (Vega *et al.,* 2008). This review, outlines the aspects of ecology and diversity of traditional fungal endophytes, addresses the functional role of some promising fungal entomopathogens as endophytes, and highlights their status as a new source of biological control strategy for future pest management programme in diverse cropped ecosystems.

THE ENDOPHYTES

Endophyte is a general term that refers to any organism that lives inside a plant, which include diverse organisms such as mycorrhizal fungi and bacteria (Azevedo *et al.,* 2000). Since the first description of symbiosis as "the living together of dissimilar organisms" (De Bary, 1879), an array of symbiotic lifestyles have been described based on fitness benefits or impacts to macroscopic hosts and microscopic symbionts (Lewis and Clements, 1986). The term endophyte was coined by the German scientist Heinrich Anton De Bary (1884), and is used to define fungi or bacteria occurring inside plant tissues without causing any apparent symptoms in the host (Wilson, 1995). Endophytes inhabit interior of plants and in some instances develop mutualistic relationships with such plants, acting as antagonists to pests and diseases (Azevedo, 1998). Mycorrhyzae and nitrogen-fixing bacteria also live in an intimate relationship with their hosts, and could be considered as endophytic microorganisms (Azevedo *et al.,* 2000), and are distinguished from other root endophytes which possess specfialized external structures like hyphae or mantels. Likewise, nitrogen-fixing endophytic bacteria such as *Rhizobium,* which form external structures called nodules, are distinguished from other root endophytic bacteria. Similar to these organisms, endophytic fungi colonize internally, either intercelluarly or intracellularly, at least for one period of their life cycle, and cause asymptomatic infections in heathy plant tissues (Suryanarayanan *et al.,* 2003). However, the definition of an endophyte is now broadened by many researchers and can include any organisms that live in plant tissue whether neutral, beneficial or detrimental (Sikora *et al.,* 2007). Cultivable endophytic communities can be isolated after surface sterilization of the plant material. Ulrich *et al.* (2008) stated that propagules of endophytes exist in a range of tissue types within a broad range of plants, colonizing the plant systemically, residing latently in intercellular spaces inside the vascular tissue or within cells.

HISTORICAL PERSPECTIVES OF FUNGAL ENDOPHYTES

The fossil records indicate that plants have been associated with endophytic fungi for more than 400 million years and were likely associated when plants colonized land (Krings *et al.,* 2007). Datebacks to 1970s, endophytes were initially considered neither neutral not causing benefits, nor showing any detrimental effect to plants, but later on they started to be better studied (Azevedo *et al.,* 2000).

Although, endophytic microorganisms were already described in the past, they only received considerable attention in the last three decades, when their capacity to protect their hosts against insects-pests and plant pathogens was recognized (Rodriguez *et al.,* 2009). As fungal endophytes started to be better analysed, it became clear that they could confer other important characteristics to plants, such as greater resistance to multiple stress conditions, alteration in physiological properties, production of phytohormones and other compounds of biotechnological interest.

Fungal Endophytes in Crop Protection

An important factor favoring endophyte-infected crop plants in natural and agricultural communities is their enhanced resistance to insect herbivores, nematodes and plant pathogens (Clay, 1996). A great deal of data has been collected on the effects of endophytic microorganisms towards crop pests. In the early 1980s the specialized literature published the first reports showing that endophytic microorganisms, in this case fungi could play an important role inside plants. It was demonstrated that the presence of endophytes in their respective hosts could result in the reduction of insect attacks (Pereira *et al.,* 1999). Landmark reports on the subject started in 1981, less than three decades ago. From 1981 to 1985, which may be considered a historical period to the field, it demonstrated the existence of natural protection against herbivore insects given by endophytic microorganisms (Azevedo *et al.,* 2000; Vicari *et al.,* 2002). Recently, many researchers focus on the study of endophytes as biocontrol agent of plant pests and diseases, and successfully applied endophytes to a better crop protection.

Webber (1981) was probably the first researcher to report an example of crop protection offered by an endophytic fungus, in which the endophyte *Phomopsis oblonga* protected elm trees against the beetle pest *Physocnemum brevilineum*. The author observed that the repellent effect towards the insect was due to toxic metabolites produced by the fungi. This was confirmed four years later by Claydon *et al.* (1985), who showed that endophytic fungi belonging to the *Xylariacea* family synthesize secondary metabolites in hosts of the genus *Fagus* and that these substances affect the beetle larvae. Other earlier observations on the control of insects-pests by endophytic fungi are those of Funk *et al.* (1983) and Lewis and Vaugham (1997) showing protection of the perennial ryegrass *Lolium perenne* L. against the sod webworm. Gaynor and Hunt (1983) and Barker (1987) observed in several ryegrasses that high fungi infection is correlated with a decrease in the attack frequency of the Argentine stem weevil, *Listronotus bonariensis*. They confirmed that these associations may be complex and factors like the level of nitrogen fertilizers may as well affect attack frequency. Lasota *et al.* (1983) had correlated in the white spruce *Picea glauca*, death rate of the *Adelges abietis* (Homoptera) when galls are infected with the endophytic fungus *Cladosporium sphaerospermum*.

With the same insect-plant interaction in New Zealand, Barker *et al.* (1984) and Prestidge *et al.* (1984) observed that plants free of endophytic fungi of the genus *Acremonium* (=*Neotyphodium)* are severely attacked. In *Lolium perenne* and members of the genus *Cyperus*, weight gain and survival of spruce budworm *Spodoptera frugiperda* (Noctuidae), were affected by endophytic fungi like *Balansia cyperi* (Clay *et al.,* 1985; Hardy *et al.,* 1985). Studies related to this insect-pest affecting different species of *Lolium* were also conducted by Latch *et al.* (1985) with two endophytic fungi *Acremonium lolii* and a member of the genus *Gliocadium*. Latch *et al.* (1985) discussed the benefits and risks resulting from the increase in toxin production in relation to insect control and resulting harm to domestic animals that would feed on these plants. Ahmad *et al.* (1985) verified similar effects of the same fungi over the grasshopper, *Acheta domesticus*. Johnson *et al.* (1985), using choice tests showed that insects would feed on endophyte-free *Festuca* plants rather than on infected samples. The role of endophytic fungi in the control of insect pests has been the subject of several reviews (Clay, 1989; Saikkonen *et al.,* 1998; Carroll, 1991; Breen, 1994). These examples of earlier reports in the literature show not only the importance of micro-organisms in controlling insects-pests in agriculture but also the diversity of control mechanisms displayed by endophytes.

More than 100 years of research suggests that most, if not all, plants in natural ecosytems are symbiotic with fungal endophytes (Carroll, 1995; Azevedo *et al.,* 2000). Unlike mycorrhizal fungi that colonize plant roots and grow into the rhizosphre, endophytic fungi reside entirely within plant tissues (Suryanarayanan *et al.,* 2003) and may grow within roots, stems and/or leaves or host tissue senescence (Stone *et al.,* 2000). New discoveries suggest that fungal endophytes can have profound effects on ecology and fitness to assist their host plants for protection against insect herbivores in diverse agro-ecosystems (Vega *et al.,* 2008). In addition to the economical aspects, the study of endophytic microorganisms has strong academic interests, concerning the discovery of new microbial species, mainly when tropical hosts are investigated.

Table 17.1. Criteria to discriminate and characterize fungal endophytes

Criteria	Clavicipitaceous (C)	Nonclavicipitaceous (NC)		
	Class 1	Class 2	Class 3	Class 4
Host range	Narrow	Broad	Broad	Broad
Tissue (s) colonized	Shoot, rhizome, seed	Shoot, root, rhizome	Shoot, flower, fruits	Root
In planta colonization	Extensive	Extensive	Limited	Extensive
In planta biodiversity	Low	Low	High	Unknown
Transm-ission	Vertical, horizontal	Vertical, horizontal	Horizontal	Horizontal
Fitness benefits*	Nonhabitat-adapted	NHA and Habitat-adapted	Nonhabitat-adapted	Nonhabitat-adapted
Endophytic genera	*Balansia*	*Arthrobotrys*	*Phyllosticta* sp.	*Chloridium*
	Cordyceps	*Colletotrichum*	*Phytophthora*	*Leptodontidium*
	Claviceps	*Curvularia*	*Ustilago*	*Phialocephala*
	Neotyphodium sp.	*Phoma* sp.		*Phialophora*

*Nonhabited-adapted (NHA) benefits from such as drought tolerance and growth enhancement are common among endophytes regardless of the habitat of origin. Habitat adapted (HA) benefits result from habitat specific selective pressures such as temperature, pH and salinity

Source: Modified from Rodriguez *et al.* (2009).

DIVERSITY OF FUNGAL ENDOPHYTES

Endophytic fungi are taxonomically and biologically diverse but all share the character of colonizing internal plant tissues without causing apparent harm to their host (Wilson, 1995). To date, mainly two discrete groups of fungal endophytes have been recognized based on phylogeny and life history traits: the clavicipitaceous (C) and the nonclavicipitaceous (NC) endophytes (Rodriguez *et al.,* 2009). They reflect differences in evolutionary relatedness, taxonomy, plant hosts, and ecological functions (Table 17.1). The clavicipitaceous endophytes (C-endophytes) infect some grasses, and the nonclavicipitaceous endophytes (NC-endophytes) can be recovered from nonvascular plants, ferns and allies, conifers and angiosperms (Rodriguez *et al.*, 2009). In all these cases, fungal growth is characterized by an implicit irregular development due to the filamentous nature of mycelia, whose complexity increases in structural substances, such as the interior of plant tissues (Macia-Vicente *et al.*, 2008). These obstructions are enhanced due to uneven distribution of nuclei (from zero to several thousands per cell), physiological activity or vitality of hyphae (Gamper *et al.*, 2008), and influence of competition amoung species.

To date, most studies on the effects of fungal endophytes on insect herbivores have concentrated on C-endophytes which systemically infect mostly temperate grasses in the Poaceae, Juncaceae, and Cyperaceae (Clay, 1989; Breen, 1994); or have examined C- and NC-endophytes together (Arnold and Lutzoni, 2007). In contrast, entomopathogenic fungal endophytes associated with tropical cultivated plants of both monocotyledonous and dicotyledonous angiosperms are poorly investigated.

HOST RANGE OF FUNGAL ENDOPHYTES

Fungal endophytes are ubiquitous and are dominated by Ascomycota (Arnold and Lutzoni, 2007). Fungal endophytes have been detected in hundreds of plants, including many important agricultural commodities such as wheat (Larran *et al.*, 2002a), rice (Tian *et al.*, 2004), maize (Cherry *et al.*, 2004; Arnold and Lewis, 2005), bananas (Pocasangre *et al.*, 2000; Akello *et al.*, 2007), date palms (Go´mez-Vidal *et al.*, 2006), soybeans (Larran *et al.*, 2002b), potato, tomatoes (Larran *et al.*, 2001), cocao (Posada and Vega, 2006), coffee (Posada *et al.*, 2007), cashew (Ambethgar *et al.*, 2010) and fodder grasses. Several roles have been ascribed to fungal endophytes, including various effects on plant performance and different levels of activity by providing increased drought tolerance (Arechaveleta *et al.*, 1989), deterance of insect herbivores (Breen, 1994; Saikkonen *et al.*, 1998; Clement *et al.*, 1992), protection against plant parasitic nematodes (West *et al.*, 1988; Elmi *et al.*, 2000), and resistance against plant pathogens (Wicklow *et al.*, 2005). However, most reports on the effects of endophytes on insect herbivores have concentrated on turf and agronomic grasses infected with endophytic clavicipitalean fungi (Ascomycota: Hypocreales: Clavicipitaceae), which systemically infect mostly grasses in the Poaceae, Juncaceae, and Cyperaceae (Clay, 1989; Breen, 1994). For example, *Neotyphodium* 3-infected perennial ryegrass (*Lolium perenne*) (Prestidge and Gallagher, 1988; Lewis and

Vaugham, 1997) and tall fescue (*Festuca arundinacea* Schreb.), have been shown to have negative effects on over 40 insect species in six orders (Clement *et al.,* 1995; Arnold *et al.,* 2003). Clement *et al.* (2005) reported different effects on two aphids (bird-cherry oat aphid, *Rhopalosiphum padi* and rose grass aphid, *Metopopophium dirhodum* and the wheat stem sawfly (*Mayetiola destructor*) exposed to different wild barleys infected with *Neotyphodium.* Fewer studies have explored this relationship in nongrass systems. Jallow *et al.* (2004) reported drastic negative effects on larvae of *Helicoverpa ar*migera Hubner reared on tomato plants infected with a nongrass endophyte, *Acremonium stric*tum W. Gams.

MODE OF ACTION AND TRANSMISSION

The traditional mode of infection of fungal entomopathogens involves spore deposition on the insect cuticle followed by formation of a germ tube, which through enzymatic and mechanical action penetrates the cuticle (Hajek and St. Leger, 1994) and reaches the hemocoel to differentiate into hyphal bodies. Hyphal growth in hemocoel causes tissue damage and nutrient depletion. Some fungi are also known to produce metabolites which cause toxicosis in insects.

Conversely, fungal endophytes invade the plants via the epidermis, and live as mycelium in the intercellular spaces inside the plants. Fungal endophytes colonize systemically the tissues of various organs through the entire growing season. Within the aerial plant body, endophyte hyphae are most dense in the leaf bases surrounding the intercalary meristems. Under a microscope, the mycelium appears to twist as the fungus grows between the cells. Mycelium can be, found in leaf sheaths, in developing seedheads and seeds. But, they do not reside in roots although they can be found in rhizomes and stolons of several hosts (Luttrell and Bacon, 1977).

The mode of action of most endophytic fungi was based on the capability to render the plant unpalatable to several types of pests like aphids, grasshoppers and beetles (Carroll, 1988). The capacity of endophytic fungus to repel insects, induce weight loss, growth and development reduction and even to increase pest death rate was correlated with toxin production (Bacon *et al.,* 1977; Clark *et al.,* 1989). The other possible mode of action of endophytic fungi such as *B. bassiana* could be result of feeding deterrence or antibiosis (Vega *et al.,* 2008).

Dynamics of Endophytic Fungal Pathogenesis

Changes in levels of endophyte infection occur in many host populations over time. Early workers noted the tendency for *Atkinsonella* and *Epichloe* to be more prevalent in older populations (Large, 1954). The infection levels of *Acremonium* endophyte in *Lolium* species were higher in older pastures than in younger pastures, and in natural populations compared to domesticated varieties (Cunningham, 1958). Increasing levels of endophyte infection with increasing age of the host population may be due to contagious spread of the fungus, increased survival and/or reproduction of infected plants, or both, depending upon the ability of the endophyte to sporulate and the host plant to produce seed (Lewis and Clements, 1986). For seed-borne *Acremonium* endophytes contagious spread is impossible; the fungus does not produce spores and never occurs outside the interior plant body (Neill, 1941). Increasing infection over time implies a fitness benefit to infected plants, which must live longer or produce more offspring than do endophyte-free plants (Latch *et al.,* 1985).

Reproduction and Dispersal

Poineering research on reproduction, transmission and dispersal of fungal endophytes has been reviewed earlier (Large, 1954). Sexual endophytes produce fruiting bodies (stromata) on the leaves or inflorescences of host plants coincident with the hosts' flowering period. Fruiting bodies generally produce both conidia and ascospores (Clay, 1986). Most of these fungal endophytes are dispersed in one of three general ways. Sexual species that fruit on their hosts can infect new plants contagiously. Contagious spread is not epidemic as in many plant pathogens, but it has been observed in the field (Large, 1954; Clay, 1989). Dispersal through the seeds of its host is the only means of spread for asexual *Acremonium* endophytes, but it also occurs in plants infected by *Atkinsonella* and *Epichloe* (Clay, 1986). Both sexual and asexual endophytes can also be disseminated through vegetative structures of host plants including rhizomes, tillers, tubers, and viviparous plantlets (Clay, 1989). Population dynamics of endophyte infection within populations could be modeled simply by equating infection with a maternally inherited gene with two alternative alleles of different selective value.

Transmission of Fungal Endophytes

The route of endophytic fungal transmission could be mainly in two ways: horizontal (non-systemic) and vertical (systemic) transmission. Systemic endophytes are transmitted vertically via the seeds, while non-systemic endophytes are transmitted horizontally with host colonization arising from the surrounding environment (Dai *et al.*, 2010). Endophytic fungi from the members of Clavicipitaceae (Ascomycota) are transmitted vertically, from mother to off-spring through seeds, as reviewed by Clay and Schardl (2002). The plant benefits by establishing an insect defence system. Fungal endophytes grow as elongate, convoluted, sparsely branched hyphae running through intercellular spaces parallel to the long axis of leaves and stems (Clay, 1986). As a common way of transmission, endophytic fungal vertical transmission could be described as seed reproduction, which is the same as the reproduction of most plants. At flowering, endophytes (e.g., *Acremonium*) grow into ovules and become incorporated into seeds. Most endophytes are able to absorb nutrients from freely available materials found in the intercellular spaces (Clay, 1989).

EFFECTS AND FUNCTIONS OF FUNGAL ENDOPHYTES

Recently, biology and function of fungal endophytes in nature have been extensively reviewed (Rodriguez *et al.*, 2009). Prior studies have shown that endophyte infected plants are more vigorous, more adaptable to severe conditions, more resilient after stressful situations and more resistant to insects and fungal pathogens than noninfected plants. Other benefits of these associations might include nitrogen fixation from the air, enhanced nutrient uptake, production of plant hormones that stimulate growth, enhanced grain quality, heightened resistance to abiotic stresses. According to Bacon *et al.* (1975), alkaloids produced by endophytes within plant tissues are in part responsible for the plant resistance to insect herbivores.

Toxins Production by Endophytic Fungi

Endophyte-infected graminaceous plants, particularly grasses contain a variety of alkaloids. These alkaloids include ergot alkaloids, loline alkaloids, lolitrems, and peramine (Johnson, *et al.*, 1985). Most of these alkaloids, or their precursors, have been detected in liquid cultures of some endophytic fungi (Ahmad, *et al.*, 1985). Further, feeding trials using pure alkaloids have revealed insecticidal activity against several herbivores (Johnson, *et al.*, 1985). Alkaloid content of endophyte-infected grasses is known to vary among host species, plant parts (leaf blades, sheaths, flowering culms), within the same parts of different age, at different times of the growing season, and with fertilization and grazing history (Luttrell and Bacon, 1977). In adult plants of tall fescue, alkaloid concentrations are higher in young than in old leaves and in leaf sheaths than in leaf blades (Hardy, *et al.*, 1985). In tall fescue the greatest concentration of loline alkaloids is in the seeds, suggesting a role in preventing seed predation (Cheplick and Clay, 1988). These patterns coincide with endophyte hyphal density, as well as with the nutritive value of the tissues.

Bacon *et al.* (1977) established for the first time a correlation between an endophytic fungi, *Epichloë typhina* and the toxicity of its host, *F. arundinacea* to herbivorous domestic mammals. It is now a fact that endophytic fungi produce several toxins and that these substances confer host protection against different herbivores (Bacon and Hills, 1996). For instance, Miller (1986) showed that the protection of Canadian fir against the spruce budworms resulted from the production of toxic secondary metabolites by endophytic fungi. Prestidge and Gallagher (1988) established a relationship between the presence of the fungus *A. lolii* in *Lolium perenne* and the growth, survival and feeding behaviour of *Listronotus bonariensis* larvae. In this case, the reduction in insect attacks towards infected plants was due to production of a strong toxin, lolitrem B, also toxic to mammals. This toxin, once added to insect diets, reduced insect growth and survival. Its assimilation occurs by ingestion but not by absorption through the insect integument.

In subsequent studies, Siegel *et al.* (1989) verified the presence of the alkaloids N-formil, N-acetil loline, peramine, lolitrem B and, ergovaline during plant attack by aphids. Grasses infected with fungal endophytes such as *Acremonium* spp. and *E. typhina* contained alkaloids, mainly peramine and ergovaline. Peramine, lolitrem B and ergovaline were found in *Lolium* and *Festuca* infected with *A. coenophialum* and *A. lolii* and in *Festuca longiflora* infected with *E. typhina*. Individuals of *Rhopalosiphum padi* and *Schizaphis graminum* did not survive in grasses containing the alkaloid loline. On the other hand, ergovaline did not affect both insect species. Riedell *et al.* (1991) observed that the methanolic extracts of *F. arundinacea* infected with *A. coenophialum* contain lolines of fungal origin, which are able to alter feeding behaviour and weight of insect-pests. Diets amended with extracts containing loline derivatives reduced weight and altered certain behaviours of *Spodoptera frugiperda* and *Ostrinia nubilalis*, notably in the first insect.

Patterson *et al.* (1992) observed in *Lolium* and *Festuca* that the production of alkaloids by *Acremonium* reduced attacks of the Japanese beetle, *Popilla japonica*. Other works find that ambient variations may affect toxin production by endophytic fungi. Breen (1992) verified that changes in temperature and seasons modify toxin levels in the plant. This is the case of the peramine toxin, produced in *L. perenne* by *A. lolii*. The antixenosis towards the aphid *S. graminum* is dependent on peramine and the endophyte concentration so, a natural consequence is that the effect on the insect-pest will also be modified. The specificity between certain endophytic fungi and host plants led Leuchtmann (1992) to suggest a further study on the occurrence of physiological races in endophytic fungi that could be important to the development of new biological controls.

Findlay *et al.* (1997) carried out surveys aiming to the discovery of new toxins useful to insect control. Two new active toxins against the Spruce budworm, *Choristoneura fumiferana* were found in an unidentified endophytic fungus infecting the wintergreen *Gaultheria procumbent*. Alkaloids from *Neotyphodium lolii* and *L. perenne* are capable of altering insect behaviour. Several of these alkaloids were added to the diet of adults of the coleopteran *Heteronychus aratur*. Peramine, lolitrem B, lysergol-type alkaloids, festuclavine and lisergic acid showed no effects on the insect. Ergovine showed moderate effects whereas ergotamine, ergovaline from the ergot-type alkaloid family seem to be responsible for the plant resistance (Ball *et al.,* 1997). Endophytic isolates of *Neotyphodium* sp. produced N-formilonine and a paxiline analogous in the host *Echinopogon ovatus* (Miles *et al.,* 1998) and these compounds show insecticidal activity against *L. bonariensis* and other insects.

Majority of works related to toxin production was performed in grasses. However, Calhoun *et al.* (1992), for the first time, identified toxic products synthesized by endophytic fungi in woody plants and that were able to modify growth and death rates in larvae of the spruce budworm, *C. fumiferana* feeding on balsam fir. The endophytes in this case were identified as *Phyllosticta* and *Hormonema dematioides* and the toxic compounds were mainly heptelidic acid and rugulosine. Bills *et al.* (1992) also detected the existence of tremorgenic toxins in tropical woody plant infected with an endophytic fungus from the genus *Phomopsis*.

Fungal strains also differ in the amounts and kinds of alkaloids produced in pure culture (Rowan and Gaynor, 1986) suggesting that between-plant variation could exist in alkaloid content resulting from genetic differences between endophyte strains. The bitter taste of many alkaloids may allow herbivores to discriminate among plants (Bush *et al.,* 1997) and avoid endophyte-infected individuals, where a species is completely endophyte infected herbivores may shift to other species (Cheplick and Clay, 1988). Populations differing historically in the level of herbivory would provide evidence as to whether selection favors endophyte-infected plants or plants infected with high alkaloid-producing endophyte strains where herbivory are common or chronic.

ECOLOGICAL MODIFICATIONS AND ENDOPHYTIC INTERACTIONS

The expression of resistance in endophyte endowed plants against herbivore insects may be affected by complex of factors, covering soil fertility and endophyte genotype (Christensen *et al.,* 1991), endophyte concentration (Breen, 1992), active amounts of allelochemicals, plant genotype, etc.. Hydric stress, temperature, soil pH, insect-pest resistance and other factors may also affect the endophyte concentration and toxin production (Breen, 1992). In plants of *F. arundinacea* and other grasses infected or not with *A. coenophialum,* attacks of two coleopterans were less frequent in the first insect instars, but afterwards there were variations (Potter *et al.,* 1992), indicating that many factors play a role in insect control. Breen (1994) idealized a model with all of these interactions affecting the increase or decrease in plant resistance to insect-pests in the presence of endophytes.

Impact of Soil Nutrients in Relation to Endophytism

Concerning alterations due to nutrients in the soil, especially nitrogen fertilizers, there are studies showing the existence of this interaction, which is very significant. Arechavaleta *et al.* (1989) verified the interaction between nitrogen fertilization, pests and the endophytic fungus *A. coenophialum,* which showed that in a general manner, insects like *S. frugiperda* developed better in nitrogen fertilized plants not infected by the endophyte. However, considering this and other insects, the occurrence of *A. coenophialum* blocks insect development. Bultman and Conard (1998) verified interactions among several factors like nutrient levels and plant damage during endophytic fungi control of *S. frugiperda* in the host *F. arundinacea.* The insect pupal mass was negatively influenced by the endophyte in low nutrient levels but the effect was negligible in plants highly fertilized (Davidson and Potter, 1995; Miller *et al.,* 2002). This study showed again that these interactions are complex and influenced by several factors, in this case by fertilizer application.

Richardson *et al.* (1999) verified that in hydroponic conditions, the Chewings fescue *F. rubra* subspecies *commutata* showed differences in alkaloid contents in respect of endophyte colonization and amounts of several nitrogen sources. Fluctuation in the contents of the alkaloids ergovaline and peramine in colonized plants was verified, which consecutively alter insect resistance. In general, insect attacks were reduced in hosts colonized by endophytes, because the plants were less chosen by the pests. However, different degrees of insect attacks were observed according to the plant associations (Azevedo *et al.,* 2000). These studies demonstrated that plant interactions would change the effects of endophytic fungi in insect control.

Host Plant Physiology and Endophytism

In certain plants, toxin production is not solely a fungal function. There are fungus-plant interaction factors acting on this production and toxins may be produced and dissipated throughout the host tissues (Azevedo *et al.,* 2000). Evidently, it may result in insect control even if the endophyte is not present in the attacked sites. The development and nutrition of the diamondback moth, *Plutella xylostella* larvae were studied in brassicas. Larvae were affected by the ingestion of leaves, before the endophytic fungus *Acremonium alternatum* reached these leaves. There were also differences in feed efficiency of males and females. The results may be related with the phytoesterol metabolism in the plants and indicate that endophytic fungi in the soil, even through a tenuous interaction, alter insect-plant interactions (Raps and Vidal, 1998). In addition, levels of alkaloids and other toxins may be altered not only quantitatively but also qualitatively, depending on the plants' physiological state (Barker, 1987). In *L. perenne* infected by *Neotyphodium lolii,* peramine production plays a major role in controlling the Argentine stem weevil *Listronotus bonariensis.* Ball *et al.* (1997) verified that with plant ageing, the amount of peramine decreases in leaves and reaches the lower levels during the inflorescence phase in eight fungus-plant interactions. White *et al.* (1993) reported that interactions may occur like the increase of stroma induced by the endophytic fungus *E. typhina* in host plants, resulting in an increase in stroma-mediated transpiration and consecutively reduction in herbivorous insect attacks.

RECOMBINANT DNA TECHNOLOGY

The presence of endophytic microorganisms in plants may be of extreme importance to host protection in the control of plant diseases caused by bacteria and fungi (Azevedo *et al.*, 2000). In relation to insects, some uses of endophyte inoculation have been mentioned before, such as the introduction into seeds or plants. It is also possible to monitor the existence of toxins inside plants by observing the presence of insects, especially aphids through molecular techniques (Eichenser and Dahlman, 1992). The recombinant DNA technology has been applied as a tool to improve endophytic microorganisms (Pereira *et al.*, 1993), aiming to the introduction of new characteristics of agronomic interests as biological control of pests. Fahey (1991) described the first work directed at the introduction of a heterologous gene in an endophytic microorganism with the purpose of insect control. He described the major steps in the construction of an endophytic bacterium with the purpose of insect control. Research concerning engineered endophytes for biocontrol is concentrated on bacteria, mainly with the endophyte *Clavibacter xyli* subsp. *cynodontis*, a gram positive, xylem-inhabiting bacterium, capable of colonizing several plant species. This was achieved through the secretion of an insecticidal toxin in the host plant. In endophytic fungi, this type of study is restricted to the development of transformation systems, introduction of reporter genes and inoculation into plants for colonization studies.

At the beginning of the 90s, Van-Heeswijck and McDonald (1992), in a review on endophytic fungi from the genus *Acremonium*, launched the idea of using engineered endophytic fungi in the control of insects and diseases affecting *L. perenne.* Murray *et al.* (1992) introduced into a lolitrem B-deficient strain of *Acremonium* (designated 187B) a ß-glucuronidase (*Gus*) gene. The transformation protocol yielded 700-800 transformants/mg DNA and transformants were successively introduced into *Lolium* where *Gus* activity was observed. Similarly, Tsai *et al.* (1992) developed a transformation system for *A. coenophialum.* Several genes from endophytic fungi and related to toxin production are now being cloned and studied in depth. Wang *et al.* (1999) cloned a gene coding for a DMATSynthase (dimetilail tryptophane synthase) that is responsible for the first step in the synthesis of ergot incited by the fungus *Claviceps purpurea*. Genes related to the latter were identified in the endophytic fungi *Balansia obtecta* and *Neotyphodium* spp., known to produce ergobalansine and ergovaline, respectively. Tudzynski *et al.* (1999) also cloned a gene (DMATS) of *C. purpurea* involved in the ergot synthetic pathway *in planta*. Young *et al.* (1999) found in *Epichloë* a cluster containing genes for the synthesis of mycotoxins like paxiline and lolitrem B. Efficient transformation systems have been established for some of these endophytic microorganisms as well as protocols for reintroducing these microorganisms into plants (Longo, 1996; Pamphile *et al.,* 1997; Glienke-Blanco, 1999; Pereira *et al.,* 1999). These studies are among the first involving genetic manipulation of endophytic microorganisms from tropical plants.

THE ENDOPHYTIC FUNGAL ENTOMOPATHOGENS

Various genera of fungal entomopathogens have been isolated as endophytes in several different plants (Table 17.2). Some fungal entomo-pathogens have been reported as naturally occurring endophytes while others have been introduced into the plant using different techniques. For example, fungal species in the genera of *Cladosporium* containing insect pathogenic species (Abdel-Baky and Abdel-Salam, 2003), have been reported as natural endophyte in *Festuca* (An *et al.,* 1993), several Ericaceae (Okane *et al.,* 1998), various grasses (Dugan and Lupien, 2002), mangroves (Suryanarayanan *et al.,* 1998; Ananda and Sridhar, 2002), wheat (Larran *et al.,* 2002a), oak (Gennaro *et al.,* 2003) and in apples (Cammatti-Sartori *et al.,* 2005). However, the investigators did not discuss the possible mode of action for these endophytic entomopathogenic fungi. The white muscardine fungus, *Beauveria bassiana* (Balsamo) Vuillemin (Ascomycota: Hypocreales) has been reported as an endophyte in maize (Bing and Lewis, 1991; Lomer *et al.,* 1997; Cherry *et al.,* 2004; Wagner and Lewis, 2000; Arnold and Lewis, 2005), potato, cotton, and tomato (Ownley *et al.,* 2004), on the cocoa relative *Theobroma gileri* (Evans *et al.,* 2003), in the bark of *Carpinus caroliniana* Walter (Bills and Polishook, 1991), in seeds and needles of *Pinus monticola* Dougl. ex. D. Don (Ganley and Newcombe, 2005), in opium poppy (Quesada-Moraga *et al.,* 2006), on date palm (Go´mez-Vidal *et al.,* 2006), in bananas (Akello *et al.,* 2007), and in coffee (Posada *et al.,* 2007). Other entomopathogenic fungi have also been reported as artificial endophytes: *Verticillium* (=*Lecanicil-lium*) *lecanii* (Zimm.) Viegas in an Araceae (Petrini and Dreifuss, 1981); *V. lecanii* and *Paecilomyces farinosus* (Holmsk) Brown & Smith (=*Isaria farinosa*) in the bark of *C. caroliniana* (Bills and Polishook, 1991); *Paecilomyces* sp. in *Musa acuminate* (Cao *et al.,* 2002) and in rice (Tian *et al.,* 2004); and *Paecilomyces varioti* Bain. in mangroves (Ananda and Sridhar, 2002).

Table 17.2. Examples of some fungal endophytes from entomopathogenic genera isolated from various plants against insects pests

Fungal endophyte	Host plant	Target pest	Selected references
Acremonium alternatum	*Coffea arabica*	*Hypothenemus hampei*	Posada *et al.* (2007)
Acremonium lolii	Tomato, *Festuca* species	*Helicoverpa armigera*, aphids	Jallow *et al.* (2004), Latch *et al.* (1985)
Acremonium spp	*Lolium* spp. & *Festuca* sp.	Grasshopper, *Acheta domesticus*	Ahmad *et al.* (1985)
Gliocadium	Ryegrass, *Lolium perenne* L	Sod webworm, Argentine stem weevil	Gaynor and Hunt (1983)
Cladosporium sphaerospermum	White spruce, *Picea glauca*	Gall insect, *Adelges abietis*	Lasota *et al.* (1983)
Balansia cyperi	Cyperus, *Cyperus* species	Spruce budworm, *Spodoptera frugiperda*	Clay *et al.* (1985)
Beauveria bassiana	Wheat, Corn, Bananas, Cacao	Stem borers, nematode pests	Cherry *et al.* (2004)
Beauveria bassiana	*Coffea arabica*	*Hypothenemus hampei*	Posada *et al.* (2007)
Beauveria brongniartii	*Coffea arabica*	*H. hampei*	Posada *et al.* (2007), Vega *et al.* (2009)
Cladosporium spp	*Coffea arabica*	*H. hampei*	Posada *et al.* (2007), Vega *et al.* (2009)
Clonostachys rosea	*Coffea arabica*	*H. hampei*	Posada *et al.* (2007), Vega *et al.* (2009)
Lecaniciallum lecanii	Arecanut	Spindle bug	Petrini and Dreifuss (1981)
Metarhizium anisopliae	Sugarbeet	Scarabeid grubs	Fuller-Schaefer *et al.* (2005)
Paecilomyces fumosoroseus	*Coffea arabica*	*H. hampei*	Posada *et al.* (2007)
Paecilomyces farinosus	*Carpinus caroliniana*	*H. hampei*	Bills and Polishook (1991)
Paecilomyces spp.	Rice, Annual grasses	Stem borers	Tian *et al.* (2004)
Phomopsis oblonga	Elm trees	Leaf beetle, *Physocnemum brevilineum*	Webber (1981)
P. oblonga	Ryegrass, *Lolium perenne* L.	Sod webworm	Funk *et al.* (1983)

Research on *Beauveria bassiana* as a maize endophyte suggests that the reduced tunneling of maize stem borer, *Ostrinia nubilalis* could be due to the presence of fungal metabolites that cause feeding deterrence or antibiosis (Vega *et al.*, 2008). Previous studies by Bing and Lewis (1993) that *O. nubilalis* individuals that feed on endophytic maize plants showed mere 2.5% mycosis and 1.7% mycoses on plants with no *B. bassiana* indicating lack of conidia within the maize plants. The lack of *B. bassiana* conidia in endophytic plants also suggests a mode of action involving feeding deterrence or antibiosis rather than direct fungal infection (Wagner and Lewis, 2000). A study by Cherry *et al.* (2004) in Africa supports the feeding deterrence/antibiosis hypothesis. They observed reduced tunneling of the maize stem borer *O. nubilalis* in plants treated with *B. bassiana.* In plants injected with *B. bassiana*, the larvae were lighter than in the control plants suggesting they did not feed as much, supporting the deterrence/antibiosis hypothesis.

Furthermore, a number of recent discoveries suggest that current approaches in insect control with fungal entomopathogens require revision. For instance, rhizosphere competence by strains of *Metarhizium anisopliae* is dependent on the plant community and not necessarily the presence of an insect host (Hu and St. Leger, 2002; Fuller-Schaefer, *et al.*, 2005), and strains of *Beauveria bassiana* exist as endophytes in various plant species (Ownley *et al.*, 2004) and exibit the potential for pest suppression. Vega *et al.* (2009) provides increasing evidence that *Beauveria, Metarhizi*um and related mitosporic genera can act as mycoparasites and plant endophytes, as well as interact with plant roots.

INOCULATION OF FUNGAL ENTOMOPATHOGENS INTO PLANTS

It is widely known that the existence of entomopathogens under natural environments is able to cause diseases in insects. Fungal entomopathogens such as *Beauveria bassiana, Metarhizium anisopliae* and others are often used in the biological control of agricultural insect-pests. Examples may be found in Hajek and St. Leger (1994), Melo and Azevedo (1998), Ambethgar (2001) and Vega *et al.* (2009). Many of these fungal entomopathogens can be placed inside plants as artificial endophytes. For an endophyte strain to be good antagonist against target pests, it must be present in the plant tissues at the time the plants are attacked by the pests. Endophytes, therefore, need to occur at high frequencies and be able to persist in the plant after inoculation. Establishment of endophytes will change the growth, physiology and metabolism of other beneficial microbes, which could break the balance between endophyte and host plant (Gao *et al.*,

2010). Even in situations where they are not present in the plant, artificial inoculation techniques may circumvent this obstacle. Endophytic fungi that do not colonize insects and cause disease, may be used as a natural insect control as well as an effective insect control if they are artificially inoculated into plants and display characteristics against the pests.

Development of Artificial Endophyte Inoculation Techniques

Until 1990s, works concerning artificial inoculation and establishment of endophytes were scarce and, however, Clay (1989), in a review on the potential of insect control by endophytic fungi, stressed that insect biocontrol may be improved by the development of artificial inoculation techniques, as those developed by Leuchtman *et al.* (1985) and Latch *et al.* (1985). As endophytic fungi, notably members of the genus *Acremonium*, are able to control some insect-pests, it is important to know if they can be inoculated in endophyte-free plants and, if they can transmit to the host the capacity to resist to certain pests. That was done with success by Koga *et al.* (1997), by infecting *F. arundinacea* and *L. perenne* seeds with the endophytic fungus *Acremonium* and, as a result, obtained plants resistant to the bluegrass webworm, *Parapediasia teterrella*.

In corn plants, this procedure was done after stem injections (Bing and Lewis, 1991) to verify the potential of entomopathogenic species, notably *B. bassiana* as endophytes to control the European corn borer *Ostrinia nubilalis*. In this case, *B.bassiana* was inoculated by injection and aspersion and was detected endophytically, colonizing several corn tissues and blocking the insect attack in some of its developmental stages (Bing and Lewis, 1993). Injection resulted in 95% colonized plants and aspersion in 98.3%. An interesting finding was that in 33.3% of the control plants (not treated), the fungus was found, indicating that the fungus is a natural endophyte. However, the process by which the fungus invades the plant was not determined, but the movement was shown to be passive, occurring in the vascular system (Bing and Lewis, 1993). It was also shown that the infected insect does not transmit the fungus because plants free of attacks could contain the fungus in an endophytic state.

Bing and Lewis (1993), also refer to the fact that the entomopathogenic fungus *B. bassiana* has been used for a long time in the control of the European corn borer and this fungus is isolated as endophyte from corn. This could explain the pest suppression in certain seasons, correlated with the presence of *B. bassiana* in the corn senescent phase (Bills and Polishook, 1991). Futhermore, in artificial inoculation, the amount of inoculum required to protect the host on a per hectare basis is minimized with the precise targeting of endophyte treatments (Gao *et al.*, 2010). This should make the appr-oach attractive to the biotechnology industry looking for alternatives to traditional pesticides, as targeting the insect and nematode infections assures improved efficacy. An important point, at least in grasses, resides in the fact that several fungi species may be transmitted by seeds, where the biocontrol becomes able to be passed to the next generation. There are works showing the importance of endophytic fungi colonizing seeds in insect control. Knoch *et al.* (1993) observed in *F. arundinacea*, in which seeds, that ants often prefer to collect endophyte-free seeds, transmit the endophytic fungus *A. coenophialum* avoiding those infected. Even seeds infected with the fungus are later discarded, favouring seed dispersion and germination.

Model System 1: Maize-*Beaveria bassiana*-European Corn Borer

The pioneering work on entomopathogenic endophytes was conducted using maize (*Zea mays* L.), *B. bassiana,* and the European corn borer, *Ostrinia nubilalis* as a model system (Lewis and Cossentine, 1986). Season-long suppression of this insect, measured as reduced tunneling of *O. nubilalis*, was achieved by applying *B. bassiana* in an aqueous (Lewis and Cossentine, 1986) or granular (Lewis and Bing, 1991) formulation on corn plants. Lewis and Bing (1991) suspected that *O. nubilalis* suppression was due to establishment of *B. bassiana* in the plant. In a subsequent study, Bing and Lewis (1991) used granular formulations and injections of *B. bassiana* spore suspensions to obtain season-long suppression of *O. nubilalis* and concluded that this was due to *B. bassiana* becoming established as an endophyte in maize. Bing and Lewis (1993) report surface-sterilization of tissues and subsequent recovery of *B. bassiana*, thus providing evidence of an endophytic relationship. However, no mycosed insects were reported (Bing and Lewis, 1993), which suggests that the reduced tunneling reported in all the papers (Lewis and Cossentine, 1986; Lewis and Bing, 1991; Lewis *et al.*, 1996) could be a result of feeding deterrence or antibiosis. The possible mode of action for endophytic *B. bassiana* against *O. nubilalis* was not discussed in any of these studies or in subsequent studies (Lewis *et al.*, 1996). Further, it is not clear how the insects that exhibited mycosis became infected since no conidia were observed internally in the plants

(Wagner and Lewis, 2000) thus indicating that the infective propagulum was not available to infect insects. Wagner and Lewis (2000) have shown that, following conidia germination and germ tube development, *B. bassiana* enters maize tissues directly through the plant cuticle. Subsequent hyphal growth occurs within the apoplast, but only occasionally extending into the xylem elements.

According to Lewis *et al.* (1996), the introduction of endophytic *B. bassiana* in maize is compatible with other pest management strategies. It has been shown that endophytic *B. bassiana* is compatible with both *B. thuringiensis* (Bt) and carbofuran applications used to suppress *O. nubilalis* (Lewis *et al.,* 1996). Lewis *et al.* (2001) observed that use of *Bt* transgenic corn did not have any detectable effect on the establishment of *B. bassiana* as a corn endophyte. Pingel and Lewis (1996) found that strains of endophytic *B. bassiana* caused no mortality to *Coleomegilla maculata*, a predator of *O. nubilalis* eggs and larvae.

Model System 2: Banana- *Beaveria bassiana*-Banana Stem Weevil

Akello *et al.* (2007) have worked out a model system for establishment of *B. bassiana* as artificial endophyte in tissue culture banana. Laboratory and screenhouse studies have revealed great potential of this fungus for use against the banana weevil, *Cosmopolites sordidus*, in banana. But, impractical field delivery methods and high costs associated with the application of *B. bassiana* against *C. sordidus* prevent its use and commercialisation in banana fields. The research has revealed that *B. bassiana* can colonise internal banana tissues for at least four months after tissue-cultured plantlets are dipped in a spore suspension. The type of banana cultivar did not affect colonisation by *Beauveria bassiana* and, even when elevated *B. bassiana* doses were used, plant growth was not reduced. In a set of three screenhouse experiments, larval mycosis rates in *B. bassiana*-treated plants were 23.5–88.9% and the presence of the fungus inside treated plants led to a reduction in larval damage of up to >50%. Application of *B. bassiana* as an artificial endophyte inside banana plants could circumvent bottlenecks associated with its application as a conventional biopesticide. This is because (i) it kills the damaging larval stages inside the plant, (ii) it is protected from adverse biotic and abiotic factors, (iii) little inoculum is required, drastically reducing its cost, and (iv) farmers do not need to apply the biological control organism themselves, as the technology is easily transferable to a commercial tissue culture producer.

In previous studies, Pereira *et al.* (1999) inoculated banana plants artificially an endophytic fungus tagged with a genetic marker for resistance to fungicides. It was observed that not only the inoculation worked well, but also the mutant fungus was kept in competition with wild type ones. Plant tissue colonization varied by cultivar and isolate combinations. These results demonstrated that inoculation of valuable strains for biological control will not only be used with success but, modified microorganisms, once inoculated into a host, will be able to coexist and even overcome wild type strains of the same species.

Model System 3: Coffee-*Beaveria bassiana*-Coffee Berry Borer

Isolates of *B. bassiana* have been established in coffee seedlings after fungal spore suspensions were applied as foliar sprays, stem injections, or soil drenches (Posada *et al.,* 2007). The results indicate that the four *B. bassiana* isolates used in this study have different capabilities for infection and colonization of coffee plant tissues even though there were not clear tendencies between isolates in their ability to persist as endophytes. Direct injection yielded the highest post-inoculation recovery of endophytic *B. bassiana*. Establishment, based on percent recovery of *B. bassiana*, decreased as time post-inoculation increased in all treatments. The isolate ARSEF 1480 (Brazil) could be detected in the injected plants four months post-inoculation, in contrast to all the other isolates, none of which was detectable at four months. Strain suitability for endophytism is perhaps an aspect that needs to be taken into account when evaluating candidates for plant inoculation. Several other endophytes were isolated from coffee seedlings and could have negatively influenced establishment of *B. bassiana*. The recovery of *B. bassiana* from sites distant from the point of inoculation indicates that the fungus has the potential to move throughout the plant. In prior studies, Santamarıa and Bayman (2005) reported presence of *Xylaria, Colletotrichum, Guignardia, Botryosphaeria* and *Pestalotia* as coffee leaf endophytes in Puerto Rico.

PRODUCTION AND FORMULATION STRATEGIES

Generalized approach to the formulation and application of traditional entomopathogenic fungi has led to inconsistency in control due to existence of diversity and variations within agro-ecosystems (Vega *et al.,* 2009). Furthermore, the formulation of propagules of fungal biocontrol agents has been guided by the need to improve product shelf life, biocontrol efficacy and the physical characteristics of the product for application (Wraight *et al.*, 2001). Formulations that improve spore desiccation tolerance or shelf life such as cryoprotectants or oils may inhibit spore germination or intimate contact of the spore with the insect host, resulting in reduced biocontrol efficacy. A direct approach to formulate an agent for improved biocontrol efficacy should include an understanding of the fungal spore-insect surface chemistries and how they interact to enhance adhesion and fungal infection.

Understanding how the insect pest or the endophytic fungi interacts or survives in a given ecological environment is critical in directing the use of appropriate formulations. Therefore, a clear understanding of host (insect)-pathogen (endophyte)-ecosystem interactions is imperative to improve the fungus's ability to become endophytic. This would enhance the consistency level and biocontrol efficacy of introduced endophytes to establish within plant systems. Jaronski and Jackson (2008) demonsrated that *M. anisopliae* was shown to be capable of producing sclerotia in fermented liquid culture. Sclerotia are over wintering structures formed by many plant pathogenic fungi that sporulate to produce infective conidia when environmental conditions are condusive for infection of their host plant (Vega *et al.,* 2009). The use of sclerotial preparations for granular application of *M. anisopliae* in soil and the use of conidia or blastospores in foliar applications for phylloplane insects are examples of how the ecology of the fungus-insect interaction directs the production and use of appropriate infective propagules. Awareness of these ecological factors will guide the development of production and formulation technologies that deliver optimally infective fungal propagules to establish as endophyte in introduced crops.

DELIVERY SYSTEM FOR ENDOPHYTIC FUNGI

Like many other microbial inoculants, propagules of endophytic entomopathogenic fungi can be used directly adopting different delivery system such as seed dressing, soil application, stem injection and foliar spray to establish endophytism in desired hosts (Carroll, 1988). For example, the strains of *B. bassiana* have been established as an artificial endophyte in various plants using different inoculation methods. For example, in corn leaves after applying a 20 ml conidia suspension to the abaxial side of leaves (Wagner and Lewis, 2000); in corn plants after granular applications to the whorl (Lewis and Bing,

1991); in tomatoes after coating seeds with a *B. bassiana* conidia (Ownley *et al.*, 2004); in opium poppies after spraying leaves or coating seeds with *B. bassiana* spore suspensions (Quesada-Moraga *et al.*, 2006), in cocoa (Posada and Vega, 2005) and coffee seedlings (Posada and Vega, 2006) after placing spore suspensions on the radicle, in rhizomes of Dioscorea zingiberensis after soil applications (Xu *et al.*, 2007) and in cashew trees after inoculating *B. bassiana* in perimeter zone of soil (Ambethgar, 2010) and cashew root-stocks after drenching the soils (Ambethgar *et al.*, 2010) help in establishment of the fungus. In general, most endophytic fungi show a certain degree of tissue specificity because they are adapted to particular conditions present in a given organ (Fisher *et al.*, 1992; Bertoni, 1994). Gold *et al.* (2001) felt that the high cost of inoculum production and delivery is a serious bottleneck precluding delivery to farmers.

Seed Dressing

In Africa, Cherry *et al.* (2004) studied *B. bassiana* endophytism when attempting to control the maize stem borer (*Sesamia calamistis*) by treating maize seeds with dry *B. bassiana* conidia, by spraying conidial suspensions in the leaf axils, and by injecting conidial suspension in the stem. In this context, Backman and Sikora (2008) opined that seed inoculation method has the advantage of substantially limiting the side effects of abiotic and biotic factors on the biological agent by almost immediately protecting them within plant tissue.

Soil Application

Vega *et al.* (2008) reported that for subterranean insect pests harbouring rhizosphere, fungal biological control agents are typically applied as granules containing hyphae or spore-hyphae preparations. For example, *B. bassiana* was applied as either an oil palm kernel cake-based formulation or as conidial powder to planting holes in soil and suckers to control banana weevil (Akello *et al.*, 2007). The cake based formulation caused 42% mortality among adult banana weevils, while conidial powder caused only 6% mortality in adults. Granules may contain infective conidia or rely on primary growth and *in situ* secondary sporulation for the formation of infective conidia (Jaronski and Jackson, 2008). The conidia-containing granules must be adequately dispersed and remain viable in the soil to insure contact with foraging insect pests.

Dipping Propagation Material

Lewis *et al.* (2001) found no differences in pant growth between *B. bassiana* treated and control plants when maize plants were dipped in a conidial suspension. Akello *et al.* (2007) also found that *B. bassiana* can form an endophytic relationship with tissue culture banana plants. Dipping the banana plants in a suspension containing *B. bassiana* conidia did not affect plant growth. Hyphal penetration of *B. bassiana* from banana plant surfaces into the tissues seemed not to be a limiting factor, because colonization in plants dipped in a conidial suspension was at par or higher than colonization in those injected conidial suspension. Further, they found that dipping was best method for delivery of *B. bassiana* into tissue culture plants, since endophytic colonization, and plant survival and growth was highest and no adverse effects were noted in this delivery system.

Injecting Plants

Studies have shown that when *B. bassiana* is injected into the corn stem, it colonizes and moves within the plant, and the fungus was subsequently isolated from the node above the injection site (Wagner and Lewis, 2000). These data support the idea proposed by Bartlett and Lefebvre (1934) that the succulent pith inside corn is ideal for growth of *B. bassiana*. Wagner and Lewis (2000) observed that fungal movement within corn was aided through passive movement of *B. bassiana* within the vascular elements of xylem via interconnecting xylem tissues to colonize the entire plant. In another study, Gomez-Vidal *et al.* (2006) found that isolates of *B. bassiana* (No. 119 and 33), *Lecanicillium dimorphum* and *L. c.f. psalliotae* were able to colonize date palm (*Phoenix dactylifera*) petiole tissues by spore injection method. The fungi colonized inoculated petioles endophytically and were recovered at least 3 cm from the injection sites from both young and adult palms after 30 days incubation in moist chambers. The fungi were detected microscopically inside the parenchyma and sparsely within vascular tissues. Inoculation of date palm petioles with these fungal entomopathogens did not show disease symptoms as compared to non-inoculated wound controls. The necrotic and hydropic zones were never longer than 1 cm after incubation for 30 days indicating evidence that important entomopathogenic fungi colonized living palm tissue and, therefore, may act as palm endophytes.

Foliar Spray

Wagner and Lewis, (2000) performed endophytic colonization of corn plants by foliar spraying method using *B. bassiana* spore suspension at 10^4 conidia/ml against *O. nubilalis* and for control of insect pests of the phylloplane. When *B. bassiana* was sprayed to plants, it colonized, translocated and persisted within the plants to provide seasonlong suppression of *O. nubilalis*. Foliar spray method was also useful against insect pests of phyllophane when numerous, discrete, infective propagules provided by spore forms satisfy the requirement for complete coverage of the foliar surface to ensure contact and infection of the insect host. Blastospores in foliar applications for phylloplane insects are examples of how the ecology of the fungus-insect interaction directs the production and use of appropriate infective propagules.

CONCLUSIONS

The role of fungal endophytes protecting plants against insect herbivores has been quite well studied from the 1980s onwards. Many species of fungal endophytes are reported to cause physiological modifications in their hosts, such as making them more resistant pests and diseases. The energy spent by the plant in the production of endophyte biomass is adequately compensated for by the improvements in plant health derived from the mutualistic microorganisms. The amount of inoculum required to protect the host on a per hectare basis is minimized with the precise targeting of endophyte treatments (Vega *et al.,* 2009). This should make the approach attractive to the biotechnology industry looking for alternatives to traditional pesticides, as targeting the insect and nematode infections assures improved efficacy.

The future use of biological-chemical combinations of endophytes in combination with commercial pesticides applied to the crop plants could lead to synergistic effects on one or multiple disease causing agents. Thus, fungal endophytes are attractive potential biocontrol agents for reducing pest damage in different crops. Additional research on endophytic entomopathogenic fungi should elucidate on the ecology of artificially introduced fungi in a context that focuses on their roles as endophytes, plant disease antagonists, rhizosphere colonizers, and plant growth promoters. These areas could lead to (i) a better understanding of the ecological niches occupied by entomopathogenic fungi; (ii) improved deployment for better pest control; and

(iii) improved production and formulation to enhance their biocontrol efficacy. Comprehensive researches are also needed for successful application of promising endophytes with wider adaptability and acceptability involving many other agri-horticultural crops for growth promotion and management of many other destructive crop pests. Insights gained from these studies will result in the effective use of promising endophytes as an integral part of agricultural systems. New discoveries suggest that endophytic entomopathogenic fungi could be utilized to "vaccinate" crop plants against a range of pests if toxins did not contaminate commodities, and they could also be used as vectors for the genetic engineering of crop plants.

REFERENCES

Abdel-Baky, N.F. and Abdel-Salam, A.H. (2003) Natural incidence of *Cladosporium* spp. as a biocontrol agent against whiteflies and aphids in Egypt. *J. Appl. Entomol.*, **127**, 228–235.

Ahmad, S., Govindarajan, S., Funk, C.R. and Johnson-Cicalese, J.M. (1985) Fatality of house crickets on perennial ryegrass infected with a fungal endophyte. *Entomol. Exp. Appl.*, **39**, 183–190.

Akello, J.T., Dubois, T., Gold, C.S., Coyne, D., Nakavuma, J. and Paparu, P. (2007) *Beauveria bassiana* (Balsamo) Vuillemin as an endophyte in tissue culture banana (*Musa* spp.). *J. Invertebr. Pathol.*, **96**, 34–42.

Ambethgar, V. (2001) Exploitation of entomogenous fungi in biological control of crop pests. In: R.K. Upadhyay, K.G. Mukerji and B.P. Chamola (eds.), *Biocontrol Potential and its Exploitation in Sustainable Agriculture, Vol. 2, Insect Pests*, Kluwer Academic/Plenum Publishers, New York, pp. 39–55.

Ambethgar, V. (2010) Field assessment of fungal and insecticidal delivery methods on cashew stem and root borer, *Plocaederus ferrugineus* L. (Cerambycidae: Coleoptera). *J. Biopestic.*, **3**, 121–125.

Ambethgar, V., Azevedo, J.L., Araujo, J.M. and Glaucia, M.S. (2010) Fungal endophytes of cashew (*Anacardium occidentale* L.): A first report. *Electr. J. Biotechnol.* (in press).

An, Z.Q., Siegel, M.R., Hollin, W., Tsai, H.-F., Schmidt, D. and Schardl, C.L. (1993) Relationships among non-*Acremonium* sp. fungal endophytes in five grass species. *Appl. Environ. Microbiol.*, **59**, 1540–1548.

Ananda, K. and Sridhar, K.R. (2002) Diversity of endophytic fungi in the roots of mangrove species on the west coast of India. *Can. J. Microbiol.*, **48**, 871–878.

Arnold, A.E. and Lewis, L.C. (2005) Ecology and evolution of fungal endophytes and their roles against insects. In: F.E. Vega, and M.

Blackwell (eds.), *Insect-Fungal Associations: Ecology and Evolution*, Oxford University Press, New York, pp. 74–96.

Arnold, A.E. and Lutzoni, F. (2007) Diversity and host range of foliar fungal endophytes: Are tropical leaves biodiversity hotspots? *Ecology*, **88**, 541–549.

Arnold, A.E., Mejia, L.C., Kyllo, D., Rojas, E.I., Maynard, Z., Robbins, N. and Herre, E.A. (2003) Fungal endophytes limit pathogen damage in a tropical tree. *Ecology,* **100**, 15649–15654.

Arechavaleta, M, Bacon, C.W., Hoveland, C.S. and Radclife, D.E. (1989) Effects of tall fescue endophyte on plant response to environmental stress. *Agronomy J.,* **81**, 83–90.

Azevedo, J.L. (1998) Endophytic microorganisms. In: I.S. Melo and J.L. Azevedo (eds.), *Ecological Microbiana*, Editoria EMPRAPA, Sao Paulo, Brazil, pp. 117–137.

Azevedo, J.L., Maccheroni Jr. W., Pereira Welington, J.O. and de Araújo, L. (2000) Endophytic microorganisms: a review on insect control and recentadvances on tropical plants. *J. Biotechnol.*, **3**, 40–65.

Backman, P.A. and Sikora, R.A. (2008) Endophytes: An emerging tool for biological control. *Biol. Contr.*, **46**, 1–3.

Bacon, C.W. and Hills, N.S. (1996) Symptomless grass endophytes: products of coevolucionary symbioses and their role in the ecological adaptation of grasses. In: S.C. Redlin and L.M. Carris. (eds.), *Endophytic Fungi in Grasses and Woody Plants*, American Phytopathologycal Society Press, St. Paul, pp. 155–178.

Bacon, C.W., Porter, J.K. and Robbins, J.D. (1975) Toxicity and occurrence of Balansia on grasses from toxic fescue, pastures. *Appl. Microbiol.,* **29**, 553–556.

Bacon, C.W., Porter, J.K., Robins, J.D. and Luttrell, E.S. (1977) *Epichloë typhi* from toxic tall fescue grasses. *Appl. Environ. Microbiol.,* **34**, 576–581.

Ball, O.J.P., Miles, C.O. and Prestidge, R.A. (1997) Ergopeptide alkaloids and *Neotyphodium lolli*-mediated resistance in perennial ryegrass agains adult *Heteronynchus arator* (Coleoptera: Scarabaeidae). *J. Econ. Entomol.,* **90**, 1382–1391.

Barker, G.M. (1987). Mycorrhizal infection influences Acremoniuminduced resistance to Argentine stem weevil in grasses. *Proceedings of the New Zealand Weed and Pest Control Conference*, **40**, 199–203.

Barker, G.M., Pottinger, R.P., Addison, P.J. and Prestidge, R.A. (1984) Effect of *Lolium* endophyte fungus infection on behaviour of adult Argentine stem weevil. *New Zealand J. Agric. Res.,* **27**, 271–277.

Bartlett, K.A. and Lefebvre, C.L. (1934) Field experiments with *Beauveria bassiana* (Bals.) Vuill., a fungus attacking the European corn borer. *J. Econ. Entomol.,* **27**, 1147–1157.

Bertoni, M.D. (1994) New record of *Ceratopycnidium baccharidicola* (endophyte on *Baccharis cordifolia*) in Brazil. *Mycotaxon,* **52**, 193–196.

Bills, G.F. and Polishook, J.D. (1991) Microfungi from *Carpinus caroliniana. Can. J. Bot.,* **69**, 1477–1482.

Bills, G.F., Giacobbe, R.A., Lee, S.H., Pelaez, F. and Tkacz, J.S. (1992) Tremorgenic mycotoxins paspalitrem A and C from a tropical *Phomopsis. Mycol. Res.,* **96**, 977–983.

Bing, L.A. and Lewis, L.C. (1991) Suppression of *Ostrinia nubilalis* (Hübner) (Lepidoptera: Pyralidae) by endophytic *Beauveria bassiana* (Balsamo) Vuillemin. *Environ. Entomol.,* **20**, 1207–1211.

Bing, L.A. and Lewis, L.C. (1993) Occurrence of the entomopathogen *Beauveria bassiana* (Balsamo) Vuillemin in different tillage regimes and in *Zea mays* L. and virulence towards *Ostrinia nubilalis* (Hübner). *Agric. Ecosyst. Environ.,* **45**, 147–156.

Breen, J.P. (1992) Temperature and seasonal effects on expression of *Acremonium* endophyte-enhanced resistance to *Schizaphis graminum* (Homoptera: Aphididae). *Environ. Entomol.,* **21**, 68–74.

Breen, J.P. (1994) *Acremonium*-endophyte interactions with enhanced plant resistance to insects. *Annu. Rev. Entomol.,* **39**, 401–423.

Bultman, T.L. and Conard, N.J. (1998) Effects of endophytic fungus, nutrient level and plant damage on performance of fall armyworm (Lepidoptera: Noctuidae). *Environ. Entomol.,* **27**, 631–635.

Bush, L.P., Wilkinson, H.H. and Schardl, C.L. (1997) Bioprotective alkaloids of grass-fungal endophyte symbioses. *Plant Physiol.,* **114**, 1–7.

Butin, H. (1992) Effect of endophytic fungi from oak (*Quercus rubor*, L.) on mortality of leaf inhabiting gall insects. *Eur. J. Forest Pathol.,* **22**, 237–246.

Calhoun, L.A., Findrlay, J.A., Miller, J.D. and Whitney, N.J. (1992) Metabolites toxic to spruce budworm from balsam fir needle endophytes. *Mycol. Res.,* **96**, 281–286.

Cammatti-Sartori, V., da Silva-Ribeiro, R.T., Valdebenito-Sanhueza, R.M., Pagnocca, F.C., Echeverrigaray, S. and Azevedo, J.L. (2005) Endophytic yeasts and filamentous fungi associated with southern Brazilian apple (*Malus domestica*) orchards subjected to conventional, integrated or organic cultivation. *J. Basic Microbiol.,* **45**, 397–402.

Cao, L.X., You, J.L. and Zhou, S.N. (2002) Endophytic fungi from *Musa acuminata* leaves and roots in South China. *World J. Microbiol. Biotechnol.,* **18**: 169–171.

Carroll, G. (1988) Fungal endophytes in stems and leaves: from latent pathogens to mutualistic symbionts. *Ecology,* **69**, 2–9.

Carroll, G. (1991) Fungal associates of woody plants as insect antagonists in leaves and stems. In: P. Barbosa, V.A. Krischik and C.G. Jones (eds.),

Microbial Mediation of Plant-Herbivore Interactions, John Wiley and Sons, New York, pp. 253–271.

Carroll, G. (1995) Forest endophytes: pattern and process. *Can. J. Bot.*, **73**, 316–324.

Cheplick G.P. and Clay, K. (1988) Acquired chemical defenses in grasses: The role of fungal endophytes. *Oikos,* **52**, 309–318.

Cherry, A.J., Banito, A., Djegui, D. and Lomer, C. (2004) Suppression of the stem-borer *Sesamia calamistis* (Lepidoptera; Noctuidae) in maize following seed dressing, topical application and stem injection with African isolates of *Beauveria bassiana. Int. J. Pest Manage.,* **50**, 67–73.

Christensen, M.J., Latch, G.C.M. and Tapper, B.A. (1991) Variation within isolates of *Acremonium* endophytes from perennial ryegrass. *Mycol. Res.,* **95**, 988–923.

Clark, C.L., Miller, J.D. and Whitney, N.L. (1989) Toxicity of conifer needle endophytes to spruce budworm. *Mycol. Res.,* **93**, 508–512.

Clay, K. (1986) Induced vivipary in the sedge Cyperus virens and the transmis sion of the fungus *Balansia cyperi* (Clavicipitaceae). *Can. J. Bot.,* **64**, 2984–2988.

Clay, K. (1989) Clavicipitaceous endophytes of grasses: their potential as biocontrol agents. *Mycol. Res.*, **92**, 1–12.

Clay, K. (1996) Interactions among fungal endophytes, grasses and herbivores. *Res. Popul. Ecol.,* **38**, 191–201.

Clay, K. and Schardl, C.L. (2002) Evolutionary origins and ecological consequences of endophyte symbiosis with grasses. *Am. Nat.,* **160**, S99–S127.

Clay, K., Hardy, T.N. and Hammond Jr., A.M. (1985). Fungal endophytes of *Cyperus* and their effect on the insect herbivore. *Am. J. Bot.,* **72**, 1284–1289.

Claydon, N., Grove, J.F. and Pople, M. (1985) Elm bark beetle boring and feeding deterrents from *Phomopsis oblonga. Phytochemistry,* **24**, 937–943.

Clement, S.L., Lester, D.G., Wilson, A.D. and Pike, K.S. (1992) Behaviour and performance of *Diuraphis noxia* (Homoptera: Aphididae) on fungal endophyte-infected and uninfected perennial ryegrass. *J. Econ. Entomol.,* **85**, 583–588.

Clement, S.L., Elberson, L.R., Bosque-Pe´rez, N.A. and Schotzko, D.J. (2005) Detrimental and neutral effects of wild barley-*Neotyphodium* fungal endophyte associations on insect survival. *Entomol. Exp. Appl.,* **114**, 119–125.

Crutchfield, B.A. and Potter, D.A. (1994) Preferences of Japanese beetle and southern masked chafer (Coleoptera: Scarabaeidae) grubs among cool-season turfgrasses. *J. Entomol. Sci.,* **29**, 398–406.

Cunningham, J.J. (1958) Non-toxicity to animals of ryegrass endophyte and other endophytic fungi of New Zealand grasses. *New Zealand J. Agric. Res.*, **1**, 487.

Dai, C., Chen, Y., Tian, L. and Shi, Y. (2010) Correlation between invasion by endophytic fungus *Phomopsis* sp. and enzyme production, *Afr. J. Agric. Res.*, **5**, 1324–1330.

Davidson, A.W. and Potter, D.A. (1995) Response of plant-feeding predatory and soil-inhabiting invertebrates to *Acremonium* endophyte and nitrogen fertilization in tall fescue turf. *J. Econ. Entomol.*, **88**, 367–379.

De Bary, H.A. (1879) Die Erscheinug der Symbiose. In: K.J. Trubner (ed.), *Vortrag auf der Versammlung der Naturforscher und Artze zu Cassel.* Strssburg, Germany: Verlag, pp. 1-30.

De Bary, H.A. (1884) *Vergleichende Morphologie und Biologie der Pilze Mycetozoen and Bacterien.* Verlag von Wilhelm Engelmann, Leipzig.

Dreyfuss, M. and Petrini, O. (1984) Further investigations on the occurrence and distribution of endophytic fungi in tropical plants. *Bot. Helv.*, **94**, 33–40.

Dugan, F.M. and Lupien, S.L. (2002) Filamentous fungi quiescent in seeds and culm nodes of weedy and forage grass species endemic to the Palouse Region of Washington and Idaho. *Mycopathologia*, **156**, 31–40.

Eisencher, H. and Dahlman, D.L. (1992) Antibiotic and deterrent qualities of endophyte-infected tall fescue to two aphid species (Homoptera: Aphididae). *Environ. Entomol.*, **21**, 1046–1051.

Elmi, A.A., West, C.P., Robbins, R.T. and Kirkpatrick, T.L. (2000) Endophyte effects on reproduction of a root-knot nematode (*Meloidogyne marylandi*) and osmotic adjustment in tall fescue. *Grass Forage Sci.*, **55**, 166–172.

Evans, H.C., Holmes, K.A. and Thomas, S.E. (2003) Endophytes and mycoparasites associated with an indigenous forest tree, *Theobroma gileri*, in Ecuador and a preliminary assessment of their potential as biocontrol agents of cocoa diseases. *Mycol. Progr.*, **2**, 149–160.

Fahey, J.W., Dimock, M.B., Tomasino, S.F., Taylor, J.M. and Carlson, P.S. (1991) Genetically engineered endophytes as biocontrol agents: A case study in industry. In: J.H. Andrews and S.S. Hirano (eds.), *Microbial Ecology of Leaves*, Springer-Verlag, New York, pp. 402–411.

Findlay, J.A., Buthelezi, S., Li, G.Q., Seveck, M. and Miller, J.D. (1997) Insect toxins from an endophytic fungus from wintergreen. *J. Nat. Prod.*, **60**, 1214–1215.

Fisher, P.J., Petrini, O. and Lappin-Scott, H.M. (1992) The distribution of some fungal and bacterial endophytes in maize (*Zea mays* L.). *New Phytol.*, **122**, 299–305.

Fuller-Schaefer, C., Jung, K. and Jaronski, S. (2005) Colonization of sugarbeet roots by entomopathogenic fungi. In: *Proceedings of the 38th*

Annual Meeting of the Society for Invertebrate Pathology, Anchorage, Alaska, p.49.

Funk, C.R., Halisky, P.M., Johnson, M.C., Siegel, M.R., Stewart, A.V., Ahmad, S., Hurley, R.H. and Harvey, I.C. (1983) An endophytic fungus and resistance to sod webworms: association in *Lolium perenne. Bioetchnology,* 1, 189–191.

Gamper, H.A., Young, J.P.W., Jones, D.L. and Hodge, A. (2008) Real-time PCR and microscopy: are the two methods measuring the same unit of arbuscular mycorrhizal fungal abundance?. *Fungal Gen. Biol.,* **45**: 581–596.

Ganley, R.J. and Newcombe, G. (2005) Fungal endophytes in seeds and needles of *Pinus monticola. Mycol. Res.,* **110**, 318–327.

Gao, F., Dai, C and Liu, X. (2010) Mechanisms of fungal endophytes in plant protection against pathogens. *Afr. J. Microbiol. Res.,* **4**, 1346–1351.

Gaynor, D.L. and Hunt, W.F. (1983) The relationship between nitrogen supply, endophytic fungus and Argentine stem weevil resistance in ryegrass. *Proc. New Zealand Grassland Assoc.,* **44**, 257–263.

Gennaro, M., Gonthier, P. and Nicolotti, G. (2003) Fungal endophytic communities in healthy and declining *Quercus robur* L. and *Quercus cerris* L. trees in northern Italy. *J. Phytopathol.,* **151**, 529–534.

Glienke-Blanco, C. (1999) *Guignardia citricarpa* Kiely: Análise genética, cariotípica e interação com o hospedeiro. PhD. Thesis. Escola Superior de Agricultura "Luiz de Queiroz", Universidade de São Paulo, Piracicaba, São Paulo, Brazil. 200p.

Gold, C.S., Pena, J.E. and Karamura, E.B. (2001) Biology and integrated management for the banana weevil, Cosmoplites sordidus (Germer) (Coleoptera: Curculionidae). *Integr. Pest Manage. Rev.,* **6**: 79–155.

Gomez-Vidal, S., Lopez-Llorca, L.V., Jansson, H.B. and Salinas, J. (2006) Endophytic colonization of date palm (*Phoenix dactylifera* L.) leaves by entomopathogenic fungi. *Micron,* **37**, 624–632.

Gurr, G.M., Wratten, S.D. and Altieri, M.A. (2004) Ecological engineering: a new direction for agricultural pest management. *AFBM J.,* **1**, 28–35.

Hajek, A.E.and St. Leger, R.J. (1994) Interactions between fungal pathogens and insect hosts. *Annu. Rev. Entomol.,* **39**, 293–322.

Hardy, T.N., Clay, K. and Hammond Jr., A.M. (1985) Fall armyworm (Lepidoptera: Noctuidae): A laboratory bioassay and larval preference study for the fungal endophyte of perennial ryegrass. *J. Econ. Entomol.,* **78**, 571–575.

Hu, G. and, St Leger, R.J. (2002) Field studies using recombinant mycoinsecticide (*Metarhizium anisopliae*) reveal that it is rhizosphere competent. *Appl. Environ. Microbiol.,* **68**, 6383–6387.

Jallow, M.F.A., Dugassa-Gobena, D. and Vidal, S. (2004) Indirect interaction between and unspecialized endophytic fungus and a polyphagous moth. *Basic Appl. Ecol.,* **5**, 183–191.

Jaronski, S.T. and Jackson, M.A. (2008) Efficacy of *Metarhizium anisopliae* microsclerotial granules. *Biocontr. Sci. Technol.,* **18**, 849–863.

Johnson, M.C., Dahlman, L.D., Siegel, M.R., Bush, L.P., Latch, G.C.M., Potter, D.A. and Varney, D.R. (1985) Insect feeding deterrents in endophyte-infected tall fescue. *Appl. Environ. Microbiol.*, **49**, 568–571.

Knoch, T.R., Faeth, S.H. and Arnott, D.L. (1993) Endophytic fungi alter foraging and dispersal by desert seed-harvesting ants. *Oecologia,* **95**, 470–473.

Koga, H., Hirai, Y., Kanda, K., Tsukiboshi, T. and Uematsu, T. (1997) Successive transmission of resistance to bluegrass webworm to perennial ryegrass and tall fescue plants by artificial inoculation with *Acremonium* endophytes. *Jap. Agric. Res. Quat.,* **31**, 109–115.

Krings, M., Taylor, T.N., Hass, H., Kerp, H., Dotzler, N. and Hermsen, E.J. (2007) Fungal endophytes in a 400-million-yr-old land plant: infection pathways, spatial distribution, and host responses. *New Phytol.,* **174**, 648–657.

Large, E.C. (1954) Surveys for choke: (*Epichloe typhina*) in cocks foot seed crops. *Plant Pathol.,* **3**, 6–11.

Larran, S., Monaco, C. and Alippi, H.E. (2001) Endophytic fungi in leaves of Lycopersicon esculentum Mill. *World J. Microbiol. Biotechnol.,* **17**, 181–184.

Larran, S., Perello, A., Simon, M.R. and Moreno, V. (2002a) Isolation and analysis of endophytic microorganisms in wheat (*Triticum aestivum* L.) leaves. *World J. Microbiol. Biotechnol.,* **18**, 683–686.

Larran, S., Rollan, C., Bruno Angeles, H., Alippi, H.E. and Urrutia, M.I. (2002b) Endophytic fungi in healthy soybean leaves. *Invest. Agraria Prod. Prot. de Veg.*, **17**, 173–177.

Lasota, J.A., Waldvogel, M.G. and Shetlar, D.J. (1983) Fungus found in galls of *Adelges abietis* (L.) (Homoptera: Adelgidae): identification within tree distribution and possible impact on insect survival. *Environ. Entomol.,* **12**, 245–246.

Latch, G.C.M., Christensen, M.J. and Gaynor, D.L. (1985) Aphid detection of endophyte infection in tall fescue. *New Zealand J. Agric. Res.,* **28**, 129–132.

Leuchtmann, A. (1992) Systematics, distribution, and host specificity of grass endophytes. *Nat. Toxins,* **1**, 150–162.

Leuchtmann, A. and Clay, K. (1985) Experimental infection of host grasses and sedges with *Atkinsonella hypoxylon* and *Balansia cyperi* (Balansiae, Clavicipitacea). *Mycologia,* **80**, 192–199.

Lewis, G.C. and Clements, R.O. (1986) A survey of ryegrass endophyte (*Acremonium loliae*) in the UK and its apparent in effectuality on a seedling pest. *J. Agric. Sci.,* **107**, 633–638.

Lewis, G.C. and Vaugham, B. (1997) Evaluation of a fungal endophyte (*Neotyphodiuum lolii*) for control of leather jackets (*Tipula* spp.) in perennial rye grass. *Tests Agrochem. Cutliv,* **18**, 34–35.

Lewis, L.C. and Cossentine, J.E. (1986) Season long intraplant epizootics of entomopathogens, *Beauveria bassiana* and *Nosema pyrausta* in a corn agroecosystem. *Entomophaga*, **31**, 36–69.

Lewis, L.C. and Bing, L.A. (1991) *Bacillus thuringiensis* Berliner and *Beauveria bassiana* (Balsamo) Vuillemin for European corn borer control: program for immediate and season long suppression. *Can. Entomol.,* **123**, 387–393.

Lewis, L.C., Berry, E.C., Obrycki, J.J. and Bing, L.A. (1996) Aptness of insecticides (*Bacillus thuringiensis* and carbofuran) with endophytic *Beauveria bassiana* in suppressing larval populations of the European corn borer. *Agric. Ecosys. Environ.,* **57**, 27–34.

Lewis, L.C., Bruck, D.J., Gunnarson, R.D. and Bidne, K.G. (2001) Assessment of plant pathogenicity of endophytic *Beauveria bassiana* in Bt transgenic and non-transgenic corn. *Crop Sci.*, **41**, 1395–1400.

Lomer, C.J., Cherry, A. and Denis, D. (1997) Systemic *Beauveria* isolates for control of maize stem borers in Africa. In: *Proceedings of the 30th Annual Meeting of the Society for Invertebrate Pathology,* Banff, Canada, p.44.

Longo, A.C. (1996) Transformação genética e variabilidade detectada por RAPD em isolados endofíticos de *Colletotrichum musae*. PhD Thesis. Escola Superior de Agricultura "Luiz de Queiroz", Universidade de São Paulo, Piracicaba, São Paulo, Brazil.

Luttrell, E. S. and Bacon, C.W. (1977) Classification of *Myriogenospora* in the Clavicipitaceae. *Can. J. Bot.,* **55**, 2090–2097.

Maciá-Vicente, J.G., Jansson, H-B., Mendgen, K. and Lopez-Llorca, L.V. (2008) Colonization of barley roots by endophytic fungi and their reduction of take-all caused by *Gaeumannomyces graminis* var. *tritici*. *Can. J. Microbiol.,* **54**, 600–609.

Melo, I.S. and Azevedo, J.L. (1998) *Controle Biológico I. Editora EMBRAPA*. Jaguariuna, São Paulo, Brazil.

Miles, C.O., diMena, M.E., Jacobs, S.W.L., Garthwaite, I., Lane, G.A., Prestidge, R.A., Marshal, S.L., Wilkinson, H.H., Schardl, C.L., Ball, O.J.P. and Latch, C.M. (1998) Endophytic fungi in indigenous australasian grasses associated with toxicity to livestock. *Appl. Environ. Microbiol.,* **64**, 601–606.

Miller, J.D. (1986) Toxic metabolites of epiphytic and endophytic fungi of conifer needles. In: N.J. Fokkema and Van den Heuvel (eds.), *Microbi-*

ology of the Phyllosphere, Cambridge University Press, Cambridge, pp. 221–231.

Miller, J.D., MacKenzie, S., Foto, M., Adams, G.W. and Findlay, J.A. (2002) Needles of white spruce inoculated with rugulosin-producing endophytes contain rugulosin reducing spruce budworm growth rate. *Mycol. Res.*, **106**, 471–479.

Murray, F.R., Latch, G.C.M. and Scott, D.B. (1992) Surrogate transformation of perennial ryegrass *Lolium perenne* using genetically modified *Acremonium* endophyte. *Mol. Gen. Genet.*, **233**, 1–9.

Neill, J.C. (1941) The endophytes of *Lolium* and *Festuca*. *New Zealand J. Sci. Technol.*, **23**, 85–93.

Okane, I., Nakagiri, A. and Ito, T. (1998) Endophytic fungi in leaves of ercaceous plants. *Can. J. Bot.*, **76**, 657–663.

Ownley, B.H., Pereira, R.M., Klingeman, W.E., Quigley, N.B. and Leckie, B.M. (2004) *Beauveria bassiana*, a dual purpose biocontrol organism, with activity against insect pests and plant pathogens, In: R.T. Lartey and A.J. Caesar (eds.), *Emerging Concepts in Plant Health Management*, Research Signpost, India, pp. 255–269.

Pamphile, J.A., Azevedo, J.L., Languin, T. and Dabousii, M.J. (1997) Transformação de *Fusarium moniliforme* endofítico com o gene *gus*A e seu uso em estudos de colonização de raizes de milho. *Braz. J. Genet.*, **20** (supplement), 296.

Patterson, C.G., Potter, D.A. and Fannin, F.F. (1992) Feeding deterrency of alkaloids from endophyte-infected grasses to Japanese beetle grubs. *Entomol. Exp. Appl.*, **61**, 285–289.

Pereira, J.O., Carneiro-Vieira, M.L. and Azevedo, J.L. (1999) Endophytic fungi from *Musa acuminata* and their reintroduction into axenic plants. *World J. Microbiol. Biotechnol.*, **15**, 37–40.

Pereira, J.O., Azevedo, J.L. and Petrini, O. (1993) Endophytic fungi of *Stylosanthes*: A first report. *Mycologia*, **85**, 362–364.

Petrini, O. and Dreifuss, M. (1981) Endophytische pilze in epiphytischen Aracea, Bromeliaceae and Orchidaceae. *Sydowia*, **34**, 135–148.

Pingel, R.L. and Lewis, L.C. (1996). The fungus *Beauveria bassiana* (Balsamo) Vuillemin in a corn ecosystem: its effect on the insect predator *Coleomegilla maculata* De Geer. *Biol. Contr.*, 6: 137–141.

Pocasangre, L., Sikora, R.A., Vilich, V. and Schuster, R.P. (2000) Survey of banana endophytic fungi from Central America and screening for biological control of the burrowing nematode (*Radopholus similis*). *Info. Musa.*, **9**, 3–5.

Posada, F. and Vega, F.E. (2005) A new method to evaluate the biocontrol potential of single spore isolates of fungal entomopathogens. *J. Insect Sci.*, **5**, 1–10.

Posada, F. and Vega, F.E. (2006) Establishment of the fungal entomopathogen *Beauveria bassiana* (Ascomycota: Hypocreales) as an endophyte in cocoa seedlings (*Theobroma cacao*). *Mycologia,* **97**, 208–213.

Posada, F., Aime, M.C., Peterson, S.W., Rehner, S.A. and Vega, F.E. (2007) Inoculation of coffee plants with the fungal entomopathogen *Beauveria bassiana* (Ascomycota: Hypocreales). *Mycol. Res.,* **111**, 749–758.

Potter, D.A., Patterson, C.G. and Redmond, C.T. (1992) Influence of turfgrass species and tall fescue endophyte on feeding ecology of Japanese beetle and southern masked chafer grubs (Coleoptera: Scarabeidae). *J. Econ. Entomol.,* **85**, 900–909.

Prestidge, R.A. and Gallagher, R.T. (1988) Endophyte conifers resistance to ryegrass: Argentine steem weevil larval studies. *Ecol. Entomol.,* **13**, 429–435.

Prestidge, R.A., Zijpp, S. and Badan, D. (1984) Effects of Argentine stem weevil on pastures in Central Volcanic Plateau. *New Zealand J. Exp. Agric.,* **12**, 323–331.

Quesada-Moraga, E., Landa, B.B., Munoz-Ledesma, J., Jime´nez-Dı´az, R.M. and Santiago-A´ lvarez, C. (2006) Endophytic colonization of opium poppy, *Papaver somniferum* by an entomopathogenic *Beauveria bassiana* strain. *Mycopathologia,* **161**, 323–329.

Raps, A. and Vidal, S. (1998) Indirect effects of an unspecialized endophytic fungus on specialized plant-herbivorous interactions. *Oecologia,* **114**, 541–547.

Richardson, M.D., Cabrera, R.I., Murphy, J.A. and Zaurov, D.E. (1999) Nitrogen form and endophyte-infection effects on growth, nitrogen uptake and alkaloid content of Chewings fescue turfgrass. *J. Plant Nutr.*, **22**, 67–79.

Riedell, W.E., Kieckhefer, R.E., Petroski, R.J. and Powell, R.G. (1991) Naturally-occurring and synthetic loline alkaloid derivatives: Insect feeding behaviour modification and toxicity. *J. Entomol. Sci.,* **26**, 122–129.

Rodriguez, R.J., White, J.F., Arnold, A.E. and Redman, R.S. (2009) Fungal endophytes: diversity and functional roles. *New Phytol.,* **176**, 1–17.

Rowan, D.D. and Gaynor, D.L. (1986) Isolation of feeding deterrents against Argentine stem weevil from ryegrass infected with the endophyte Acremoniumloliae. *J. Chem. Ecol.,* **12**, 647–658.

Saikkonen, K., Faeth, S.H., Helander, M. and Sullivan, T.J. (1998) Fungal endophytes: A continuum of interactions with host plants. *Annu. Rev. Ecol. Syst.,* **29**, 319–343.

Sikora, R.A., Schafer, K. and Dababat, A.A. (2007) Modes of action associated with microbially induced in planta suppression of plantparasitic nematodes. *Aust. Plant Pathol.,* **36**, 124–134.

Santamarıa, J. and Bayman, P. (2005) Fungal epiphytes and endophytes of coffee leaves (*Coffea arabica*). *Microb. Ecol.*, **50**, 1–8.

Seigel, M.R., Dahlaman, D.I. and Bush, L.P. (1989) The role of endophytic fungi in grasses: new approaches to biological control of pests. In: A.R. Leslie and R.L. Metcalf (eds.), *Integrated Pest Managrment for Turfgrass and Ornamentals*, EPA, Washington, DC, pp. 169–186.

Stone, J.K., Bacon, C.W. and White, Jr J.F. (2000) An overview of endophytic microbes: endophytism defined. In: C.W. Bacon and J.F. White (eds.), *Microbial Endophytes*. Marcel Dekker, New York, USA, pp. 3–29.

Suryanarayanan, T.S., Kumaresan, V. and Johnson, J.A. (1998) Foliar fungal endophytes from two species of the mangrove *Rhizophora*. *Can. J. Microbiol.,* **44**, 1003–1006.

Suryanarayanan, T.S., Murali, T.S. and Venkatesan, G. (2003) Endophytic fungal communities in leaves of tropical forest trees: diversity and distribution patterns. *Curr. Sci.*, **85**, 489–493.

Tian, X.L., Cao, L.X., Tan, H.M., Zeng, Q.G., Jia, Y.Y., Han, W.Q. and Zhou, S.N. (2004) Study on the communities of endophytic fungi and endophytic actinomycetes from rice and their antipathogenic activities in vitro. *World J. Microbiol. Biotechnol.,* **20**, 303–309.

Tsai, H.F., Siegel, M.R. and Schardl, C.L. (1992) Transformation of *Acremonium coenophialum*, a protective fungal symbiont of the grass *Festuca arundinacea*. *Curr. Genet.,* **22**, 399–406.

Tudzynski, P., Holter, K., Correia, T., Arntz, C., Grammell, N. and Keller, U. (1999) Genetics of alkaloid biosynthesis in *Claviceps purpurea*: evidence for a gene cluster. *Fungal Genet. Newslett.*, **46** (Supplement), 140.

Ulrich, K., Ulrich, A. and Ewald, D. (2008) Diversity of endophytic bacterial communities in poplar grown under field conditions. *FEMS Microb. Ecol.,* **63**, 169–180.

Van-Heeswijck, R. and McDonald, G. (1992) *Acremonium* endophyte in perennial ryegrass and other pasture grasses in Australia and New Zealand. *Aust. J. Agric. Res.,* **43**, 1683–1709.

Vega, F.E., Posada, F., Aime, M.C., Pava-Ripoll, M., Infante, F. and Rehner, S.A. (2008) Entomopathogenic fungal endophytes. *Biol. Contr.,* **46**, 72–82.

Vega, F.E., Goettel, M.S., Blackwell, M., Chandler,D., Jackson, M.A., Keller, S., Koike, M., Maniania, N.K., Monzon, A., Ownley, B.H., Pell, J.K. Rangel. D.E.N. and Roy, H.E. (2009). Fungal entomopathogens: new insights on their ecology. *Fungal Ecol.,* **2**, 149–159.

Vicari, M., Hatcher, P.E. and Ayres, P.G. (2002) Combined effect of foliar and mycorrhizal endophytes on an insect herbivore. *Ecology*, **83**, 2452–2462.

Wagner, B.L. and Lewis, L.C. (2000) Colonization of corn, *Zea mays*, by the entomopathogenic fungus *Beauveria bassiana*. *Appl. Environ. Microbiol.,* **66**, 3468–3473.

Wang, J., Machado, C., Panaccione, D. and Schardl, C. (1999) Ergot alkaloid biosynthesis genes cloned from *Claviceps* and *Balansia*. *Fungal Genet. Newslett.*, **46** (Supplement), 120.

Webber, J. (1981) A natural control of Dutch elm disease. *Nature*, **292**, 449–451.

White Jr., J.F., Glenn, A.E. and Chandler, K.F. (1993) Endophyte-host associations in grasses. XVIII- Moisture relations and insect herbivory of the emergent stromal leaf of *Epichloe*. *Mycologia*, **85**, 195–202.

Welsh, R., Hubbell, B., Ervin, D.E. and Jahn, M. (2002) GM crops and the pesticide paradigm. *Nat. Biotechnol.*, **20**, 548–549.

West, C.P., Izekor, E., Oosterhuis, D.M. and Robbins, R.T. (1988) The effect of *Acremonium coenophialum* on the growth and nematode infestation of tall fescue. *Plant Soil*, **112**, 3–6.

Wicklow, D.T., Roth, S., Deyrup, S.T. and Gloer, J.B. (2005) A protective endophyte of maize: *Acremonium zeae* antibiotics inhibitory to *Aspergillus flavus* and *Fusarium verticillioides*. *Mycol. Res.*, **109**, 610–618.

Wilson, D. (1995) Endophyte: the evolution of a term, and clarification of its use and definition. *Oikos*, **73**, 274–276.

Wraight, S.P., Jackson, M.A. and de Kock, S.L. (2001) Production, stabilization, and formulation of fungal biocontrol agents. In: T.M. Butt, C.W. Jackson and N. Magan (eds.), *Fungi as Biocontrol Agents: Progress, Problems and Potential*, CAB Intyernational, Wallingford, United Kingdom, pp. 253–288.

Xu, L., Zhou, L., Li, X. And Wang, J. (2007) Fungal endophytes from Dioscorea zingiberensis rhizomes and their antibacterial activity. *Appl. Microbiol.*, **46**, 68–72.

Young, C., McMillan, L. and Scott, B. (1999) Molecular cloning of an indole-diterpenoid gene cluster from *Penicillium paxilii*. *Fungal Genet. Newslett.*, **46** (Supplement), 121.

18

Biofungicides: Current Scenario and Future Prospects in the Management of Plant Diseases

JASPAL KAUR AND RAMA S. SINGH*

Department of Plant Pathology, Punjab Agricultural University
Ludhiana- 141 004, India
*E-mail: ramassingh5@yahoo.com

INTRODUCTION

Nowadays, the use of toxic chemicals for the management of plant diseases is of great concern due to their harmful effects on the non-target organisms including human beings and the environment. The chemical toxicity persists in food chains, pollutes the environment, accumulates in soil and contaminates groundwater. Intensive use of chemicals has also led to the increase in pathogen resistance to these chemicals (Haggag and Mohamed, 2007; Kim and Hwang, 2007). Moreover, certain soil borne diseases are difficult to control by the use of fungicides. In several cases the chemical pesticides that are registered for use may not be available and registrations are being lost because of pressure and concerns of general public. The alternatives such as cultural practices, resistant varieties, and biological protections are suggested to get rid of hazardous effects of toxic chemicals. However, resistant varieties and cultural methods have several limitations. Breeding for resistance, which continues to be the most practical and feasible method to control plant diseases, is not able to keep pace with the development of more virulent pathogens. Therefore, the biological control is highly preferred as an alternative of chemicals and to obtain high quality produce of agricultural and horticultural crops (Kulkarni *et al.*, 2007). Biological methods mainly consist of using microorganisms to control harmful microorganisms causing plant diseases without disturbing the ecological balance. Overall, the biocontrol or use of microbial biofungicides is the eco-friendly method for disease control and has been widely accepted by the scientists, farmers and consumers in developed and developing countries.

Biological control in plant pathology has been defined by various scientists in their own words. Garret (1965) defined biological control as, "any condition under which or practice whereby survival activity of a pathogen is reduced through the agency of any other living organisms (except man himself) with the result that there is reduction in incidence of diseases caused by pathogen". Baker and Cook (1974) defined biological control as *the reduction of inoculum density or disease producing activities of a pathogen or parasite in its active or dormant state by one or more organisms, accomplished naturally or through manipulation of the environment, host or antagonist or by mass introduction of one or more antagonists.* In simple words, biocontrol involves the use of saprophytic living micro- organisms to manage other pathogenic microorganisms. Thus, it is desired eco-friendly method of disease management. As the demand of chemical free food products is increasing, the area under organic farming is increasing day by day, as a result the demand of biocontrol/ bioformulation is also increasing.

BIOCONTROL AGENTS AND THEIR FORMULATIONS

Thousands of saprophytic microorganisms, fungi, bacteria have been identified as antagonists of microbial population inciting plant diseases. Further, the most effective microorganisms selected out of antagonistic population and used to manage plant diseases are termed as biological control agents (BCAs). The formulations of BCAs are known as biofungicides/biopesticides. A good number of biocontrol agents belonging to fungi and bacteria are being used as biofungicides against various plant pathogens (Table 18.1).

The effectiveness of biocontrol depends on the choice of the antagonist, methods for introducing and maintaining their optimal population density to the site required protection. The species of *Trichoderma*, *Gliocladium*, *Coniothyrium*, non pathogenic *Fusarium*, fluorescent *Pseudomonas*, *Bacillus*, etc. are most commonly used as potential BCAs. Only brief account on commonly used bacterial and fungal biocontrol agents is given here.

Table 18.1. Commonly used biological control agents in the management of various plant pathogens

Genera	Common species	Genera	Common species
Fungi			
Ampelomyces	*A. quisqualis*	*Peniophora*	*P. gigantea*
Arthrobotrys	*A. dactyloides, A. oligospora*	*Pythium*	*P. oligandrum, P. nunns*
Ascocoryne	*A. sarcoides*	*Sporidesmium*	*S. sclerotivorum*
Candelabrella	*C. javanica,C. musiformis*	*Trichothecium*	*T. roseum*
Catenaria	*C. anguillulae, C. auxiliaris*	*Trichoderma*	*T. viride, T. hamatum,*
Chaetomium	*C. cochliodes, C. globosum*		*T. polysorum, T. koningii*
Cladosporium	*C. herbarum,*		*T. harzianum*
	C. cladosporioides	*Tuberculina*	*T. maxima*
Coniothyrium	*C. minitans*	*Verticillium*	*V. lecanii, V. biguttatum,*
Dactylaria	*D. vermicola, D. haptotyla*		*V. nigrescens*
Dactylella	*D. oviparasitica,*	**Bacteria**	
	D. doedycoides, D. lobata	*Agrobacterium*	*A. radiobacter*
Fusarium	*F. roseum, F. lateritium*	*Bacillus*	*B. subtilis, B. cereus,*
	F. oxysporum		*B. penetrans*
Gliocladium	*G. virens, G. roseum*	*Bdellovibrio*	*B. bacteriovorus*
	G. catenulatum	*Erwinia*	*E. herbicola,*
Glomus	*G. fasciculatum, G. mosseae*		*E. uredovora*
Laetisaria	*L. arvalis*	*Pseudomonas*	*P. fluorescens, P. cepacia,*
Myrothecium	*M. verrucaria*		*P. putida*
Nematophthora	*N. gynophila*		
Penicillium	*P. lilacinum, P. nigricans,*	***Streptomyces***	*S. griseus, S. praecox,*
	P. frequentens, P. oxalicum		*S. lavendulae*
	P. chrysogenum		

BACTERIAL ANTAGONISTS

Agrobacterium

Agrobacterium radiobacter is a naturally occurring bacterium which is present in many soil types and plant rhizospheres. *A. radiobacter* is one of the best examples of a successful microbial control agent that has been isolated from nature (first isolated from soil surrounding a peach tree in Australia in the early 1970s) and used in agriculture. The use of *A. radiobacter* is targeted at mitigating pathogenic strains of other closely related bacteria, *A. tumefaciens* causing crown gall infections in certain fruit, nut, and ornamental nursery stock. A few bioformulation based on it are available in the market for their commmercial use (Table 18.2).

Table 18.2. *Agrobacterium*-based biofungicides for the control of various plant diseases

Name of the formulation	Active ingredient	Effective against
Gallatrol	*Agrobacterium radiobacter* (strain K-84)	Crown gall
No Gall	*Agrobacterium radiobacter* (strain K-1026)	Crown gall
Norbac 84C	*Agrobacterium radiobacter*(strain K-84)	Crown gall

Pathogen is susceptible to Agrosin 84, the bacteriocin produced by *A radiobacter*. Strain K84 of *Agrobacterium* (formerly called *A. radiobacter*) has been a successful biocontrol agent and most effective against crown gall disease for almost 30 years all over the world. In spite of its demonstrated efficiency, the most important risk of failure when using strain K84 is the possibility of transfer of plasmid pAgK84 to pathogenic *Agrobacterium* strains. pAgK84 codifies production of and immunity to agrocin 84, the main factor involved in crown gall biocontrol by strain K84. Then, a second generation of strain K84 was obtained and the genetically engineered strain was called K1026. It contains a deletion in the transfer region of pAgK84. A number of studies have been performed to compare both strains in their ability to control crown gall, plasmid transfer, antibiotic production, root colonization and survival in the rhizosphere. Now-a-days, strain K1026 is in use as a biopesticide to safeguard biocontrol of crown gall wherever strain K84 was employed. Except for a small portion of DNA removed from K84, the two strains are essentially identical and have the same characteristics. Removal of this genetic material prevents strain K1026 from transferring a piece of DNA to the bacteria that cause crown gall disease and reduces the likelihood of resistance. A toxic compound produced by both K1026 and K84 controls certain other *Agrobacterium* species that cause crown gall disease.

Bacillus

Bacillus belongs to Eubacteria. They are spore forming rods, mesophilic, gram positive, endospores formed in cells, which remain dormant for long periods. Several saprophytic species of bacterium have been well known antagonist of various plant pathogens and widely used as biocontrol agents in the management of plant diseases (Table 18.3).

Table 18.3. *Bacillus* - based biofungicides for the control of various plant diseases

Formulation	Active ingredient	Effective against
Campanion	*Bacillus subtilis GB03*	*Rhizoctonia, Pythium, Fusarium, Phytophthora*
Ecogaurd	*Bacillus licheniformis* SB3086	Fungi causing dollar spots
Green Releaf	*Bacillus licheniformis* B3086	Foliar fungi
KodiakTM, KodiakHB, Epic, Concentrate Quantum 4000 and System 3 TM	*B. subtilis* GB03	*Rhizoctonia solani, Fusarium* spp, *Alternaria* spp, and, *Aspergillus* spp
Bio Yield	*B. subtilis* and *B. amyloliquefaciens*	Broad spectrum action against greenhouse pathogens
Sernade Max Sernade ASO	*B. subtilis* QST 713 WP	Powdery mildew, Downy mildew, Cercospora leaf spot, early blight, late blight, brown rot, fire blight
GB34 Biological fungicide	*Bacillus pumilus GB34*	*Rhizoctonia* and *Fusarium*
Rhizo-Plus	*B. subtilis* strain FZB24	*R. solani, Fusarium* spp., *Alternaria* spp., *Sclerotinia* and *Verticillium*
Rhapsody	*B. subtilis* strain QST713.	Powdery mildew, sour rot, downy mildew, and early leaf spot, early blight, late blight, bacterial spot, and walnut blight diseases.
Subtilex	*B. subtilis* MB1600	*Fusarium* spp., *Rhizoctonia* spp. and *Pythium* spp
GB 34 Concentrate	*B. pumilus*	*Rhizoctonia* and *Fusarium*
SonataTM	*B. pumilus* strain QST 2808	Fungal pests such as molds, mildews, blights, rusts and to control Oak death syndrome Used in nurseries, landscapes, oak trees and green house crops

In general, bacilli have received less attention as biocontrol agents of plant pathogens as compared to pseudomonads. However, some of the species of *Bacillus* are particularly effective for practical use as they produce endospores which give better shelf life to the formulation and can tolerate hot and desiccation conditions that may usually be faced by the biocontol agents (Handelsman and Stabb, 1996). Some of strains of species *B. subtilis, B. cereus, B. penetrans, B. licheniformis* and B. *pumilus* have been commonly used for the control of soil-borne plant pathogens (Emmert and Handelsman, 1999).

Pseudomonas

Non-phytopathogenic fluorescent pseudomonads include species such as *P. fluorescens, P. putida, P. cholorophis, P. aureofaciens* and *P. aeruginosa.* It is a common habitant of soil specially rhizosphere and plant debries. The species suppress the phytopathogens by various methods such as competition for nutrients, systemic acquired resistance (SAR) antibiosis, etc. They produce yellow-green fluorescent siderophores which chelate iron and make it available to the plants. These species are known to produce several secondary metabolites with antagonistic characteristics (Dawling and O'Lara, 1994). Among these are phenazines and indole derivatives, which are nitrogen containing heterocyclic compounds (Schipper *et al.*, 1989). Several other antibiotic substances are also produced by pseudomonads. Different commercial formulations are available in market based on pseudomonads for the management of plant diseases (Table 18.4). Mostly, these bioformulations are effective against several soil-borne plant pathogens including *Pythium*, *Rhizoctonia, Fusarium* species, etc.

Table 18.4. *Pseudomonas* - based bioformulations for the control of plant diseases

Formulation	Active ingredient	Effective against
Atzee	*P. chlororaphis 63-28*	Wilts, root rot and stem rot
Bio-save 10LP, 110	*P. syringae*	*Botrytis, Penicillium, Mucor* etc.
Deny	*Burkholderia cepacia*	*Rhizoctonia, Fusarium, Penicillium*
Blight ban A 506	*P. fluorescens* A506	*Erwinia*
Frostban A,B,C,D	*P. fluorescens A506* *P. fluorescens 1629RS* *P. syringe 742RS*	Frost protection agents
Conqueror	*P. fluorescens*	Bacterial blotch
Biocoat	*P. fluorescens*	Fusarial wilts
Biosield	*P. fluorescens*	Various fungal pathogens

Pantoea agglomerans (Erwinia herbicola, Enterobacter agglomerans)

Pantoea agglomerans is a bacterium that is ubiquitous in nature and occurs naturally on fruit trees. *Erwinia herbicola* is highly effective against *Erwinia amylovora* causing apple and pear fire blight and act against the pathogen by competing for space and nutrients. *P. agglomerans* strain C9-1 is able to reproduce on the blooms for several days, effectively colonizing the tree and competitively displacing the pathogen. Ishimaru *et al.* (1988) identified two antibiotics produced by strain C9-1, herbicolin O and herbicolin I, that may contribute to the protective effect. Some of its commercial formulations are widely used to control fire blight (Table 18.5).

Table 18.5. *Pantoea*-based biofungicides for the control of various plant diseases

Name of the formulation	Active ingredient	Effective against
Bloomtime Biological FD	*Pantoea agglomerans* strain E325	Fire blight in apples and pears
Blightban	*Pantoea agglomerans* strain C9-1	Fire blight in apples and pears

Streptomyces

Streptomyces is a ubiquitous and naturally-occurring bacterium that is commonly found in soil environments. It acts against the pathogens mainly through competition, antibiosis and degradation of phytotoxins and production of extracellular enzymes (Doumbou *et al.*, 2002; Crawford *et al.*, 1993) *S. lydicus* WYEC is a ubiquitous, naturally-occurring soil microorganism that colonizes root tips and acts as a mycoparasites. Some of its formulations are widely used to control plant pathogens specially the soil borne (Table 18.6).

Table 18.6. *Streptomyces* - based biofungicides for the control of various plant diseases

Name of the formulation	Active Ingredient	Effective against
Actinovate	*Streptomyces lydicus* WYEC 108	Root rot and damping-off fungi
Mycostop	*Streptomyces griseoviridis*	Wilt and root rot

FUNGAL ANTAGONISTS

Ampelomyces

The fungus is specialised hyperparasite of Erysiphaceae. One species, *A. quisqualis* has been widely used against fungi causing powdery mildew diseases. *A. quisqualis* strain M10 was found in the year 1984 in Israel. *Ampelomyces* acts directly by invasion and destruction of host cytoplasm. *Ampelomyces* kills the parasitized powdery mildew cells by causing a rapid degeneration of the cytoplasm (Hashioka and Nakai, 1980). These mycoparasites suppress both asexual and sexual sporulation of the attacked powdery mildew mycelia by colonizing and destroying the conidiophores, and the immature ascocarps, respectively. The early stage of mycoparasitism is apparently biotrophic, but the invaded cytoplasm then begins to die and a necrotrophic interaction results (Hashioka and Nakai, 1980; Sundheim and Krekling, 1982). Its commercial formulation AQ10TM, produced by Ecogen, Inc. (USA) obtained EPA registration in 1995 for use in the control of grape-vine powdery mildew. It is available as a conidial preparation dry-formulated in a stable water-dispersible granule (Daoust and Hofstein, 1996). The early problems related to reduced efficacy at lower relative humidity could be overcome by spraying in the early morning or late afternoon and by incorporation of a mineral oil-based surfactant as a wetting agent (Daoust and Hofstein, 1996). AQ10 was subsequently registered for use on a number of other fruits and vegetables in conjunction with the wetting agent, Add-Q (Paulitz and Be'langer, 2001; Be'langer and Labbe', 2002).

Aspergillus niger

Aspergillus niger is a member of the genus *Aspergillus* which includes a set of fungi that are generally considered asexual, although perfect forms (forms that reproduce sexually) have been found. Aspergilli are ubiquitous in nature. They are geographically widely distributed, and have been observed in a broad range of habitats because they can colonize a wide variety of substrates. *A. niger* is commonly found as a saprophyte growing on dead leaves, stored grain, compost piles, and other decaying vegetation. The spores are wide-spread, and are often associated with organic materials and soil. A rhizosphere isolate of *A. niger* AN-27 was found effective as biocontrol agent against various diseases caused by *Fusarium oxysporum* fsp. *ciceri, Macrophomina phaseolina, Pythium aphanidermatum* and

Rhizoctonia solani (Sen, 1997, 1998). It acts by mycoparasitism and also promotes the plant growth.

Coniothyrium

The genus *Coniothyrium* belongs to the class "Coelomycetes" within the subdivision "Deuteromycotina". The species *C. minitans* has been recognised as strong sclerotial parasite. *C. minitans* grows profusely on the culture medium and produces abundant single celled conidia (Fig. 18.1). It is well known parasite of the sclerotial fungi, i.e. *Sclerotinia*, *Sclerotium*, etc. Following penetration of the hyphal walls of the sclerotium by hyphae of *C. minitans*, the cytoplasm of the host disintegrates and the walls collapse.

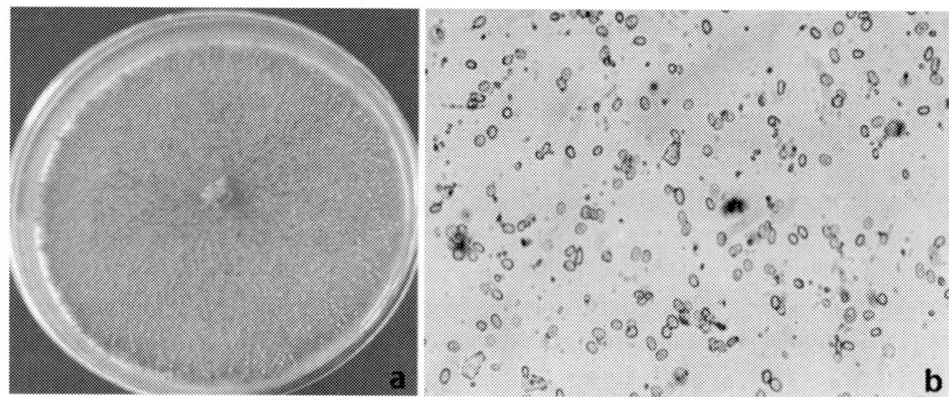

Fig. 18.1. *Coniothyrium minitans (a)* Colony on culture medium (b) Conidia.

Table 18.7. *Coniothyrium minitans* - based biofungicides for the control of plant diseases

Name of the formulation	Registered in country	Method of application	Effective against
Contans WG, Intercept WG	Germany Switzerland, Austria, Hungary, Luxembourg and Poland	Soil application	Sclerotial fungi
Coniothyrin	Russia	Soil Application	Sclerotial fungi
Koni WG	Hungary	Soil application and spray	Sclerotial fungi

Conidia of *C. minitans* germinate in contact with sclerotia, infect the rind, cause lysis of the cytoplasm and then grow into the medulla. The hydrolytic enzymes β-1, 3-glucanase and chitinase appear to be involved in the infection of sclerotia by *C. minitans*. Sporulation of the mycoparasite occurs on the surface of and inside the sclerotium, thus providing the inoculum required for further infections of sclerotia in the soil. The commercial formulations *of C. minitans* are widely used for the biocontrol of sclerotinia diseases of oilseed and other crops in developed countries (Table 18.7).

Candida

It is a naturally occurring yeast on various food commodities. One species *Candida oleophila* has been widely used against post-harvest rots due to *Penicillium digitatum, P. italicum, Geotrichum candidum* and *Botrytis* species (Droby *et al.*, 1998). It act through competition for nutrients (Jijakli *et al.,* 1993) and pre-colonization of plant wound sites. Two formulations based on it are available in the market (Table 18.8).

Table 18.8. *Candida* - based biofungicides for the control of plant diseases

Name of the formulation	Active ingredient	Effective against
Nexy	*Candida oleophila*	*Botrytis cineria, Penicillium expansum*
Aspire	*Candida oleophila*	*Botrytis cineria, Penicillium expansum*

Chaetomium

The application of *Chaetomium* as a biological control agent to control plant pathogens first commenced in about 1954 when Martin Tviet and M.B. Moor found *C. globosum* and *C. cochliodes* occurring on oat seeds and that these taxa provided some control of *Helminthosporium victoriae* (Tviet and Moore, 1954). *Chaetomium globosum* and C. *cochlioides* are antagonistic to species of *Fusarium* and *Helminthosporium* (Tveit and Moore, 1954). Many species of *Chaetomium* with potential to be biological control agents suppress the growth of bacteria and fungi through competition (for substrate and nutrients), mycoparasitism, antibiosis, or various combinations of these (Marwah *et al.*, 2007; Zhang and Yang, 2007), are effective against different diseases caused by Phytophthora on tangerine and strawberry, wilt of tomato caused by *Fusarium oxysporum* f.sp. *lycopersici* and basal rot of corn caused by *Sclerotium rolfsii.* Biocontrol technology of plant diseases with this BCA has successfully been demonstrated, either completely or partially, when integrated with other control measures for appropriate disease management.

Chaetomium based formulations are used for biocontrol of soil born as well as aerial plant pathogens (Table 18.9).

Table 18.9. *Chaetomium* based biofungicides for the control of various plant diseases

Name of the formulation	Active ingredient	Effective against
Ketomium (R)	*Chaetomium cupreum* CC01-CC10 and *Chaetomium globosum* CG01-CG12	*Phytophthora palmivora, Phytophthora parasitica, Colletotrichum gloeosporioides, Fusarium oxysporum* f.sp. lycopersici, *Fusarium moniliform, Pyricularia oryzae Sclerotium rolfsii, Drechslera maydis*
Ketocin	*Chaetomium cupreum*	*Fusarium oxysporum* f.sp. *lycopersici*

Gliocladium

It is commonly occurring soil saprophyte and known to suppress various soil borne plant pathogens, viz. *Rhizoctonia, Pythium,* etc. It acts against pathogens mainly through mycoparasitism, competition and antibiosis. Recently, molecular evidence indicates that *G. virens* is more closely related to *Trichoderma*. This supports suggestions that this taxon should be referred to as *Trichoderma virens* (Hebber and Lumsden, 1999; Paulitz and Belanger, 2001; Punja and Utkhede, 2004). Two formulations based on *Gliocladium* are available in the market (Table 18.10).

Table 18.10. *Gliocladium*-based biofungicides for the control of various plant diseases

Name of the formulation	Active ingredient	Effective against
SoilGard® 12G	*Gliocladium virens* strain GL-21	damping-off and Root rot pathogens
Primastop	*Gliocladium catenulatum* strain J1446	damping-off, seed rot, root and stem rot, and wilt diseases

Non-pathogenic *Fusarium*

Smith and Synder (1971) and Toussoum (1975) were first to demonstrate the role of non-pathogenic *Fusarium* in suppressive soil and indicated that soil suppressive to Fusarium wilts supported a large population of non-pathogenic *Fusarium* spp. *Fusarium* spp. act against pathogens mainly through induced systemic resistance (Larkin and Fravel, 2002; Shimizu *et al.*, 2000) and competition (Mandeel and Baker, 1991; Alabouvette and Couteadier, 1992; Benhamou and Grand, 2001). It is used against plant pathogens alone as well as in combination with *Pseudomonas* spp. to obtain an effective control (Kaur *et al.,* 2007; Elmer, 2008; Lemanceau *et al.,* 1992). Two formulations based on non-pathogenic *Fusarium* are available in the market (Table 18.11).

Table 18.11. Non-pathogenic *F. oxysporum*-based formulations for control of Fusarial wilts

Name of the formulation	Active ingredient	Registered in country	Method of application	Effective against
Biofox C (Dust/alginate granules)	Non-pathogenic *F. oxysporum*	Italy	Seed/Soil treatment	*F. oxysporum* and *F. moniliforme*
Fusaclean (Liquid)	Non-pathogenic *F. oxysporum*	France	Potting mixture and in rows/soil treatment	*F. oxysporum*

Pythium

Pythium oligandrum Drechs. is a soil borne fungus belonging to the order Oomycetes. It is able to live as a saprophyte, but is antagonistic to and parasitic on many fungi from the same or other orders namely, *Botrytis, Fusarium, Pythium, Phytophthora, Gaeumannomyces* spp., etc. (Brozovo, 2002). The different interactions between *Pythium oligandrum* and the various plant pathogens show that the antagonism of *P. oligandrum* is not a simple process mediated by one simple antimicrobial metabolite. Instead, the studies suggest that each interaction is mediated by highly specific events that ultimately determine the outcome of the interaction. From the scope of ultrastructural and cytochemical changes, it is evident that the antagonism is a multifaceted process and dependent on the target species involved (Benhamou *et al.,* 1999). It acts as hyperparasite and also produces protein oligandrin and other compounds which stimulate plant cell walls to fend off pathogen invasion and stimulate plant defence mechanism. *P. oligandrum* based formulations are widely used to protect damping off and other root diseases (Table 18.12).

Table 18.12. *Pythium* - based biofungicides for the control of various plant diseases

Name of the formulation	Active ingredient	Effective against
Polygandron	Oospores of *P. oligandrum*	damping-off
Polyversum	Oospores of *P. oligandrum*	*Pythium ultimum Rhizoctonia solani,* and *Fusarium*

Peniophora gigantea

Phlebiopsis gigantea (= *Peniophora*) is a common saprophytic fungus, found abundant on rotten conifer logs and stumps. It is used for biological control of annosum root rot, caused by *Heterobasidion* spp. in several countries (Table 18.13).

Table 18.13. *Phlebiopsis gigantean* - based biofungicides for the control of various plant diseases

Name of the formulation	Active ingredient	Effective against	Registered in country
Rotstop	*Phlebiopsis gigantea*	*Heterobasidion* spp	Finland
PG Suspension	*Phlebiopsis gigantea*	*Heterobasidion* spp	UK
PG IBL	*Phlebiopsis gigantea*	*H. annosum*	Poland

The hyphae of this fungus antagonize the hyphae of *Heterobasidion* spp. (and some other fungi) on contact - a phenomenon termed hyphal interference. Any hypha of *Heterobasidion* that makes contact with a hypha of *P. gigantea* shows rapid, localized disruption, the protoplasm becomes disorganized and its membrane integrity is affected.

Trichoderma

Trichoderma is classified under subdivision Deuteromycotina with its perfect stage *Hypocrea* in Ascomycotina. It produces typically greenish colonies on medium and branched conidiophores bearing apical phialides, which bear conidia in a mucilage at the tip (Fig. 18.2). It is most widely exploited biocontrol agent. Weindling (1934) first time reported *T. viride* as an antagonist for control of soil-borne pathogens. Earlier the genus was recognised with 9 species (Rifai, 1969), however, revised classification includes several other species. The common species which are known as BCA are *T. viride, T. harzianum, T. hametum, T. polysorum, T. koningii*, etc. It contributes almost 60% of the fungal BCA market mostly as growth enhancers (Whipps and Lumsden, 2001). It is effective against wide range of plant pathogens. It acts against plant pathogens mainly by mycoparasitism, however, other modes of actions such as antibiosis and competition have also been well documented. Several studies have been conducted on the kind of lytic enzymes and secondary metabolites produced by *Trichoderma* spp. These direct or indirect modes of action may be operating simultaneously and biocontrol efficacy depends on the *Trichoderma* strains, host, pathogen, crop and the environmental conditions including pH, temperature, nutrient availability, iron concentration, etc. Different formulations/biofungicides based on *Trichoderma* spp. are available in the market for the control of wide range of plant pathogens specially soil-borne plant pathogens (Table 18.14).

Table 18.14. *Trichoderma*-based biofungicides for the control of plant diseases

Formulation	Active ingredient	Effective against
Root shield, Plant shield, Biotrek 22g, T-22G, T-22HB, T22 Planter Box, Supresivit	*T. harzianum*	*Pythium, Rhizoctonia, Fusarium, Phytophthora, Macrophomina, Sclerotinia, Sclerotium*, etc.
Monitor SD, WP	*Trichoderma* spp.	
Trichodex	*T. harzianum*	
Trichoderma 2000	*T. harzianum*	
TUSAL	*T. harzianum*	
Binab	*T. harzianum+ T. polysorum*	
Trichopel, Trichojet, Trichodowels, Trichoseal	*T. harzianum + T. viride*	
Ecofit, Trieco	*T. viride*	
Promot, Biofungus	*Trichoderma* spp.	
TrianumG/Bioten Wp	*T. gamsii + T. asperellum*	

T. harzianum based formulations are abundant in market for their commercial use. T22 Planter Box, a conidial formulation of *T. harzianum* can be applied on seeds already treated with fungicides (Harman and Bjorkman, 1998). The biocontrol inoculants grow and colonise spermosphere, developing root system and protect against damping of diseases. Similarly, Root Shield, a granular formulation is used as a green house soil amendments. Another product Root Shield drench consist of conidia used as a water suspensible drench to protect wide range of root diseases. In India, Ecofit, talc based *T. viride* formulation is sold to control root rot, damping off, wilts in cotton, chickpea, pigeonpea, groundnut, soybean, and vegetable crops.

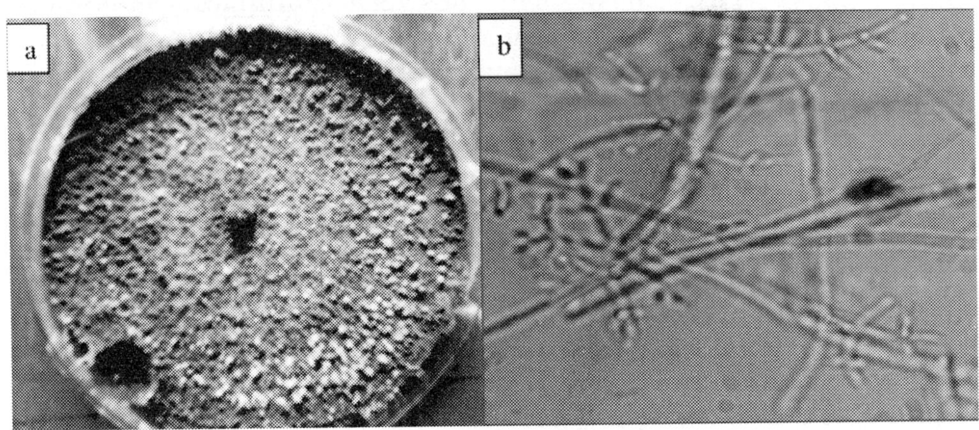

Fig. 18.2. *Trichoderma* (a) colony on artificial medium, (b) Conidiophores and conidia

MODE OF ACTION

The biological control process may be operated by different mechanisms: These BCAs can act against pathogens by one or combination of following mechanisms :

Competition

The pathogen and the antagonist compete for food/nutrients, space and infection sites, which results in lesser or retarded growth of the pathogen. It is generally believed that competition for nutrients is most important among the microbial population, i.e. pathogenic and non-pathogenic for the colonization of rhizosphere for success of biological control. The best example may be cited for the control of fire blight, caused by *Erwinia amylovora* and non-pathogenic *Pseudomonas fluorescens* A506 (Lindow and Leveau, 2002). The siderophores produced by the fluorescent pseudomonads deprive the pathogen of iron. Kloepper *et al.* (1980) firstly demonstrated the role of siderophores in biological control of *Erwinia caratovora* by several plant growth promoting *Pseudomonas fluorescens* strains A1, Bk1, TL3B1 and B10. Other bacteria like *Streptomyces* spp., *A. tumefaciens, Erwinia herbicola, E. amylovora, E. carotovora, Rhizobium leguminosorum, E. chrysanthemi, Pseudomonas* spp. and *Bradyrhizobium japonicum* are capable of producing siderophores (Datroff *et al.*, 2007).

Competition for space and nutrients is also the main mechanism of action in case of non-pathogenic *Fusarium oxysporum*, *Rhizoctonia* and *Pythium* spp. (Larkin and Fravel, 1999; Lemanceau, 1992; Olivain *et al.*, 2006; Kaur *et al.*, 2010). Kessel *et al.* (2005) reported the colonisation of necrotic leaf tissues by *Ulocladium atrum* for the management of *Botrytis cineria*.

Mycoparasitism/Hyperparasitism

The process in which one fungal organism attacks and parasitizes other fungal strain is known as mycoparastism. The concept of biological control of soil-borne pathogens and abundance of mycoparasites was born with discovery of *Trichoderma lignorum* parasitizing a number of soil-borne fungi in culture (Weindling, 1934). The term has been widely used with hyperparasitism, direct parasitism or interfungus parasitism. The process of mycoparasitism involves coiling of hyphae and penetration, production of haustoria and lysis of host hyphae. In laboratory, gradual overgrowth of *Trichoderma* colony may be observed on colony growth of *S. sclerotiorum* in dual culture (Fig. 18.3). Four major groups of hyperparasites have been identified as hypoviruses, facultative parasites, obligate bacterial pathogens and predators. Biological control agents are able to lyse hyphae of pathogens by releasing the lytic enzymes and this is an important and powerful tool for control of plant disease (Harman *et al.*, 2004; Viterbo *et al.*, 2007) such as chitinases, proteases and β-1,3- glucanases (Whipps, 2001).

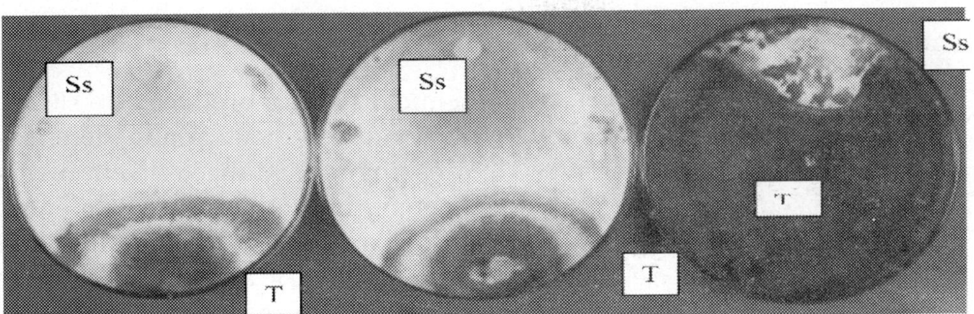

Fig. 18.3. *Trichoderma* (T) showing mycoparasitism against *Sclerotinia sclerotiorum* (Ss) in dual culture

Antibiosis

It includes the inhibition of pathogen growth due to the production of toxic metabolites or antibiotics by the biocontrol agents.

Antibiotics are microbial toxins produced at low concentrations and are able to kill microorganisms (Fravel, 1988). Thousands of antibiotic substances produced by bacteria and fungi are simply demonstrated *in vivo* following the dual culture method (Fig. 18.4). One species of antagonist may be producing several antibiotics. Forty three antibiotic substances are reported to be produced by the *Trichoderma* spp. (Sivasithamparam and Ghisalberti, 1998). Out of these, alkyl pyrones, isonitriles, polyketides, peptaibols, diketopiperazines, sesquiterpenes and steroids are important. Pal and Gardner (2006) enlisted different antibiotics produced by biocontrol agents. *Agrobacterium radiobacter* produces agrocin 84, against *Agrobacterium tumefaciens* (Kerr, 1980). *Pseudomonas fluorescens* 2–79 and 30–84 produce phenazines against *Gaeumannomyces graminis* var. *tritici* (Thomashow *et al.*, 1990).

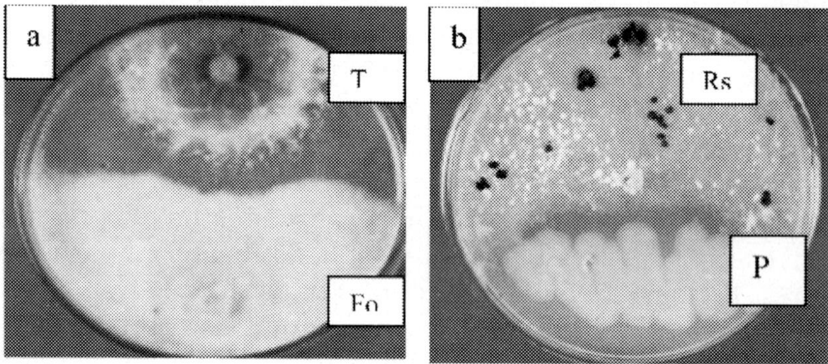

Fig. 18.4. Zone of inhibition produced in dual culture due to antibiosis of (a) *Trichoderma* (T) against *Fusarium oxysporum* (Fo), (b) fluorescent *Pseudomonas* (P) against *Rhizoctonia solani* (Rs).

Induced Resistance

Plants respond to a variety of chemical stimuli produced by BCAs. Such stimuli can either induce or condition host plant defences through biochemical changes expressing resistance mechanisms against subsequent infection by pathogens. Induction of host defences can be localised and/or systemic in nature. The determinants and pathways of induced resistance stimulated by BCAs and other non-pathogenic microbes have being occasionally characterized.(Tjamos *et al.*, 2010) According to Harman *et al.* (2004) strains of *Trichoderma* added to the rhizosphere protect plants against many pathogens including viruses, bacteria, and fungi, because of the induction of resistance mechanisms similar to the hypersensitive response (HR), systematic acquired resistance (SAR), and induced systematic resistance (ISR) in plants. A non-pathogenic strain of *Fusarium*,

Pythium ultimum, and *Rhizoctonia* could induce plant resistance to pathogenic stains (Harman *et al.*, 2004; Kaur *et al.*, 2010).

Hypovirulence

The term hypovirulence is used to describe the reduced pathogenic ability of a pathogenic strain. Basically, the hypovirulence is a type of hyperparasitic relationship between virus and fungal pathogen. The best example is the virus which infects *Cryphonectria parasitica*, a fungal pathogen causing chestnut blight, resulting hypovirulence, which reduced the disease-producing efficacy of the pathogen. This phenomenon has been known to control the chestnut blight in large scale at several places (Milgroom and Cortesi, 2004). Overall, the interaction of virus, fungus, tree, and environmental conditions determines the success of hypovirulence.

COMMERCIAL BIOPESTICIDES

The Environmental Protection Agency (EPA) defines a biopesticide as a pesticide derived from natural materials such as animals, plants, bacteria, and certain minerals. Micro- organisms such as a bacterium, virus, fungus, or microscopic nematode worms can be used in agricultural practices to control or eliminate pests that can inflict damage to a plant. Commercialization of biofungicides is a multi-step process involving a wide range of activities starting from isolation of microorganisms, their evaluation, study of mechanisms, strain improvement, mass production, formulation, delivery, compatibility and registration in sequence. First biocontrol agent was registered in 1979, but now-a-days there are several. First *Trichoderma* biofungicide registration was issued for suppressing *Verticillium malthousei* (*fungicola*), a soil-borne fungal pathogen in 1976 (De Trogoff and Ricard, 1976). In 2005, 26 microbial products for biological disease control with proven effects were marketed in the US by commercial companies (Fravel, 2005). In Europe, 14 microbial products for disease control were registered in 2008 and listed in Annex 1 of Directive 91/414 EEC (Ehlers, 2010). According to Köhl *et al.* (2010) the development of new biocontrol products against plant diseases requires screening of high numbers of candidate antagonists. Antagonists for commercial use have to fulfil many different requirements. Besides being active against the specific targeted plant pathogens (laboratory virulence, field performance, genetic stability, productivity, stability of conidia in storage, stability in formulation, field persistence including tolerance to environmental factors such as

UV, temperature extremes and desiccation (Jenkins *et al.*, 1998; Spadaro and Gullino, 2005), they must be safe and cost-effective. Important criteria besides pathogenic efficacy are market size, ecological characteristics, production costs, safety, environmental risks, and possibilities for intellectual property protection.

Reduced environmental damage and less risk to human health are the main advantages of use of biopesticides. Since the public awareness has increased about ill effects of chemicals, there is more and more demand of organic agricultural produce. It resulted overall decreased plant protection chemicals and increased use of biopesticides in crop protection. According to the US Environmental Protection Agency (USEPA) at end of 1998 there were approximately 175 registered active ingredient of biopesticides and 700 products (Rathod, 2009).According to a new technical market research report, Biopesticides: The Global Market (CHM029C) from BCC Research (www.bccresearch.com), the global market for biopesticides was worth an estimated $1.6 billion in 2009, but is expected to increase to $3.3 billion in 2014, for a 5-year compound annual growth rate (CAGR) of 15.6%. Biopesticides represent a strong growth area in the global pesticide market.

Table 18.15. Antagonistic fungi and bacteria included under the Insecticides Act, 1968

Fungi	Bacteria
Gliocladium spp.	*Bacillus subtilis*
Trichoderma spp.	*Bacillus* spp.
Ampelomyces quisqualis	*Streptomyces griseoviridis*
Candida oleophila	*Streptomyces lydicus*
Pythium oligandrum	*Pseudomonas fluorescens* Migula
Aspergillus niger-strain AN27	*Burkholderia cepacia*
VAM-Fungi	*Agrobactarium radiobacter* strain 84
Phlebia gigantea	*Agrobactarium tumefaciens*
Piriformospora indica	*Erwinia amylovora* (Hairpin protein)
Coniotyrium minitans	*Serratia marcescens* GPS 5
Penicillium islanidicum	*Alcaligenes* spp.

In India, biopesticides are included under the Insecticide Act, 1968. As per the Gazette Notifications No.GSR 200 (E) dated 29/03/78, GSR 839 (E) dated 14/11/85, GSR 13 (E) dated 05/01/90, GSR 621 (E) dated 27/10/97, G.S.R. 224 (E) dated 26/03/99, G.S.R. 868 (E) dated

Haggag, W.M. and Mohamed Abdel-Latif (2007) Biotechnological aspects of microorganisms used in plant biological control. *Am.-Eur. J. Sustain. Agric.*, **1**, 7–12.

Handelsman, J. and Stabb,, E.V. (996) Biocontrol of soil borne plant pathogens. *Plant Cell*, **8**, 1855–1869.

Harman, G.E. and Bjorkman, T. (1998) Potential and existing uses of *Trichoderma* and *Gliocladium* for plant disease control and plant growth enhancement. In: G.E. Harman and C.P. Kubicek (eds.), *Trichoderma and Gliocladium: Enzymes, Biological Control and Commercial Applications*, Vol 2, Taylor & Francis, London, pp. 229–265.

Harman, G.E., Howell, C.R., Viterbo, A., Chet, I. and Lorito, M. (2004) *Trichoderma* species opportunistic, avirulent plant symbionts. *Nat. Rev. Microbiol.*, **2**, 43–56.

Hashioka, Y. and Nakai, Y. (1980) Ultrastructure of pycnidial development and mycoparasitism of *Ampelomyces quisqualis* parasitic on Erysiphales. *Trans. Mycol. Soc. Jap.*, **21**, 329–338.

Hebbar, P.K. and Lumsden, R.D. (1999) Biological control of seedling diseases. In: R.H. Frinklin and J.M. Julius (eds.), *Methods in Biotechnology, Biopesticides: Use and Delivery*, Vol. 5, Humana Press, New York, pp. 103–116.

Ishimaru, C.A., Klos, E.J. and Brubaker, R.R. (1988) Multiple antibiotic production by *Erwinia herbicola*. *Phytopathology*, **78**, 746–750.

Jenkins, N.E., Heviefo, G., Langewald, J., Cherry, A.J. and Lomer, C.J. (1998) Development of mass production technology for aerial conidia for use as mycopesticides. *Biocontr. News Inf.*, **19**, 21N–31N.

Jijakli, M.H., Lepoivre, P., Tossut, P. and Thornard, P. (1993) Biological control of *Botrytis cinerea* and *Penicillium* sp. on post-harvest apples by two antagonistic yeasts. *Med. Fac. Landbouw. Univ. Gent.*, **53**, 1349–1358.

Kaur, R., Kaur, J., Singh, R.S. and Alabouvette, C. (2007) Biological control of *Fusarium oxysporum* f.sp. *ciceri* by non-pathogenic *Fusarium* and fluorescent *Pseudomonas*. *Int. J. Bot.*, **3**, 114–117.

Kaur R., Kaur J. and Singh, R.S. (2010) Nonpathogenic *Fusarium* as a biological control agent. *Plant Pathol. J.*, **9**, 79–91.

Kerr A. (1980) Biological control of crown gall through production of agrocin 84. *Plant Dis.*, **64**, 25–30.

Kessel, G.J.T., Köhl, J., Powell, J.A., Rabbinge, R. and Werf, W. van der (2005) Modeling spatial characteristics in the biological control of fungi at the leaf scale: Competitive substrate colonization by *Botrytis cinerea* and the saprophytic antagonist *Ulocladium atrum*. *Phytopathology*, **95**, 439–448.

Kim, B.S. and Hwang, B.K. (2007) Microbial fungicides in the control of plant diseases. *J. Phytopathol.*, **155**, 641–653.

Kloepper, J.W., Leong, J., Teintze, M. and Schroth, M.N. (1980) *Pseudomonas* siderophores: A mechanism explaining disease suppressive soils. *Curr. Microbiol.*, **4**, 317–320.

Köhl P J., Nicot, P., Ruocco, M. and Blum, B. (2010) Stepwise screening of microorganisms for commercial use in biological control of plant-pathogenic fungi and bacteria. *Biol. Contr.*, **57**, 11–12.

Kulkarni, M., Chaudhari, R. and Chaudhari, A. (2007) Novel tensio-active microbial compounds for biocontrol applicatins. In: A. Ciancio and K.G. Mukerji (eds.), *General Concepts in Integrated Pest and Disease Management,* Springer, Dordrecht, pp. 295–304.

Kiss, L. (2003) A review of fungal antagonists of powdery mildews and their potential as biocontrol agents. *Pest Manage. Sci.*, **59**, 475–483.

Larkin, R.P. and Fravel, D.R. (1999) Field efficacy of selected nonpathogenic *Fusarium* spp. and other biocontrol agents for the control of Fusarium wilt of tomato 1997–1998. *Biol. Cult. Tests*, **14**, 116.

Larkin, R.P. and Fravel, D.R. (2002) Effects of varying environmental condit-ions on biological control of *Fusarium* wilt of tomato by nonpathogenic *Fusarium* spp. *Phytopathology*, **92**, 1160–1166.

Lemanceau, P. (1992) Effets bénéfiques de rhizobactéries sur les plantes: exemple des *Pseudomonas* spp. fluorescents. *Agronomie*, **12**, 413–437.

Lemanceau, P., Baker, P.A.H.M., De Kogel, W.J., Alabouvette, C. and Schippers, B. (1992) Effect of pseudobactin 358 production by *Pseudomonas putida* WCS358 on suppression of Fusarium wilt of carnation by nonpathogenic *Fusarium oxysporum* Fo47. *Appl. Environ. Microbiol.*, **58**, 2978–2982.

Lindow, S.E. and Leveau, J.H. (2002) Phyllosphere microbiology. *Curr. Opin. Biotechnol.*, **13**, 238–243.

Marwah, R.G., Fatope, M.O., Deadman, M.L., Al-Maqbali, Y.M. and Husband, J. (2007) Musanahol: A new aureonitol-related metabolite from a *Chaetomium* sp. *Tetrahedron*, **63**, 8174–8180.

Mandeel, Q. and Baker, R. (1991) Mechanisms involved in biological control of *Fusarium* wilt of cucumber with strains of non-pathogenic *Fusarium oxysporum*. *Phytopathology*, **81**, 462–469.

Milgroom, M.G. and Cortesi, P. (2004) Biological control of chestnut blight with hypovirulence: A critical analysis. *Annu. Rev. Phytopathol.*, **42**, 311–338.

Olivain, C., Humbert, C., Nahalkova, J., Fatehi, J., L'Haridon, F. and Alabouvette, C. (2006) Colonization of tomato root by pathogenic and non-pathogenic *Fusarium oxysporum* strains inoculated together and separately into the soil. *Appl. Environ. Microbiol.*, **72**, 1523–1531.

Paulitz, T.C. and Be´ Langer, R.R. (2001) Biological control in greenhouse systems. *Annu. Rev. Phytopathol.*, **39**, 103–133.

Pal, K. and Gardener, B.M. (2006) Biological control of plant pathogens. The plant health instructor. doi: 10.1094/PHI-A-2006-1117-02. APS net: 1–25.

Punja, Z.K. and Utkhede, R.S. (2004) Biological control of fungal diseases on vegetable crops with fungi and yeasts. In: D.K. Arora (ed.), *Fungal Biotechnology in Agricultural, Food, and Environmental Applications*, Marcel Dekker Inc., New York/ Basel, pp. 157–171.

Rifai, M.A. (1969) A revision of the genus *Trichoderma*. *Mycol. Papers*, **116**, 1–116.

Rathod, J. (2009) Use of biopesticide for sustainable agriculture : Green alternatives in pest management. http://www.suite101.com/content/biopesticides-a-step-towards-sustainable-agric-a110263.

Schipper, B., Lugtenberg, B. and Wiesbeck, P.J. (1987) Plant growth control by fluorescent pseudomonads. In: I. Chet (ed.), *Innovative Approaches to Plant Disease Control*, John Wiley & Sons, New York, pp. 19–39.

Sen, B. (1997) *Aspergillus niger* strain AN-27, A successful BCA for soil borne fungal pathogens of agricultural crops. *FAO Regional Expert Consultation on the Application of Biotech for Pest Management*. Feb. 1997, New Delhi.

Sen, B. (1998) Bioformulation Kalisena-A boon to the farmers. *Pestic. Inf.*, **24**, 29–30.

Shimizu, M., Nakagawa, Y., Sato, Y., Furumai, T., Igarashi, Y., Onaka, H., Yoshida, R. and Kunoh, H. (2000). Studies on endophytic actinomycetes. (I) *Streptomyces* sp. isolated from *Rhododendron* and its antifungal activity. *J. Gen. Plant Pathol.*, **66**, 360–366.

Sivasithamparam, K. and Ghisalberti, E.L. (1998) Secondary metabolism in *Trichoderma* and *Gliocladium*. In: G.E. Harman and C.P. Kubicek (eds.), *Trichoderma and Gliocladium*, Vol. 1, Taylor and Francis, London, pp. 139–191.

Smith, S.N. and Snyder, W.C. (1971) Relationship of inoculums density and soil types to severity of *Fusarium* Wilt of sweet potato. *Phytopathology*, **61**, 1049–1051.

Spadaro, D. and Gullino, L.M. (2005) Improving the efficacy of biocontrol agents against soilborne pathogens. *Crop Prod.*, **24**, 601–613.

Sundheim, L. and Amundsen, T. (1982) Fungicide tolerance in the hyperparasite *Ampelomyces quisqualis* and integrated control of cucumber powdery mildew. *Acta Agric. Scand. Vica*, **32**, 349–355.

Thomashow, L.S., Weller, D.M., Bonsall, R. F. and Pierson, III L.S. (1990) Production of the antibiotic phenazine-lcarboxylicacid by fluorescent *Pseudomonas* species in the rhizosphere of wheat. *Appl. Environ. Microbiol.*, **56**, 908–912.

Tjamos, E.C., Tjamos, S.E. and Antoniou, P.P. (2010) Biological management of plant diseases: Highlights on research and application. *J. Plant Pathol.*, **92**(Supplement), S4.17–S4.21.

Toussoun, T.A. (1975) *Fusarium*-Suppressive Soils. In: G.W. Bruehl (ed.), *Biology and Control of Soil-Borne Plant Pathogens,* American Phytopathological Society, St Paul, MN, pp. 145–151.

Tveit, M. and Moore, M.B. (1954) Isolates of *Chaetomium* that protect oats from *Helminthosporium victoriae. Phytopathology,* **44**, 686–689.

Viterbo, A., Inbar, J., Hadar, Y. and Chet, I. (2007) Plant disease biocontrol and induced resistance via fungal mycoparasites. In: C.P. Kubicek and I.S. Druzhinina (eds.), *Environmental and Microbial Relationships*, 2nd Editon, Springer-Verlag, Berlin Heidelberg, pp. 127–146.

Whipps, J.M. (2001) Microbial interactions and biocontrol in the rhizosphere. *J. Exp. Bot.*, **52**, 487–511.

Whipps, J.M. and Lumsden, R.D. (2001) Commercial use of fungi as plant disease biological control agents: Status and prospects. In: T.M. Butt, C.W. Jackson and N. Magan (eds.), *Fungi as Biocontrol Agents: Progress,Problems and Potential*, CAB International, Wallingford, pp. 9–22.

Weindling, R. (1934) Studies on a lethal principle effective in the parasitic action of *Trichoderma lignorum* on *Rhizoctonia solani* and other soil fungi. *Phytopathology*, **24**, 1153–1179.

Zhang, H.Y. and Yang, Q. (2007) Expressed sequence tags-based identification of genes in the biocontrol agent *Chaetomium cupreum. Appl. Microbiol. Biotechnol.,* **74**, 650–658.

19

Three Decades of Research on Botanical Pesticides in Canada : 1980-2010

RITA SEFFRIN AND MURRAY B. ISMAN*

Faculty of Land and Food Systems,
University of British Columbia Vancouver, Canada V6T1Z4
*E-mail: murray.isman@ubc.ca

INTRODUCTION

Over 34,000 tonnes of pesticides are used annually across Canada. According to the National Research Council, expenditures on pesticides doubled between 1980 and 1990 and increased eightfold since 1970. According to Canadian studies, DDT and DDE residual levels in the breast milk to Inuit women in the Eastern Arctic average 1210 ppb, compared to 171 ppb in women in Southern Ontario. The extreme levels of persistent pesticides in Arctic people and animals high on the food chain are largely a result of intercontinental transport and bioaccumulation. Similar build-ups of other harmful pesticides have been recorded in various parts of Canada including the Great Lakes, the St. Lawrence River and the Prairies. The negative effects of pesticides in the Canadian environment are well documented. For example, autopsies performed on a group of Canada Geese that died inexplicably near Simcoe Ontario in September 1996 revealed that the grass and clover found in their digestive systems contained diazinon –an organophosphate insecticide commonly used on turfgrass (http://www.sierraclub.ca/national/programs/health-environment/pesticides/index.shtml).

Worldwide pesticide use increased steadily during the 1970s and 1980s at around 5 percent per year. The Canadian pesticide industry consists primarily of products for the protection of agricultural crops, but insecticides account for only 9% of overall pesticide sales in Canada (Crop Protection Institute 1996-97 Annual Report). The major market for pesticides in Canada is for those herbicides used on the major field crops – wheat, barley and canola. In 2002, the world pesticide market was valued at about USD 27.7 billion, with Canada accounting for about 3% of this amount. From a global perspective, the pesticide industry in Canada is considered to be modest in profit potential for a manufacturer. The main reasons for this are: the large acreage crops in Canada (wheat, barley, canola) are grown using extensive agricultural practices requiring modest quantities of pesticides on a per hectare basis; slower and higher cost of pesticide registration in comparison to large global markets; relatively small market size for pesticides in comparison to the U.S., Europe and Japan; and typically higher corporate tax rates. The majority of pesticide research in Canada consists of field trials of products developed in other countries by multinational agrochemical companies, and government sponsored research on the environmental fate and impact of pesticides used in agriculture and forestry. Efforts aimed at development of novel pest management products in Canada have been very modest in comparison.

Globally, the rapid increase in the human population and limited availability of arable land constitute the key factors stimulating market growth for agrochemicals. The worldwide pesticides market reached USD 52 billion in 2008, and is expected to grow to by up to 16% per annum by 2013. This is in spite of global economic instability and consumer pressure for food safety and environmental protection. Although pesticide regulations differ from country to country, pesticides are traded across international borders. The use of pesticides in Canada is regulated by Health Canada through the Pest Control Products Act. In Canada, 154 municipalities and the entire provinces of Quebec, Ontario and New Brunswick have now placed restrictions or outright bans on the cosmetic use of pesticides as a result of health and environmental concerns. Over 250 products will be banned for sale and more than 95 pesticide ingredients will be banned for cosmetic uses (http://en.wikipedia.org/wiki/Pesticides_in_Canada). Since 1990, the world market for organic products has grown, reaching US$ 55 billion in 2009 according to Organic Monitor (www.organicmonitor.com). This demand has driven a similar increase in organically managed farmland. Approximately 37 million hectares worldwide were farmed organically in 2009, representing approximately 0.9% of total world farmland (Willer and Kilcher 2011). Pressure from consumers for organic produce and other foods are leading even conventional growers to reduce their use of synthetic pesticides and consider alternatives, creating greater market space for botanical insecticides.

The majority of research on botanical insecticides conducted in Canada has taken place at the University of British Columbia in the laboratory of Prof. Murray B. Isman, collaborating for a number of years with the late Prof. G.H. Neil Towers, and at the University of Ottawa in the laboratory of Prof. John T. Arnason, collaborating with Prof. Bernard Philogene. The research summarized in the following sections represents their contributions to our knowledge of botanical insecticides over the past 30 years.

NEEM AND AZADIRACHTIN

Some of the earliest work on botanical insecticides in Canada focused on the development of products based on extracts from the seeds of the Indian neem tree, *Azadirachta indica* (Meliaceae) (Fig. 19.1), a plant used for centuries in India for pest control (Saxena *et al.*, 1989) and studied since the early 1960's for use in modern agriculture. (Pradhan *et al.*, 1962).

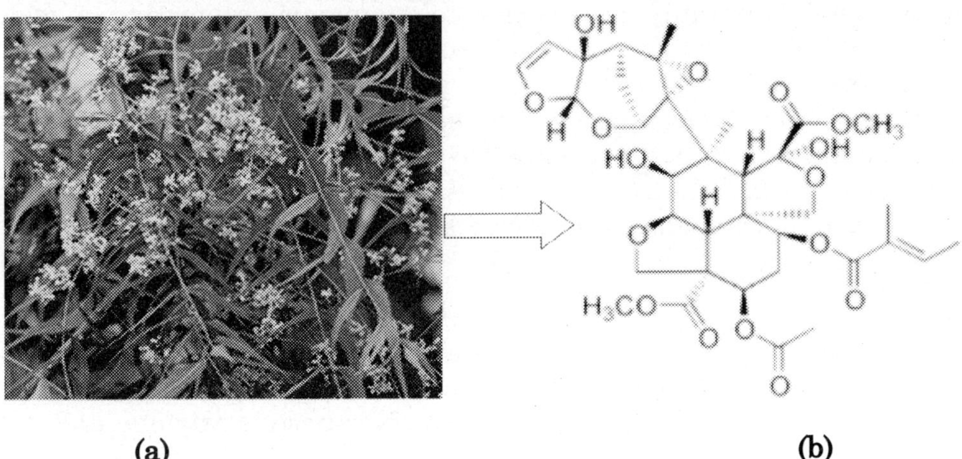

Fig. 19.1. Neem tree (a), and structure of azadirachtin (b)

The correct structure of azadirachtin, a complex triterpenoid responsible for the outstanding biological properties of neem, was not established until the mid 1980s. The attributes of neem-based insecticides are widely knows. They include both behavioural (antifeedant) and physiological effects on insects, efficacy against a broad spectrum of insects pests, systemic action in some plants, minimal disruption of natural enemies, and low mammalian toxicity (Isman, 2004). Research at the University of British Columbia demonstrated that azadirachtin is clearly the most important active constituent in commercial preparations (Isman *et al.*, 1990). Examination of acute and sublethal effects of neem on the biocontrol agents *Phytoseiulus persimilis* and *Amblyseius cucumeris* (Acari: Phytoseiidae) demonstrated that neem favored these natural enemies compared to their prey the twospotted spider mite *Tetranychus urticae* (Acari: Tetranychidae) making it compatible with most integrated pest management progròams (Spollen and Isman, 1996). Other studies at UBC demonstrated the efficacy of neem and azadirachtin to aphids (Lowery and Isman 1995), the relative safety of neem to foraging honeybees (Naumann and Isman, 1996), the safety of neem to juvenile salmonids (Stroh *et al.*, 1998) and the systemic action of neem in pine trees against bark beetles (Naumann *et al.*, 1995). An important finding was that green peach aphids (*Myzus persicae*) could evolve up to 42-fold resistance to pure azadirachtin, but not to a neem extract containing the same absolute amount of azadirachtin (Feng and Isman, 1995). This study indicated the value of using botanical insecticides in that as natural mixtures, resistance in insects can be mitigated. Neem insecticides have minimal impact in the environment, and essentially no toxicity to mammals, conferring a wide margin of safety to the user and consumers. No surprisingly then, neem has been considered a modern paradigm for botanical pesticides (Isman, 1997). Unfortunately the pesticide regulatory authority in Canada did not facilitate registration of neem insecticides even though efficacious formulations with commercial potential were developed by two Canadian companies and growers lobbied for this use.

AGLAIA

Aglaia is a genus of more than 100 species belonging to the mahogany family (Meliaceae). These trees occur in the tropical and subtropical forests of Southeast Asia, Northern Australia and the Pacific, and several species demonstrate bioactivity against insects (Satasook *et al.*, 1993). Limonoid triterpenes, characteristic secondary metabolites of the Meliaceae, are normally responsible for bioactivity in plants of this family, but in the case of *Aglaia*, the insecticidal principles are modified benzofurans (Isman, 1997). More than a third of 19 *Aglaia* species screened, mostly of Indo-Malaysian origin, were significantly active in bioassays against the polyphagous lepidopteran *Peridroma saucia* (Noctuidea). Extracts of at least seven of these species significantly reduced larval growth of *P. saucia; A. odorata* yielded the most inhibitory extracts. The insecticidal principle of *A. odorata* was isolated and identified as the highly substituted benzofuran, rocaglamide, a potent natural insecticide discovered in 1993 (Janprasert *et al.*, 1992). The mode of action of rocaglamide is inhibition of protein synthesis (Ohse *et al.*, 1996), which accounts for its potent but slow action against insects (Satasook *et al.*, 1993).

TRICHILIA

The mahogany family Meliaceae has proven to be a profitable taxon to investigate for natural products active against pests. This plant family is also of interest because it includes some economically important timber species among which are species in the genus *Trichilia*. With 230 recognized species in the New World Tropics the probability of finding bioactivity species was high (Isman *et al.*, 1997). Screening of six Costa Rican species for biological activity against larvae of *Spodopetra litura* (Lepidptera : Noctuidae) was conducted in 2001. Extracts from twigs from six species showed that the most active species was *T. americana* and the least was *T. glabra* (Wheeler *et al.*, 2001). Several investigators have isolated limonoids from different *Trichilia* species and demonstrated biological effects on insects, including feeding deterrence. Crude wood extracts of *T. americana* were effective at protecting cabbage plants (Wheeler and Isman, 2000).

TALL OIL

Tall oil is a viscous yellow-black odorous liquid obtained as a by-product of the Kraft process of wood pulp manufacture when pulping mainly coniferous trees.(Lars, 2000; Gullichsen *et al.*, 2000). The name originated as an anglicization of the Swedish "tallolja" ("pine oil") (Biermann, 1993). Tall oil is the third largest chemical by-product in a Kraft mill after lignin and hemicellulose; the yield of crude tall oil from the process is in the range of 30–50 kg/ton pulp (Stenius, 2000). Crude tall oil and two of its derivatives were assessed for antifeedant and growth inhibitory effects, in the variegated cutworm (*Peridroma saucia* Hübner) (Xie and Isman, 1992). The substances tested are both toxic to neonate *P. saucia* and inhibitory to larval growth. The results suggest that an environmentally sound, low cost natural pest control agent could be developed based on tall oil. Comparative toxicity and deterrence of depitched tall oil, a by-product of the kraft process for pulping softwood, and Superior 70 oil, a common horticultural oil, was evaluated in the green peach aphid, *Myzus persicae* (Sulzer). Leaf disc choice bioassays using second-instar *M. persicae* indicated that both depitched tall oil and Superior 70 oil deter aphids in a dose-dependent manner. The deterrent effect of depitched tall oil on *M. persicae* persisted for at least 3 days under greenhouse conditions (Xie and Isman, 1995). The active principles in tall oil against the variegated cutworm, *Peridroma saucia* Hübner were investigated abietic, dehydroabietic, and isopimaric acids, the major resin acid components of crude tall oil and depitched tall oil (Fig. 19.2). When crude tall oil samples of differing resin acid composition were incorporated into artificial diet at a concentration of 2.0% fresh weight, they suppressed larval growth by 45–60% compared to controls. This suppression was significantly correlated with the equivalent contents of abietic, dehydroabietic, isopimaric, and total resin acids. These results were also evident from a diet choice test, showing that the second-instar larvae obviously selected diets with low levels of resin acids. Bioassays with pure resin acids (abietic, dehydroabietic, and isopimaric acids) demonstrated that all individual chemicals have similar bioactivity against this insect. Comparison of the bioactivities of depitched tall oil and an equivalent mixture of pure resin acids in the*Peridroma* chronic growth bioassay indicated that pure resin acids and depitched tall oil share a common mode of action to this insect. This study confirms that resin acids are major active principles in tall oil against the variegated cutworm, but other chemicals likely also contribute to the bioactivity of tall oil (Xie *et al.*, 1993).

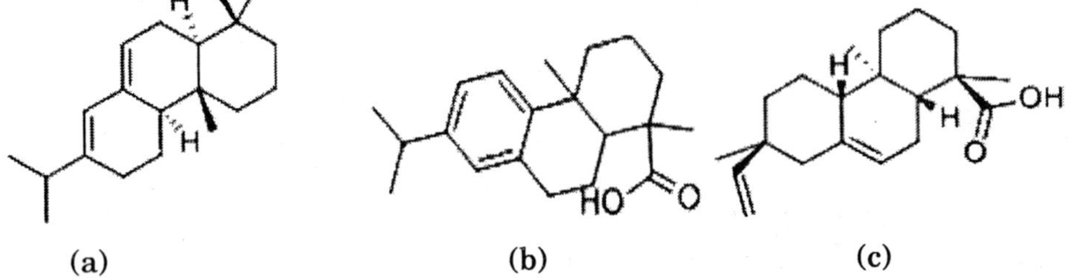

OH H H

(a)

HO O

(b)

H OH H O

(c)

Fig. 19.2. Resin and components of tall oil, (a) Abietic acid, (b) Dehydroabietic acid, and (c) Isopimaric acid

CEDRELA

The genus *Cedrela* includes some important timber species of neotropical forests (Arnason, 2004). The common tropical cedar tree of the Americas, *C. odorata*, contains the limonoid gedunin and closely related derivatives in substantial amounts in the bark and wood. Gedunin (Fig. 19.3a) and *C. odorata* extract showed antifeedant properties, and significantly reduced growth and delayed development of lepidopteran larvae (Arnason *et al.*, 1987; Ewete *et al.*, 1996).

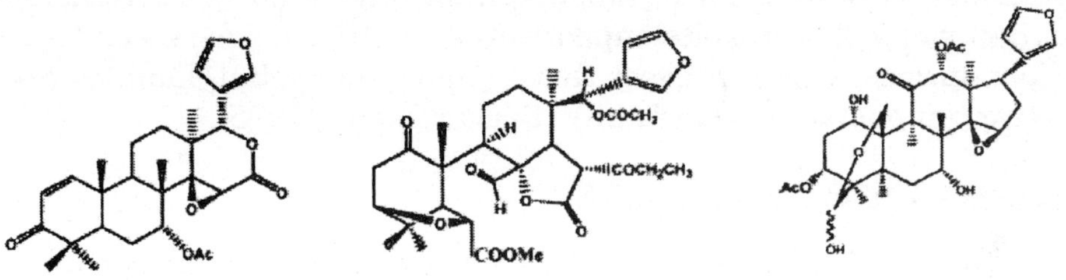

(a) (b) (c)

Fig. 19.3. Structure of gedunin (a), cedrelanolide (b), and toosendanin (c)

Investigation of a related species, *C. salvadorensis* yielded the limonoid cedrelanolide (Fig. 19.3b), which is comparable in activity to gedunin and to toosendandin (Fig. 19.3c), a commercial insecticide derived from *Melia azedarach,* in the European corn borer, *Ostrinia nubilalis*. When incorporated into artificial diets of neonates at 50 ppm, all compounds caused larval mortality as well as growth reduction and increased the development time of survivors in a concentration-dependent manner. Additional effects observed in many of the limonoid-treated groups included a significant delay in time to pupation and adult emergence (Jimenez *et al.*, 1997).

PIPER LIGNANS

The tropical plant family Piperaceae has proven to be a source of diverse medicines and culinary spices. The secondary plant compounds that produce these desired qualities function also as chemical defenses for many species in the genus *Piper*. The compounds with the greatest insecticidal activity in this genus are the isobutylamides. Many studies have shown the effectiveness of *Piper* spp. extracts for the control of stored products pests and studies from the University of Ottawa have evaluated efficacy of extracts of *Piper nigrum*, *P. guineense* and *P. tuberculatum* against insect pests of the home and garden. *Piper* extracts offer a unique and useful source of biopesticides (Scott *et al.*, 2007). Insecticidal and growth-reducing properties of extracts of 14 species of American neotropical Piperaceae were investigated by inclusion in diets of a polyphagous lepidopteran, the European corn borer, *Ostrinia nubilalis*. Nutritional indices suggested most extracts acted via postdigestive toxicity. *Piper aduncum, P. tuberculatum*, and *P. decurrens* were among the most active species and were subjected to bioassay-guided isolation of the active components. Dillapiol was isolated from the active fraction of *P. aduncum*, piperlonguminine from *P. tuberculatum*, and a novel neolignan from *P. decurrens*. The results support other studies on Asian and African *Piper* species, which suggest that lignans and isobutylamides are the active compounds in this family (Bernard *et al.*, 1995).

ANNONA SEED EXTRACTS

Various *Annona* species (Annonaceae) produce edible fruits that are directly consumed or used to produce juice. The bitter tasting seeds that are not consumed are well known as a source of insecticidal long-chain acetogenins that are toxic to insects via the same mode of action as the rotenoids from *Derris* and other tropical legumes (McLauglin *et al.*, 1998). Screening of seeds of tropical fruits grown in Ambon and adjacent islands in eastern Indonesian indicated that extracts of seeds of the sweetsop, *A. squamosa*, were the most toxic to the tobacco cutworm *Spodoptera litura* (Leatemia and Isman, 2004). A crude seed extract was shown to protect cabbage plants in the greenhouse from attack by the diamondback moth (*Plutella xylostella*) with efficacy comparable to the widely used botanical insecticide, rotenone (Leatemia and Isman, 2004). A similar study indicated that extracts of *A. squamosa* and *A. atemoya* from Brazil protected cabbage in the greenhouse from attack by the cabbage looper, *Trichoplusia ni* (Seffrin *et al.*, 2010). Crude methanolic seed extracts deterred feeding of third instar *T. ni* larvae in a leaf disc choice bioassay. *A. squamosa* was 10 times more active as a feeding deterrent than *A. atemoya*. *A. squamosa* was also three times more active as a growth inhibitor than *A. atemoya*. Methanolic seed extracts of *A. squamosa* and *A. atemoya* were toxic to third instar *T. ni* larvae both through topical and oral application. The activity in these plants suggests the potential for development as botanical insecticides, especially for local use in Brazil.

PLANT ESSENTIAL OILS

Certain plant essential oils, widely used as fragrances and flavors in the perfume and food industries, have long been reputed to repel insects. Recent investigations in several countries demonstrate that some plant essential oils not only repel insects, but have contact and fumigant insecticidal actions against many pests. Various essential oils are documented to exhibit acute toxic effects against insects. Monoterpenoids (terpenes and biogenically-related phenols) commonly found in plant essential oils were tested for acute toxicity via topical application to tobacco cutworms (*Spodoptera litura*). The most toxic among ten such compounds was thymol (LD_{50} = 25.4 μg/larva) from garden thyme, *Thymus vulgaris* (Hummelbrunner and Isman, 2001). Twenty-one essential oils were tested for insecticidal activity via topical administration to third instar larvae of the cutworm, *S. litura*. Oils of *Satureia hortensis, Thymus serpyllum* and *Origanum creticum* produced >90% larval mortality at a dose of 100 μg per larva. The LD_{50} value for *S. hortensis* (48.4 μg) was comparable to that for *Thymus vulgaris* (46.9 μg). The monoterpenoid phenols thymol and carvacrol, major constituents of the oils of *Thymus* and *Satureia* species, likely account for the insecticidal action of the species (Isman *et al.*, 2001). *Rosmarinus officinalis* (rosemary) oil caused complete mortality of *Tetranychus urticae* (Acari: Tetranychidade) (Miresmailli *et al.*, 2006). Machial *et al.* (2010) recently screened a selection of 17 essential oils against two lepidopteran pests, the cabbage looper, *Trichoplusa ni* and the oblique banded eafroller, *Choristoneura rosaceana*. Among these oils, patchouli oil (*Pogostemon cablin* Benth.) and thyme oil (*Thymus vulgaris* L.) were the most toxic against C. rosaceana, whereas patchouli oil, garlic oil (*Allium sativum* L.) and lemongrass oil (*Cymopogon nardus* L.) were the most toxic against *T. ni*.

One of the most interesting and important findings has been the repeated demonstration of internal synergy among terpenoid constituents of plant essential oils. Based on their toxicity to the twospotted spider mite, *Tetraynychus urticae* in isolation, major constituents of rosemary could be categorized as putatively very active, moderately active, or inactive. Mixtures of actives alone did not reproduce the toxicity of the intact essential oil to mites – admixture with the putatively inactive constituents was required to reproduce full toxicity observed with the natural oil (Miresmailli *et al.*, 2006). This phenomenon of internal synergy was demonstrated further based on toxicity of *Litsea* essential oils to the cabbage looper, *T. ni* (Jiang *et al.*, 2009).

Most essential oil and their major constituents are relatively non-toxic to laboratory animals and fish in toxicological tests, and meet the criteria for "reduced risk" pesticides. These products have been well received by consumers for use against home and garden pests, and have made significant inroads in urban and industrial pest management where human safety is placed at a premium. In addition, they can also prove effective in certain agricultural situations, particularly for organic food production (Isman *et al.*, 2010).

THIOPHENES

Plants have a set of physical and chemical traits that operate as defense mechanisms against herbivores (Harborne, 1993; Ingrouille, 1992). Chemical defenses involve the manufacture and storage of organic substances in different plant parts, which repel, act as toxic substances, or have antinutritional effects on herbivores (Harborne, 1991). Thiophenes and their derivatives are among these chemical defenses in plants. They are sulfurous heterocyclic compounds derived from polyacetylenes, which may be stored in plant tissues or released into the soil (Tang *et al.*, 1987).

Polyacetylenes and their thiophene derivatives, characteristic secondary metabolites of the Asteraceae, were examined for their effects on herbivorous insects. Three thiophenes (a monothiophene, a bithiophene, and α-terthienyl) and four polyacetylenes (phenylheptatriyne, phenylheptadiynene, phenylheptadiyene acetate, and matricaria lactone) were studied for their phototoxicity and light-independent toxicity to the polyphagous lepidopteran, *Ostrinia nubilalis*, whose host range includes a number of phototoxic Asteraceae, a polyphagous lepidoteran, *Euxoa messoria*, whose host range includes very few species of Asteraceae, and an oligophagous lepidopteran, *Manduca sexta*, which is a specialist on Solanaceae. Several compounds were phototoxic to *M. sexta* and *E. messoria* even at very low irradiance levels, but behavioral adaptations, including spinning silk and boring into diet, allowed *O. nubilalis* to avoid photosensitization. Light-independent activity of the compounds to all three species involved feeding deterrence and long term metabolic toxicity in the form of impaired nutrient utilization. The thiophenes were more toxic than their acetylenic precursors, and toxicity increased with increasing number of thiophene rings (Champagne,1986).

The secondary compound α-terthienyl derived from the common marigold, *Tagetes* and related molecules are especially toxic to mosquito larvae as photoactivated insecticides. Trials under tropical conditions indicate a very high level of activity against the malaria mosquito, *Anopheles gambiae* (Arnason, 1989). The thiophenes, α-terthienyl and methyl- α -terthienyl are found in many plant species of the family Asteraceae. These compounds and a synthetic analogue, cyano- α-terthienyl, controlled *Aedes intrudens* Dyar (Diptera: Culicidae) larvae at application rates between 10 and 40 g per hectare in field trials (Fields *et al.*, 1991).

OTHERS SPECIES OF MELIACEAE

Encouraged by the commercialization of neem, many other plant species from the Meliaceae family havae been evaluated and developed as botanical insecticides and antifeedants. Closely related to the India tree neem is the chinaberry tree, *Melia azedarach.* The seeds of this tree contain insecticidal constituents. The bark of this tree and the *Melia toosedan* contain limonoid triterpenes that can be used to produce botanical pesticide (Chiu, 1995). In the early 1990s a botanical insecticide produced in China was based on an extract of bark of *Melia toosendan*, a tree considered by most taxonomists to be synonymous with *M. azedarach*. The extract contains a number of triterpenoids based on toosendanin, a substance reported to be a stomach poison for chewing insects (Chiu, 1988). Later studies suggest that this substance acts primarily as a feeding deterrent but can also serve as a synergist for conventional insecticides (Chen *et al.*, 1995; Feng, 1995). An insecticide containing toosendanin and nicotine (2.5%) as active ingredients is currently sold in China for pest management (Isman, 2006).

Extracts of *Azadirachta indica* (neem), *A. excelsa* (sentang), *Melia volkensii*, *M. azedarach* (chinaberry) and *Trichilia americana*, (all belonging to the family Meliaceae) along with commercial botanical insecticides ryania, pyrethrum, rotenone and essential oils of rosemary and clove leafs were tested against the cabbage looper, *Trichoplusia ni* and the armyworm, *Pseudaletia unipuncta*. Most of the extracts and botanicals tested proved to be strong growth inhibitors, contact toxins and significant feeding deterrents to both lepidopteran species (Akhtar *et al.*, 2007). Thirty-one species in twenty genera of the plant family Meliaceae were assayed for the production of growth-inhibiting phytochemicals, using the generalist herbivore *Peridroma saucia*. Most species were inhibitory when methanolic extracts were incorporated into artificial diets at concentrations at or below those occurring naturally. In general members of the subfamily Melioideae were more inhibitory than members of the Swietenioideae (Champagne *et al.*, 1993).

CONCLUSIONS

Research on botanical insecticides in Canada over the past thirty years has led to the discovery of a wide array of novel principles from plants with useful bioactivity against insect pests. In addition to the isolation and elucidation of novel natural insecticides, other studies in Canada have determined or confirmed the modes of action of some insecticidal natural products, and demonstrated selectivity favoring natural enemies and pollinators. Although commercial formulations have been developed, only in the case of the essential oils have the products proceeded into production and use, and only then outside of Canada. To date the regulatory climate in Canada has not been encouraging for companies wishing to register natural pesticides in the country and, therefore, they have developed markets elsewhere.

REFERENCES

Arnason, J.T., Guillet, G. and Durst, T. (2004) Phytochemical diversity of insect defenses in tropical and temperate plant families. *In* J.T. Arnason, B.J.R. Philogene and M. Abou-Zaid (eds.), *Chemical Ecology and Phytochemistry of Forest Ecosystems*, Plenum Press, New York, pp. 1–18.

Arnason, J.T., Philogène, B.J.R., Morand, P., Imrie, K., Iyengar, S., Duval, F., Soucy- Breau, C., Scaiano, J.C., Werstiuk, N.H., Hasspieler, B. and Downe, A.E.R. (1989) Naturally occurring and synthetic thiophenes as photoactivated insecticides. In: J.T. Arnason, B.J.R. Philogene and P. Morand (eds.), *Insecticides of Plant Origin*, Amer. Chem. Soc. Symposium Series 387, Washington DC, pp. 164–172.

Arnason, J.T., Philogene, B. J.R., Donskov, N. and Kubo, I. (1987) Limonoids of the Meliaceae and Rutaceae reduce feeding, growth and development of *Ostrinia nubilalis*. *Entomol. Exp. Appl.,* **43**, 221–226.

Akhtar,Y, Yeoung,Y,R. and Isman, M.B. (2007) Comparative bioactivity of selected extracts from Meliaceae and some commercial botanical insecticides against two noctuid caterpillars, *Trichoplusia ni* and *Pseudaletia unipuncta*. *Phytochem. Rev.*, **7**, 77–88.

Bernard, C.B., Krishnamurty, H.G., Chauret, D., Durst, T., Philogène, B.J.R., Sánchez-Vindas, P., Hasbun, C., Poveda, L., San Román, L. and Arnason, J.T. (1995) Insecticidal defenses of Piperaceae from the neotropics. *J. Chem. Ecol.*, **21**, 801–814.

Biermann, C.J. (1993) *Essentials of Pulping and Papermaking.* Academic Press, Inc., San Diego.

Champagne, D.E., Isman, M.B., Downum, K.R. and Towers, G.H.N. (1993) Insecticidal and growth-reducing activity of foliar extracts from Meliaceae. *Chemoecology*, **4**,165–173.

Champagne, D.E,. Arnason, J.T., Philogène, B.J.R., Morand, P. and Lam, J. (1986) Light-mediated allelochemical effects of naturally occurring polyacetylenes and thiophenes from Asteraceae on herbivorous insects. *J. Chem. Ecol.*, **12**, 835–858.

Chauret, D.C., Bernard, C.B., Arnason, J.T., Durst, T., Krishnamurty, H.G., Sanchez-Vindas, P., Moreno, N., Roman, L.S. and Poveda, L. (1996) Insecticidal neolignans from *Piper decurrens. J. Nat. Prod.*,**59**,152–155.

Chen, W., Isman, M.B. and Chiu, S.F. (1995) Antifeedant and growth inhibitory effects of the limonoid toosendanin and *Melia toosendan* extracts on the variegated cutworm, *Peridroma saucia* (Lepidoptera: Noctuidae). *J. Appl. Entomol.*, **119**, 367–370.

Chiu, S.F. (1988) Recent advances in research on botanical insecticides in China. In: J.T. Arnason, B.J.R. Philogene and P. Morand (eds.), *Insecticides of Plant Origin.* Amer. Chem. Soc. Symposium Series 387, American Chemical Society, Washington DC, pp. 69–77.

Chiu, S.F. (1995) *Melia toosendan* Sieb. & Zucc. In: H. Schumutterer (ed.), *The Neem Tree: Source of Unique Natural Products for Integrated Pest Management, Medicine, Industry and Other Purposes*, VCH, Weinheim, pp. 642–646.

Ewete, F., Nicol, R.W., Hengsawad, V., Sukumalanand, P., Satasook, C., Wiriyachritra, P., Isman, M.B., Kahn, Y., Duval, F., Philogene, B.J.R. and Arnason, J.T. (1996) Insecticidal activity of *Aglaia odorata* extract and the active principle, rocaglamide, to the European corn borer, *Ostrinia nubilalis* Hubner (Lepidoptera: Pyralidae). *J. Appl. Entomol.*, **120**, 483–488.

Feng, R., Chen, W. and Isman, M.B. (1995) Synergism of malathion and inhibition of midgut esterase activities by an extract from *Melia toosendan* (Meliaceae). *Pestic. Biochem. Physiol.*, **53**, 34–41.

Fields, P.G., Arnason, J.T., Philogène, B.J.R., Aucoin, R.R., Morand, P. and Soucy-Breau, C. (1991) Phototoxins as insecticides and natural plant defenses. *Mem. Entomol. Soc. Can.*, **159**, 29–38.

Gullichsen, J. and Paulapuro, H. (2000) Chemical pulping. *Papermaking Sci. Technol. Finland*, **6B**, 378–388.

Harborne, J.B. (1991) The chemical basis of plant defense. *In* R.T. Palo and C.T. Robbins (eds.), *Plant Defense against Mammalian Herbivory*, CRC Press, Boca Raton, FL, pp. 45–59.

Harborne, J.B. (1993) Insect feeding preferences. *In* J.B. Harborne (ed.), *Introduction to Ecological Biochemistry*, Academic Press, London, pp. 128–158.

Hummelbrunner, L.A. and Isman. M.B. (2001) Acute, sublethal, antifeedant, and synergistic effects of monoterpenoid essential oil compounds on the tobacco cutworm, *Spodoptera litura* (Lepidoptera:Noctuidae). *J. Agric. Food Chem.*, **49**, 715–720.

Ingrouille, M. (1992) Competition, herbivory and dispersal: The limiting biotic environment. In: M. Ingrouille (ed.), *Diversity and Evolution of Land Plants*, Chapman & Hall, London, pp. 263–290.

Isman, M.B. (2006) Botanical insecticides, deterrents and repellents in modern agriculture and an increasingly regulated world. *Annu. Rev. Entomol.*, **51**, 45–66.

Isman, M.B., Koul, O., Luczynski, A. and Kaminski, J. (1990) Insecticidal and antifeedant bioactivities of neem oils and their relationship to azadirachtin content. *J. Agric. Food Chem.*, **38**, 1406–1411.

Isman, M.B. (2004) Factors limiting commercial success of neem insecticides in North America and Western Europe. In: O. Koul and S. Wahab (eds.), *Neem: Today and in the New Millennium*, Kluwer Academic, Dordrecht, pp. 33–41.

Isman, M.B. (1997) Neem insecticides. *Pestic. Outlook*, **8**, 57–63.

Isman, M.B., Gunning, P.J. and Spollen, K.M. (1997) Tropical timber species as sources of botanical insecticides. In: P.A. Hedin, R.M. Hollingworth, E.P. Masler, J. Miyamoto and D.G. Thompson (eds.), *Phytochemicals for Pest Control*, Amer. Chem. Soc. Symp. Ser. 658, American Chemical Society, Washington DC, pp. 27–37.

Isman, M.B., Miresmailli, S. and Machial, C. (2011) Commercial opportunities for pesticides based on plant essential oils in agriculture, industry and consumer products. *Phytochem. Rev.,* **10**, 197–204

Isman. M.B., Wan, A.J. and Passreiter, C.M. (2001) Insecticidal activity of essential oils to the tobacco cutworm, *Spodoptera litura. Fitoterapia,* **72**, 65–68.

Janprasert, J., Satasook, C., Sukumalanand, P., Champagne, D.E,, Isman, M.B.; Wiriyachitra, P. and Towers, G.H.N. (1992) Rocaglamide, a natural benzofuran insecticide from *Aglaia odorata. Phytochemistry*, 32, 67–69.

Jiang, Z., Akhtar, Y., Bradbury, R., Zhang, X. and Isman, M.B. (2009) Comparative toxicity of essential oils of *Litsea pungens* and *Litsea cubeba* and blends of their major constituents against the cabbage looper, *Trichoplusia ni. J. Agric. Food Chem.,* **57**, 4833–4837.

Jimenez, A., Mata, R., Pereda-Miranda, R., Calderon, J., Isman, M.B., Nicol, R. and Arnason, J.T. (1997) Insecticidal limonoids from *Swietenia humilis* and *Cedrela salvadorensis. J. Chem. Ecol.,* **23**, 1225–1234.

Lars, H.N. (2002) "*Tall Oil*" *in Ullmann's Encyclopedia of Industrial Chemistry*. Wiley-VCH, Weinheim.

Leatemia, J.A. and Isman, M.B. (2004) Insecticidal activity of crude seed extracts of *Annona* spp., *Lansium domesticum* and *Sandoricum koetjape* against lepidopteran larvae. *Phytoparasitica*, **32**, 30–37

Leatemia, J.A. and Isman, M.B. (2004) Efficacy of crude seed extracts of *Annona squamosa* against diamondback moth, *Plutella xylostella* L. in the greenhouse. *Int. J. Pest Manage.*, **50**, 129–133.

Lowery, D.T. and Isman, M.B. (1995) Toxicity of neem to natural enemies of aphids. *Phytoparasitica*, **23**, 297–306

McLaughlin, J.L., Zeng, L., Oberlies, N.H., Alfonso, D., Johnson, H.A. and Cummings, B. (1997) Annonaceous acetogenins as new natural pesticides: Recent progress. In: P.A. Hedin, R.M. Hollingworth, E.P. Masler, J. Miyamoto and D.G. Thompson (eds.), *Phytochemicals for Pest Control*, ACS Symposium Series 658, American Chemical Society, Washington, DC, pp. 117–133.

Machial, C.M., Shikano, I., Smirle, M.J., Bradbury, R. and Isman, M.B. (2010) Evaluation of the toxicity of 17 essential oils against *Choristoneura rosaceana* (Lepidoptera:Tortricidae) and *Trichoplusia ni* (Lepidoptera: Noctuidae). *Pest Manage. Sci.*, **66**, 1116–1121.

Miresmailli, S., Bradbury, R. and Isman. M.B, (2006) Comparative toxicity of *Rosmarinus officinalis* L. essential oil and blends of its major constituents against *Tetranychus urticae* Koch (Acari: Tetranychidae) on two different host plants. *Pest Manage. Sci.*, **62**, 366–371.

Naumann, K. and Isman, M.B. (1995) Evaluation of neem (*Azadirachta indica* A. Juss) seed extracts and oils as oviposition deterrents to noctuid moths. *Entomol. Exp. Appl.*. **76**, 115–120.

Naumann, K. and Isman, M.B. (1996) Toxicity of neem (*Azadirachta indica* A. Juss) insecticide to larval honey bees. *Amer. Bee J.*, **136**, 518–520.

Ohse, T., Ohba, S, Yamamoto, T., Koyano, T. and Umezawa, K. (1996) Cyclopentabenzofuran lignin protein synthesis inhibitors from *Aglaia odorata*. *J. Nat. Prod.*, **59**, 650–652.

Pradhan, S., Jotwani, M.C. and Rai, B.K. (1962) The neem seed deterrent to locusts. *Indian Fmg.*, **12**, 650–652.

Satasook, C., Isman, M.B. and Wiriyachitra, P. (1993) Toxicity of rocaglamide, an insecticidal natural product, to the variegated cutworm, *Peridroma saucia* (Lepidoptera: Noctuidae). *Pestic. Sci*, **36**, 53-58.

Satasook, C., Isman, M.B., Ishibashi, F., Medbury, S., Wiriyachitra, P. and Towers, G.H.N. (1994) Insecticidal bioactivity of crude extracts of *Aglaia* species (Meliaceae). *Biochem. Syst. Ecol.*, **22**, 121–127.

Saxena, R.C. (1989) Insecticides from neem. In: J.T. Arnason, B.J.R. Philogene and P. Morand (eds.), *Insecticides of Plant Origin*, Amer.

Chem. Soc. Symp. Ser. 387, American Chemical Society, Washington DC, pp. 164–172.

Scott, I.M., Jensen, H.R., Philogène, B.J.R. and Arnason, J.T. (2007) A review of *Piper* spp. (Piperaceae): Phytochemistry, insecticidal activity and mode of action. *Phytochem. Rev.*, **7**, 65–75.

Seffrin, R.C., Shikano, I., Akhtar, Y. and Isman, M.B. (2010) Effects of crude seed extracts of *Annona atemoya* and *A. squamosa* L. against the cabbage looper, *Trichoplusia ni* in the laboratory and greenhouse. *Crop Prot.,* **29**, 20–24.

Spollen, K.M. and Isman, M.B. (1996) Acute and sublethal effects of neem insecticide on commercial biological control agents *Phytoseiulus persimilis* and *Amblyseius cucumeris* (Acari: Phytoseiidae) and *Aphidoletes aphidimyza* (Diptera: Cecidomyiidae). *J. Econ. Entomol.,* **89**, 1379–1386.

Stenius, P. (2000) Forest products chemistry. *Papermaking Sci. Technol, Finland,* **3**, 73–76.

Stroh, J., Wan, M.T., Isman, M.B. and Mou, D.J. (1998) Evaluation of the acute toxicity to juvenile Pacific coho salmon and rainbow trout of some plant essential oils, a formulated product, and the carrier. *Bull. Environ. Contam. Toxicol.,* **60**, 923–930.

Tang, C.S., Wat, C.K. and Towers, G.H.N. (1987) Thiophenes and benzofurans in the undisturbed rhizosphere of *Tagetes patula* L. *Plant Soil*, **98**, 93–97.

Wheeler, D., Isman, M.B., Sanchez-Vindas, P. and Arnason, J.T. (2001) Screening of Costa Rican *Trichilia* species for biological activity against the larvae of *Spodoptera litura* (Lepidoptera: Noctuidae). *Biochem. Syst. Ecol.*, **29**, 973–978.

Wheeler, D.A. and Isman, M.B. (2000) Effect of *Trichilia americana* extract on feeding behavior of Asian armyworm, *Spodoptera litura*. *J. Chem. Ecol.*, **26**, 2791–2800.

Whiller, H. and Kilcher, L. (2011) *The World of Organic Agriculture. Statistics and Emerging Trends.* FiBl, Frick, IFOAM, Bonn.

Xie, Y.S. and Isman, M.B. (1995) Toxicity and deterrency of depitched tall oil to the green peach aphid, *Myzus persicae*. *Crop Prot..*, **14**, 51–56.

Xie, Y.S. and Isman, M.B. (1992) Antifeedant and growth inhibitory effects of tall oil and derivatives against the variegated cutworm, *Peridroma saucia* Hùbner (Lepidoptera: Noctuidae). *Can. Entomol.*, **124**, 861–869.

Xie, Y.S., Isman, M.B., Feng, Y. and Wong, A. (1993) Diterpene resin acids: Major active principles in tall oil against Variegated cutworm,*Peridroma saucia* (Lepidoptera: Noctuidae). *J. Chem. Ecol.*, **19**, 1075–1084.

Xie, Y.S. and Isman. M.B. (1995) Tall oil: enhancement of neem and azidarachtin toxicity to the variegated, *Peridroma saucia* (Lepidoptera : Noctuidae). *J. Appl. Entomol.*,**119**, 361–365.

20

Biopesticides in Africa: Current Status and Future Potential

S. FACKNATH[1*] AND B. LALLJEE[2]

[1]Faculty of Agriculture, University of Mauritius, Réduit, Mauritius
[2]Consultancy and Contract Research Centre,
University of Mauritius, Réduit, Mauritius
*E-mail: sunif@uom.ac.mu

INTRODUCTION

Biopesticides provide an alternative tool for the management of agricultural pests, and contribute to the reduction in use of the toxic synthetic chemical pesticides that represent an increasing cause for environmental and health concerns in many parts of the world. While having comparable efficacies to the conventional chemical pesticides, biopesticides exhibit the significantly important advantage of being more eco-friendly and less hazardous to non-target organisms, including humans. In particular, they are safer to use in environmentally sensitive areas, such as national parks or near water bodies. Furthermore, many of these can be mass produced fairly easily in artificial media, and can be applied using conventional spraying equipment and techniques. On the other hand, their popularity and use suffers due to their high cost, and relatively slow/ delayed action (between 2 and 15 days to kill the pest, disease or weed).

Biopesticides include bacteria, viruses, fungi, protozoa, nematodes, plant based preparations, as well as semiochemicals. Hence, there are two types of biopesticides - biochemical and microbial. Biochemical (also known as biorational pesticides) pesticides are similar to naturally occurring chemicals, such as insect pheromones, plant allelochemicals, fermentation products, etc., while microbial biopesticides come from microscopic organisms that cause pathogenicities in the pest. The latter can occur naturally or may be genetically engineered. Biopesticidal agents are all particles and sprayable formulations consist of adequately stable suspensions of these particles.

BIOPESTICIDE MARKET IN AFRICA

Compared to the rest of the world, commercial biopesticide use in Africa is very limited, with an annual market of only about 8.1 million USD for Africa and Middle East combined, compared to the 74.6 million USD for Latin America, 127 million USD for North America (2007 figures), 132.5 million for Asia and Australasia (2007 figures) and about 54.5 million USD for Europe (CPLBC, 2010). Of the various biological organisms that have been commercialized for control of insect and other pests, *Bacillus thuringiensis* is the one that is relatively better known in parts of Africa. Until very recently, there was very little use being made of fungal-based, virus-based, protozoan-based, or nematode-based preparations.

The main causes of this relatively poor use in Africa is the mainly subsistence nature of agricultural production on the continent, high cost of such preparations as compared to the synthetic chemical formulations, lack of knowledge and awareness among the farming community, lack of access of the farmers to these products, lower efficacy compared to toxic chemical pesticides, slow/delayed action of biological pesticides, narrower spectrum of activity, absence of local production of biopesticides, absence of legislation for registration and use of biopesticides. In many African countries, biopesticides are still in the experimental stage. Unlike in Latin America, there is negligible local production of microbial pesticides in Africa. There is no import of biopesticides in Brazil, Cuba and Colombia, and practically all the biopesticides (mainly fungal and viral-based products) utilized in these countries come from local production through the support of Government agencies and NGOs. This is lacking in the African continent, and is one of the major causes of the paucity of use of microbials in African countries.

While there is a general trend towards promoting IPM on the continent, pressure is not yet adequate in all markets to drive adoption of alternatives to synthetic pesticides. Global trends suggest that biopesticides tend to address high value niche markets, therefore, they may not produce direct or immediate benefits to the resource-poor African farmer. Among the African countries, South Africa and Kenya are demonstrating growth in this sector, and their market for microbial products is on the increase. Development of an export market for horticultural produce could stimulate the creation of a biopesticide industry due to the stringent requirements of this market. It is interesting to note that there appears to be a direct relationship between the increase in research on microbial preparations and the rising market in these products.

POTENTIAL OF BIOPESTICIDES IN AFRICA

There is substantial room for growth in the market for microbial biopesticides in Africa with currently available products. Africa, in particular, is almost totally undeveloped and with increasing trade in horticultural products into the EU and NAFTA regions, the market has the potential to reach $150 million by 2015. While Bt-based products represent only about 40% of the biopesticide market in Latin America (mainly Cuba and Argentina), 77% of the overall microbial biopesticide market in Africa and the Middle East is dominated by Bt (CPLBC, 2010).

Africa has produced some highly promising pioneering scientific data on the potential of some viruses, fungi and plant extracts for field pest control and for post harvest protection, but has not been very successful in commercializing the results of these studies. Must studies do not go beyond the stage of scientific publications, and there is hardly any commercial product emanating from the research work. Lack of expertise in product development, lack of expertise in patenting and copyright protection, insufficient market information, inadequate funds for commercial development and marketing, inability to attract private investment, lack of local legislation and inability to meet western registration requirements, are some of the reasons for this failure to exploit what could be a very lucrative economic sector. Africa is one of the last great unknowns for obtaining natural bioactive organisms and compounds, and commercialisation is the key implementation pathway for ensuring availability and utilisation of biopesticides in the region and also as a source for export-based revenue.

Legislation for Biopesticides

While many African countries have pesticide regulatory frameworks, few have regulatory frameworks specific for biopesticides, for which the data requirements can be significantly different. The production and promotion of biocontrol agents for the wider farming community can be implemented only when the appropriate legislation on sale and use of biopesticides is in place. Kenya set the example in 2003, when the Kenya Agricultural Research Institute and the Pest Control Products Board prepared drafted legislation for registration of biopesticides in Kenya, following a landmark workshop organised with the support of the DFID Crop Protection Programme.

Several countries in sub-Saharan Africa have already used new guidelines to develop systems for registering biological controls for pests and diseases - bacteria, viruses, nematodes, fungi, predators and parasitoids. Both South Africa and Kenya now have laws - based on the guidelines - that allow biological control agents to be registered, sold and used. Many other countries, including Tanzania, Ghana and Benin, are also using the guidelines to draw up similar laws to help their booming exports of fresh produce. Ghana is on the verge of having biological control agents commercially available in the country for the first time.

Microbial Pesticides

Although use of commercial biopesticides in Africa is still far below that of the developed nations, it is interesting that microbial pesticides are showing great promise in the fight against the two most deadly agricultural insect pests in Africa, namely locusts and armyworms. Until very recently, locust swarms and armyworm outbreaks have been controlled using huge amounts of toxic synthetic insecticides. Rising concern about the associated health and environmental side effects led to research on safer biological alternatives, culminating in a couple of highly promising microbial, and microbial+semiochemical combination treatments.

One of these is GREEN MUSCLE® (GREEN GUARD in Australia) – a formulation of an entomopathogenic fungus, *Metarhizium anisopliae* (earlier *M. anisopliae* var. *acridum* and *M. flavoviride*). This fungal variety has specialised on short-horned grasshoppers and represents an important weapon in the fight against the devastating locust swarms that develop periodically. Developed under the LUBILOSA[1] project of 1989 (following environmental concerns over

[1] LUtte BIologique contre les LOcustes et les SAuteriaux: a collaborative, multi-disciplinary research and development programme funded by the Governments of

the 1.5 million litres of toxic synthetic chemicals used to control the 1987-1989 locust plague), this biopesticide avoided the problem of high humidity requirements of entomopathogenic fungi by developing an oil based formulation of the fungal spores. Hence, GREEN MUSCLE® can be used to kill locusts even under arid conditions. This enables its application to hoppers living under desert conditions, far from cropping areas, where the almost two week delay in insect death would not cause appreciable crop loss. Furthermore, it allows birds, parasitic and predatory wasps, parasitic flies and beetles to contribute better to the pest control since the infected insects become sluggish and easy to catch.

Desert Locust Outbreak

Swarming desert locusts cause significant crop losses and exacerbate the problem of food insecurity. During the twentieth century, desert locust plagues occurred in 1926-1934, 1940–1948, 1949–1963, 1967–1969, 1987–1989 and 2003-2005. The 1987-1989 one took several years and more than US$ 300 million before it was brought to an end. The 2003-2005 one was brought under control at an estimated cost of over US$ 400 million with harvest losses valued at about US$ 2.5 billion.

Image Source : Wikimedia Commons

In 2003, small independent upsurges in Mauritania, Mali, Niger and Sudan spread to the region between Dakar, Senegal to Morocco as a result of unusually heavy rains, and thereafter moved to crops in the Sahel in West Africa. The swarms migrated over the continent causing devastation, and by 2004, had spread to northern Egypt, Jordan and Israel for the first time in 50 years, and also into Cape Verde, the Canary Islands, southern Portugal, and Crete. One swarm in Morocco was estimated as being 230 km long, at least 150 m wide, and contained an estimated 69 billion locusts. Several countries lost significant portions of their crops to the locusts, particularly Mauritania, which lost as much as half of its harvest. These had disastrous effects on the food security situation in the eleven or so Africa nations affected by the plague. Experts estimate that the swarm consumed vegetation equivalent to the food needs of about 200 million people.

The formulations consist of a dry powder (TC formulations) and an oil miscible concentrate (OF formulations). It is applied as an oil suspension using ultra low volume (ULV) spinning disk spray equipment. The biopesticide has no "knock down" effect, and grasshoppers and locusts take 6 to 10 days to die, however, they are slowly incapacitated during this period, eating and moving much less than healthy insects after only 3 or 4 days. When a spore of the live fungus lands on an insect, it bores a hole in the insect's exoskeleton and invades the tissue of the insect's entire body. Because locusts and grasshoppers often prey on deceased members of their own kind, the spores can attack new insects. This helps to reduce the amount of the microbial pesticide that needs to be sprayed, translating

Canada (CIDA), the Netherlands (DGIS), Switzerland (SDC) and the UK (DfID).

into lower costs for farmers, and less likelihood of resistance developing in the insect pest.

Depending on environmental conditions the spores can survive several weeks and continue to be a source of infection to healthy grasshoppers (Langewald, 1999). The speed at which the fungal infection spreads through the insect is primarily determined by temperature. Infected locusts exhibit an interesting behavioural adaptation – staying out in the sun to raise their body temperature and thereby slowing fungal spread. This delays the killing time, sometimes even doubling it, and is unacceptable to farmers whose crops are being damaged by the insects. It is, therefore, recommended that the biopesticide be used proactively to treat breeding sites, rather than reactively to control outbreaks. This could be done most effectively by establishing community-based early warning systems. Organizing these, with the involvement of wildlife rangers and farmers, is one of the next challenges for locust control.

GREEN MUSCLE® was launched in South Africa in March 1998, and registered for use there. For the Sahel, GREEN MUSCLE® was registered in 2001. Niger launched GREEN MUSCLE® in August 2000 and is the first country which integrated GREEN MUSCLE® into it's grasshopper control operations, with 2000 ha sprayed against rice grasshoppers. After the completion of the LUBILOSA programme, production and use of GREEN MUSCLE® stagnated somewhat, but was revived in Senegal during the late 2003s when the first of the 2003–2005 locust swarms started developing in West Africa. With hindsight, it is now apparent that the high cost of GREEN MUSCLE®, due to the high dosages recommended at the time for locust control (50-100 g/ha at a cost of \$10–20, as compared to \$5–10, respectively for chemical pesticides), was one of the main causes for the insufficient interest shown in this product in Africa at the time.

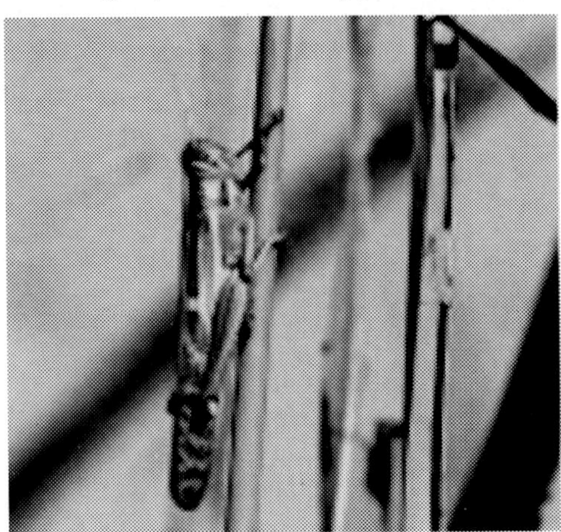

Red locust
(Image source: Wikimedia Commons)

Recently, in 2009, a red locust outbreak in Tanzania was successfully controlled, and a potential disaster averted in Eastern and Southern Africa, through large scale applications of GREEN MUSCLE® mixed with a mineral oil. The rapid interventions in Tanzania, Malawi, Mozambique and Zambia, organized and coordinated by the FAO together with the International Locust Control Organization for Central and Southern Africa (IRLCO-CSA), and funded by FAO and the UN's Central Emergency Response Fund, significantly reduced red locust infestations, thereby preventing a full-blown invasion that could have affected the food crops of around 15 million people in the region (FAO, 2009). In Tanzania, around 10,000 ha infested with adult locusts, in the environmentally sensitive areas of the Iku-Katavi National Park, the Lake Rukwa and the Malagarasi Basin, and home to large wildlife such as elephants, hippos, and giraffes were successfully treated with this microbial.

The microbial is now produced in commercial quantities in Senegal and South Africa. It is cultured in partially boiled rice, and the microscopic spores collected by harvesting machines (Mycoharvester) that evolved through many variations, culminating in an industrial-scale version. Most biological pesticides have a limited shelf-life, but the packaged product of GREEN MUSCLE® has a shelf life of 18 months at 30°C, and up to 5 years under refrigeration. In 2008, more than 7000 ha were sprayed in Senegal against Sahelian grasshoppers. The biopesticide has also been effectively used in Niger and Mauritania.

Another devastating insect pest in Africa is the African armyworm, *Spodoptera exempta*. It is a migratory pest of pasture and grain crops. The unusual migratory behaviour of the adult moths, which arrive in huge numbers and lay billions of eggs in a very short time results in masses ('armies') of voracious, black caterpillars that devastate any Graminae crop (maize, sorghum, millet, rice and wheat), as well as pasture grasses. The resulting damage is not only the direct crop loss, but also the indirect impact on livestock production through starvation and poisoning of the animals (Rose *et al.*, 2000).

Armyworm larvae

(Image Source : Mushobozi, 2011)

Outbreaks start in southern and central Tanzania and spread to other parts of Tanzania and eastern, northern and southern Africa every year, being seriously damaging in some years, when conditions are favourable, and less so in others. Tanzania and Kenya are among the hardest hit countries in Africa (Haggis, 1984). Following a particularly severe outbreak in 2006 that attacked 500,000 ha in Tanzania and against which conventional pesticide control failed due to lack of availability of pesticides to farmers, R & D projects funded by international donor agencies were set up in Tanzania in order to develop and evaluate biopesticides for armyworm control.

A microbial *S. exempta* nucleopolyhedrovirus (SpexNPV), and a naturally occurring, plant-based biochemical pesticide, namely neem, were tested as alternatives to the environmentally damaging and expensive synthetic chemical pesticides commonly used. Armyworm larvae become infected when they ingest vegetation contaminated with the virus occlusion bodies (OBs). These OBs break down in the alkaline conditions of the insect's mid-gut and the virus replicates within host cells, generating millions of new OBs. Within 3-5 days the larvae die and they exhibit a characteristic inverted-V shape as they hang from vegetation. Rupturing of the cuticle liberates the OBs which can then drift and settle on nearby vegetation where they can be taken up by other armyworm larvae. Neem allelochemicals act as antifeedants and growth regulators, causing feeding and reproductive disturbances. The success of both the virus and neem in small-plot trials and large-scale aerial application (the first use of aerial application for any NPV biopesticide in Africa) led to the adoption of SpexNPV as National Policy in Tanzania, with the use of local DIY preparations of neem being promoted in remote villages and local communities who lack access to other resources. Neem pesticide is also available under the brand name Neemrin and is registered in Tanzania to control various caterpillars including armyworms, but production is still small scale, and needs stepping up to meet the demand.

Heavy infestation of armyworm larvae
(Image Source : Mushobozi, 2011)

SpexNPV is produced according to a modified Brazilian production protocol in the form of a low-cost clay formulation. Simple homogenates of infected insects are filtered, absorbed onto pulverised local china clay and then air dried within 12 h at ambient temperature. These are then ground and stored for use as a bioinsecticide. The system does not require expensive centrifugation, freeze drying or deep freeze storage and gives a stable product (> 6 months at 20°C, > 2 years at 4°C) that can be supplied directly to farmers as a substitute for chemical insecticides. It is sold to farmers in affordable 50-100 g bottles, and is effective at only half the cost of chemical pesticides. Being non-toxic to humans, it requires no specialist safety gear and represents no danger from stockpiles, unlike with chemical pesticides. Production of this biopesticide is expected to be stepped up by Eco Agri Consult Ltd in the company's DFID RIU-funded baculovirus production facility located in Arusha, Tanzania. The recent January 2011 outbreak of armyworms destroyed over 300 ha of maize, sorghum and other crops in Southern Tanzania and in Zimbabwe. However, in the case of Tanzania, the outbreak helped to provide the raw material for the production of SpexNPV. Although there are still registration issues to be overcome for the biopesticide, the recent registration of GREEN MUSCLE in Tanzania has set a useful precedent.

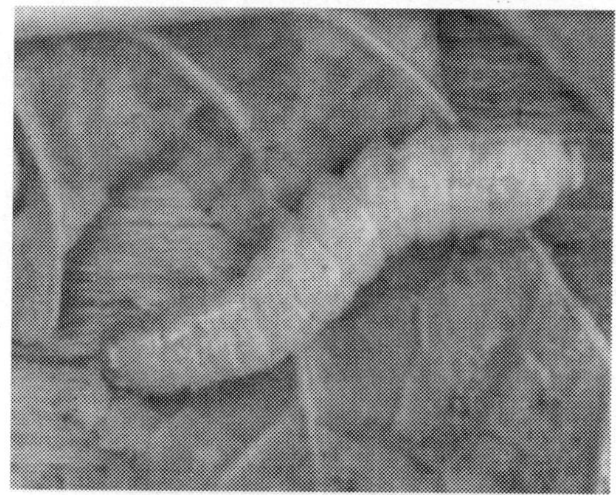

Diamondback moth larva

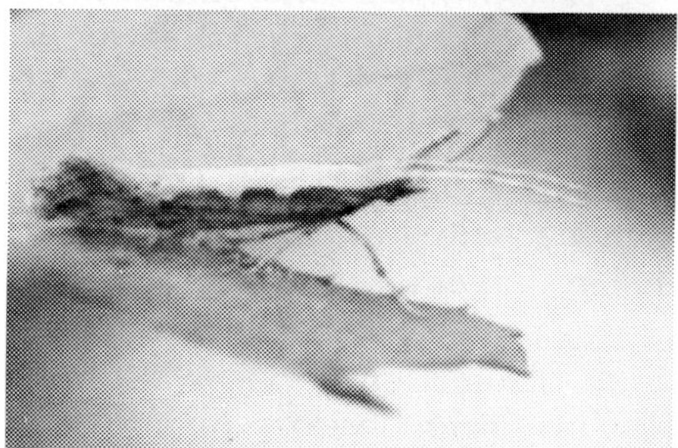

Diamondback moth adult
(Image source: Wikimedia Commons)

DFID is supporting an integrated approach which combines establishing community-based armyworm forecasting based on Straight Chain Lepidopteran Pheromones (SCLP) baited traps, with the use of SpexNPV and neem in a programme involving Tanzania, Kenya, and Ethiopia. Malawi too has expressed its interest in adopting this approach, and it also has potential for use in other African countries, where armyworm epidemics are a problem (Uganda, Mozambique, South Africa, etc.).

The International Institute of Tropical Agriculture (IITA) in Benin has developed a fungal based biopesticide against the notorious diamondback moth (DBM), *Plutella xylostella* using isolates of *Beauveria bassiana.*

This insect pest is a serious pest of cruciferous crops and can cause complete destruction of a marketable crop, if not controlled in time.

Heavy insecticide use throughout the world has resulted in this pest exhibiting resistance to practically most of the synthetic chemicals used so far. The fungus, *B. bassiana* shows great promise as a cost-effective solution for the sustainable control of DBM when used in an integrated pest management approach. *Beauveria bassiana* grows naturally in soils throughout the world, is generally specific to target pests, persists in the environment, is ecologically-friendly and easy to mass produce. It can remain active in the field for several months after initial application and hence will need fewer applications. Inspite of its success, commercial production of the fungus awaits the interest of a private investor for manufacture and marketing of this biopesticide.

Under the DFID's Crop Protection Programme (CPP), isolates of the bacterium *Pasteuria penetrans* and the fungus *Pochonia* (formerly *Verticillium*) *chlamydosporium* were extracted from Kenyan soils and their efficacy in suppressing root-knot nematode populations was confirmed. Partnership between a commercial producer, Dudutech (Kenya) Ltd. for mass production of these microbials, and CENSA Cuba for their expertise in microbial production technologies (with technical support from Rothamsted Experimental Station and Reading University at UK and financed by DFID) led to weekly production figures of over 100 kg of the fungal biopesticide.

SCLPs

SCLPs are lepidopteran pheromones that are naturally occurring compounds, or identical or substantially similar synthetic compounds, designated by an unbranched aliphatic chain (between 9 and 18 carbons) ending in an alcohol, aldehyde or acetate functional group and containing up to 3 double bonds in the aliphatic backbone. Most countries have developed simplified registration procedures for introduction and use of the SCLPs. (US EPA, 2009) http://www.epa.gov/opp00001/biopesticides/ingredients/tech_docs/brad_129109.pdf).

In Kenya, the chemical identity of the *S. exempta* sex pheromone was determined in the 1970s (Beevor *et al*., 1975), and a synthetic version of the pheromone was developed for use as a lure in moth traps. Over 400 pheromone baited traps, operated by extension agents for forecasting purposes, provide armyworm data to the Plant Protection Services Migrant Pest Unit each week during the cropping season. However, the recent trend in several African countries is to develop community-based forecasting systems (as opposed to the central government extension officers) using SCLP baited traps.

International Centre for Insect Physiology and Ecology (ICIPE) has developed combinations of the entomopathogenic muscardine fungi, *Beauveria bassiana* and *Metarhizium anisopliae*, for the control of the hard tick, *Amblyomma variegatum,* a major pest of cattle and other livestock animals (Maranga *et al.,* 2005), while extracts from the Fabaceae plant, *Calpurnia aurea*, are being developed for use as attractants of the ticks *Rhipicephalus appendiculatus* and *R. pulchellus* (Nana *et al.,* 2010).

Botanical Pesticides

Unlike microbial pesticides, plant based pesticidal preparations have been in use in Africa since a very long time. In fact, two of the major groups of synthetic pesticides widely used today, namely carbamates and pyrethroids, originate from plants known since ancient times to medicine men and farmers in Africa, and used either as medicines or poisons. Methyl carbamate insecticides are modeled on bioactive compounds from the calabar bean (*Physostigma venenosum)* which was traditionally used as an ordeal poison in West Africa. Similarly, African tribes controlled insect pests using extracts of the flowers of *Chrysanthemum cinerariaefolium*; the isolation and identification of the pyrethrins in these flowers led to the extraction of pyrethrums and synthesis of pyrethroids, which today is a multibillion dollar global industry.

There is a common tradition among subsistence farmers in many parts of Africa to use natural products such as plant powders/ashes, leaf/root/shoot decoctions, whole fresh/dried leaves, seed/kernel oils, etc. for protection of field crops and stored harvest. For instance, as much as 20–50% of the harvested maize and pulses can be lost due to infestations from weevils, bruchids and other insects (FAO, 1985). Chemicals for fumigation or seed treatment are sometimes available but their high cost precludes their widespread use. Farmers, therefore, make use of local plants, placing whole fresh or dried leaves among layers of the grains or mixing powdered plant parts directly with the grain. Such botanicals have advantages over synthetic pesticides in terms of their cost for the resource-poor farmers. In addition, it represents an easily available and environmentally sustainable alternative to the chemical insecticides. Many farmers believe that botanicals are safer to use and are more flexible in their usage. Plants affect insects in three main ways, repellency/feeding deterrency, toxicity, and growth disturbance. There are also no legal restrictions to their preparation and use. The use of plant materials is ethnically and culturally biased, sometimes restricting the knowledge within a given community. However, all communities possess knowledge of local plants that can be used as agricultural, medical, and veterinary pesticides. Given that many of the plants used as pesticides have alternative uses as medicines or food spices, their possible toxicity to man or other vertebrates would have been established through the gaining of ethnobotanical knowledge. There is a wealth of indigenous and traditional knowledge in Africa about such pesticidal plants, about the appropriate times and locations of their collection in the wild, and about their preparation and application.

In sub-Saharan Africa, small-scale farmers face the problem of high grain losses in storage due to insects and other stored product pests. The age old grain storage structures are not sufficient to protect grain from insect infestation, which can represent a substantial threat to household incomes and food security. Farmers adopt one of two main options – they avoid the need to store their grain by selling the produce soon after harvest. But this also represents a monetary loss since the glut of grains on the market results in significantly lowered prices on the market. Storing the grain for later selling requires protecting the storage from insect and other pests, which is not available to subsistence level farmers. Use of cheap, inappropriate insecticides can result in pesticide residues in human food and lead to health risks, and may also promote insecticide resistance. The other option is the use of low cost, easily available, safer using local pesticidal plants. Botanicals do not need to be purchased and are, therefore, low-cost (requiring labour inputs but no direct financial outlay). Many farmers also believe that botanicals are safer to use and are more flexible in their usage. Botanicals are also often much more available than commercial products as they grow in the local environment. Some of the plants traditionally used by farmers for stored product protection offer similar levels of protection as compared to commercial synthetics.

Plant species show different levels and durations of efficacy depending on the pest species and the commodity affected. Variations in plant efficacy can also arise due to differences in variety of the plant, the soil type, the location, the time of year and time of day the collection of the plant part was done, the method of preparation of the extracts, mode of application, dosage, etc. Farmer acceptability is based not only on demonstrated efficacy, but also on other factors such as availability, familiarity and versatility.

In most parts of Africa, synthetic chemical pesticides are not sufficiently regulated, and several banned and outdated pesticides are still widely available and used. To make matters worse, the products are often adulterated by dilution, mixed incorrectly and sold beyond their expiry date and the use of these products can exacerbate the problem the farmers are trying to address by encouraging the development of resistance in the target pests (SAPP, 2010). Furthermore, the instructions provided for the correct use and safe handling of the pesticides are usually given in English rather than the local language, and since many users do not read or understand English or are illiterate, there are a large number of instances of incorrect application. Given such a situation and the high cost of the synthetic commercial pesticides, pesticidal plants offer these farmers a real and effective alternative to the synthetic products.

One of the best studied and most widely known botanical is *Azadirachta indica* (neem), an evergreen tree belonging to the Meliaceae family. The International Centre for Insect Physiology and Ecology (ICIPE), based in Kenya, studied the potential of this tree as a source of biopesticides for several years. In 1993, it partnered with the German Development Agency, GTZ and a Kenyan private investor to explore the potential of promoting neem use in East Africa. The first commercial neem pesticides in Africa (Neemros® and Neemroc®) were developed by this research-industry collaboration under the name of Saroneem Ltd. The scientific and technical inputs of ICIPE coupled with the financial backing of the private investor resulted in capturing a good portion of the market and made a healthy profit for several years. After the withdrawal of ICIPE from this venture in 2001, Saroneem Biopesticides Ltd. continues to manufacture the two neem pesticides for the African market. In 2002, ICIPE started a new indigenous company called Biop Ltd which developed and registered several competitive products, including pesticides for horticultural pest management, and oils for the preparation of soaps, skin creams and shampoos for which there is a big demand in supermarkets and drug stores (ICIPE, 2011).

A large number of other local plants have been studied and are known to have promising potential as natural pesticides. This can contribute to alleviating one of the major constraints to food production (namely pests) and increasing agricultural production and food security. The possibility of making raw extracts from plants grown in the neighborhood where the farmer lives assists in ensuring self sufficiency and gives him/her a cheap alternative to conventional pesticides which often are imported using foreign exchange. Large-scale production of natural pesticides based on crude or purified extracts could also provide an income for people in the countryside. ICIPE has partnered with other organisations and established a programme to help communities that live in or near biodiversity-rich forested areas to generate income through the collection and sale of the raw materials needed for the production of the biopesticides and at the same time to conserve the biodiversity–rich forests by relieving some of the economic pressure on the forest resources. The programme covers communities based in Kenya, Tanzania and Uganda. The main insecticidal plant that is being promoted in the programme in Kenya is *Ocimum kilimandscharicum* in the area adjacent to the Kakamega forest. This plant is used by the community for repelling mosquitoes and protecting stored grains, and is also used to treat colds, cough and flu in the form of a decoction or as the active ingredient in a medicinal preparation (marketed under the name of Naturub®) (ICIPE, 2011). Similarly, neem and aloe cultivation and processing is being supported under this programme for empowering communities living adjacent to the Shimba Hills forest in Kenya (ICIPE, 2011).

The three-year Southern African Pesticidal Plants Project (Jan 2007–Dec 2009) included detailed studies of pesticidal plants in Malawi, Zimbabwe and Zambia with technical expertise from the Natural Resources Institute, University of Greenwich, UK. Outcomes of the project have been incorporated into a new project called the African Dryland Alliance for Pesticidal Plant Technologies (ADAPPT), a network for optimising and promoting the use of indigenous botanical knowledge for food security and poverty alleviation in Africa, with 12 partner institutions and eight African countries (Malawi, Kenya, Zimbabwe, Zambia, Ghana, Namibia, Tanzania and South Africa) coordinated by Natural Resources Institute, University of Greenwich, UK (ADAPPT, 2011).

The excessive removal of a pesticidal plant without any replacement runs the risk of threatening survival of the plant species. In order for such plants to be used sustainably, their propagation and cultivation potential as well as their conservation in the wild needs to be addressed. Promotion of the use of botanicals will assist biodiversity conservation by increasing perceived value of the plant species concerned. In addition, the indigenous and traditional knowledge gathered by generations of farmers must be protected and given due recognition and credit.

INDIGENOUS AND TRADITIONAL KNOWLEDGE IN AFRICA

Indigenous and Traditional Knowledge (ITK) sums up all the community practices, institutions, relationships and rituals of the people; it has developed over generations of farmers and englobes centuries of experimentation on adapting to local conditions. It embodies all the skills and innovations of the community and represents the collective wisdom and resourcefulness of its people. It is part of the local knowledge and is deeply rooted in a particular community based on its local historical experience. As such, it is unique to every culture and society, and plays an important role in defining the identity of the community.

ITK has made it possible for African indigenous communities to live in harmony with their environment for generations, to utilise their natural resources sustainably using a variety of innovations to deal with pest and disease outbreaks, crop health issues, environmental conservation and natural disaster management. For most subsistence and resource-poor farmers in Africa, naturally occurring, free pesticidal preparations are the only affordable and accessible farm input for crop protection and animal health care (Facknath and Lalljee, 2005). In keeping with local tradition, ITK is passed down orally from generation to generation, and there is little documented literature. The UNEP website has a page for Indigenous Knowledge in Africa which contains information on ITK gathered from a study of four African countries, namely Kenya, Tanzania, Swaziland and South Africa. However, there is a wealth of information in all the

The Nagoya Protocol on Access to Genetic Resources and the Fair and Equitable Sharing of Benefits Arising from their Utilization

This is an international agreement which aims at sharing the benefits arising from the utilization of genetic resources in a fair and equitable way, including by appropriate access to genetic resources and by appropriate transfer of relevant technologies, taking into account all rights over those resources and to technologies, and by appropriate funding, thereby contributing to the conservation of biological diversity and the sustainable use of its components. It was adopted by the Conference of the Parties to the Convention on Biological Diversity at its 10th meeting on 29th October 2010 in Nagoya, Japan. The fair and equitable sharing of the benefits arising out of the utilization of genetic resources is one of the three objectives of the Convention on Biological Diversity.

Source : CBD website

other African countries as well, but which has not yet been inventoried.

The promotion of ITK practices and technologies is associated with the issues of property and copyrights, and the need to protect and preserve this knowledge for the benefit of all. Most African countries do not have specific laws to protect their nation's ITK. Some have isolated national policies that fit within international frameworks and agreements (e.g. the Nagoya Protocol) that help to combat biopiracy and share benefits from national resources research fairly. The African Regional Inte-llectual Property Organization protects the interest of holders of ITK from any infringement of their rights and the misappr-opriation, misuse or exploitation of their knowledge. The Access and Benefit-Sharing (ABS) agreements under the Convention on Biological Diversity protects a country's and/or a communities' autonomous rights over the genetic resources on their territory and recognizes their collective rights. According to this agreement, local stakeholders of any genetic resource must be informed about any bioprospecting activities, must give their formal consent to it, and have to be compensated if future revenues are to be achieved. On their part, the local resource providers have to commit themselves to biodiversity conservation. Most of the cases of bioprospecting and biopiracy in Africa come from the field of traditional medicines and pharmaceuticals.

The principle is that such deals can help to finance conservation efforts, and also ensure a fair and equitable distribution of the benefits of the use of genetic resources among all those who invest in their continued existence. However, such agreements do

> One of the most publicized agreements between Costa Rica's National Biodiversity Institute (INBio) a private, non-profit organization and the U.S.-based pharmaceutical firm Merck & Co., Ltd. in 1991 opened the way for bilateral deals for bioprospecting in other countries. In exchange for exclusive rights to develop and patent products from species found in Costa Rican rainforests for use in its drug-screening program, Merck would give INBio a two-year research and sampling budget of $1,135,000 and royalties on any resulting commercial products. At the national level, INBio agreed to contribute 10% of the budget and 50% of royalties to the government's National Park Fund for the conservation of national parks in Costa Rica, while Merck agreed to provide technical assistance and training to help establish drug research capacity in Costa Rica (Aldhous, 1991).
>
> A similar deal was signed between Nigeria and Shaman over the period 1990 to 1996 (Kate and Laird, 1999). Kenya too has explored similar possibilities for biodiversity prospecting activities.

not necessarily ensure that countries where the bioactive compounds originated benefit from a share of the profits accruing from the development of any commercial product. Since the success rate for deve-loping a commercial marketa-ble product is exceedingly low, royalties and other similar financial benefits are either highly unlikely (McChesney, 1996) or long time in coming (upto 10–12 years) (DiMasi *et al.*, 2003). Thus this model, with the source country offering biological materials and the developed country supplying the research, may not always give the expected results. McGown (2006) cites several cases from Africa suggestive of biopiracy. A few include:

- Two nematophage strains of the fungus *Arthrobotrys conoides* Dreschsler originating from Burkina Faso are patented by a Swiss agrichemical multinational for use as biopesticides for the control of gall nematodes which attack lentils and maize. More recent patent applications cover large scale fungus production methods and formulation of biopesticide granules from this fungus.
- Slug Barrier, a molluscide has been patented in the UK. The active compound is from myrrh and other African plants (e.g. *Commiphora molmol*) common in Somalia, Ethiopia, Egypt, etc.
- Insect resistance and feed quality of tall fescue (*Festuca arundinacea*) grass (used for animal pasture, hay and silage) have been improved by the inoculation of three North African fungal endophytes from the genus *Neotyphodium*, by a company in New Zealand in 2000. The endophytes produce peramine, which enhances the plant's resistance to insects.

CONCLUSIONS

Africa has immense potential in the field of biopesticides, not only as a potential market for commercial biopesticides manufactured outside the continent, but also as a source of novel bioactive (pesticidal) molecules from natural (either plant or animal) sources, or new microbial species that may have pest controlling properties. This represents an interesting area of research for agricultural scientists.

The huge internal market for Africa biopesticides within the continent justifies the need for Africa to develop is own standards and norms for registration of its biopesticides, and its own registration and certification bodies. With the increasing dismantling of technical barriers to trade (TBTs) under the SPS regulations of the WTO protocol, there is a need to standardize and harmonise registration and certification norms and standards, not only for the biopesticides themselves, but also for agricultural commodities treated with such biopesticides, e.g. organic produce.

The above, coupled with the appropriate IPR protection, can pave the way to make biopesticides an important source of safer and more environmentally friendly and sustainable options for pest management, and also as a significant source of revenue for African nations.

REFERENCES

ADAPPT (2011) *African Dryland Alliance for Pesticidal Plant Technologies.* http://www.nri.org/projects/adappt/

Aldhous, P. (1991) Hunting license for drugs. *Nature*, **353**, 290.

Beevor, P.S., Hall, D.R., Lester, R., Poppi, R.G., Read, J.S. and Nesbitt, B.F. (1975) Sex pheromones of the armyworm moth *Spodoptera exempta. Experientia*, **31**, 22–23.

CBD (2011) *Convention on Biological Diversity.* United Nations Environment Programme, Geneva.

CPLBC (2010) *Latin America, Africa and The Middle East: Biopesticides Market.* CPL Business Consultants, Wallingford, UK.

DiMasi, J.A., Hansen, R.W. and Grabowski, H.G. (2003) The price of innovation: New estimates of drug development costs. *J. Health Econ.*, **22**, 51–185.

Facknath, S. and Lalljee, B. (2005) Indigenous/traditional knowledge from Mauritius for sustainable agriculture. In: A. Bandyopadhyay, K.V. Sundaram, M. Moni, P.S. Kundu and M.M. Jha (eds.), *Sustainable Agriculture-Issues in Production, Management, Agronomy and ICT*, Northern Book Centre, New Delhi, India, pp. 147–164.

FAO (1985) *Prevention of Post-Harvest Food Losses: Fruits, Vegetables and Root Crops.* A Training Manual, Food and Agriculture Organization, Rome.

FAO (2009) *Red Locust Disaster in Eastern Africa Prevented. Biopesticides Being Used on a Large Scale.* FAO Media Centre, 24th June 2009, Food and Agriculture Organization, Rome.

FAO TECA (2011) *Sustainable Approaches for Management of Root-Knot Nematodes on Vegetables in Kenya.* http://teca.fao.org.

FAO TECA (2011) *Development of Biopesticide Legislation in Africa.* http://teca.fao.org

Haggis, M.J. (1984) *Distribution, Frequency of Attack and Seasonal Incidence of the African Armyworm, Spodoptera exempta (Walk.) (Lepidoptera:*

Noctuidae), with Particular Reference to Africa and Southwestern Arabia. Tropical Development Research Institute, London.

Hassanali, A. and Lwande, W. (1989) Antipest secondary metabolites from African plants. In: J.T. Arnason, B.J.R. Philogene and P. Morand (eds.), *Insecticides of Plant Origin*. ACS Symposium Series 387, American Chemical Society, Washington, DC, pp. 78–96.

Kate, K. and Laird, S.A. (1999) *The Commercial Use of Biodiversity. Access to Genetic Resources and Benefit-sharing*. Earthscan, London.

Langewald, J. (1999) *Lubilosa Green Muscle® User Handbook*, Ver 4, International Institute of Tropical Agriculture, Cotonou, Benin.

Maranga R.O., Kaaya G.P., Mueke J.M. and Hassanali A. (2005) Effects of combining the fungi *Beauveria bassiana* and *Metarhizium anisopliae* on the mortality of the tick *Amblyomma variegatum* (Ixodidae) in relation to seasonal changes. *Mycopathologia,* **159**, 527–532.

McChesney, J. (1996) Biological diversity, chemical diversity and the search for new pharmaceuticals. In: M.J. Balick, T. Elisabetsky and S. Laird (eds.), *Medicinal Resources of the Tropical Forest: Biodiversity and its Importance to Human Health*, Columbia University Press, New York, pp. 63–74.

McGown, J. (2006) *Out of Africa: Mysteries of Access and Benefit Sharing*. Edmonds Institute in cooperation with African Centre for Biosafety. West Edmonds, Washington.

Musebe R., Day, R., Kipkoech, S., Musavi, F., Kimani, M., Opiyo, P. and Hassan, N. (2010) Putting research into use: community based armyworm forecasting in Kenya. *12th Kenya Agricultural Research Institute (KARI) Biennial Scientific Conference*, 8–12 November 2010,The KARI Headquarters, Kaptagat Road, Loresho, Nairobi, Kenya.

Mushoboz (2011) *Armyworm Outbreak Allows Harvest of Biological Control Agent*. Ecoagri Consult Ltd., Tanzania.

Nana, P., Maniania, N.K., Maranga, R.O., Kutima, H.L., Boga, H.I., Nchu, F. and Eloff, J.N. (2010) Attraction response of adult *Rhipicephalus appendiculatus* and *Rhipicephalus pulchellus* (Acari: Ixodidae) ticks to extracts from *Calpurnia aurea* (Fabaceae). *Vet. Parasitol.,* **174**, 124–130.

Rose, D.J., Dewhurst, C.F. and Page, W.W. (2000) *The African Armyworm Handbook: The Status, Biology, Ecology, Epidemiology and Management of Spodoptera exempta (Lepidoptera : Noctuidae)*. Second Edition. Natural Resources Institute, Chatham, UK.

US EPA (2009) *Straight Chain Lepidopteran Pheromones (SCLP)*. Office of Pesticide Programs, Biopesticides Registration Action Document, US Environmental Protection Agency, USA.

21

Biopesticides for Pest Management in Thailand: Prospects and Constraints

VASAKORN BULLANGPOTI[1]* AND OPENDER KOUL[2]**

[1]Department of Zoology, Faculty of Science, Kasetsart University
Bangkok 10900, Thailand
[2]Insect Biopesticide Research Centre, 30 Parkash Nagar,
Jalandhar 144003, India, and Adjunct Professor, Faculty of Science
Kasetsart University, Bangkok 10900, Thailand
* E-mail: fscivkb@ku.ac.th; **okoul@koulresearch.org

INTRODUCTION

In the mid 1980s Thai agriculture faced severe problems due to insecticide resistance of key insect pests that made production of cotton, vegetables and fruit increasingly expensive (Jones *et al.*, 1993). Another aspect was the health hazards due to pesticide poisoning because of their overuse (Harris, 2000). In view of these problems, Thailand initiated a national program to develop biocontrol alternatives and various research programs were undertaken from late 1980s to develop local products based on bacteria, viruses, fungi and nematodes. Some emphasis was given to predators and parasitoids as well. On the basis of this program NPV production was established by the Department of Agriculture (Jones *et al.*, 1998). However, Thailand faces problems due to limited market for biopesticides and the poor quality of some of the non-commercial biopesticide production units (Warburton *et al.*, 2002). Illegal importations into Thailand add ineffective products and needs streamlining and proper regulation.

With the expanding gross domestic product (GDP) in Thailand (1.2% in the fourth quarter of 2010 alone over the previous quarters), average quarterly GDP Growth was seen to be maximum (7.1%) in September of 1993 and lowest (–5.1%) in March 1998 (Trading Economics, 2010). However, agricultural product has contributed significantly to this expanding GDP, which has been possible because Thailand is one of the largest consumers of synthetic pesticide in South East Asia (Ratanasatien *et al.*, 2005) and follows almost the global trend in pesticide consumption (Ketunuti, 2003). The major crops grown in Thailand are rice, cassava, rubber, coconut, cotton, sugar cane and oil palm. Most pesticides are imported, and foreign companies hold the biggest market share such as Monsanto, Bayer, and Du pont. However, due to indiscriminate use of these pesticides, various environmental hazards and residual problems have prompted the promotion of organic agriculture, which provides environment, social and economic benefits. The use of bio-rational pesticides like biopesticides in integrated pest management (IPM) has been one of the strategies employed in order to obtain safe system of crop protection.

As mentioned above that Thailand is one of major markets for pesticide in South East Asia, the Thailand government, specifically the Ministry of Agriculture has prompted to develop new and effective methods to use pesticides. The researchers have tried to discover and develop biopesticides for farmer use under field conditions. However, as in other parts of the world the registration rate of biopesticide products has remained quite low in comparison to imports of synthetic pesticides.

Pesticide market in Thailand seems to increase in the same proportion as that of the increase of export value of rice, food and fruit. Quantitative value of pesticide imports in 2009 and 2010 depict this (Table 21.1). The import quantity and value of the herbicide was the highest, 97,956,856.41 kg and 33,783,149.58 kg, respectively in 2009 and 2010. The data from 1997 to 2000 suggest that biopesticide import seems to be decreasing dramatically (Poramacom, 2001) and goes further down by 2010 (Table 21.1). In fact, the import was 44 tons in the year 2000.

BIOPESTICIDES IN THAILAND

Research and development of biopesticides in Thailand started in 1971 in order to control the diamondback moth, *Plutella xylostella* (Lepidoptera: Yponomeutidae), and cabbage looper, *Trichoplusia ni*

(Lepidoptera : Noctuidae). At the beginning, all the research on biopesticides emphasized testing the efficacy for controlling various insect pests. It is reported that in the early days farmers and others involved in crop protection took little interest because biopesticides were slower acting in controlling insect pests compared to synthetic chemical insecticides. It is only after biopesticides were proven operationally to control some arthropod pests that researchers, universities, government agencies and the private sector became interested in this new, promising group of organisms. Field surveys conducted throughout the country by the Department of Agriculture revealed the presence of several promising biopesticide agents.

Table 21.1. Type, quantity, value and percent of pesticide imports in 2009 and 2010

Type	2009			2010		
	Quantity (kg)	Values (baht)	Percentage	Quantity (kg)	Values (baht)	Percentage
Insecticide	24,680,168.90	3,972,447,198.25	23.59	8,094,452.98	1,640,803,728.03	26.21
Fungicide	10,366,987.08	2,967,933,290.16	17.63	2,442,567.60	883,269,165.16	14.11
Herbicide	97,956,856.41	9,338,498,815.25	55.46	33,783,149.58	3,554,250,336.21	56.78
Acaricide	622,097.24	133,273,180.30	0.79	129,572.00	36,337,917.00	0.58
Rodenticide	222,490.01	25,784,622.08	0.15	89,000.00	14,309,855.50	0.23
PGR	2,107,459.63	183,017,953.30	1.09	711,734.00	67,822,501.03	1.08
Molluscide	659,387.50	47,170,497.87	0.28	162,000.00	12,421,752.00	0.20
Fumigant	942.946.50	147,643,520.35	0.88	300,911.00	47,388,977.46	0.76
Biopesticide	144,998.00	21,404,651.00	0.13	20,682.00	2,558,510.00	0.04
Total	112,080,275.90	16,837,173,729.00	-	45,734,069.16	6,259,162,742.00	-

Source: Office of Agricultural Economics, Thailand (2010)

Botanicals

There are no reports that show when the botanical insecticides were first used in Thailand (Wongsiri, 1980). However, there are many plant species such as *Nicotiana tabacum, Azadriachta indica, Azadrachta indica* var. *siamensis* or *Derris elliptica* that Thai farmers knew have insecticidal efficiencies and used them for control of insect pests for a long time. In addition, the Thai government promoted pyrethrum cultivation and production of pyrethrum extracts in Chiang Mai (Wongsiri, 1980; Jeerapong, 1999). In fact, Thailand is a tropical rainforest country having diversity of plant species and Thai farmers have tried to make botanical extracts for use in their farms using self-created formulation devices.

There are many reports of research and development of botanicals as pest control agents in Thailand, which have been comprehensively reviewed recently (Bullangpoti, 2009). Many commercial botanical products, mostly neem base product, are now available as insecticide formulations and are evaluated against number of insect species. Although botanical insecticides encounter obstacles in research, development and commercialization, yet they still have a good future in Thailand due to the Thai government's campaigns to promote national "Food Safety" and international recognition of "Thailand: Kitchen of the World" by GAP and integrated pest management programs that emphasize utilization of biopesticide compounds (Bullangpoti, 2009). In addition, knowledge of botanical insecticides in Thailand has also been progressively increasing. It is noteworthy that many Thai plants with insecticidal effects such as *Azadiracta indica* var *siamensis*, *Momordica charantia* (Cucurbitaceae), *Cassia fistula* and *Senna siamea* (Leguminosae, Caesalpinioideae), *Derris elliptica* (Leguminosae, Papilionoideae), *Lansium domesticum* (Meliaceae) and *Nicotiana tabacum* (Solanaceae) are being exploited. The odoriferous plants such as *Citrus hystrix and Citrus reticulata* (Rutaceae) have various essential oil components that tend to be repellents rather than insecticides. Dried leaves of *Coriandrum sativum* (Apiaceae) have been used by South Asian farmers against stored insects, and some essential oil components of this plant are reported to be effective as fumigants and repellents against the red flour beetle *Tribolium castaneum* (Koul *et al.*, 2008). *Bixa orellana* (Bixaceae), *Jatropha curcas* (Euphorbiaceae) and *Ricinus communis* (Euphorbiaceae) also have potency as bioinsecticides, but their extracts show high toxicity to mammals, so before developing them as commercial products or recommending them to users, scientists must consider their toxicity to other organisms (Bullangpoti, 2009).

Microbials

It was in 1969 that the first strain of *Bacilus thuringiensis* var *kurstaki* was used in Thailand for lepidopteran larval control (Prasetphol *et al.*, 1969; Vattanatangum, 1989) and substantial control was achieved for chemical pesticide resistant pests, like *Spodoptera exigua* and *Helicoverpa armigera.* Today, *Btk* is used in many vegetable grow areas (Ratanasatien *et al.*, 2005). Some farmers also accept these microbials as combination products with synthetic chemicals of various groups, including organophosphates. There is, however, no quantitative static usage of this microbial in Thailand and depends on the proportion of synthetic chemicals used (Ratanasatien *et al.*, 2005). If a new effective synthetic chemical is introduced, the use of biopesticide decreases. The importation of biopesticides into Thailand doubled in 1986 compared to 1985. However, the amount of *Btk* imported decreased in 1987 compared to that of 1986 due to the introduction of a new and effective group of pyrethroid insecticides. It has been reported that the percentage of *Btk* applications sharply increased by at least 15-fold between 1985 1987(Ratanasatien *et al.*, 2005). At present, there are more than 30 pest species in Thailand that can be controlled by *Btk*. In the past decade, Department of Agriculture, Ministry of Agriculture and Cooperative (DOA) researchers have discovered over 100 new isolates of *Bt*. To promote the use of *Bt*, DOA has built a pilot plant with a production capacity of at least 1000 liters per year to serve the northern vegetable growing areas and hopefully to reduce the dependence on toxic chemical insecticides in the northern part of Thailand. Research on formulation of *Btk* has also been carried out to prolong the efficacy and stability in the environment. Furthermore, the private sector has also developed *Bt* strains (*Btk* and *Bta)* which can control the important pest of cruciferous crops (Ratanasatien *et al.*, 2005).

In case of entomopathogenic fungi, *Metarhizium anisopliae* (Hypocreales: Clavicipitaceae) and *Beauveria bassiana* (Deuteromycotina: Hypomycetes or Moniliales: Hyphomycetes) are other microbials tried as alternative pest control strategy in Thailand. The research has focused on the use of green muscardine fungus, *M. anisopliae*, for the control of many pests such as *Oryctes rhinoceros* (Coleoptera: Scarabaeidae), which causes considerable economic damage to both coconut and oil palms. A large scale application of this fungus has been conducted in the coconut plantation in the southern part of Thailand. Some research has also been done to demonstrate the efficacy of this entomopathogenic fungus against *Coptotermes gestroi* (Ruchinarong, 2002), *Musca domestica* (Youngcha-itrakul, 1999), and *Boophilus microplus* (Tungrabieb, 2009)

Nucleopolyhedrovirus (NPV) is recognized as the most effective virus for controlling several insect pests (Marasri, 1994). Three NPVs were developed by DOA researchers such as *Helicoverpa armigera* NPV(HaNPV), *Spodoptera exigua* NPV (SeNPV) and *Spodoptera litura* NPV (SlNPV) (Ratanasatien *et al.*, 2005). HaNPV provides effective control of the cotton bollworm on cotton, mungbean, sorghum, okra, asparagus, tomato, tobacco, grape vine, tangerine, rose and marigold. SeNPV is a promising agent to control the beet armyworm on onion, shallot, okra, asparagus, grape vine, rose, marigold and orchids. SlNPV is produced for the control of common cutworm on some vegetables such as cabbage, yard long bean and orchids. The DOA has built and has been operating a nucleopolyhedrovirus pilot plant since 1994 (Ratanasatien *et al.*, 2005).

Overall, in Thailand, there are several large and very competent groups conducting research, development and extension work on microbial control. For instance, at Mahidol University, a group is working on *Bt* subsp. *israelensis* for control of the *Aedes* mosquitoes that transmit dengue, a very important arbovirus disease in Thailand. Similarly, at Kasetsart University the focus is on the three-dimensional structure, mode of action and insect specificity of Bt Cry toxins. There is also a smaller group working on *Bt* subsp. *israelensis*. The Biocontrol Section of the Thai Department of Agriculture conducts applied research on nuclear polyhedrosis viruses (NPVs), fungi and Bt. The extension service of the same ministry has a project to educate farmers on the use of microbial .pesticides, and to help in the establishment of small-scale biopesticide production units. The production efforts are currently focusing on the fungi *Trichoderma harzianum*, *Beauverria bassiana* and, to a lesser extent, *Metarhizium* (Skovmand, 2007)

Entomopathogenic Nematodes

Research on entomopathogenic nematodes for controlling insect pests in Thailand started in 1986 (Ratanasatien *et al.*, 2005). The first field trials were conducted to determine the effectiveness of *Steinernema carpocapsae* against *Cossus* sp. (Lepidoptera: Cossidae), *Microchlora* sp. (Lepidoptera: Pyralidae), *Phyllotreta sinuata* (Coleoptera: Chrysomellidae), *Cylas formicarius* (Coleoptera: Curculionidae) and *S. exigua*. Since 1999, the nematode production technology was transferred to a private company for marketing and distribution (Somsuk *et al.*, 1999). As of today, the research efforts are on to develop entomopathogenic nematodes as new biopesticide in Thailand. The prominent ones are the studies of *Steinernema siamkayai* for the control of *Plutella xylostella* (Sasnarukkit, 2003) and *Steinernema siamkayai* for *Helicoverpa armigera* (Chongchitmate, 2005)

Protozoans

Microsporan protozoans have been investigated extensively as possible components of integrated pest management programs. Microsporidia are ubiquitous, obligatory intracellular parasites that are disease agents for several insect species. Two genera, *Nosema* and *Vairimorpha* have some potential as they attack lepidopteran and orthopteran insects and seem to kill hoppers more than any other insect (Lewis, 2002). Earliest report of evaluation of the microbial control potential of a protozoan in Thailand is that of *Helicosporidium* sp. (Protozoa : Helicosporida) from *Aedes aegypti* and *Culex quinquefasciatus* (Hembree, 1981). However, there is no other record to show that protozoans have been used in pest management in Thailand. In fact, only commercial product known globally is based on *Nosema locustae* (known as Nolo bait), which infect grasshoppers. However, in Thialand new rodenticide based on protozoan *Sarcocystis singaporensis* is one good example (Jaekel *et al.*, 1997). The cooperative research of this protozoa was conducted with the University of Hohenheim, Bayer A.G. and the Agricultural Zoology Research Group, DOA, Thailand. These protozoa were developed to control rats in rice fields, oil palm plantations and chicken farms. Present research on *Sarcocystis* includes the study of the virulence of the parasites, viability, immunological selection, and mass production techniques of this organism. The research group currently has a cooperative project with Uniseeds Co., Ltd. to develop this parasite as a commercial product. The Department of Agriculture and TGZ-project (German Technical Cooperation Agency) are pushing and encouraging the ASEAN countries to use such poison baits against rats, both in the fields and in households. So far, considerable success using this organism to control rats has been confirmed in both Thailand and the People's Democratic Republic of Laos (Promkird, 2004).

Entomophagous Insects

Thailand has a total land area of 51.3 million ha; 41% of this is used for agricultural production while another 30% is forest. The country is divided into 6 regions made up of 77 provinces. Approximately 40% of the population (65 million) is farmers. The main agricultural crops are rice, rubber, oil palm, sugarcane, vegetables and fruit trees. The key pests for these are brown planthopper, leaf folder, blast and dirty pinnacle for rice; stemborer, stem boring grub and white grub for sugarcane; thrips, whitefly, wilt, anthracnose and mosaic virus for chilli; diamondback moth, armyworm, aphid, rootrot and softrot for crucifer vegetables; root and stem rot, fruitfly, thrips, mite, mealybug, and stemborer for fruit trees. In order to reduce the hazards of chemical pesticides, in 1995 it was decide to promote biological control. Accordingly, nine Pest Management Centers were established in different regions in Thailand, i.e. Chiangmai and Phitsanulok in the northern region; Khon Kaen and Nakhon Ratchasima in the north-eastern region; Chonburi in the eastern region; Suphanburi in the western region; Chainat in the central region; and Songkha and Suratthani in the southern region. Each of these centers was rendered responsible for 7-10 provinces to mass rear biological control agents like *Trichogramma* sp. and *Cotesia* sp. to manage sugarcane stemborer. They were suggested to arrange preliminary field trials to develop skills, confirm the results and benefits of the biological control agents, and subsequently transfer this knowledge to farmers using the farmer field school approach. This was good beginning and now there are some projects where studies are being conducted on using entomophagous insects in Thailand. For example, the project conducted by DOA on *Anastatus sp. nr. Japonicus* for control *Tessaratoma papillosa*. The studies on *Trichogramma confusum viggini* for control of *Heliothis amigera* or *Chilo infuscatelus, Chilo sacchariphagus, Sesamia inferens* have been taken up by DOA. The control of *Heteropsylla cubana* by *Curinus coeruleus* and the use of *Eocanthecona furcellata* for control of *Spodoptera exigua* are also on record. However, mostly parasitoids and predators used in Thailand have not been commercialized but mostly DOA or government agencies promote them and supply to the farmer directly. For example, *Brontispa longissima* Gestro for control of *Plesispa reichei* Chapuis is provided to farmers in Samutsongkram province by Samutsongkram Provincial Agricultural Extension Office. However, efforts of DOA or researchers to promote biological control through natural enemies during last decade has not really picked up because farmers do not accept the use of natural enemies against the conventional insecticides. It may be because slow control efficiency and least field application demonstrations (Nanta, 2002), which need to be promoted by extension agencies on larger scale.

Pheromones

There are three main uses of pheromones in the integrated pest management of insects. The most important application is in monitoring a population of insects to determine if they are present or absent in an area or to determine if enough insects are present to warrant a costly treatment. This monitoring function is the keystone of integrated pest management. A second major use of pheromones is to mass trap insects to remove large numbers of insects from the breeding and feeding population. Massive reductions in the population density of pest insects ultimately help to protect resources such as food or fiber for human use. Mass trapping has been explored with pine bark beetles and has resulted in millions of insects attracted specifically into traps and away from trees. Mass trapping has also been used successfully against the codling moth, a serious pest of apples and pears. A third major application of pheromones is in the disruption of mating in populations of insects. This has been most effectively used with agriculturally important moth pests. In this scenario, synthetic pheromone is dispersed into crops and the false odor plumes attract males away from females that are waiting to mate. This causes a reduction of mating, and thus reduces the population density of the pests.

There are some projects in Thailand where pheromones have been used under the Thai-German Program for Enterprise Competitiveness. This program prepares the ground for introduction of pheromone mass trapping technology in orchards to combat insect pests and strengthen organic production. The current introduction and testing of a sex pheromone to control the lychee fruit borer (*Conopomorpha sinensis*) in organic longan and lychee plantations in Thailand is a follow-up of a very successful program of GTZ in Indonesia, where a new pheromone for mass trapping of a sister species, the notorious pest cocoa pod borer (*Conopomorpha cramerella*), was tested (Zhang *et al.*, 2008) and was subsequently commercialized. Preliminary results from trials on organic longan farms in Chiang Mai show that the pheromone is trapping lychee fruit borer effectively. Should it be possible to reduce damage to longan and lychee fruits significantly, sales of organic longan and lychee would be facilitated, because presence of the pest's larva in infested fruits deterred customers of buying it previously (Poonuchaphai and Jäkel, 2009).

BIOPESTICIDE PRODUCTION AND REGULATION

Although research and development of biopesticides in Thailand started in 1971 with a goal to control *Plutella xylostella* and *Trichoplusia ni* (Ratanasatien *et al.*, 2005), the promising organisms used as biopesticide in Thailand include *Bacillus thuringiensis, B. subtilis, nucleopolyhedrovirus, Metarhizium anisopliae, Beauveria bassiana, Trichoderma harsianum, Steinernema siamkayai,* and

Sarcocystis singaporensis and some botanical products such as neem (Table 21.2) Currently, the DOA has built a small pilot factory for producing Bt, NPV and neem. The main objectives are promoting the use of biopesticides in Thailand. However, the interest at farmer level has not been very encouraging, apparently due to slow action of biopesticides compared to synthetic chemicals. Although many pest organisms have been intensively studied and tested, in both the laboratory and field, for efficacy; yet the application technology, mass production methods or mechanism still need to be comprehensively investigated.

Regulatory systems for biopesticides were established to cover commercialization of these products, but non-commercial production by farmers, NGOs, extension services, research institutes and products distributed as part of IPM initiatives could not be stopped (Warburton *et al.*, 2002). The system allowed imports of some biopesticides subjected to local registration. In country efficacy trials were supervise by DOA and production units are required to go through proper regulatory process (Grzywacz, 2003)

There are 3 stage registration procedure followed in Thailand for biopesticides:

1. The preliminary test to determine the acute and residual toxicity.
2. The field evaluation to demonstrate its efficiency and analyze chronic toxicity.
3. The 2 years data for human and environment safety and also chronic toxicity to test animals.

A key focus for regulation is to ensure that all commercial products meet acceptable performance and quality standards. There are proposed standards for a number of such biopesticides. In determining appropriate protocols for field efficacy test, we now have considerable consensus on acceptable practices for field trials though no recommended guidelines have been published for most biopesticides. In developing these systems, active dialogue between producer scientists and regulators is important in order to balance the sometimes-conflicting needs of regulation and commerce.

One problem for registration authorities can be how to judge the validity of submitted data? This becomes especially acute where the data are of a type unfamiliar to regulators whose technical expertise may be only in chemistry. We believe that recognized referral laboratories are important to eliminate such discrepancies.

Table 21.2. Some of biopesticides intensively used for pest control in Thailand

Biopesticide type	Pests to be controlled
Botanical Pesticides	
Neem; *Derris* sp.	Lepidoptera species
Lemon grass oil	Mosquitoes (Repellent)
Mixed botanical pesticide	Various used such as Lepidoptera species or termite control
Microbial Biopesticides	
Bacillus thuringiensis var. *kurstaki; B. thuringiensis* var. *aizawai; B. thuringiensis* var. *isaelensis*	Lepidoptera such as *S. exigua, P. xylostella*
B. subtilis	Plant disease eg, *Pyricularia grisea, Curvularia lunata* and water treatment in Shrimp industries.
NPV	Lepidoptera sp. such as *Helicoverpa armigera, S. litura, S. exigua*
Bipolaris oryzae	*Cerocospora oryzae*
Chaetomium globosum or C. cupreum	*Pyricularia oryzae, Puthium altimum, Sclerotium rolfsii, Fusarium oxysporum, Phytophthera palmivora*
Beauvaria bassiana	*Eutetranychus africanus*
Entomophthora grylli	*Patanga succincta*
Metarhizium anisopliae	*Oryctes rhinoceros*
Paecilomyces lilacinus	*Meloidogyne*
Nematodes and Protozoa	
Steinèrnema carpocapsae; S. siamkayai	Flea beetles, *Spodoptera exigua, S. litura, Plutella xylostella, Cossus* sp., *Microchlora* sp.
Trichoconis padwickii	*Helmintosporium oryzae, Fusarium semitectum*
Trichoderma harzianum	*Rhizoctonia* spp., *Sclerotium* spp., *Pytium* spp.
Sacocystis Singaporensis	*Rattus* sp., *Banticola* sp.
Entomophagous Insects	
Anastatus sp. nr. *Japonucus*	*Tessaratoma papillosa*
Trichogramma spp.: *Trichogramma chilonis, Trichogramma confusum,*	*Helicoverpa armigera, Chilo infuscatellus, Chilo tumidicostalis,*

Biopesticide type	Pests to be controlled
Trichogramma japonicum, Trichogrammatoidea bactreae	*Trihoplusia ni, Plutella xylostella, Ostrinia furnacalis, Papilio demoleus malayanus, Scirpophaga incertulas*
Diachasmimorpha longicaudata	*Bactocera dorsalis*
Mallada basalis	*Trialeurodes vaporiorum, Tetranychus urticae*
Ankylopteryx octopunctata	*Trialeurodes vaporiorum, Tetranychus urticae*
Eocanthecona furcellata	*Helicoverpa armigera, Spodoptera litura, Spodoptera exigua.*
Pheromones	Lepidoptera sp.

BIOPESTICIDE MARKET

Biopesticide suppliers can be classified in two groups in view of the scenario existing in Thailand.

Non-profit Organizations

This group provides help and technical support to farmers to reduce the use of synthetic chemicals and their harmful effects to the environment. This group also ensures to reduce the levels of toxic residues in the agricultural products. For example, many researchers in universities of Thailand look for new and promising organisms. They conduct research on their efficacy, mode of action and also provide some technical assistance and impart technical knowledge to farmers.

Department of Agriculture (DOA), Ministry of Agriculture and Cooperatives are other agencies, which are slightly different from university approach. DOA mostly focuses on technical aspects and investigates the effectiveness of biopesticides through laboratory and field experiments. They produce the biopesticides in small scale and distribute them to farmers, at no cost. Once the effectiveness of the biopesticides is demonstrated, the technology is then transferred to interested private sector for production and sale. For example, DOA in Thailand has established *Bt* and NPV production units or neem production factories in Chiangmai Experiment Station and at the Entomology and Zoology centers at Bangkok.

Department of Agriculture Extension Service (DOAE) is another agency having at least eight centers throughout the country. Quite different from DOA, these centers provide information and technical support to farmers by demonstrating, training, and supervising the use of biopesticides. Sometimes DOAEs also produce and distribute biopesticides to farmers or occasionally sells it to farmers or private sectors at production cost.

Non-governmental Organizations (NGOs) also play important role, who assist farmers on the basis of the knowledge obtained from DOA, DOAEs or Universities by providing either technical information or assisting with funding. The NGOs are mostly non-profit organizations whose budgets come from donations from developed countries or from the Thai government.

Commercial Organizations

Commercial organizations are important because they produce, import and sell the products on larger scale. Usually, these products are distributed directly to farmers by local representatives of the companies, or through local dealers who charge farmers at market price. Some companies import and sell both synthetic chemicals and biopesticides. They are usually representative of the multinational companies. In addition, there also have local producers who produce and sell the products directly to farmers. Under the national project on 'Hygienic Food and Safety', it is expected that the production and use of biopesticides will increase. Mostly technical assistance usually comes from universities and government sectors. To promote and encourage the use of biopesticides, the DOA has transferred the production technology of both *Btk* and nematodes to the private sector for commercialization. At present, there are five local private companies producing the three varieties of *Bt* and *S. siamkayai* and their combined production is about 20 tons a year (Ratanasatien *et al.*, 2005). This does not cover the total demand in Thailand, and the rest has to be imported each year. The imported biopesticides (27.7 tons) represent less than 1% of all pesticide imports (42,248 tons) (Office of Agriculture Economics, 2010) and local production of *Btk*, *Bti* and *Bta* will double by the year 2010.

In fact, the initiative, "Commercialization of Biopesticides in Southeast Asia" was implemented by the German government agency GTZ (German Technical Cooperation) in 2003 in Thailand. Based in Bangkok, Thailand, the goal was to bring biopesticides (microbial control products, pheromones, natural enemies) to the Southeast Asian market place by providing technical support to private companies and government institutions who share this goal. Technical expertise is available in areas such as mass production, formulation, product evaluation, regulatory strategies and business and marketing efforts (Skovemand, 2007).

Right now in Thailand, the biopesticides are required to be registered with DOA. The products based on predators, parasitoids and nematodes, etc. are provided free to farmers by DOA and DOAE (Table 21.3).

Table 21.3. Some registered biopesticide products in Thailand

Type	Trade name and source[1]	Active ingredient
Microbial	Mogin (Nontrisia co. Ltd./Nontrisia Co. Ltd.)	*Bacillus thuringiensis* var. *aizawai* 55,000 SU/mg (5% WP)
Microbial	Suyaki (Nontrisia Co. Ltd./ Nontrisia co. Ltd)	*Bacillus thuringiensis* var. *aizawai* 55,000 SU/mg (5% WP)
Microbial	Moyas (Nontrisia co, Ltd./ Agrsia inter trade Co. Ltd.)	*Bacillus thuringiensis* var. *aizawai* 55,000 SU/mg(5% WP)
Microbial	Proback K (ProCrop Co. Ltd./ ProCrop Co. Ltd.)	*Bacillus thuringiensis* var. *kurstaki* 55,000 SU/mg (5% WP)
Microbial	GSRI Biocin (Strong Crop Co. Ltd.-GSRI network Co. Ltd.)	*Chaetomium cupreum* 1.5×10^6 cfu/gm
Microbial	Set Point (Applied Chem. Co. Ltd.)	*Bacillus thuringiensis* var. *kurstaki* 16,000 IU/mg (3.2% WP)
Microbial	*Bt* Gold (Oscar Agro Co. Ltd.)	*Bacillus thuringiensis* var. *aizawai* 1×10^9 CFU/ml (12% W/V SC)
Microbial	Tricosia (Applied Chem. Co. Ltd./Lectic Chem. Co. Ltd.)	*Trichoderma harzianum* 2×10^8 cfu/gm WP
Microbial	Balensia (Applied Chem. Co. Ltd./ Lectic Chem Co. Ltd.)	*Bacillus subtilis* var. AP-01 1×10 cfu/gm WP
Microbial	B&T (Agro Biomed Co. Ltd.-Siam Know How Co. Ltd.)	*Bacillus thuringiensis* var. *kurstaki* 1×10 cfu/g WP
Microbial	MV (Farmer Supply Co. Ltd.-Marwell Major Co. Ltd.)	*Bacillus thuringiensis* var. *kurstaki* 55000 SU/mg (5% WP)
Microbial	Sanback (Chem. Fac Co. Ltd./ Santhai Co. Ltd.)	*Bacillus thuringiensis* var. *kurstaki* 32000 IU/mg (15% WP)
Microbial	KU BT 32,000 (chem. Fac. KU53 co. Ltd.- chemi kaset Co. Ltd.)	*Bacillus thuringiensis* var. *kurstaki* 32000 IU/mg (15% WP)
Microbial	Gardang (Sharp Formulator Co. Ltd./Pisulin Co. Ltd.)	*Bacillus thuringiensis* var. *kurstaki* 64,000 IU/mg (15% WG)

Microbial	Diamond Tox (Project Field Co. Ltd./World Agro Group Co. Ltd.)	*Bacillus thuringiensis* var. *kurstaki* 55000 SU/mg (5% WP)
Microbial	World Bact F C (PF Factory Co. Ltd./Power First Agro Co. Ltd.)	*Bacillus thuringiensis* var. *aizawai* 8500 IU/mg SC
Microbial	Honma (Project Field Co. Ltd./ S Team Inter Trade Co. Ltd.)	*Bacillus thuringiensis* var. *aizawai* 550,000 SU/mg (5% WP)
Microbial	P Fresh (Applies Chem. Co. Ltd./Rungsipan Intertrade Co. Ltd.)	*Bacillus subtilis* var. AP-01 1×10 cfu/gm WP
Microbial	Bacto (TFI Green Biotech Co. Ltd./Kast Mahachon Co. Ltd.)	*Bacillus thuringiensis* var. *aizawai* 1×10^9 cfu/ml (12% W/V SC)
Microbial	Metazan (Applied Chem. Co. Ltd.)	*Metarhizium anisopliae* 1×10^9 cfu/gm WP
Microbial	Beauverin (Applied chem. Co. Ltd., Applied chem. Co. Ltd.)	*Beauveria bassiana* 1×10^9 cfu/gm WP
Microbial	K B T (K-mer Co. Ltd.)	*Bacillus thuringiensis* var. *kurstaki* 55000 SU/mg (5% WP)
Microbial	Cesar F. C. (Thep Watana Co. Ltd./Planter United Co. Ltd.)	*Bacillus thuringiensis* var. *kurstaki* 10,600 IU/mg SC
Microbial	Pro V-Z (Project Field Co. Ltd./Pro Crop Co. Ltd.)	Nuclear Polyhedrosis Virus of *Helicoverpa zea* 0.64% w/v SL
Microbial	Pro V-X (Project Field Co. Ltd./Pro Crop Co. Ltd.)	Nuclear Polyhedrosis Virus of *Spodoptera exiqua* 0.64% w/v SL
Microbial	Climax (Sharp Formulator Co. Ltd.)	*Bacillus thuringiensis* var. *kurstaki* 64,000 IU/mg (15% WG)
Microbial	Basic (Applied Chem. Co. Ltd./SV Biotech Co. Ltd.)	*Beauveria bassiana* 1×10^9 cfu/gm WP
Botanicals	Sadao-Fight (Bio Seven)	Azadirachtin 0.1% w/v SL
Botanicals	Azatin XL (Certis USA/Chem Trading Co. Ltd.)	Azadirachtin 3 % w/v EC
Botanical	Strong Black (Strong Crop Co.Ltd.)	*Azadirachtin* 0.1% w/v SL

[1]the record is in Thai name

Source: Department of Agriculture (DOA), Ministry of Agriculture and Cooperative, Government of Thailand, Bangkok (2011)

CONSTRAINTS

Thailand still faces some problems in liberal use of biopesticides. There are several of reasons why farmers are reluctant to adopt biopesticides to replace synthetic chemicals. Biopesticides generally act slower than synthetic pesticides, killing arthropod pests over a longer time rather than having an immediately apparent "knock-down" effect, so the pests can still damage the crops after application. Because of this, farmers still prefer to use and rely on the quick acting synthetic pesticides, causing high pest mortality shortly after application. Therefore, cost-competitiveness and product quality and performance of biopesticides are the two major barriers standing in the way of their adoption. Inexpensive conventional pesticides and all of their associated problems will continue to dominate until these barriers are addressed. To do so will require more effort in technology transfer, so that research results are more readily translated into field use. Even more importantly, it will require additional funding for optimization of local production strategies, farmer education, continuing research and even subsidization of biopesticide retail costs so that their availability to farmers is ensured.

Technical knowledge on the proper application methods, both number and size of spray droplets per m^2 foliage, is very important for enhancing the efficacy of biopesticides. Applying too much spray volume causes loss of product, while applying too little spray volume can cause uneven coverage of the target crops and lower pest mortality. Researches have proved that conventional applications of *Btk* (high spray volume, ca. 750 L/ha) are less effective than the reduced volume application (of less than 250 L/ha with droplet density 198-209 droplets/cm^2) in controlling *S. exigua* (Tantichodok *et al.*, 2003).

Generating adequate, local biopesticide capacity to support local producers is difficult in a limited market. The poor quality of some of the non-commercial biopesticide production is also a cause for concern (Warburton *et al.*, 2002). Other reason is the illegal importation of unregistered, often in-effective biopesticide products from China and Vietnam. These products may contain extremely low levels of active agents and sometime a cocktail of several agents so that their use is highly voluminous. In addition, some commercial products, which do not pass the registration process, still sell in the market that creates doubts of ineffective products for control and thus ignored to use them.

Overall scenario in Thailand suggests that issues perceived by biopesticide suppliers are lack of demand, knowledge, and extension.

There is competition with old and new conventional chemicals. The biopesticides are costly having slow action with shelf life constraints. There is also lack of quality with technical problems and moderate availability of dealers. Production cost is also one of the major constraints, which is further complicated by government policies and regulatory procedures. Last but not the least, there is substantial gap between biopesticide producers and the farmers.

CONCLUSIONS

Overall market for *Bacillus thuringiensis*-based products in Asia contribute to about 55.3% including Australia. The remainder of the market comprises other bacteria (including *Bacillus subtilis*, *Pseudomonas fluorescens*, various *Bacillus* spp for plant disease control, a number of fungal-based products (especially *Trichoderma*, *Beauveria* and *Metarhizium* products) and entomopathogenic viruses (notably *Helicoverpa armigera* NPV and *Spodoptera litura* NPV). China is the largest biopesticide market in the region, followed by India and Japan and Thailand is still in its growing stages of development of products. The majority of the products available for use in Thailand have been researched, developed and manufactured locally and mostly on the advice and assistance by NGOs or government and controlled by DOA. However, there are a number of critical barriers to product introduction in the region. These include an ability to pay, availability of cheap chemicals, the policy of many of the governments to support local production and the need for often difficult and long-term product support. In case of botanical insecticide development, these are hindered by research and developmental barriers; they still have a promising future due to bolstering campaigns by the Thai government. Therefore, many Thai plant species are analyzed for their insecticidal efficiency.

However, botanical insecticides require additional research and developmental efforts in order to eliminate the key limitations. Only then could Thailand's acceptance of botanical insecticides as effective pest control agents be widespread in Thai agricultural and commercial sectors.

ACKNOWLEDGEMENT

Authors thank De. Suprada Na Patarung from Department of Agriculture for providing information on natural enemies in Thailand.

REFERENCES

Bullangpoti V. (2009) Botanical pesticides in Thailand: Hope and hindrances. *Biopestic. Int.*, **5**, 75–99.

Chongchitmate P. (2005) Bionomics of entomopathogenic nematode *Steinernema siamkayai* Stock, Somsook and Reid (n. sp.) and Its efficacy against *Helicoverpa armigera* Hubner (Lepidoptera: Noctuidae). Thesis Kasetsart University, Bangkok.

Department of Agriculture, Ministry of Agriculture and Cooperative (2011) http://m.doa.go.th/ard/stat.php (in Thai)

Harris J. (2000) *Chemical Pesticides Markets, Health Risks and Residues.* CABI Publishing, Wallingford, UK

Hembree, S.C. (1981) Evaluation of the microbial control potential of a *Helicosporidium* sp. (Protozoa: Helicosporida) from *Aedes aegypti* and *Culex quinquefasciatus* from Thailand. *Mosq. News*, **41,** 770–783.

Islam, M.S., Hasan, M.M., Xiong, W., Zhang, S.C. and Lei, C.L. (2009) Fumigant and repellent activities of essential oil from *Coriandrum sativum* (L.) (Apiaceae) against red flour beetle *Tribolium castaneum* (Herbst) (Coleoptera: Tenebrionidae). *J. Pest Sci.*, **82,** 171–177.

Jaekel, T., Khoprasert, Y., Sorger, I., Kliemt, D., Seehabutr, V., Suasa-ard, K. and Hongnark, S. (1997) Sarcosporidiasis in rodents from Thailand. *J. Wild. Dis.*, **33,** 860–867.

Jeerapong, L. (1999) *The Method and Use of Herbs for Control of Pests.* Ministry of Agriculture, Thailand.

Jones, K.A,, Westby, A., Reilly, P.J.A. and Jeger, M.J. (1993) Exploitation of microorganisms in the developing countries of the tropics. In: D.G. Jones (ed.), *Exploitation of Microorganisms*, Chapman and Hall. London, pp. 343–370.

Jones, K.A., Zalazny, B., Ketunuti, U., Cherry, A. and Gryzwacz, D.A. (1998) World survey: South east Asia and Western Pacific. In: F.R. Hunter-Fuijita, P.F. Entwistle, H.F. Evans and N.E. Crook (eds.), *Insect Viruses and Pest Management*, Wiley, Chichester, pp. 244– 257

Grzywacz, D. N.D. (2003) Development and registration of biopesticides in Asia. *Biopestic. Asia*, 101–109.

Ketunuti, U. (2003). Survey and study of the biopesticides use in controlling arthropods pests Thailand. Reported Paper submitted to Natural Resources Institute, Greenwich University U.K. (in Thai) (unpublished report).

Koul, O., Walia, S. and Dhaliwal, G.S. (2008) Essential oils as green pesticides: Potential and constraints. *Biopestic. Int.*, **4,** 63–84.

Lewis, L.C. (2002) Microbial control potential of a protozoan. In: D. Pimentel (ed.), *Encyclopedia of Pest Management*, Taylor & Francis, UK, pp. 673–676.

Marasri P. (1994) *Using NPV, Bt and Natural Enemy for Control of Vegetable Crop Pests*. KURDI Report, Bangkok, Thailand.

Nanta P. (2002) *Natural Enemies as the Key Component of IPM*. Department of Agriculture, Ministry of Agriculture and Cooperative, Bangkok, Thailand.

Nongnooch Poramacom (2001) Pesticide markets and related situations in Thailand. *Kasetsart J.*, **22**, 205–211.

Prasetphol, S., Areekul, P., Buranarerk, A. and Kritpitayaavuth, M. (1969) *Life History of Orange Dog Butterfly and its Microbial Control*. Technical Bulletin No. 10, Department of Agriculture, Bangkok, Thailand.

Promkird, P. (2004) *Evaluation of Rat Population in Luang Prabang and the Impact of Protozoa, Sarcocystis singaporensis on other Vertebrate Pests*. GTZ (German Technical Cooperation Agency), (in Thai).

Office of Agricultural Economics (2010) *Agricultural Statistics of Thailand Crop Year 2008/09*. Ministry of Agriculture and Cooperatives, Bangkok, Thailand.

Poonuchaphai, W. and Jäkel, T. (2009) From production to the consumer: What German Technical Cooperation offers to establish a successful organic farming business. *As. J. Food Ag-Ind.*, Special Issue, S271–S278.

Poramacom, N. (2001) Pesticide markets and related situations in Thailand. *Kasetsart J.*, **22**, 205–211.

Ratanasatien, P., Ketunuti, U. and Tantichodok, A. (2005) *Positioning of Biopesticides in Thailand*. 6th Pacific Rim Conference on the Biotechnology of *Bacillus thuringiensis* and its Environmental Impact, Victoria, BC.

Ruchinarong, P. (2002) Conidia production of a green muscadine fungus, *Metarhizium anisopliae* on solid substrates and its efficacy to control the termites, *Coptotermes gestroi* (Wasmann). Thesis Kasetsart University, Bangkok.

Sasnarukkit, A. (2003) Efficacy of an entomopathogenic nematode, *Steinernema siamkayai* (Stock, Somsook, and Reid) on controlling diamondback moth, *Plutella xylostella* (linnaeus). Thesis, Kasetsart University, Bangkok.

Skovmand, O. (2007) Microbial control in Southeast Asia. *J. Invertebr. Pathol.*, **95**, 168–174.

Somsuk, W., Somsuk, S. and Nanta, P. (1999) Comparison on production cost of parasitic nematode, *Steinernema carpocapsae* for controlling lepidopterous pests. In: *Research and Development on Entomology and Zoology Proceeding*, Department of Agriculture, Bangkok, Thailand (in Thai), pp. 104–108.

Tantichodok, A., Chudee, P., Ratanasatien, P. and Ketunuti, U. (2003) Efficacy of *Bacillus thuringiensis* by using high volume, low volume and ultra low volume applications for controlling the beet armyworm, *Spodoptera exigua* (Hubner) on asparagus. *Ent. Zool. Gas.*, **25**, 80–93 (in Thai).

Thai German Partnership Program. (2010) *Regional Programme on Commercialization of Biopesticides*. www.thai-germen-cooperation.info.

Trading Economics (2010) http://www.tradingeconomics.com/Economics/GDPGrowth.aspx?Symbol=THB

Tungrabieb, M. (2009) Effect of *Metarhizium anisopliae* against *Boophilus microplus*, (in Thai). Thesis, Kasetsart University, Bangkok.

Vattanatangum, A. (1989) Diamondback moth control with *Bacillus thuringiensis* products in Thailand. *Isr. J. Entomol.*, **23**, 131–139.

Youngchaitrakul, S.(1999) An investigation on the effect of using the fungus, *Metarhizium anisopliae* for the control of the housefly, *Musca domestica* L. (Diptera: Muscidae). Thesis, Kasetsart University, Bangkok.

Wongsiri, S. (1980) *Insecticides*. Se-Education, Thailand

Warburton, H., Ketunuti, U. and Grzywacz, D. (2002) *Survey of the Supply, Production and use of Microbial Pesticides in Thailand*. NRI Report 2723. National Resource Institute, Univesity of Greenwitch, Chatham, UK.

Zhang, A., Kuang, L.F., Maisin, N., Karumuru, B., Hall, D.R., Virdiana, I., Lambert, S., Purung, H.B., Wang, S. and Hebbar, P. (2008) Activity evaluation of cocoa pod borer sex pheromone in cacao fields. *Environ. Entomol.*, **37**, 719–724.

Index